Edited by
Theo Mang and
Wilfried Dresel

Lubricants and Lubrication

Edited by Theo Mang and Wilfried Dresel

Lubricants and Lubrication

Third, Completely Revised and Enlarged Edition

Volume 1

Verlag GmbH & Co. KGaA

Editors

Prof. Dr. Theo Mang
Fuchs Petrolub, Mannheim, Germany
Holzweg 30
69469 Weinheim
Germany

Dr. Wilfried Dresel
Fuchs Petrolub, Mannheim, Germany
Treppenweg 4
67063 Paul - Martin - Ufer 51
Mannheim, Germany

Cover: Pictures from (c) fotolia/sveta (gears) and (c) fotolia/stuartbur (oil can).

Library of Congress Card No.: applied for

British Library Cataloguing-in-Publication Data
A catalogue record for this book is available from the British Library.

Bibliographic information published by the Deutsche Nationalbibliothek
The Deutsche Nationalbibliothek lists this publication in the Deutsche Nationalbibliografie; detailed bibliographic data are available on the Internet at http://dnb.d-nb.de.

Print ISBN: 978-3-527-32670-9
ePDF ISBN: 978-3-527-64558-9
ePub ISBN: 978-3-527-64557-2
Mobi ISBN: 978-3-527-64559-6
oBook ISBN: 978-3-527-64556-5

Cover Design Grafik-Design Schulz, Fußgönheim, Germany
Typesetting Thomson Digital, Noida, India
Printing and Binding Markono Print Media Pte Ltd, Singapore

Printed on acid-free paper

Contents

List of Contributors *XXXVII*
A Word of Thanks *XXXIX*
Preface to the 3rd Edition *XLI*
Abbreviations *XLIII*

Volume 1

1 Lubricants and Their Market *1*
Theo Mang and Apu Gosalia
1.1 Introduction *1*
1.2 Lubricants Demand *2*
1.3 Lubricants Competitor Landscape *5*
1.4 Lubricant Systems *7*
References *9*

2 Lubricants in the Tribological System *11*
Theo Mang and Christian Busch
2.1 Lubricants as Part of Tribological Research *11*
2.2 The Tribological System *12*
2.3 Friction *12*
2.3.1 Types of Friction *13*
2.3.1.1 Sliding Friction *13*
2.3.1.2 Rolling Friction *14*
2.3.1.3 Static Friction *16*
2.3.1.4 Kinetic Friction *16*
2.3.1.5 Stick–Slip *16*
2.3.2 Friction and Lubrication Conditions *17*
2.3.2.1 Solid Friction (Dry Friction) *17*
2.3.2.2 Boundary Friction *17*
2.3.2.3 Fluid Friction *17*
2.3.2.4 Mixed Friction *18*
2.3.2.5 Solid Lubricant Friction *18*
2.3.2.6 Stribeck Diagram *19*
2.3.2.7 Hydrodynamic Lubrication *19*

2.3.2.8 Elastohydrodynamic Lubrication (EHD Regime) *20*
2.3.2.9 Thermo-Elasto-Hydrodynamic Lubrication (TEHD) *21*
2.4 Wear *21*
2.4.1 Wear Mechanisms *21*
2.4.1.1 Abrasion *22*
2.4.1.2 Adhesion *22*
2.4.1.3 Tribochemical Reactions *23*
2.4.1.4 Surface Fatigue *23*
2.4.1.5 Erosion *24*
2.4.1.6 Fretting *24*
2.4.1.7 Cavitation *24*
2.4.1.8 Corrosive Wear *24*
2.4.2 Types of Wear *25*
2.4.3 The Wear Process *25*
2.4.4 Tribomutation *25*
2.4.5 Nanotribology *27*
2.4.6 Tribosystems of Tomorrow *29*
References *29*

3 Rheology of Lubricants *31*
Theo Mang
3.1 Viscosity *31*
3.2 Influence of Temperature on Viscosity (*V–T* Behaviour) *33*
3.2.1 Viscosity Index *34*
3.3 Viscosity–Pressure Dependency *34*
3.4 The Effect of Shear Rate on Viscosity *37*
3.5 Special Rheological Effects *38*
3.5.1 Greases *38*
3.6 Viscosity Grades *39*
3.6.1 ISO Viscosity Grades *39*
3.6.2 Other Viscosity Grades *40*
3.6.2.1 Engine Oils *40*
3.6.2.2 Automotive Gear Oils *40*
3.6.2.3 Industrial Gear Oils *40*
3.6.2.4 Viscosity Grades for Base Oils *40*
3.6.2.5 Comparison of Viscosity Grades *40*
3.7 Viscosity Measurements *42*
Nael Zaki
3.7.1 Measurement of Viscosity by Capillary Tubes *42*
3.7.1.1 Newtonian Fluids *42*
3.7.2 2 Non-Newtonian Fluids *43*
3.7.2.1 Lubricating Greases Viscosity Measurements by the Standard Oil Development (SOD) Viscometer *44*
3.7.3 High Shear Rate Capillary Viscometers *44*
3.7.4 Rotational Viscometers *46*

3.7.4.1 Coaxial Cylinder Viscometer *46*
3.7.4.2 Cone/Plate Viscometer *46*
3.8 Viscosity Measurements at High Pressure *47*
References *49*

4 Base Oils *51*
Theo Mang and Georg Lingg
4.1 Base Oils: A Historical Review and Outlook *51*
4.2 Chemical Characterization of Mineral Base Oils *52*
4.2.1 Rough Chemical Characterization *52*
4.2.1.1 Viscosity–Gravity Constant (VGC) *52*
4.2.1.2 Aniline Point *52*
4.2.2 Carbon Distribution *53*
4.2.3 Hydrocarbon Composition *53*
4.2.4 Polycyclic Aromatics in Base Oils *53*
4.2.4.1 Aromatics in White Mineral Oils *55*
4.3 Refining *55*
4.3.1 Distillation *56*
4.3.2 De-Asphalting *57*
4.3.3 Traditional Refining Processes *58*
4.3.3.1 Acid Refining *58*
4.3.3.2 Solvent Extraction *58*
4.3.4 Solvent Dewaxing *60*
4.3.5 Finishing *61*
4.3.5.1 Lube Crudes *62*
4.4 Base Oil Manufacturing by Hydrogenation and Hydrocracking *62*
4.4.1 Manufacturing Naphthenic Base Oils by Hydrogenation *63*
4.4.2 Production of White Oils *65*
4.4.3 Lube Hydrocracking *66*
4.4.4 Catalytic Dewaxing *67*
4.4.5 Wax Isomerization *70*
4.4.6 Hybrid Lube Oil Processing *70*
4.4.7 All-Hydrogen Route *71*
4.4.8 Gas-to-Liquids Conversion Technology *72*
4.5 Boiling and Evaporation Behaviour of Base Oils *73*
4.6 Base Oil Categories and Evaluation of Various Petroleum Base Oils *78*
References *81*

5 Synthetic Base Oils *83*
Wilfried Dresel
5.1 Synthetic Hydrocarbons *84*
5.1.1 Polyalphaolefins *85*
5.1.2 Polyinternalolefins *87*
5.1.3 Polybutenes *88*

5.1.4 Alkylated Aromatics *89*
5.1.5 Other Hydrocarbons *90*
5.2 Halogenated Hydrocarbons *91*
5.3 Synthetic Esters *92*
5.3.1 Esters of carboxylic acids *92*
5.3.1.1 Dicarboxylic Acid Esters *93*
5.3.1.2 Polyol Esters *94*
5.3.1.3 Other Carboxylic Esters *95*
5.3.1.4 Complex Esters, Complex Polymer Esters *96*
5.3.1.5 Fluorinated Carboxylic Acid Esters *97*
5.3.2 Phosphate Esters *97*
5.4 Polyalkylene Glycols *98*
5.5 Other Polyethers *100*
5.5.1 Perfluorinated Polyethers *101*
5.5.2 Polyphenyl Ethers *102*
5.5.3 Polysiloxanes (Silicone Oils, Silicones) *103*
5.6 Other Synthetic Base Oils *105*
5.7 Comparison of Synthetic Base Oils *109*
5.8 Mixtures of Synthetic Lubricants *109*
References *110*

6 Additives *117*
Jürgen Braun
6.1 Antioxidants *118*
6.1.1 Mechanism of Oxidation and Antioxidants *118*
6.1.2 Compounds *120*
6.1.2.1 Phenolic Antioxidants *120*
6.1.2.2 Aromatic Amines *121*
6.1.2.3 Compounds Containing Sulfur and Phosphorus *122*
6.1.2.4 Organosulfur Compounds *123*
6.1.2.5 Organophosphorus Compounds *123*
6.1.2.6 Other Compounds *123*
6.1.2.7 Synergistic Mixtures *123*
6.1.3 Testing of the Oxidation Stability *123*
6.2 Viscosity Modifiers *124*
6.2.1 Physical Description of Viscosity Index *124*
6.2.2 VI Improvement Mechanisms *125*
6.2.3 Structure and Chemistry of Viscosity Modifiers *127*
6.3 Pour Point Depressants (PPD) *129*
6.4 Detergents and Dispersants *130*
6.4.1 Mechanism of DD Additives *130*
6.4.2 Metal-Containing Compounds (Detergents) *131*
6.4.2.1 Phenates *131*
6.4.2.2 Salicylates *131*
6.4.2.3 Thiophosphonates *132*

6.4.2.4 Sulfonates *133*
6.4.3 Ashless Dispersants (AD) *134*
6.5 Antifoam Agents *136*
6.5.1 Silicon Defoamers *136*
6.5.2 Silicone-Free Defoamers *137*
6.6 Demulsifiers and Emulsifiers *137*
6.6.1 Demulsifiers *137*
6.6.2 Emulsifiers *138*
6.7 Dyes *138*
6.8 Antiwear (AW) and Extreme Pressure (EP) Additives *138*
6.8.1 Function of AW/EP Additives *138*
6.8.2 Compounds *139*
6.8.2.1 Phosphorus Compounds *139*
6.8.2.2 Compounds Containing Sulfur and Phosphorus *140*
6.8.2.3 Compounds Containing Sulfur and Nitrogen *141*
6.8.2.4 Sulfur Compounds *142*
6.8.2.5 PEP Additives *144*
6.8.2.6 Chlorine Compounds *144*
6.8.2.7 Solid Lubricating Compounds *144*
6.9 Friction Modifiers (FM) *144*
6.10 Corrosion Inhibitors *145*
6.10.1 Mechanism of Corrosion Inhibitors *146*
6.10.2 Antirust Additives (Ferrous Metals) *146*
6.10.2.1 Sulfonates *146*
6.10.2.2 Carboxylic Acid Derivatives *147*
6.10.2.3 Amine-Neutralized Alkylphosphoric Acid Partial Esters *148*
6.10.2.4 Vapour Phase Corrosion Inhibitors *148*
6.10.3 Metal Passivators (Non-Ferrous Metals) *149*
References *151*

7 Lubricants in the Environment *153*
Rolf Luther
7.1 Definition of 'Environment-Compatible Lubricants' *153*
7.1.1 CEN Technical Report 16227 – Standard Designation of the Term Biolubricant *155*
7.2 Current Situation *156*
7.2.1 Statistical Data *156*
7.2.2 Economic Consequences and Substitution Potential *158*
7.2.3 Agriculture, Economy and Politics *160*
7.2.4 Political Initiatives *161*
7.3 Tests to Evaluate Biotic Potential *162*
7.3.1 Biodegradation *163*
7.3.2 Ecotoxicity *163*
7.3.3 Emission Thresholds *164*
7.3.4 Water Pollution *164*

7.3.4.1 The German Water Hazardous Classes *164*
7.3.4.2 German Regulations for Using Water-Endangering Lubricants (VAwS) *166*
7.4 Environmental Legislation 1: Registration, Evaluation and Authorization of Chemicals (REACh) *169*
7.4.1 Registration *170*
7.4.2 Evaluation *172*
7.4.3 Authorization *172*
7.4.4 Registration Obligations *173*
7.5 Globally Harmonized System of Classification and Labelling (GHS) *175*
7.6 Environmental Legislation 2: Classification and Labelling of Chemicals *177*
7.6.1 Dangerous Preparations Directive (1999/45/EC) *177*
7.6.2 Globally Harmonized System of Classification and Labelling of Chemicals (GHS) *179*
7.7 Environmental Legislation 3: Regular use *180*
7.7.1 Environmental Liability Law *181*
7.7.2 The Chemicals Law, Hazardous Substances Law *181*
7.7.3 Transport Regulations *182*
7.7.4 Disposal (Waste and Recycling Laws) *182*
7.7.5 Disposal Options for 'Not Water Pollutant' Vegetable Oils *183*
7.8 Environmental Legislation 4: Emissions *184*
7.8.1 Air Pollution *184*
7.8.2 Water Pollution *184*
7.8.3 German Law for Soil Protection *185*
7.8.4 German Water Law *186*
7.8.5 Wastewater Charges *187*
7.8.6 Clean Air: German Emissions Law *187*
7.8.7 Drinking Water Directive *187*
7.9 Standardization of Environment-Compatible Hydraulic Fluids *188*
7.9.1 The German Regulation VDMA 24568 *188*
7.9.2 ISO Regulation 15380 *188*
7.10 Environmental Seal *194*
7.10.1 Global Eco-Labelling Network *194*
7.10.2 European Eco-Label for Lubricants (EEL) *195*
7.10.3 The German 'Blue Angel' *200*
7.10.4 Nordic Ecolabel (Norway, Sweden, Finland, Iceland): 'White Swan' *204*
7.10.4.1 Requirements Concerning Renewable Resources *206*
7.10.4.2 Requirements Concerning Re-Refined Oil *206*
7.10.4.3 Requirements Concerning Environmentally Harmful Components *206*
7.10.4.4 Requirements for Hydraulic Fluids, Mould Oil, Metalworking Fluids *206*

7.10.5 Sweden Standard 207
7.10.6 The Canadian 'Environmental Choice' (Maple Leaf) 208
7.10.7 Other Eco-Labels 209
7.10.7.1 Austria 209
7.10.7.2 France 210
7.10.7.3 Japan 210
7.10.7.4 USA 211
7.10.7.5 The Netherlands 211
7.10.7.6 US-Regulation VGP and Definition of Environmentally Acceptable Lubricants (EAL) 212
7.11 Base Fluids 214
7.11.1 Biodegradable Base Oils for Lubricants 214
7.11.2 Synthetic Esters 214
7.11.3 Polyglycols 215
7.11.4 Polyalphaolefins 215
7.11.5 Relevant Properties of Ester Oils 215
7.11.5.1 Evaporation Loss 215
7.11.5.2 Viscosity–Temperature Behaviour 216
7.11.5.3 Boundary Lubrication 216
7.12 Additives 216
7.12.1 Extreme Pressure/Antiwear Additives 217
7.12.2 Corrosion Protection 217
7.12.3 Antioxidants 217
7.13 Products (Examples) 218
7.13.1 Hydraulic Fluids 218
7.13.2 Metal Working Oil 218
7.13.3 Oil-Refreshing System 219
7.14 Safety Aspects of Handling Lubricants (Working Materials) 220
7.14.1 Toxicological Terminology and Hazard Indicators 220
7.14.1.1 Acute Toxicity 220
7.14.1.2 Subchronic and Chronic Toxicity 221
7.14.1.3 Poison Categories 221
7.14.1.4 Corrosive and Caustic 221
7.14.1.5 Explosion and Flammability 221
7.14.1.6 Carcinogenic 222
7.14.1.7 Teratogens and Mutagens 222
7.14.2 MAK (Maximum Workplace Concentration) Values 222
7.14.3 Polycyclic Aromatic Hydrocarbons (PAK, PAH, PCA) 223
7.14.4 Nitrosamines in Cutting Fluids 224
7.14.5 Law on Flammable Fluids 225
7.15 Skin Problems Caused by Lubricants 225
7.15.1 Structure and Function of the Skin 225
7.15.2 Skin Damage 226
7.15.2.1 Oil Acne (Particle Acne) 226
7.15.2.2 Oil Eczema 227

7.15.3 Testing Skin Compatibility *228*
7.15.4 Skin Function Tests *230*
7.15.5 Skin Care and Skin Protection *231*
Further Reading *232*

8 Disposal of Used Lubricating Oils *237*
Theo Mang
8.1 Possible Uses of Waste Oil *237*
8.2 Legislative Influences on Waste Oil Collection and Reconditioning *239*
8.3 Re-Refining *240*
8.3.1 Sulfuric Acid Refining (Meinken) *241*
8.3.2 Propane Extraction Process (IFP, Snamprogetti) *241*
8.3.3 Mohawk Technology (CEP–Mohawk) *243*
8.3.4 KTI Process *243*
8.3.5 PROP Process *243*
8.3.6 Safety Kleen Process *244*
8.3.7 DEA Technology *245*
8.3.8 Other Re-Refining Technologies *245*
References *246*

9 Lubricants for Internal Combustion Engines *249*
Manfred Harperscheid
9.1 Four-Stroke Engine Oils *249*
9.1.1 General Overview *249*
9.1.1.1 Fundamental Principles *250*
9.1.1.2 Viscosity Grades *251*
9.1.1.3 Performance Specifications *253*
9.1.1.4 Formulation of Engine Oils *254*
9.1.1.5 Additives *255*
9.1.1.6 Performance Additives *255*
9.1.1.7 Viscosity Improvers *256*
9.1.2 Characterization and Testing *256*
9.1.2.1 Physical and Chemical Testing *256*
9.1.2.2 Engine Testing *257*
9.1.2.3 Passenger Car Engine Oils *257*
9.1.2.4 Engine Oil for Commercial Vehicles *261*
9.1.3 Classification by Specification *262*
9.1.3.1 MIL Specifications *262*
9.1.3.2 API and ILSAC Classification *262*
9.1.3.3 CCMC Specifications *265*
9.1.3.4 ACEA Specifications *266*
9.1.3.5 Manufacturers' Approval of Service Engine Oils *268*
9.1.3.6 Future Trends *272*
9.1.3.7 Fuel Efficiency *274*

9.1.3.8 Long Drain Intervals *276*
9.1.3.9 Low Emission *277*
9.2 Two-Stroke Oils *278*
9.2.1 Application and Characteristics of Two-stroke Oils *278*
9.2.2 Classification of Two-stroke Oils *280*
9.2.2.1 API Service Groups *280*
9.2.2.2 JASO Classification *280*
9.2.2.3 ISO Classification *281*
9.2.3 Oils for Two-Stroke Outboard Engines *282*
9.2.4 Environmentally Friendly Two-stroke Oils *283*
9.3 Tractor Oils *283*
9.4 Gas Engine Oils *285*
9.4.1 Use of Gas Engines: Gas as a Fuel *286*
9.4.2 Lubricants for Gas Engines *286*
9.5 Marine Diesel Engine Oils *287*
9.5.1 Low-speed Crosshead Engines *288*
9.5.2 Medium-Speed Engines *288*
9.5.3 Lubricants *288*
References *290*

10 Gear Lubrication Oils *293*
Thorsten Bartels
10.1 Requirements of Gear Lubrication Oils *294*
10.2 Gear Lubrication Oils for Motor Vehicles *297*
10.2.1 Driveline Lubricants for Commercial Vehicles *298*
10.2.2 Driveline Lubricants for Passenger Cars *302*
10.2.3 Lubricants for Automatic Transmissions and CVTs *306*
10.2.3.1 Fluid Requirements for Hydrodynamic Transmissions *308*
10.2.3.2 Fluid Requirements for Wet Clutches and Brakes *309*
10.2.3.3 Fluid Requirements for CVT Applications *311*
10.2.3.4 B-CVT Push Belt and Link Chain Drives *312*
10.2.3.5 T-CVT Traction Drives *313*
10.2.3.6 H-CVT Hydrostatic Dynamic Powershift Drives *314*
10.2.4 Multifunctional Fluids in Vehicle Gears *315*
10.3 Gear Lubricants for Industrial Gears *316*
Wolfgang Bock
10.3.1 Introduction *316*
10.3.2 Industrial Lubricants: Industrial Gear Oils – Statistics *319*
10.3.2.1 Market Shares/Market Situation *319*
10.3.3 Composition of Industrial Gear Oils *320*
10.3.3.1 Mineral Oils (CLP-M) *321*
10.3.3.2 Synthetic Hydrocarbons: Polyalphaolefins (CLP-PAO) *321*
10.3.3.3 Polyglycols (CLP-PG) *322*
10.3.3.4 Synthetic Esters (CLP-E) *324*
10.3.3.5 Gear Oil Additives *324*

10.3.4 Classification of Gear Oils 326
10.3.4.1 Requirements and Specifications 326
10.3.4.2 Requirements for Industrial Gear Oils According to DIN 51517, Part 3 326
10.3.5 Temperature Ranges and Lifetime of Industrial Gear Oils 328
10.3.5.1 Lifetime of Gear Oils: Oxidation Stability (RPVOT Test) 328
10.3.6 Cost-Benefit Ratio of Industrial Gear Oils 330
10.3.7 Filtration Behaviour: Electrical Conductivity of Gear and Lubricating Oils 330
10.3.8 Oil and Water: Saturation Values of Dissolved Water in Oil 331
10.3.9 Special Industrial Gear oil Formulations 333
10.3.9.1 Special Corrosion Protection Gear Oils 333
10.3.9.2 Detergent/Dispersant Types of Gear Oils 333
10.3.9.3 'Plastic Deformation' Additive Technology 333
10.3.9.4 'Reducing Friction' by Special Industrial Gear Oils 333
10.3.10 Gear Oils for Wind Turbines: Demands and Characteristics 334
10.3.10.1 Demands on Wind Turbine Gear Oils 334
10.3.10.2 Special Tests for Wind Turbine Gear Oils 338
10.3.10.3 SKF Specifications for Wind Turbine Gear Oils 339
10.3.10.4 Low-Speed Wear Behaviour of Industrial Gear Oils 339
10.3.10.5 Low-Temperature Viscosity of Industrial Gear Oils 339
10.3.10.6 Conclusion 342
10.3.11 Summary 343
References 343

11 Hydraulic Oils 345
Wolfgang Bock
11.1 Introduction 345
11.2 Hydraulic Principle: Pascal's Law 346
11.3 Hydraulic Systems, Circuits and Components 347
11.3.1 Elements of a Hydraulic System 347
11.3.1.1 Pumps and Motors 348
11.3.1.2 Hydraulic Cylinders 348
11.3.1.3 Valves 350
11.3.1.4 Circuit Components 350
11.3.1.5 Seals, Gaskets and Elastomers 350
11.4 Hydraulic Fluids 353
11.4.1 Composition of Hydraulic Fluids: Base Fluids and Additives 353
11.4.1.1 Base Oil or Base Fluid 353
11.4.1.2 Hydraulic Fluid Additives 353
11.4.2 Primary, Secondary and Tertiary Characteristics of a Hydraulic Fluid 354
11.4.3 Selection Criteria for Hydraulic Fluids 355
11.4.4 Classification of Hydraulic Fluids: Standardization of Hydraulic Fluids 357

11.4.4.1 Classification of Hydraulic Fluids *357*
11.4.5 Mineral-Oil-Based Hydraulic Fluids *357*
11.4.5.1 H Hydraulic Oils *360*
11.4.5.2 HL Hydraulic Oils *360*
11.4.5.3 HLP Hydraulic Oils *360*
11.4.5.4 HVLP Hydraulic Oils *360*
11.4.5.5 HLPD Hydraulic Oils *370*
11.4.6 Fire-Resistant Hydraulic Fluids *370*
11.4.6.1 HFA Fluids *371*
11.4.6.2 HFB Fluids *371*
11.4.6.3 HFC Fluids *371*
11.4.6.4 HFD Fluids *372*
11.4.7 Biodegradable Hydraulic Fluids *372*
11.4.7.1 HETG: Triglyceride and Vegetable-Oil Types *373*
11.4.7.2 HEES: Synthetic Ester Types *373*
11.4.7.3 HEPG: Polyglycol Types *375*
11.4.7.4 HEPR: Polyalphaolefin and Related Hydrocarbon Products *378*
11.4.8 Hydraulic Fluids for the Food and Beverage Industry *378*
11.4.8.1 H2 Lubricants *378*
11.4.8.2 H1 Lubricants *378*
11.4.9 Automatic Transmission Fluids *379*
11.4.10 Fluids in Tractors and Agricultural Machinery *379*
11.4.11 Hydraulic Fluids for Aircraft *379*
11.4.12 International Requirements on Hydraulic Oils *380*
11.4.13 Physical Properties of Hydraulic Oils and Their Effect on Performance *380*
11.4.13.1 Viscosity and Viscosity–Temperature Behaviour *380*
11.4.13.2 Viscosity–Pressure Behaviour *383*
11.4.13.3 Density *384*
11.4.13.4 Compressibility *388*
11.4.13.5 Gas Solubility and Cavitation *389*
11.4.13.6 Air Release *390*
11.4.13.7 Foaming *391*
11.4.13.8 Demulsification *391*
11.4.13.9 Pour Point *392*
11.4.13.10 Copper Corrosion Behaviour: Copper Strip Test *392*
11.4.13.11 Water Content: Karl Fischer Method *392*
11.4.13.12 Ageing Stability: Baader Method *393*
11.4.13.13 Ageing Stability (TOST Test) *393*
11.4.13.14 Neutralization Number *393*
11.4.13.15 Steel/Ferrous Corrosion Protection Properties *394*
11.4.13.16 Wear Protection (Shell Four-Ball Apparatus; VKA, DIN 51 350) *394*
11.4.13.17 Shear Stability of Polymer-Containing Lubricants *394*
11.4.13.18 Mechanical Testing of Hydraulic Fluids in Rotary Vane Pumps (DIN EN ISO 20763 [Q]) *395*

11.4.13.19 Wear Protection (FZG Gear Rig Test; DIN ISO 14635-1 [R]) *395*
11.5 Hydraulic System Filters *395*
11.5.1 Contaminants in Hydraulic Fluids *396*
11.5.2 Oil Cleanliness Grades *397*
11.5.3 Filtration *398*
11.5.4 Requirements of Hydraulic Fluids *400*
11.6 Machine Tool Lubrication *400*
11.6.1 The Role of Machine Tools *400*
11.6.2 Machine Tool Lubrication *401*
11.6.3 Machine Tool Components: Lubricants *402*
11.6.3.1 Hydraulic Unit *402*
11.6.3.2 Slideways *404*
11.6.3.3 Spindles: Main and Working Spindles *405*
11.6.3.4 Gearboxes and Bearings *406*
11.6.4 Machine Tool Lubrication Problems *406*
11.6.5 Hydraulic Fluids: New Trends and New Developments *407*
11.6.5.1 Applications *407*
11.6.5.2 Chemistry *407*
11.6.5.3 Extreme Pressure and Anti-Wear Properties *408*
11.6.5.4 Detergent/Dispersant Properties *409*
11.6.5.5 Air Release *409*
11.6.5.6 Static Coefficient of Friction *410*
11.6.5.7 Oxidation Stability *410*
11.6.5.8 Shear Stability *410*
11.6.5.9 Filtration of Zn- and Ash-Free Hydraulic Fluids *411*
11.6.5.10 Electrostatic Charges *412*
11.6.5.11 Micro-Scratching *413*
11.6.5.12 Updated Standards *413*
11.6.5.13 Conclusion *416*
11.7 Summary *416*
References *416*
Further Reading *419*
Books *419*
Standards *419*

12 Compressor Oils *421*
Wolfgang Bock and Christian Puhl
12.1 Air Compressor Oils *421*
12.1.1 Displacement Compressors *423*
12.1.1.1 Reciprocating Piston Compressors *423*
12.1.1.2 Lubrication of Reciprocating Piston Compressors *423*
12.1.1.3 Rotary Piston Compressors: Single Shaft, Rotary Vane Compressors *424*
12.1.1.4 Lubrication of Rotary Piston Compressors *424*
12.1.1.5 Screw Compressors *424*

12.1.1.6 Lubrication of Screw Compressors *425*
12.1.1.7 Roots Compressors *426*
12.1.1.8 Lubrication of Roots Compressors *426*
12.1.2 Dynamic Compressors *426*
12.1.2.1 Turbo Compressors *426*
12.1.2.2 Lubrication of Turbo Compressors *427*
12.1.2.3 Preparation of Compressed Air *427*
12.1.2.4 Lubrication of Gas Compressors *427*
12.1.2.5 Characteristics of Compressor Oils *428*
12.1.2.6 Standards and Specifications of Compressor Oils *429*
12.2 Refrigeration Oils *436*
12.2.1 Introduction *436*
12.2.2 Minimum Requirements of Refrigeration Oils *436*
12.2.3 Classifications of Refrigeration Oils *438*
12.2.3.1 Mineral Oils (MO) – Dewaxed Naphthenic Refrigeration Oils *438*
12.2.3.2 Mineral Oils (MO) – Paraffinic Refrigeration Oils *439*
12.2.3.3 Semi-synthetic Refrigeration Oils: Mixtures of Alkylbenzenes (AB) and Mineral Oils (MO) *440*
12.2.3.4 Fully Synthetic Refrigeration Oils: Alkylbenzenes *440*
12.2.3.5 Fully Synthetic Refrigeration Oils: Polyalphaolefins (PAO) *440*
12.2.3.6 Fully Synthetic Refrigeration Oils: Polyol Esters (POE) *441*
12.2.3.7 Fully Synthetic Refrigeration Oils: Polyalkylene Glycols (PAG) for R134a and HFO-1234 yf *442*
12.2.3.8 Fully Synthetic Refrigeration Oils – Polyalkylene glycols for NH_3 *442*
12.2.3.9 Other Synthetic Fluids *443*
12.2.3.10 Refrigeration Oils for CO_2 *443*
12.2.3.11 New Refrigeration Oils for HFO Refrigerants *443*
12.2.3.12 Copper Plating *444*
12.2.4 Types of Compressor *444*
12.2.5 Viscosity Selection *445*
12.2.5.1 General Overview *445*
12.2.5.2 Mixture Concentration in Relation to Temperature and Pressure (RENISO Triton SE 55 – R134a) *447*
12.2.5.3 Mixture Viscosity in Relation to Temperature, Pressure and Refrigerant Concentration (RENISO Triton SE 55 – R134a) *447*
12.2.6 Summary *450*
References *450*

13 Turbine Oils *453*
Wolfgang Bock
13.1 Introduction *453*
13.2 Demands on Turbine Oils – Characteristics *454*
13.3 Formulation of Turbine Oils *454*
13.4 Physical and Chemical Data of Turbine Oils *455*

13.4.1 Colour According to DIN ISO 2049 455
13.4.2 Density According to DIN 51757 455
13.4.3 Kinematic Viscosity According to DIN EN ISO 3104 456
13.4.4 Flashpoint According to DIN ISO 2592 457
13.4.5 Pourpoint According to DIN ISO 3016 458
13.4.6 Foaming According to ASTM D 892 458
13.4.7 Neutralization Number According to DIN 51558 459
13.4.8 FZG Mechanical Gear Test Rig According to DIN ISO 14635-1 460
13.4.9 Air Release at 50 °C According to DIN ISO 9120 462
13.4.10 Water Content According to DIN 51777 462
13.4.11 Water Separation According to DIN 51589 463
13.4.12 Demulsifying Power at 54°C According to DIN ISO 6614 464
13.4.13 Steel/Ferrous Corrosion Protection Properties According to DIN ISO 7120 464
13.4.14 Copper Corrosion Protection Properties According to DIN EN ISO 2160 465
13.4.15 RPVOT 150 °C According to ASTM D2272 465
13.4.16 TOST Lifetime According to DIN EN ISO 4263-1 466
13.4.17 Thermal Stability 467
13.4.18 Thermal Conductivity 468
13.4.19 Specific Heat 468
13.4.20 Vapour Pressure 469
13.4.21 Surface Tension 470
13.4.22 Remaining Useful Life Evaluation Routine (RULER Method) 470
13.4.23 Membrane Patch Colorimetry (MPC) Test 472
13.5 Turbine Lubricants: Description According to DIN 51515, Parts 1 and 2 473
13.6 Turbine Lubricants: Specifications 474
13.6.1 Lubricants for Turbines: ISO 8068 Specifications 477
13.7 Turbine Oil Circuits 479
13.8 Flushing Turbine Oil Circuits 480
13.9 Monitoring and Maintenance of Turbine Oils – General 481
13.10 Turbine Oils: Evaluation of Used Oil Values – Parameters and Warning Values/Limits According to VGB Recommendation 481
13.11 Turbine Oils: Evaluation of Used Oil Values – Causes and Measures 482
13.11.1 Turbine Oils: Used Oil Values 483
13.11.2 Turbine Oils: Used Oil Values – Causes and Measures 484
13.11.3 Turbine Oils: Used Oil Values – Causes and Measures 485
13.12 Lifetime of (Steam) Turbine Oils 485
13.13 Gas Turbine Oils: Application and Requirements 486
13.14 Fire-Resistant, Water-Free Fluids for Power Station Applications 487
13.15 Lubricants for Water Turbines and Hydroelectric Plants 488
References 489

Volume 2

14 Metalworking Fluids *491*

Theo Mang, Carmen Freiler, and Dietrich Hörner

14.1 Action Mechanism and Cutting Fluid Selection *492*
14.1.1 Lubrication *493*
14.1.2 Cooling *494*
14.1.3 Significance of Cutting Fluid with Various Cutting Materials *496*
14.1.3.1 High-Speed Steels *497*
14.1.3.2 Cemented Carbide Metals *497*
14.1.3.3 Coated Carbide Metals *497*
14.1.3.4 Ceramic Materials *498*
14.1.3.5 Cubic Boron Nitride (CBN) *498*
14.1.3.6 Polycrystalline Diamond (PCD) *498*
14.1.3.7 Coatings *498*
14.1.4 Cutting Fluid Selection for Various Cutting Methods and Cutting Conditions *499*
14.2 Friction and Wear Assessment Method for the Use of Cutting Fluids *501*
14.2.1 Tool Life and Number of Parts Produced by the Tool as Practical Assessment Parameters *502*
14.2.2 Measuring Cutting Forces in Screening Tests *502*
14.2.3 Feed Rates at Constant Feed Force *503*
14.2.4 Measuring Tool Life by Fast-Screening Methods *503*
14.2.5 Cutting Geometry and Chip Flow *504*
14.2.6 Other Fast Testing Methods *504*
14.2.6.1 Temperature Measurement *504*
14.2.6.2 Radioactive Tools *504*
14.2.6.3 Surface Finish *505*
14.3 Water-Miscible Cutting Fluids *505*
14.3.1 Nomenclature and Breakdown *506*
14.3.2 Composition *507*
14.3.2.1 Emulsifiers *508*
14.3.2.2 Viscosity of Emulsions *514*
14.3.2.3 Phase Reversal, Determination of the Type of Emulsion *514*
14.3.2.4 Degree of Dispersion *516*
14.3.2.5 Stability *517*
14.3.2.6 Corrosion Inhibitors and Other Additives *519*
14.3.2.7 Cutting Fluids Containing Emulsifiers *521*
14.3.2.8 Coolants Containing Polyglycols *523*
14.3.2.9 Salt Solutions *523*
14.3.3 Corrosion Protection and Corrosion Test Methods *523*
14.3.4 Concentration of Water-Mixed Cutting Fluids *524*
14.3.4.1 Determination of Concentration by DIN 51 368 (IP 137) *525*
14.3.4.2 Concentration Measurement Using Hand-Held Refractometers *526*

14.3.4.3 Concentration Measurement Through Individual Components 526
14.3.4.4 Determination of Concentration by Titration of Anionic Components 526
14.3.4.5 Determination of Concentration Through Alkali Reserve 527
14.3.4.6 Concentration After Centrifuging 527
14.3.5 Stability of Coolants 527
14.3.5.1 Determination of Physical Emulsion Stability 527
14.3.5.2 Electrolyte Stability 528
14.3.5.3 Thermal Stability 529
14.3.5.4 Stability to Metal Chips 530
14.3.6 Foaming Properties 531
14.3.6.1 Definition and Origin of Foam 531
14.3.6.2 Foam Prevention 531
14.3.6.3 Methods of Determining Foam Behaviour 533
14.3.7 Metalworking Fluid Microbiology 534
14.3.7.1 Hygienic and Toxicological Aspects of Microorganisms 535
14.3.7.2 Methods of Determining Microbial Count 536
14.3.7.3 Determination of the Resistance of Water-Miscible Coolants Towards Microorganisms 537
14.3.7.4 Reducing or Avoiding Microbial Growth in Coolants 537
14.3.8 Preservation of Coolants with Biocides 539
14.3.8.1 Aldehydes 541
14.3.8.2 Formaldehyde Release Compounds 541
14.3.8.3 Phenol Derivatives 543
14.3.8.4 Compounds Derived from Carbon Disulfide 543
14.3.8.5 Isothiazoles 544
14.3.8.6 Fungicides 544
14.3.8.7 Hypochlorites 544
14.3.8.8 Hydrogen Peroxide, H_2O_2 544
14.3.8.9 Quaternary Ammonium Compounds 544
14.4 Neat Cutting Fluids 545
14.4.1 Classification of Neat Metalworking Oils According to Specifications 545
14.4.2 Composition of Neat Metalworking Fluids 545
14.4.2.1 Base Oils and Additives 545
14.4.2.2 Significance of Viscosity on the Selection of Neat Products 546
14.4.3 Oil Mist and Oil Evaporation Behaviour 547
14.4.3.1 Evaporation Behaviour 548
14.4.3.2 Low-Misting Oils 548
14.4.3.3 The Creation of Oil Mist 548
14.4.3.4 Sedimentation and Separation of Oil Mists 549
14.4.3.5 Toxicity of Oil Mist 550
14.4.3.6 Oil Mist Measurement 552
14.4.3.7 Oil Mist Index 553
14.4.3.8 Oil Mist Concentration in Practice 553
14.5 Machining with Geometrically Defined Cutting Edges 554

14.5.1 Turning *554*
14.5.2 Drilling *557*
14.5.3 Milling *557*
14.5.4 Gear Cutting *558*
14.5.5 Deep Hole Drilling *559*
14.5.5.1 Deep Hole Drilling Methods *560*
14.5.5.2 Tasks to be Fulfilled by the Cutting Fluid *561*
14.5.6 Threading and Tapping *562*
14.5.7 Broaching *563*
14.6 Machining with Geometric Non-Defined Cutting Edges *564*
14.6.1 Grinding *564*
14.6.1.1 High-Speed Grinding *565*
14.6.1.2 Grinding Wheel Abrasive Materials and Bondings *565*
14.6.1.3 Requirements for Grinding Fluids *566*
14.6.1.4 Special Workpiece Material Considerations *567*
14.6.1.5 CBN High-Speed Grinding *567*
14.6.1.6 Honing *568*
14.6.1.7 Honing Oils *570*
14.6.1.8 Lapping *571*
14.6.1.9 Lapping Powder and Carrier Media *571*
14.7 Specific Material Requirements for Machining Operations *572*
14.7.1 Ferrous Metals *572*
14.7.1.1 Steel *572*
14.7.1.2 Tool Steels *572*
14.7.1.3 High-Speed Steels (HSS) *572*
14.7.1.4 Stainless Steels *573*
14.7.1.5 Cast Iron *573*
14.7.2 Aluminium *573*
14.7.2.1 Influence of the Type of Aluminium Alloy *573*
14.7.2.2 The Behaviour of Aluminium During Machining *575*
14.7.2.3 Tool Materials *577*
14.7.3 Magnesium and its Alloys *577*
14.7.4 Cobalt *579*
14.7.4.1 The Health and Safety Aspects of Carbides *579*
14.7.4.2 Use of Cutting Oils in Carbide Machining Processes *580*
14.7.5 Titanium *580*
14.7.6 Nickel and Nickel Alloys *581*
14.8 Metalworking Fluid Circulation System *581*
14.8.1 Metalworking Fluid Supply *582*
14.8.1.1 Grinding *584*
14.8.2 Individually-Filled Machines and Central Systems *585*
14.8.3 Tramp Oil in Coolants *585*
14.8.4 Separation of Solid Particles *587*
14.8.4.1 Swarf Concentration and Filter Fineness *587*
14.8.4.2 Full, Partial or Main Flow Solids Separation *588*

14.8.4.3 Filtration Processes 588
14.8.4.4 Solids Separation Equipment 591
14.8.5 Plastics and Sealing Materials in Machine Tools – Compatibility with Cutting Fluids 598
14.8.6 Monitoring and Maintenance of Neat and Water-Miscible Cutting Fluids 599
14.8.6.1 Storage of Cutting Fluids 599
14.8.6.2 Mixing Water-Miscible Cutting Fluids 599
14.8.6.3 Monitoring Cutting Fluids 600
14.8.6.4 Cutting Fluid Maintenance 601
14.8.6.5 Corrective Maintenance for Neat and Water-miscible Cutting Fluids 603
14.8.7 Splitting and Disposal 605
14.8.7.1 Disposal of Cutting Fluids 605
14.8.7.2 Evaluation Criteria for Cutting Fluid Water Phases 606
14.8.7.3 Electrolyte Separation 607
14.8.7.4 Emulsion Separation by Flotation 609
14.8.7.5 Splitting of Emulsions with Adsorbents 609
14.8.7.6 Separating Water-Miscible Cutting Fluids by Thermal Methods 610
14.8.7.7 Ultrafiltration 610
14.8.7.8 Evaluation of Disposal Methods 612
14.9 Coolant Costs 613
14.9.1 Coolant Application Costs 613
14.9.1.1 Investment Costs (Depreciation, Financing Costs, Maintenance Costs) 613
14.9.1.2 Energy Costs 613
14.9.1.3 Coolant and Coolant Additives 614
14.9.1.4 Coolant Monitoring 614
14.9.1.5 Other Auxiliaries 614
14.9.1.6 Coolant Separation and Disposal 614
14.9.2 Coolant Application Costs with Constant System 614
14.9.2.1 Specific Coolant Costs 614
14.9.2.2 Optimization of Coolant use by Computer 618
14.10 New Trends in Coolant Technology 619
14.10.1 Oil Instead of Emulsion 619
14.10.1.1 Fluid Families and Multifunctional Fluids for Machine Tools 621
14.10.1.2 Washing Lines 622
14.10.1.3 De-Oiling of Chips and Machined Components 622
14.10.1.4 Future Perspectives–Unifluid 622
14.10.2 Minimum Quantity Lubrication 623
14.10.2.1 Considerations When Dispensing with Coolants 623
14.10.2.2 Minimum Quantity Lubrication Systems 625
14.10.2.3 Coolants for Minimum Quantity Lubrication 626
14.10.2.4 Oil Mist Tests with Minimum Quantity Lubrication 628

14.10.2.5 Product Optimization of a Minimum Quantity Coolant Medium for Drilling *630*
References *631*

15 Forming Lubricants *639*
Theo Mang
15.1 Sheet Metal Working Lubricants *639*
Theo Mang, Achim Losch, and Franz Kubicki
15.1.1 Processes *640*
15.1.2 Basic Terms in Forming Processes *640*
15.1.2.1 Lattice Structure of Metals *640*
15.1.2.2 Yield Strength *641*
15.1.2.3 Strain *641*
15.1.2.4 Flow Curve *641*
15.1.2.5 Efficiency of Deformation, Resistance to Forming, Surface Pressure *643*
15.1.2.6 Strain Rate *643*
15.1.2.7 Anisotropy, Texture: R Value *644*
15.1.3 Deep Drawing *644*
15.1.3.1 Friction and Lubrication in the Different Areas of a Deep Drawing Operation *645*
15.1.3.2 Significance of Lubrication Dependent upon Sheet Metal Thickness, Drawn-Part Size and the Efficiency of Deformation *648*
15.1.3.3 Assessment of the Suitability of Lubricants for Deep Drawing *650*
15.1.4 Stretch Drawing and a Combination of Stretch and Deep Drawing *652*
15.1.5 Shear Cutting *652*
15.1.5.1 Stamping *654*
15.1.5.2 Fineblanking *657*
15.1.6 Material and Surface Microstructure *659*
15.1.6.1 Material *659*
15.1.7 Tools Used in Sheet Metal Forming Operations *661*
15.1.8 Lubricants for Sheet Metal Forming *663*
15.1.8.1 Application and Types of Lubricants *664*
15.1.8.2 Lubricant Behaviour During Forming *668*
15.1.8.3 Post-Processes *670*
15.1.8.4 Trends in Sheet Metal Forming Lubrication *671*
15.1.9 Special Case: Automobile Manufacturing *672*
15.1.9.1 Prelubes *672*
15.1.9.2 Skin Passing *673*
15.1.9.3 Coil Oiling *673*
15.1.9.4 Transport and Storage of Steel Coils or Blanks *674*
15.1.9.5 Washing of Steel Strips and Blanks *674*
15.1.9.6 Additional Lubrication *674*

15.1.9.7 Pressing 675
15.1.9.8 Transport and Storage of Pressed Parts 675
15.1.9.9 Welding and Adhesive Bonding 676
15.1.9.10 Cleaning and Phosphating 677
15.1.9.11 Cataphoretic Painting 677
15.1.9.12 Savings Potential Using Prelubes 678
15.1.9.13 Dry-Film Lubricants 678
15.1.10 Special Case: Water-Based Synthetics 679
15.1.10.1 Introduction and Historical Background 679
15.1.10.2 Synthetic Lubricants Today 681
15.1.11 Testing Tribological Characteristics 682
15.1.11.1 General Considerations 682
15.1.11.2 Strip Drawing 683
15.1.11.3 Cup Drawing 685
15.1.12 Corrosion Protection 690
15.1.12.1 Corrosion Mechanisms 690
15.1.12.2 Temporary Corrosion Protection 691
15.1.12.3 Corrosion Tests 693
15.2 Lubricants for Wire, Tube and Profile Drawing 693
Theo Mang
15.2.1 Friction and Lubrication, Tools and Machines 693
15.2.1.1 Forming Classification 693
15.2.1.2 Friction and Lubrication, Machines and Tools When Wire Drawing 694
15.2.1.3 Drawing Force and Tension 695
15.2.1.4 Drawing Tool and Wear 697
15.2.1.5 Wire Cracks 699
15.2.1.6 Hydrodynamic Drawing 699
15.2.1.7 Wire Friction on Cone 700
15.2.1.8 Lubricant Feed in Wet Drawing 703
15.2.1.9 Dry Drawing 703
15.2.1.10 Applying Lubricant as Pastes or High-Viscosity Products 704
15.2.2 Drawing Copper Wire 705
15.2.2.1 Lubricants 706
15.2.2.2 Lubricant Concentration 707
15.2.2.3 Solubility of Copper Reaction Products 708
15.2.2.4 Water Quality and Electrolyte Stability 708
15.2.2.5 Laboratory Testing Methods 708
15.2.2.6 Lubricant Temperature 709
15.2.2.7 Influence of the Lubricant on Wire Enamelling 709
15.2.2.8 Circulation Systems, Cleaning and Disposal of Drawing Emulsions 710
15.2.3 Drawing of Steel Wire 710
15.2.3.1 Requirements 710
15.2.3.2 Lubricant Carrier Layers 711

15.2.3.3 Lime as a Lubricant Carrier *712*
15.2.3.4 Borax as Lubricant Carrier *712*
15.2.3.5 Phosphate as Lubricant Carrier *712*
15.2.3.6 Oxalate Coatings and Silicates *712*
15.2.3.7 Lubricants for Steel Wire Drawing *713*
15.2.4 Drawing Aluminium Wire *714*
15.2.4.1 Drawing Machines and Lubrication *715*
15.2.4.2 Lubricants for Aluminium Wire Drawing *715*
15.2.5 Wire from Other Materials *716*
15.2.5.1 Stainless Steel *716*
15.2.5.2 Nickel *716*
15.2.5.3 Tungsten *716*
15.2.6 Profile Drawing *716*
15.2.6.1 Lubricating Tasks in Profile Drawing *717*
15.2.6.2 Pretreatment and the Use of Lubricant When Profile Drawing Steel *718*
15.2.7 Tube Drawing *718*
15.2.7.1 Tube-Drawing Methods *718*
15.2.7.2 Tools and Tool Coatings *720*
15.2.7.3 Lubricants and Surface Pretreatment for Tube Drawing *721*
15.2.8 Hydroforming *723*
15.2.8.1 Process Principle *725*
15.2.8.2 Process Configuration *725*
15.2.8.3 Tribological Aspects of Hydroforming *727*
15.2.8.4 Lubricants for Hydroforming *727*
15.3 Lubricants for Rolling *728*
Theo Mang
15.3.1 General *728*
15.3.1.1 Rolling Speed *729*
15.3.1.2 Rationalization *730*
15.3.1.3 Surface and Material Quality *730*
15.3.1.4 Hygienic Commercial Requirements *730*
15.3.2 Friction and Lubrication When Rolling *730*
15.3.3 Rolling Steel Sheet *735*
15.3.3.1 Hot Rolling *735*
15.3.3.2 Sheet Cold Rolling *736*
15.3.3.3 Finest Sheet Cold Rolling *740*
15.3.3.4 Cold Rolling of High Alloy Steel Sheet *741*
15.3.4 Rolling Aluminium Sheet *742*
15.3.5 Aluminium Hot Rolling *744*
15.3.6 Aluminium Cold Rolling *745*
15.3.7 Rolling Other Materials *746*
15.4 Solid Metal Forming Lubricants: Solid Forming, Forging and Extrusion *747*
Theo Mang, Wolfgang Buss

15.4.1 Processes 747
15.4.1.1 Upsetting 747
15.4.1.2 Extrusion 747
15.4.1.3 Impression Die Forging 748
15.4.1.4 Open Die Forging 748
15.4.2 Forming Temperatures 748
15.4.2.1 Cold 749
15.4.2.2 Warm 749
15.4.2.3 Hot 749
15.4.3 Friction and Lubrication with Cold Extrusion and Cold Forging 749
15.4.3.1 Friction and Lubricant Testing Methods 750
15.4.3.2 Selection Criteria for Lubricants and Lubrication Technology 752
15.4.3.3 Lubricating Oils for Cold Extrusion of Steel: Extrusion Oils 753
15.4.3.4 Phosphate Coatings and Soap Lubricants for Cold Extrusion of Steel 755
15.4.3.5 Solid Lubricants for Cold Extrusion of Steel 758
15.4.4 Warm Extrusion and Forging 760
15.4.4.1 Temperature Range Up to 350 °C 762
15.4.4.2 Temperature Range 350–500 °C 763
15.4.4.3 Temperature Range 500–600 °C 763
15.4.4.4 Temperature Range >600 °C 763
15.4.5 Lubrication When Hot Forging 763
15.4.5.1 Demands on Hot Forging Lubricants 764
15.4.5.2 Lubricant Testing Methods 766
15.4.6 Hot Forging of Steel 766
15.4.6.1 Lubricants 767
15.4.7 Aluminium Forging 769
15.4.8 Isothermal and Hot Die Forging 769
15.4.9 Application and Selection of Lubricant 770
References 773

16 Lubricating Greases 781
Wilfried Dresel and Rolf-Peter Heckler
16.1 Introduction 781
16.1.1 Definition 781
16.1.2 History 782
16.1.3 Advantages over Lubricating Oils 782
16.1.4 Disadvantages 783
16.1.5 Classification 783
16.2 Thickeners 784
16.2.1 Simple Soaps 785
16.2.1.1 Soap Anions 785
16.2.1.2 Soap Cations 785
16.2.1.3 Lithium Soaps 785
16.2.1.4 Calcium Soaps 786

16.2.1.5 Sodium Soaps *787*
16.2.1.6 Other Soaps *787*
16.2.1.7 Cation Mixed Soaps M_1X/M_2X *788*
16.2.1.8 Anion Mixed Soaps MX_1/MX_2 *789*
16.2.2 Complex Soaps *789*
16.2.2.1 Lithium Complex Soaps *789*
16.2.2.2 Calcium Complex Soaps *791*
16.2.2.3 Calcium Sulfonate Complex Soaps *792*
16.2.2.4 Aluminium Complex Soaps *792*
16.2.2.5 Other Complex Soaps *793*
16.2.3 Other Ionic Organic Thickeners *793*
16.2.4 Non-Ionic Organic Thickeners *793*
16.2.4.1 Diureas and Tetraureas *794*
16.2.4.2 Other Non-Ionic Organic Thickeners *795*
16.2.5 Inorganic Thickeners *795*
16.2.5.1 Clays *795*
16.2.5.2 Highly Dispersed Silicic Acid *796*
16.2.6 Miscellaneous Thickeners *796*
16.2.7 Temporarily Thickened Fluids *796*
16.3 Base Oils *797*
16.3.1 Mineral Oils *798*
16.3.2 Synthetic Base Oils *798*
16.3.2.1 Synthetic Hydrocarbons *798*
16.3.2.2 Other Synthetic Base Oils *799*
16.3.2.3 Immiscible Base Oil Mixtures *799*
16.4 Grease Structure *799*
16.5 Additives *800*
16.5.1 Structure Modifiers *801*
16.5.2 Antirust Additives (Corrosion Inhibitors) *801*
16.5.3 Extreme-Pressure and Anti-Wear Additives *801*
16.5.4 Solid Lubricants *802*
16.5.5 Friction Modifiers *802*
16.5.6 Nanomaterials *803*
16.6 Manufacture of Greases *803*
16.6.1 Metal Soap-Based Greases *804*
16.6.1.1 Batch Production with Preformed Metal Soaps *804*
16.6.1.2 Batch Production with Metal Soaps Prepared In Situ *804*
16.6.1.3 Continuous Production *806*
16.6.2 Oligourea Greases *806*
16.6.3 Gel Greases *807*
16.7 Grease Rheology *807*
16.8 Grease Performance *808*
16.8.1 Test Methods *810*
16.8.2 Analytical Methods *812*
16.9 Applications of Greases *812*

16.9.1 Rolling Bearings 813
16.9.1.1 Re-Lubrication Intervals 814
16.9.2 Cars, Trucks and Construction Vehicles 815
16.9.3 Steel Mills 819
16.9.4 Mining 820
16.9.5 Railroad, Railway 820
16.9.6 Gears 821
16.9.7 Food-Grade Applications 821
16.9.8 Textile Machines 822
16.9.9 Application Techniques 822
16.9.10 Special and Lifetime Applications 822
16.9.11 Applications with Polymeric Materials 823
16.10 Grease Market 824
16.11 Ecology and the Environment 825
16.12 Grease Tribology 827
References 827

17 Solid Lubrication *843*
Christian Busch
17.1 Classification of Solid Lubricants 845
17.1.1 Class 1: Structural Lubricants 846
17.1.2 Class 2: Mechanical Lubricants 847
17.1.2.1 Self-Lubricating Substances 847
17.1.2.2 Substances with Lubricating Properties That Need a Supporting Medium 850
17.1.2.3 Substances with Indirect Lubricating Properties Based on Their Hardness (Physical Vapour Deposition, Chemical Vapour Deposition and Diamond-Like Carbon Layers) 851
17.1.3 Class 3: Soaps 852
17.1.4 Class 4: Chemically Active Substances 852
17.2 Characteristics 852
17.2.1 The Crystal Structures of Lamellar Solid Lubricants 853
17.2.1.1 Graphite – C 853
17.2.1.2 Molybdenum Disulfide 855
17.2.1.3 Tungsten(IV) Sulfide 856
17.2.1.4 Boron Nitride 856
17.2.2 Heat Stability of Lamellar Solid Lubricants 857
17.2.3 Melting Point 858
17.2.4 Thermal Conductivity 858
17.2.5 Adsorbed Films 858
17.2.6 Mechanical Properties 859
17.2.7 Chemical Stability 859
17.2.8 Purity 859
17.2.9 Particle Size 860
17.3 Products Containing Solid Lubricants 860

17.3.1 Powders *860*
17.3.1.1 Solid Lubricants in Carrying Media *861*
17.3.2 Dispersions and Suspensions *862*
17.3.3 Greases and Grease Pastes *863*
17.3.4 Pastes *864*
17.3.5 Bonded Solid Lubricants or Dry-Film Lubricants *865*
17.4 Industrial Uses of Products Containing Solid Lubricants *871*
17.4.1 Screw Lubrication *872*
17.4.2 Roller-Bearing Lubrication *874*
17.4.3 Slide Bearing, Slide Guideway and Slide Surface Lubrication *875*
17.4.4 Chain Lubrication *876*
17.4.5 Plastic and Elastomer Lubrication *876*
Further Reading *877*
Journals *877*
Standards, Reprints *879*
Books *879*

18 Laboratory Methods for Testing Lubricants *881*
Roman Müller
18.1 Introduction *881*
18.2 Density *881*
18.3 Viscosity *882*
18.3.1 Capillary Viscometers *882*
18.3.2 Rotary Viscometers *882*
18.4 Refractive Index *883*
18.5 Structural Analyses *883*
18.6 Flash Point *884*
18.7 Surface Phenomena *884*
18.7.1 Air Release *884*
18.7.2 Water Separation and Demulsibility *884*
18.7.3 Foaming Characteristics *885*
18.8 Cloud Point, Pour Point *885*
18.9 Aniline Point *885*
18.10 Water Content *886*
18.11 Ash Content *886*
18.12 Acidity, Alkalinity *886*
18.13 Ageing Tests *887*
18.14 Hydrolytic Stability *888*
18.15 Corrosion Tests *888*
18.16 Oil Compatibility of Seals and Insulating Materials *889*
18.17 Evaporation Loss *889*
18.18 Analysis and Testing of Lubricating Greases *890*
18.18.1 Consistency *890*
18.18.2 Dropping Point *890*
18.18.3 Oil Separation *890*

18.18.4 Shear Stability of Greases *891*
18.18.4.1 Prolonged Grease Working *891*
18.18.4.2 Roll Stability of Lubricating Greases *891*
18.18.5 High-temperature Performance *891*
18.18.6 Wheel Bearing Leakage *891*
18.18.6.1 Leakage Tendency of Automotive Wheel Bearing Greases *891*
18.18.6.2 Wheel Bearing Leakage Under Accelerated Conditions *892*
18.18.7 Wheel Bearing Life *892*
18.18.8 Water Resistance *892*
18.18.8.1 Water Washout Characteristics *892*
18.18.8.2 Water Spray-Off Resistance *892*
18.18.9 Oxidation Stability of Lubricating Greases by the Oxygen Pressure-Vessel Method *893*
18.18.10 Corrosion-preventive Characteristics *893*
18.18.10.1 Rust Test *893*
18.18.10.2 EMCOR Test *893*
18.18.10.3 Copper Corrosion *894*
18.19 Elemental Analyses by Spectroscopic Methods *894*
18.19.1 Atomic Absorption Spectroscopy – AAS *894*
18.19.2 Atom Emission Spectroscopy (AES) and Inductively Coupled Plasma Atomic Emission Spectroscopy (ICP-AES) *895*
18.19.3 Energy Dispersive X-Ray Fluorescence (EDXRF) *895*
18.19.4 Wavelenght Dispersive X-Ray Fluorescence (WDXRF) *895*
18.20 List of Equivalent Standardized Methods for Testing Lubricants *896*
References *901*

19 Mechanical–Dynamic Test Methods and Tribology *903*
Thorsten Bartels
19.1 Tribological System Categories within Mechanical–Dynamic Tests *906*
19.2 Simple Tribological Mechanical–Dynamic Test Machines and Test Methods *907*
19.2.1 Four-Ball Apparatus *907*
19.2.2 Brugger Apparatus and Reichert's Friction–Wear Balance *909*
19.2.3 Falex Test Machines *910*
19.2.3.1 Falex Block-on-ring Test Machine *910*
19.2.3.2 Falex Pin and Vee Block Test Machine *911*
19.2.3.3 Falex High-Performance Multispecimen Test Machine *911*
19.2.3.4 Falex Tapping Torque Test Machine *913*
19.2.4 Timken Test Machine *913*
19.2.5 High-Frequency Reciprocating Test Machines *914*
19.2.5.1 High-Frequency Reciprocating Rig (HFRR) *914*
19.2.5.2 High-Frequency, Linear-Oscillation Test Machine (SRV) *915*
19.2.6 Mini Traction Machine (MTM) *917*

19.2.7 Low-Velocity Friction Apparatus (LVFA): The Tribometer *919*
19.2.8 Diesel Injector Apparatus *919*
19.2.9 Further Standardized and Non-Standardized Test Machines and Test Methods *919*
19.3 Mechanical–Dynamic Tests for Gearbox and Transmission Application *920*
19.3.1 Variety of Gear and Transmission Types *920*
19.3.2 Gears and Roller Bearings *922*
19.3.2.1 Gears *922*
19.3.2.2 Roller Bearings *924*
19.3.3 Gear and Roller Bearing Tribo-Performance Test Rigs *925*
19.3.3.1 Older Test Rigs Used for Gear and Axle Application *925*
19.3.3.2 FZG Gear-Test Rig *927*
19.3.3.3 The Micro-Pitting Rig (MPR) *928*
19.3.3.4 The High-Temperature Bearing Tester (HTHT) *929*
19.3.3.5 The Roller Bearing Test Apparatus (FE8) *930*
19.3.3.6 The Roller Bearing Test Apparatus (FE9) *931*
19.3.4 Types of Tooth Flank and Roller Bearing Damage *933*
19.3.4.1 Wear and Wear Performance Tests *934*
19.3.4.2 Scuffing/Scoring and Performance Tests *936*
19.3.4.3 Surface Fatigue/Micro-Pitting and Performance Tests *939*
19.3.4.4 Sub-Surface Fatigue/Pitting and Performance Tests *941*
19.3.4.5 Tooth Fracture *944*
19.3.5 Gear Efficiency and Fuel Economy *944*
19.3.6 Transmission Trends with Regard to Tribology and Oil Ageing *945*
19.3.7 Synchronizer Applications and Performance Tests *948*
19.3.7.1 Area of Application *948*
19.3.7.2 Function and Tribology of the Synchronizer *948*
19.3.7.3 Standardized Test Rigs and Test Methods *952*
19.3.8 Wet Clutch/Break Applications and Performance Tests *956*
19.3.8.1 Area of Application *956*
19.3.8.2 Function and Tribology of the Friction Disks *956*
19.3.8.3 Standardized Test Rigs and Test Methods *957*
19.3.9 Variator Applications and Performance Tests *963*
19.3.9.1 Area of Application *963*
19.3.9.2 Function and Tribology of Chain, Push Belt and Fluid *964*
19.3.9.3 Standardized Test Rigs and Test Methods *965*
19.4 Mechanical–Dynamic Tests for Internal Combustion Engines *968*
19.4.1 Variety of Internal Combustion Engines *968*
19.4.2 Area of Application and Operation of Two-Stroke Engines *969*
19.4.3 Mechanical–Dynamic Tests for Two-Stroke Engines *971*
19.4.4 Area of Application and Operation of Four-Stroke Engines *973*
19.4.5 Mechanical–Dynamic Tests for Four-Stroke Engines *975*
19.4.5.1 Piston *977*

19.4.5.2 Piston Pin to Piston Contact *978*
19.4.5.3 Piston Skirt to Cylinder Block *979*
19.4.5.4 Piston Rings to Cylinder Block *981*
19.4.5.5 Crankshaft and Crankshaft Bearings *982*
19.4.5.6 Camshaft to Cam Follower, Valves, Valve Mechanisms and Valve Train *984*
19.4.5.7 Gear Trains: Timing Gear Chain and Timing Belt *987*
19.4.5.8 Fuel Supply System *987*
19.4.5.9 Oil Pump *989*
19.4.5.10 Oil Filter *990*
19.4.5.11 Single- and Dual-Mass Flywheel Systems *990*
19.4.5.12 Auxiliary Assemblies *992*
19.4.5.13 Engine Testing *992*
19.4.6 Rotary Engines *996*
19.4.6.1 Application and Operation of Rotary Engines *996*
19.4.6.2 Tribology of Rotary Engines *998*
19.4.7 Gas Turbines: High-Power Engines *1000*
19.4.7.1 Area of Application and Operation of Gas Turbines *1000*
19.4.7.2 Gas Turbine Engines *1003*
19.4.8 Future Trends *1005*
19.4.8.1 Fuel Economy *1005*
19.4.8.2 Extended Drain and Biofuel Compatibility *1006*
19.4.8.3 Emissions *1008*
19.4.8.4 Weight Reduction *1008*
19.5 Hydraulic Pump and Circuit Design *1009*
19.5.1 Hydraulic Fundamentals *1009*
19.5.2 Area of Hydraulic Application and Operation *1010*
19.5.3 Hydraulic System Components and Loop Circuits *1011*
19.5.4 Tribology of Hydraulic Pumps and Performance Tests *1012*
19.5.4.1 Gear Pumps *1014*
19.5.4.2 Vane Pumps *1015*
19.5.4.3 Piston Pumps *1018*
19.5.4.4 Hybrid Pumps *1020*
19.5.5 Valves *1022*
19.5.6 Seals, Gaskets and Elastomers *1023*
19.5.7 Fuel Efficiency and Emissions Reductions *1026*
19.5.7.1 Trends and Key Factors *1026*
19.5.7.2 The Role of Hydraulic Fluids for Energy Efficiency *1026*
19.5.7.3 Hydraulic Pump Efficiency Test Rigs and Test Methods *1027*
19.5.7.4 Test Results of Pump Efficiency Tests and Field Test Verification *1031*
19.6 Interpretation and Precision of Tribological Mechanical–Dynamic Testing *1041*
Acknowledgements *1041*
References *1041*

20 Lubrication Systems *1053*
Theo Mang
20.1 Introduction *1053*
20.2 The Taxonomy of Centralized Lubrication Systems DIN 24271 Part 1 and DIN ISO 5170 *1054*
20.3 Total-Loss Lubrication Systems *1055*
20.3.1.1 Structure of Centralized Lubrication Systems *1055*
20.3.2 Single-line System *1055*
20.3.2.1 System Description *1055*
20.3.2.2 Applications *1056*
20.3.2.3 Single-Line System with Piston Distributors *1057*
20.3.3 Dual-Line System *1057*
20.3.3.1 System Description *1058*
20.3.3.2 Applications *1059*
20.3.4 Oil + Air System *1059*
20.3.4.1 Applications *1060*
20.3.5 Oil–Mist System *1060*
20.3.5.1 System Description *1060*
20.3.5.2 Applications *1061*
20.4 Circulating Lubrication Systems *1061*
20.4.1 General Remarks *1061*
20.4.2 Circulating Lubrication Systems with Orifice Tubes *1062*
20.4.2.1 General Remarks *1062*
20.4.3 Symmetric Systems *1063*
20.4.4 Circulating Lubrication Systems with Multicircuit Pumps *1064*
20.4.4.1 General Remarks *1064*
20.4.4.2 Structure of a System *1065*
20.5 Special Applications *1065*
20.5.1 Vehicle Lubrication *1065*
20.5.1.1 General Remarks *1065*
20.5.1.2 Lubricants in the Vehicle Sector *1066*
20.5.2 Wheel Flange Lubrication for Rail Vehicles *1069*
20.5.2.1 General Remarks *1069*
20.5.2.2 Types of Systems *1071*
20.5.2.3 Systems *1072*
20.5.3 System Example, System 2 – Metering by Piston with Ring Groove *1072*
20.5.3.1 Wheel Flange Lubrication System for One Direction of Travel and Curve-Dependent Lubrication *1072*
20.5.4 Chain Lubrication *1074*
20.5.4.1 General Remarks *1074*
20.5.4.2 Chain Friction Points *1076*
20.5.4.3 Lubrication Systems *1077*
20.5.4.4 Lubrication Systems for Oil *1077*
20.5.4.5 Lubrication Systems for Grease *1077*

20.5.5 Minimal Quantity Lubrication (MQL) *1077*
20.5.5.1 General Remarks *1077*
20.5.5.2 Advantages and Disadvantages of (Cooling) Lubrication Systems *1079*
20.5.5.3 Equipment and Systems *1079*
20.5.6 Hydrostatic Lubrication *1081*
20.5.6.1 Advantages and Disadvantages *1082*
20.5.6.2 Application *1082*
20.5.6.3 Supply Systems *1083*
20.5.7 Wind Energy Systems *1084*
20.5.7.1 Lube Points *1084*
20.5.7.2 Centralized Lubrication System *1084*
20.5.8 Large Diesel Engines *1084*
20.5.8.1 General Remarks *1084*
Reference *1088*

21 Removal of Lubricants: Industrial Cleaners *1089*
Achim Losch
21.1 Introduction to Industrial Cleaning *1089*
21.2 Substrates (Workpieces) *1090*
21.3 Contamination: Soil *1091*
21.3.1 Oils and Emulsions *1091*
21.3.2 Greases, Pastes and Compounds *1092*
21.3.3 Inorganic Particles and Salts *1092*
21.4 Mechanisms of Aqueous Cleaning *1093*
21.5 Detection and Control of Cleaning Result *1095*
21.5.1 Water Break Test *1096*
21.5.2 Determination of Surface Carbon *1097*
21.5.3 Copper Cementation *1097*
21.5.4 IR Oil Film Analyser *1097*
21.5.5 UV Fluorescence Analyser *1097*
21.5.6 Surface Tension Inks *1098*
21.6 Cleaning Methods and Equipment *1098*
21.6.1 Avoiding of Cleaning *1099*
21.6.2 Pickling and Precleaning *1100*
21.6.3 Aqueous Immersion Cleaning *1100*
21.6.4 Emulsion Cleaners *1101*
21.6.5 Aqueous Spray Cleaning *1102*
21.6.6 Manual Part Washing *1103*
21.6.7 Ultrasound Cleaning *1104*
21.6.8 Electrolytic Cleaning *1105*
21.6.9 Solvent Cleaning and Vacuum Systems *1105*
21.6.10 Solvent Vapour Cleaning *1106*
21.6.11 Cleaning after Heat Treatment *1106*
21.6.12 Rinsing, Passivation and Corrosion Protection *1106*

21.6.13 Comparison of Cleaning Methods *1107*
21.7 Aqueous Cleaners *1108*
21.7.1 Builders *1110*
21.7.2 Surfactants *1110*
21.7.2.1 Non-ionic Surfactants *1111*
21.7.2.2 Cationic Surfactants *1113*
21.7.2.3 Anionic Surfactants *1113*
21.7.3 Sequestrants and Inhibitors *1114*
21.7.4 Water *1115*
21.7.5 Emulsion Cleaners *1115*
21.8 Solvent Cleaners *1115*
21.8.1 Hydrocarbon Solvents *1116*
21.8.2 Halogenated Solvents *1117*
21.8.3 Other Solvent Cleaners *1118*
21.9 Maintenance of Cleaner Systems *1119*
21.9.1 Gravity Separators *1120*
21.9.2 Dimensional Separators *1120*
21.9.3 Other Separators *1121*
21.9.4 Replenishment of Chemicals *1122*
References *1122*

Index *1125*

List of Contributors

Thorsten Bartels
Dr.-Ing., Weisenheim am. Sand, Germany
Evonik Resource Efficiency GmbH
Director Performance Testing

Wolfgang Bock
Dipl.-Ing., Weinheim, Germany
Fuchs Schmierstoffe GmbH
International Product Management Industrial Oils

Jürgen Braun
Dr. rer. nat., Speyer, Germany
Fuchs Schmierstoffe GmbH
Head of R&D for Industrial Oils

Christian Busch
Prof. Dr. Ing., Zwickau, Germany
Westsächsische Hochschule Zwickau
Vice President

Wilfried Dresel
Dr. rer. nat., Mannheim, Germany
Fuchs Petrolub SE
R&D for Lubricating Greases (International)

Carmen Freiler
Dipl.-Ing., Hüttenfeld, Germany
Fuchs Schmierstoffe GmbH
Head of R&D and Product Management for Metal Cutting Fluids

Apu Gosalia
Dipl.-Kfm/MBA, Mannheim Germany
Fuchs Petrolub SE
Vice President Sustainability & Global Competitive Intelligence

Manfred Harperscheid
Dr. rer. nat., Römerberg, Germany
Fuchs Schmierstoffe GmbH
Head of R&D for Engine Oils

Rolf-Peter Heckler
Dipl.-Ing., Rimbach, Germany
Fuchs Petrolub SE
International Product Management for Lubricating Greases

Dietrich Hörner
Dr. rer. nat., Hassloch, Germany
Fuchs Petrolub SE
International Product Management for Metal working Fluids and Quenching Oils

Franz Kubicki
Dipl.-Ing., Hockenheim, Germany
Fuchs Petrolub SE
International Product Management for Corrosion Preventives and Sheet Metal forming

Georg Lingg
Dr.-Ing., Mannheim, Germany
Fuchs Petrolub SE
Member of the Executive Board,
(until 2013)

Achim Losch
Dr. rer. nat., Westhofen, Germany
Fuchs Schmierstoffe GmbH
Head of R&D for Corrosion
Preventives, Metal forming and
Cleaners

Rolf Luther
Dipl.-Phys., Speyer, Germany
Fuchs Schmierstoffe GmbH
Head of Advanced Development

Theo Mang
Prof. Dr.-Ing., Weinheim, Germany
Fuchs Petrolub SE
Group's Executive Board,
Technology, Group Purchasing,
Human Resources (retired)

Roman Müller
Chem. Eng., Mannheim, Germany
Fuchs Petrolub SE
Vice President Know How
Transfer/Group Laboratory

Christian Puhl
Dipl.-Ing., Grünstadt, Germany
Fuchs Schmierstoffe GmbH
Product Management for
Compressor and Turbine Oils

Nael Zaki
Ph.D., Shawnee, Kansas, USA
Fuchs Lubricants Co.
Lubricating Grease R&D Manager

A Word of Thanks

We thank the Vogel-Verlag for permission to use texts and illustrations from the book titled *Schmierstoffe in der Metallbearbeitung* written by Prof. Dr. Mang, published in Würzburg in 1983.

The authors thank the following persons for their specialist and linguistic contributions:

Prof. Dr. Dieter Schmoeckel and Dirk Hortig, Institut für Produktions- und Umformtechnik, Darmstadt, Germany; Prof. Will Scott, Tribology Research Group, Queensland University of Technology, Australia; Dr. Anand Kakar, FUCHS LUBRICANTS Co., Emlenton, PA; Paul Wilson, FUCHS LUBRICANTS CO., Harvey, IL; Ted McClure, FUCHS LUBRICANTS CO., Harvey, IL; Albert Mascaro, FUCHS LUBRICANTES, Castellbisbal, Spain; Cliff Lea, FUCHS LUBRICANTS (UK), Stoke-on-Trent, UK; Paul Littley, FUCHS LUBRICANTS (UK), Stoke-on-Trent, UK; Heinz-Gerhard Theis, FUCHS, Mannheim, Germany; Mercedes Kowallik FUCHS, Mannheim, Germany; Dr. Helmut Seidel, FUCHS LUBRITECH GmbH, Weilerbach, Gisela Dressler, FUCHS PETROLUB AG, Mannheim, Germany; Ursula Zelter, FUCHS PETROLUB AG, Mannheim, Germany and Jochen Held, Bolanden-Weierhof, Germany.

Roman Müller would extend his special thanks to his former Supervisor and mentor, Siegfried Noll of FUCHS PETROLUB AG, Mannheim, who drew up the initial version of Chapter 18 in the 1st edition of *Lubricants and Lubrication*.

Dr. Thortsen Bartels, Evonik Resource Efficiency GmbH, Darmstadt, the test laboratory engineers Daniel Debus, Lucas Voigt, test assistants Ludwig Herdel, Marcus Stephan and Robert Cybert for collecting and preparing descriptions, making literature searches, providing valuable information and for preparing figures used in Chapter 19. The author also thanks Wolfgang Bock and Dr. Manfred Harperscheid for the good cooperation, and for providing detailed information and their permission to use their data and figures in Chapter 10.

Prof. Dr. Theo Mang thanks SKF Lubrication Systems Germany AG, formerly Willy Vogel AG, to use the internal papers of the Company in Chapter 20.

He thanks especially Frank Bechtloff, Jan Ruiter, Götz Mehr, Alexander Tietz and Hans Gaca. He also thanks Dr. Hermann-J. Gummert, Viersen, Germany, Dr. Kai F. Karhausen, Hydro Aluminium Rolled Products GmbH, Bonn Germany as well as Dr. Hartmut Pavelski, SMS Siemag AG, Erhard Schloemann, Düsseldorf, Germany, for important Information in the area of drawing and rolling (Chapter 15).

Preface to the 3rd Edition

Nine years after the publication of the second edition of *Lubricants and Lubrication*, its high acceptance motivated the publisher and the editors to realize a third edition. The result is this largely revised and extended version in two volumes.

The use of lubricants is as old as the history of mankind but the scientific analysis of lubrication, friction and wear as an aspect of tribology is relatively new. The reduction of friction along with the reduction or even avoidance of wear by the use of lubricants and lubrication technologies results in energy savings, the protection of resources and also fewer emissions. These benefits describe the economic and ecological importance of this field of work.

This third edition spent more attention to environmental facts, to new energies, test methods for lubricants and modern applications of lubricants with centralized lubrication systems and the removal of lubricants in two additional chapters.

Only recently have lubricants begun to be viewed as functional elements in engineering and this group of substances are also attracting increasing attention from engineers.

This book offers chemists and engineers a clear interdisciplinary introduction and orientation to all major lubricant applications. The book focuses not just on the various products but also on specific application engineering criteria.

The authors are internationally recognized experts. All can draw on many years of experience in lubricant development and application.

This book offers the following readers a quick introduction to this field of work: the laboratory technician who has to monitor and evaluate lubricants; plant maintenance people for whom lubricants are an element in process technology; research and development people who have to deal with friction and wear; engineers who view lubricants as functional elements and as media which influence service life and increasingly safety and environmental protection officers who are responsible for workplace safety, an acceptable use of resources along with the reduction or avoidance of emissions and wastes.

Theo Mang
Wilfried Dresel

October 2016

Abbreviations

AAS	Atomic Absorption Spectrometry
ACEA	Association des Constructeurs Européens d'Automobiles Association of the European Car Manufacturers
AD	Ashless Dispersant
AENOR	Asociación Espanola de Normalización y Certificación Spanish Association for Normalization and Certification
AFNOR	Association Francaise de Normation French Association for Normalization
AGMA	American Gear Manufacturers Association (USA)
API	American Petroleum Institute (USA)
ASTM	American Society of Testing Materials (USA)
ASTME	American Society of Tool and Manufacturing Engineers (USA)
ATF	Automatic Transmission Fluid
ATIEL	Association Technique de L'Industrie Européenne des Lubrifiants Technical Association of the European Lubricant Industry
AW	Anti-Wear
AWT	Almen-Wieland-Test
BAM	Bundesanstalt für Materialprüfung (D) Federal Institute for Materials Research and Testing
BOD	Biological Oxygen Demand
CAFE	Californian Act for Fuel Emission (USA)
CCEL	China Certification Committee for Environmental Labelling of products
CCMC	Committee of Common Market Automobile Constructors (EU)
CCS	Cold Cranking Simulator
CEC	Coordinating European Council for the development of performance tests for lubricants and engine fuels
CETOP	Comité Européen des Transmissions Oleohydrauliques et Pneumatiques European Oil Hydraulic and Pneumatic Committee
CFMS	Closed-Field Magnetron Sputtering
COD	Chemical Oxygen Demand
CONCAWE	The Oil Companies' Organization for the Conservation of Clean Air and Water in Europe
CRC	Coordinating Research Council (EU)
CVD	Chemical Vapor Deposition
CVT	Constantly Variable Transmission
DD	Detergent and Dispersant

DGMK	Deutsche Wissenschaftliche Gesellschaft für Erdöl, Erdgas und Kohle German Scientific Society for Mineral Oil, Natural Gas and Coal
DIN	Deutsches Institut für Normung German Institute for Normalization
DKA	Deutscher Koordinierungsausschuß im CEC German Committee for Coordination in the CEC
DLC	Diamond-like amorphous Carbon
DSC	Differential Scanning Calorimetry
EBT	Electron Beam Texturing
ECP	Environmental Choice Program (CDN)
EHD	Elastohydrodynamic
EHEDG	European Hygienic Equipment Design Group
ELGI	European Lubricating Grease Institute
EN	European Norm
EP	Extreme Pressure
FCC	Fuel Catalytic Cracker
FDA	Food and Drug Administration (USA)
FM	Friction Modifier
FTMS	Federal Test Methods Standardization (USA)
FVA	Forschungsvereinigung Antriebstechnik (D) Research Association for Drive Technology
FZG	Forschungsstelle für Zahnräder und Getriebebau (D) Research Center for Toothed Wheel and Gearing Engineering
GC	Gas Chromatography
GfT	Gesellschaft für Tribologie (D) Society for Tribology
GOST	Gossudarstwenny Obschtschessojusny Standart (former USSR) State Standard (Governmental Union Standard)
HC	Hydrocracked
HD	Heavy Duty
HLB	Hydrophilic-Lyophilic Balance
HPDSC	High Pressure Differential Scanning Calorimetry
HPLC	High Pressure Liquid Chromatography
HRC	Rockwell C Hardness
HTHS	High Temperature High Shear
HVI	High Viscosity Index
IARC	International Agency for Research on Cancer
IBAD	Ion-Beam-Assisted Deposition
ICP	Inductively Coupled Plasma Atomic Emission
IFP	Institute Francais de Pétrole French Institute of Petroleum
ILMA	Independent Lubricant Manufacturers Association (USA)
ILSAC	International Lubricant Standardization and Approval Committee
IP	Institute of Petroleum (UK)
ISO	International Standard Organisation
ISO VG	ISO Viscosity Grade
JEA	Japanese Environmental Association
JIS	Japanese Industrial Standard
LDF	Long Drain Field Test
LIMS	Laboratory Information and Management System
LVFA	Low Velocity Friction Apparatus

LVI	Low Viscosity Index
MAK	Maximale Arbeitsplatzkonzentration Maximum Workplace Concentration
MIL	Military Standard (USA)
MLDW	Mobil Lube Dewaxing
MQL	Minimum Quantity Lubrication
MRV	Mini Rotary Viscosimeter
MS	Mass Spectroscopy
MSDW	Mobil Selective Dewaxing
N-D-M	(Refractive Index) n-Density-Molecular Weight
NLGI	National Lubricating Grease Institute (USA)
NMR	Nuclear Magnetic Resonance
OECD	Organization for Economic Cooperation and Development (EU)
OEM	Original Equipment Manufacturer
PCMO	Passenger Car Motor Oil
PEP	Passive Extreme Pressure
PPD	Pour Point Depressant
PTFE	Polytetrafluorethylene
PVD	Physical Vapor Deposition
RAL UZ	Reichsausschuß für Lieferbedingungen und Gütesicherung, Umweltzeichen (D) Imperial Committee for Quality Control and Labelling, Environmental Symbol
RNT	Radionuclide Technique
RVT	Reichert-Verschleiß-Test Reichert-Wear-Test
SAE	Society of Automotive Engineers (USA)
SCDSC	Sealed Capsule Differential Scanning Calorimetry
SHPD	Super High Performance Diesel
SRE	Standard Reference Elastomer
SRV	Schwing-Reibverschleiß Gerät (translatorisches Oszillations-Prüfgerät) Translatory oscillation apparatus
STLE	Society of Tribologists and Lubrication Engineers
STOU	Super Tractor Oil Universal
SUS	Saybolt Universal Seconds
TAN	Total Acid Number
TBN	Total Base Number
TDA	Thermal Deasphalting
TEI	Thailand Environment Institute
TEWL	Transepidermal Water Loss
TG	Thermogravimetry
TGL	Technische Normen, Gütevorschriften und Lieferbedingungen (former DDR) Technical Standards, Quality Specifications and Terms of Delivery
TLV	Threshold Limit Value
TOST	Turbine Oil Oxidation Stabilty Test
TRIP	Transformation Induced Plasticity
TRK	Technische Richtkonzentration (D) Technical Guideline Concentration
TSSI	Temporary Shear Stability Index
UEIL	Union Européenne des Indépendents en Lubrifiants European Union of Independent Lubricant Manufacturers
UHPD	Ultra High Performance Diesel
UHVI	Ultra High Viscosity Index

UNITI	Bundesverband Mittelständischer Mineralölunternehmen (D) Federal Association of the Middle-class Oil Companies
USDA	United States Department for Agriculture
UTTO	Universal Tractor Transmission Oil
VAMIL	Regeling Willekeurig Afschrijving Milieu-Investeringen (NL) Regulation of discretion deductions for investments into the environment
VCI	Vapor Phase (Volatile) Corrosion Inhibitor
VDA	Verband der Automobilindustrie (D) Automobile Industry Association
VDI	Verein Deutscher Ingenieure Association of German Engineers
VDMA	Verband Deutscher Maschinen- und Anlagenbau German Engineering Federation
VDS	Volvo Drain Specification
VGB	Technische Vereinigung der Großkraftwerkbetreiber (D) Technical Association of Large Power Plant Operators
VGO	Vacuum Gas Oil
VHVI	Very High Viscosity Index
VI	Viscosity Index
VIE	Viscosity Index Extended
VII	Viscosity Index Improver
VKA	Vier-Kugel-Apparat Four-Ball-Apparatus
VOC	Volatile Organic Compounds
Vp	Viscosity-pressure
VPI	Vapor Phase Inhibitor
VT	Viscosity-Temperature
VTC	Viscosity-Temperature-Coefficient
WGK	Wassergef_hrdungsklasse Water Hazardous Class (Water Pollution Class)
XHVI	Extra High Viscosity Index
XRF	X-ray Fluorescence spectrometry
ZAF	Zinc and Ashes Free

1
Lubricants and Their Market

Theo Mang and Apu Gosalia

1.1
Introduction

The main task and most important function of lubricants are to reduce friction by lubricants and offer wear protection, which extends machine runtimes and thereby protects raw materials. In some cases, the relative movement of two bearing surfaces is possible only if a lubricant is present. At present times when sustainability has become a driving force in the industry, saving energy and resources as well as cutting emissions have become central environmental matters. Therefore, the scarcity of resources and the responsibility towards future generations are also a particular focus of corporate action. Lubricants are increasingly attracting public awareness, because they support sustainability targets in economic, ecological and social areas. Lubricants make a contribution to the sparing use of resources and thereby to sustainability. Their task of reducing friction reduces the amount of energy input required and in this way saves emissions. Their task of wear protection extends the service life of equipment and saves resources. Scientific research has shown that up to 1% of gross domestic product could be saved in terms of energy in Western industrialized countries if current tribological knowledge, that is the science of friction, wear and lubrication, was just applied to lubricated processes.

Apart from important applications in internal combustion engines, vehicle and industrial gearboxes, compressors, turbines or hydraulic systems, there are a vast number of other applications which mostly require specifically tailored lubricants. This is illustrated by the numerous types of greases or the different lubricants for chip-forming and chip-free metalworking operations which are available. About 5000–10 000 different lubricant formulations are necessary to satisfy more than 90% of all lubricant applications.

If one thinks of lubricants today, the first type that comes to mind is mineral oil-based lubricant. Mineral oil continue to constitute quantitatively most important component of lubricants. Petrochemical components and increasing

Lubricants and Lubrication, Third Edition. Edited by Theo Mang and Wilfried Dresel.

derivatives of natural, harvestable raw materials from the oleo-chemical industry are finding increasing acceptance because of their environmental compatibility and some technical advantages.

On average, lubricants consist of about 90% base oils and 10% chemical additives and other components on a volume basis, while on a value base the respective ratio is estimated to be around 80:20.

The development of lubricants is closely linked to the specific applications and application methods. As a simple description of materials in this field makes little sense, the following sections will consider both lubricants and their application.

1.2
Lubricants Demand

Lubricants today are classified into five product groups: automotive oils, industrial oils, greases, metalworking fluids (including corrosion preventatives) and process oils. Process oils are included as raw materials in processes, but above all as plasticizers for the rubber industry. Their only link with lubricants is that they are mineral oil products resulting from the refining of base oils, but they often distort lubricant consumption figures. Therefore, they will not be covered in this book.

Interestingly, the breakdown by product groups in the past 15 years only slightly changed. 56% of all lubricants still go into automotive oils (e.g. engine oils, gear oils and transmission fluids), which continue to be the prevailing product group and largely dictate what will be available (or not) for making other products. Only 26% are industrial oils, with the rest comprising process oils, lubricating greases, metalworking fluids and corrosion preventatives.

The global lube market volume (without marine oils) was at around 36 million tonnes at the turn of the millennium and more or less quite stable until 2008. Then lubricants demand on a worldwide basis plunged by more than 10% year-on-year to just around 32 million metric tonnes in 2009. Since 2010 the worldwide market consumption showed a partial recovery in light of the partly unexpected rapid economic growth, to nearly reach the 36 million tonnes level again in 2015. Thus, one could think that not much happened market volume-wise between 2000 and 2015 (Chart 1.1).

However, the underlying regional lube market dynamics of the past 15 years were enormous in terms of quantity and quality. The Asia-Pacific region together with Africa and the Middle East accounted for a little more than one-third of global volume in 2000 and now makes more than half of it, as a result of growing industrialization and motorization and consequently higher consumption. The mature markets of Western Europe and North America experienced a continuous move to more quality lubricants, which resulted in extended oil change intervals and consequently lower demand per year. Asia-Pacific today consumes twice the lubricants amount per year than North America (Chart 1.2).

Since 1975, quantitative lubricant demand has significantly detached itself from gross national product and also from the number of registered vehicles. This quantitative view, which at first glance shows a continuous decline in lubricant

LUBRICANTS. TECHNOLOGY. PEOPLE. FUCHS

Lubricants Market
Development Global Lubricants Demand (Million Tons)*

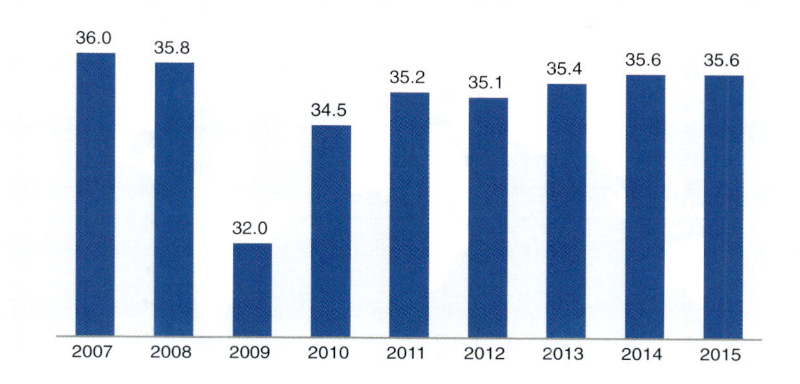

*Without Marine Oils

Chart 1.1 Development global lubricants demand [1–6].

Lubricants Market
Development Regional Lubricants Breakdown

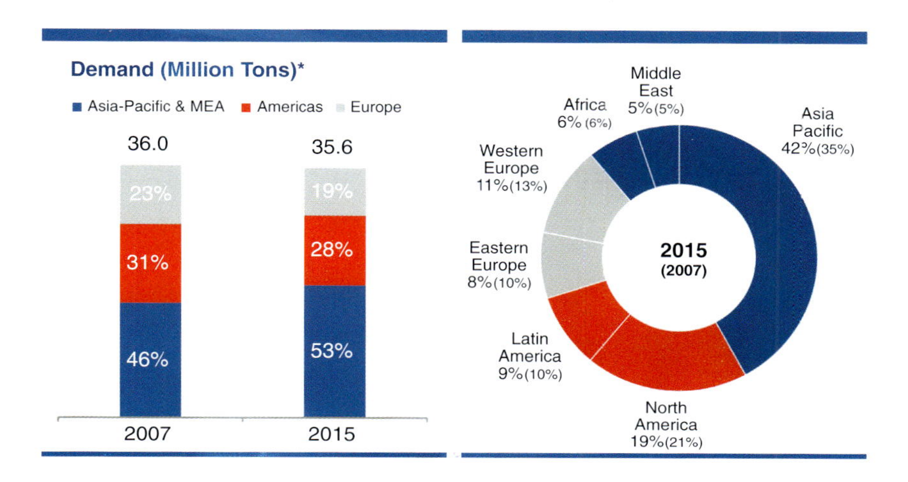

*Without Marine Oils

Chart 1.2 Development regional lubricants breakdown [1–6].

volumes, gives an inadequate impression of the significance of the lubricants business today. In almost all areas, products now have a longer life and offer greater performance, that is specific lubricant consumption has declined but specific revenues have increased noticeably. This is also confirmed by the volumetrically very important group of engine oils: The doubling of requirements with extended oil change intervals in recent years has quadrupled the cost of such oils. The efforts to increase the life of lubricants are not based on the wish to reduce lubricant costs. Much more important is the reduction of service and maintenance costs which result from periodic oil changing or regreasing.

As about 50% of the lubricants sold worldwide end in and thus pollute the environment, every effort is made to minimize spillages and evaporation. An example is diesel engine particulate emissions, about a third of which are caused by engine oil evaporation. These high lubricant losses into the environment were behind the development of environment-friendly lubricants which are thoroughly covered in this book.

A further incentive to reduce specific consumption is the ever-increasing cost of disposal or recycling of used lubricants. But this again creates new demands on lubricants because reduced leakage losses means less topping-up and less refreshing of the used oil. The new oils must therefore display good ageing stability.

Another consequence of the aforementioned developments was that global per capita consumption decreased from around 9 to 5 kg per year between 1970 and 2015, that is the increase in lubricant demand (+7%) did not keep up with the worldwide growth in population (+90%) during this period; in other words, the compounded annual growth rate (CAGR) of world population between 1970 and 2015 was 1.6% and 10 times higher than the CAGR of global lubricants demand, which amounted to just 0.16% in this time frame (Chart 1.3).

Lubricants Market

Regional Per-Capita Lubricants Demand 2015 (kg)

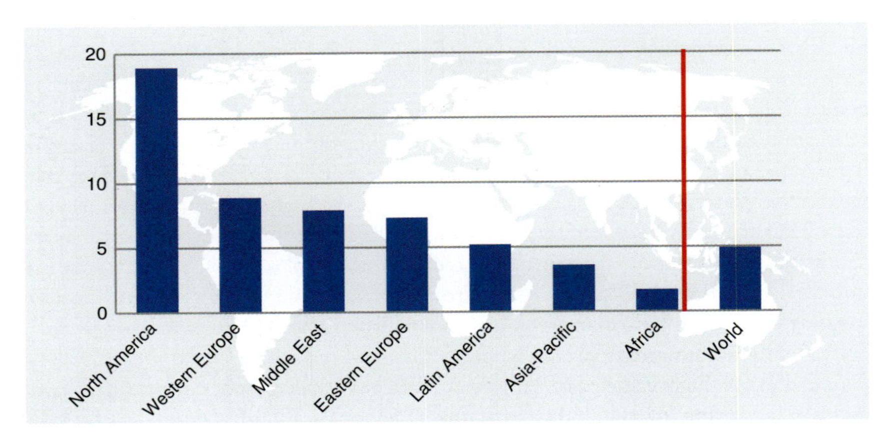

Chart 1.3 Regional per capita lubricants demand [3,4,6].

Bearing in mind the growth potential in Asia where per capita consumption in some areas is still extremely low (2015: India 1 kg) and a continuing reduction in volumes or stagnation in Western industrialized countries, overall a modest global growth is forecast. The growth in value will be more pronounced because the rapid globalization of technologies will promote high-value products even in the developing and emerging lubricant markets such as India and the machines and plants used in these countries will be similar or identical to those used in the developed industrialized countries.

1.3 Lubricants Competitor Landscape

The structure of the global lubricants industry changed significantly between the mid-1990s and 2005. Towards the end of the 1990s, the petroleum industry was affected by a wave of mergers and acquisitions (M&A). These created new and larger lubricant structures at the merged companies. The principal reasons for these mergers were economic factors in crude oil extraction and refining which resulted in lower refining margins.

The number of manufacturers (with lubricants production over 1000 tonnes per year) decreased by close to 60% or in nominal terms by around 1000 players from around 1700 to just above 700 market participants at the end of 2005.

On the one hand, there are vertically integrated petroleum companies whose main business objective is the discovery, extraction and refining of crude oil (Majors). Lubricants account for only a very small part of their oil business. In 2005, there were about 130 such national and multinational oil companies engaged in manufacturing lubricants, with the focus on high-volume lubricants such as engine, gear and hydraulic oils.

The consolidation and concentration proceeded much stronger on the level of the small-sized and independent lube manufacturers (Independents), with technological, safety-at-work and ecological considerations along with the globalization of lubricant consumers playing an important role and critical mass becoming increasingly important in company strategies. Their number halved between 2000 and 2005 to around 600 players, down from around 1200 at the beginning of the millennium. These 590 independent lubricant companies view lubricants as their core business, focusing on specialties and niches, where apart from some tailor-made lubricants, comprehensive and expert customer service is part of the package. They mainly concentrate on the manufacturing and marketing of lubricants. The independent lubricant manufacturers also generally purchase raw materials on the open market from the chemical and oleo-chemical industry and their mineral base oils from the large petroleum companies and they rarely operate base oil refineries (Chart 1.4).

Consolidation nowadays proceeds rather slowly. In case there are deals, then they are mostly on a high-value basis. However, there are other driving forces in the competitive landscape of the industry today: The vertically integrated mineral oil

Competitive Landscape
Structure Global Lubricants Industry*

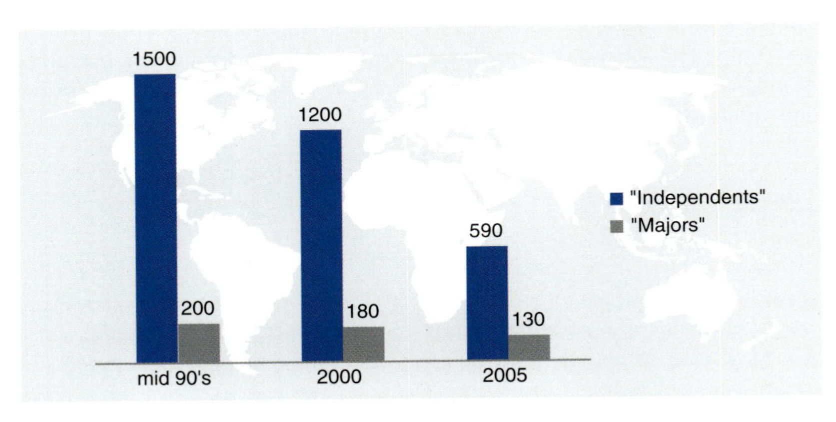

I 14

*Number of Players > 1,000 tons

Chart 1.4 Structure global lubricants industry [1,3,4,6].

companies of the 'old world' concentrate on big volume business and retract from niches. New and growing market participants enter the scene. Large oil companies restructure their lube businesses as stand-alone subsidiaries. National Oil Companies in China, South Korea and Russia, for example, go global. Companies, so far mostly known as raw material suppliers to the lubricants industry, go for vertical/forward or lateral diversification steps and Private Equity comes into play (Chart 1.5).

The aforementioned merger and acquisition activities changed the ranking of the top 15 lubricant manufacturers in the past 15 years:

- EXXON and SHELL switched leading positions, after Shell acquired PENNZOIL
- FUCHS gained 3 positions in the top 15 ranking and made it into the top 10 as number 9
- GULF OIL, PERTAMINA and PETRONAS newly came into the top 15 ranking, while INDIAN OIL, AGIP and REPSOL had to leave it.

At the end of 2015, the top 15 manufacturers share two-thirds of the worldwide lube market, while the rest of more than 700 manufacturers share the other half.

The production of simple lubricants normally involves blending processes but specialties often require the use of chemical processes such as saponification (in the case of greases), esterification (when manufacturing ester base oils or additives) or amidation (when manufacturing components for metalworking lubricants). Further manufacturing processes include drying, filtration, homogenizing, dispersion or distillation. Depending on their field of activity, lubricant manufacturers invest between 1 and 5% of their sales in research and development (Chart 1.6).

Competitive Landscape
Driving Forces Global Lubricants Industry

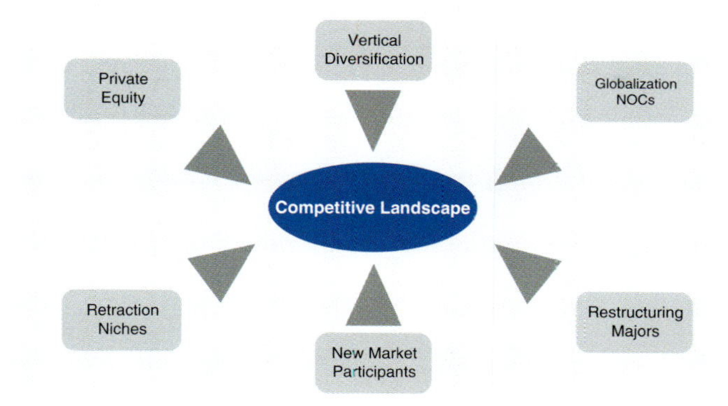

Chart 1.5 Driving forces global lubricants industry.

FUCHS

Competitive Landscape
Global Ranking Top 15 LubricantsManufacturers*

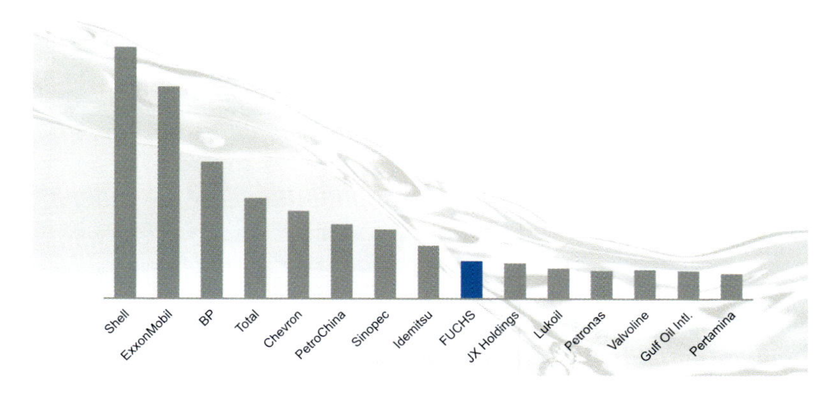

*By Volume

Chart 1.6 Global ranking top 15 lubricants manufacturers [1,3,4,6].

1.4 Lubricant Systems

Apart from the most common lube oils, the many thousands of lubricant applications necessitate a diverse number of systems which is seldom equalled in other product groups.

The group next to oils are emulsions, which as oil-in-water emulsions are central to water-miscible cutting fluids (Chapter 14), rolling emulsions and fire-resistant HFA hydraulic fluids (Chapter 11). In these cases, the lubricant manufacturer normally supplies a concentrate which is mixed with water locally to form an emulsion. The concentration of these emulsions with water is generally between 1 and 10%. The annual consumption of such emulsions in industrialized countries is about the same as all other lubricants together. From this point of view, the volumetric proportion of these products (as concentrates) is significantly underrated in lubricant statistics with regard to the application engineering problems they create and their economic significance.

The next group of lubricant systems are water-in-oil emulsions. Their most important application is in metal forming. These products are supplied ready-to-use or as dilutable concentrates. Fire-resistant HFB fluids are designed as water-in-oil emulsions too (invert emulsions).

In some special cases, oil-in-oil emulsions are developed as lubricants and these are primarily used in the field of metalworking.

Water-based solutions in the form of non-dispersed systems are sometimes used in chip-forming metalworking operations.

Greases (Chapter 16) are complex systems consisting of base oils and thickeners based on soaps or other organic or inorganic substances. They are available in semiliquid form (semifluid greases) through to solid blocks (block greases). Special equipment is required for their production (grease-making plants). A group of products closely related to greases are pastes.

Solid lubricant suspensions normally contain solid lubricants in stable suspension in a fluid such as water or oil. These products are often used in forging and extrusion as well as other metalworking processes. Solid lubricant films can also be applied as suspensions in a carrier fluid which evaporates before the lubricant has to function.

Solid lubricant powders can be applied directly to specially prepared surfaces.

In the case of dry-film lubricants (Chapter 17), solid lubricants are dispersed in resin matrices. Dry-film lubricants are formed when the solvent (principally water or hydrocarbons) evaporates.

Molten salts or glass powder are used for hot forming processes such as extrusion. These are normally supplied as dry powders and develop lubricity when they melt on the hot surface of the metal.

Polymer films are used when special surface protection is required in addition to lubricity (e.g. the pressing of stainless steel panels). Together with greases, these products are also used to some extent in the construction industry.

An intermediary field between materials and lubrication technology is the wide area of surface treatment to reduce friction and wear. While the previously mentioned dry-film lubricants are an accepted activity of the lubricants industry, chemical coatings are somewhat controversial. These coatings are chemically bonded to the surface of the metal. They include oxalation and phosphating (zinc, iron and manganese). In cases when such coatings adopt the carrier function of an organic lubricant, the entire system could be supplied by the lubricant manufacturer. If the chemical coating is not designed to be supplemented with an additional lubricant coating (e.g. dry film on phosphatized gear), it will probably be supplied by a company which specializes in surface degreasing and cleaning.

Even more different from traditional lubricants are metallic or ceramic coatings which are applied with CVD (chemical vapour deposition) or PVD (physical vapour deposition) processes. They also sometimes replace the EP functions of the lubricants (Chapter 6). Such coatings are increasingly being used together with lubricants to guarantee improved wear protection in extreme conditions and over long periods of time.

References

1 Fuchs, M. (2002) The world lubricants market: year 2001 and outlook, 13th International Colloquium Tribology on Lubricants, Materials, and Lubrication Engineering, Technische Akademie Esslingen (TAE), Esslingen.

2 Lindemann, L. (2013) Lubricant development against the background of new raw materials, OilDoc Conference, Rosenheim.

3 Gosalia, A. (2013) *The Lubricants Industry & FUCHS: A Journey along the Process & Value Chain*, FUCHS Bankeninformationsveranstaltung, Mannheim.

4 Gosalia, A. (2013) *Sustainability & Intelligence @ FUCHS*, Mannheim Business School & Tongji Executive MBA Module, Competition & Industry, Mannheim.

5 Gosalia, A. (2012) The sustainability of the European lubricants industry, The Annual Congress of the European Lubricants Industry, Lisbon.

6 Gosalia, A. (2012) Sustainability and the global lubricants industry, The 16th ICIS World Base Oils & Lubricants Conference, London.

2
Lubricants in the Tribological System

Theo Mang and Christian Busch

The development of lubricants has become an integral part of the development of machinery and its corresponding technologies. It is irrevocably and interdisciplinarily linked to numerous fields of expertise; without this interdisciplinary aspect, lubricant developments and applications would fail to achieve success.

2.1
Lubricants as Part of Tribological Research

Tribology (derived from the Greek tribein, or tribos meaning rubbing) is the science of friction, wear and lubrication. Although the use of lubricants is as old as mankind, scientific focus on lubricants and lubrication technology is relatively new. The term tribology was first introduced in 1966 and has been used globally to describe this far-reaching field of activity since 1985. Even though efforts had been made since the sixteenth century to describe the whole phenomenon of friction scientifically (Leonardo da Vinci, Amontons, Coulomb), the work always concentrated on single aspects and lubricants were not even considered. Some research work performed up to the early 1970s totally ignored the chemical processes which take place in lubricated friction processes.

Tribology, with all its facets, is only sporadically researched. Fundamental scientific tribological research takes place at universities which have engineering or materials testing departments. Naturally, lubricant manufacturers also perform research. The advantage of tribological research by engineering departments is the dominant focus on application engineering. The most common disadvantage is the lack of interdisciplinary links to other fields of expertise. Joint research projects which combine the disciplines of engineering, materials, chemistry, health and safety, and the work conducted by lubricant manufacturers themselves therefore offer the best prospects of practical results.

Lubricants and Lubrication, Third Edition. Edited by Theo Mang and Wilfried Dresel.

2.2
The Tribological System

The tribological system (commonly referred to as the tribosystem) (Figure 2.1) consists of four elements: the contacting partner, the opposing contacting partner (material pair), the interface between the two and the medium in the interface and the environment [1]. In lubricated bearings, the lubricant is located in this gap. In plain bearings, the material pair are the shaft and the bearing shells; in combustion engines, they are the piston rings and the cylinder wall or the camshaft lobes and the tappets; in metalworking, the tool and the workpiece.

The variables are the type of movement, the forces involved, temperature, speed and duration of the stress. Tribometric parameters – such as friction, wear and temperature data – can be gathered from the stress area. Tribological stress is the result of numerous criteria of surface and contact geometry, surface loading or lubricant thickness. Tribological processes can occur in the contact area between two friction partners – which can be physical, physical–chemical (e.g. adsorption, desorption) or chemical in nature (tribochemistry).

2.3
Friction

The description of friction as the cause of wear and energy losses has always posed significant problems because of the complexity of the tribological systems. There is also no internationally recognized nomenclature. In this discussion,

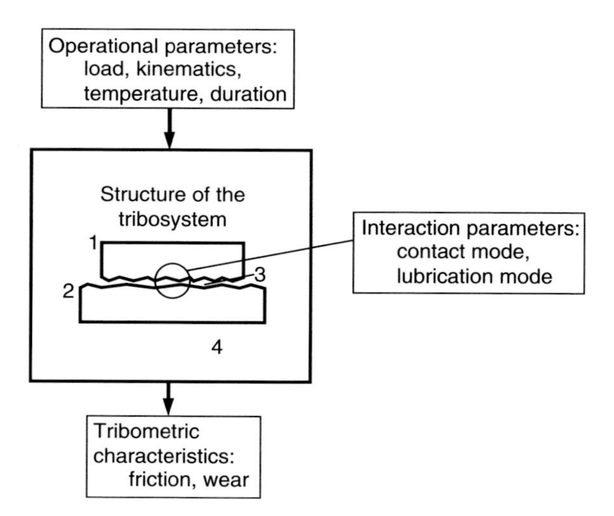

Figure 2.1 Structure of the tribosystem according to Czichos [1]. 1 and 2: material pair; 3: interface and medium in the interface (lubricant); 4: environment.

friction is described according to its type, and by the combination of friction and lubrication conditions, in line with the view taken by most experts [2,3].

2.3.1 Types of Friction

Friction is the mechanical force which resists movement (dynamic or kinetic friction) or hinders movement (static friction) between sliding or rolling surfaces. These types of friction are also called external friction.

Internal friction results from the friction between lubricant molecules; this is described as viscosity (Chapter 3).

The cause of external friction is, above all, the microscopic contact points between two sliding surfaces; these cause adhesion, material deformation and grooving. Energy which is lost as friction can be measured as heat and/or mechanical vibration. Lubricants should reduce or avoid the microcontact which causes external friction.

2.3.1.1 Sliding Friction

This is friction in a pure sliding motion with no rolling and no spin (Figure 2.2).

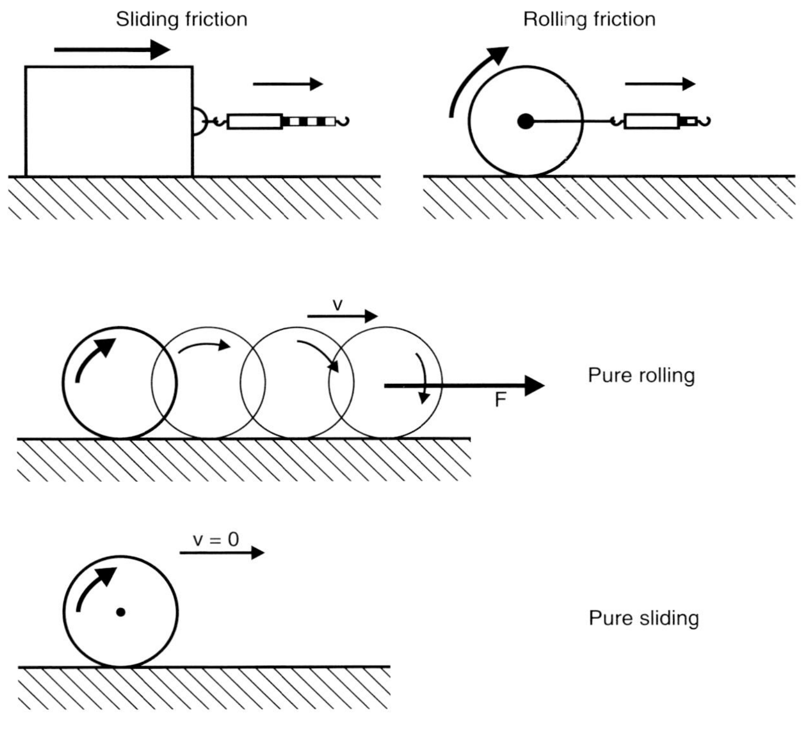

Figure 2.2 Sliding and rolling.

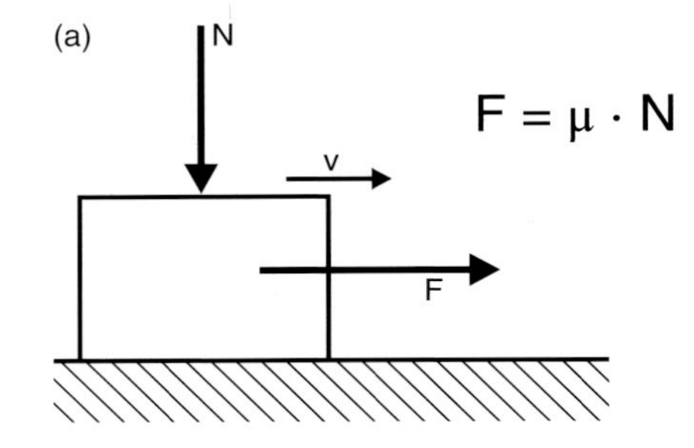

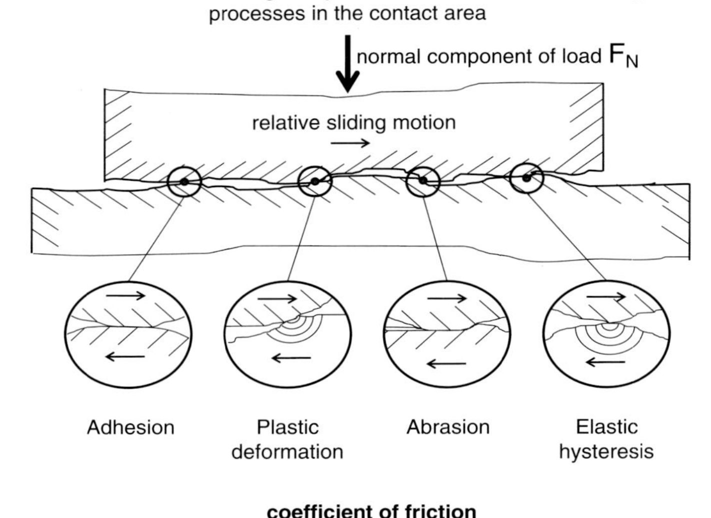

Figure 2.3 (a) Coefficient of friction. (b) The coefficient of friction in a tribological system as a result of four energy dissipation processes in the contact area.

Figure 2.3a defines the coefficient of friction as the dimensionless ratio of the friction force F and the normal force N. The proportionality between normal force and frictional force is often given under dry and boundary friction conditions but not in fluid-film lubrication Figure 2.3b.

2.3.1.2 Rolling Friction

This is the friction generated by rolling contact (Figure 2.4). In roller bearings, rolling friction mainly occurs between the rolling elements and the raceways, whereas sliding friction occurs between the rolling elements and the cage. The main cause of friction in roller bearings is sliding in the contact zones between the rolling elements and the raceways. It is also influenced by the geometry of the contacting surfaces and the deformation of the contacting elements. In addition, sliding also occurs between the cage pockets and the rolling elements.

If rolling motion and sliding motion combine to any significant extent, as for gear tooth meshing, special terminology has been created. The word 'Wälzreibung' which is derived from 'Wälzen' (rolling, e.g. steel rolling) is used in Germany. Situations in which a high sliding/rolling ratio occurs require totally different lubrication than does pure sliding. Figure 2.4 shows this 'rolling friction' during rolling and during gear meshing.

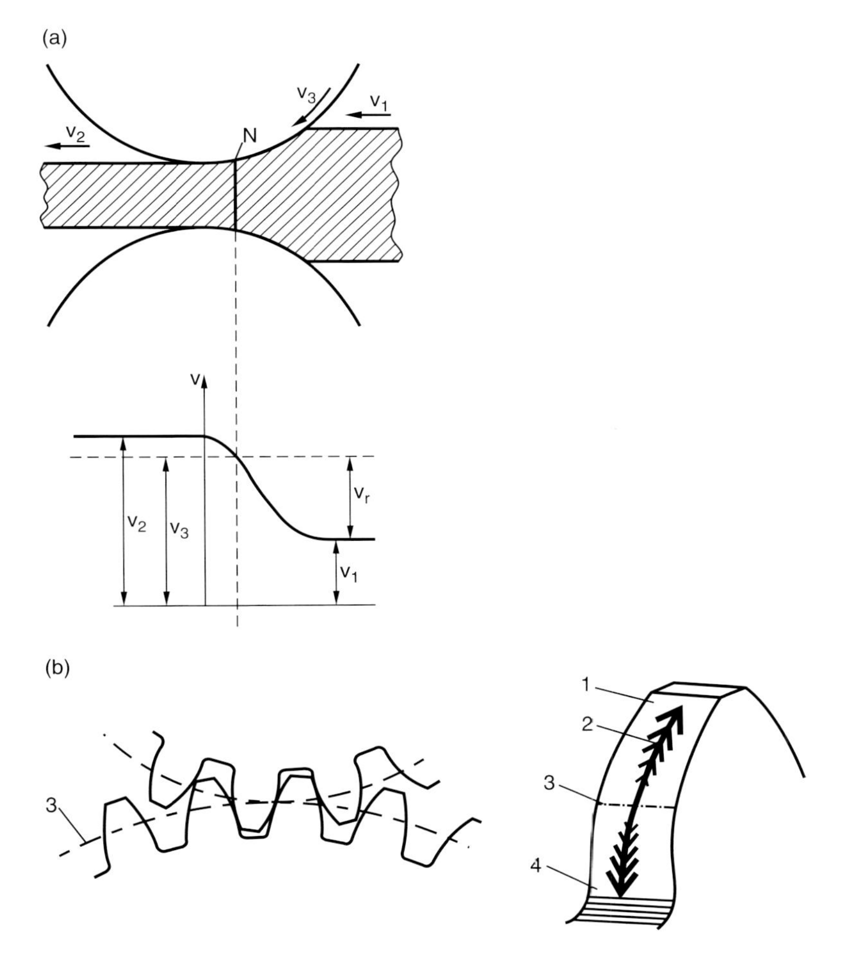

Figure 2.4 'Wälzreibung', mixing of rolling and sliding motions. (a) Rolling in metal forming; v_1, initial speed of the sheet metal; v_2, final speed of the sheet metal; v_3, speed of the roller; v_r, speed difference in the roll gap (sliding part); *N*, neutral point (non-slip point, pure rolling). (b) Engagement of gear teeth, 1, 2, 4; high sliding/rolling ratio; 3: pitch circle (pure rolling, no slip).

2.3.1.3 **Static Friction**

The static coefficient of friction is defined as the coefficient of friction corresponding to the maximum force that must be overcome to initiate macroscopic motion between two bodies (ASTM).

2.3.1.4 **Kinetic Friction**

Different from static friction, kinetic friction occurs under conditions of relative motion. ASTM defines the kinetic coefficient of friction as the coefficient under conditions of macroscopic relative motion of two bodies. The kinetic coefficient of friction sometimes called dynamic coefficient of friction is usually somewhat smaller than the static coefficient of friction.

2.3.1.5 **Stick–Slip**

Stick–slip is a special form of friction which often results from very slow sliding movements when the friction partners are connected to a system which can vibrate. The process is influenced by the dependence of the coefficient of sliding friction on speed. This generally occurs when the static coefficient of friction (f_{stat}) is larger than the dynamic coefficient of friction (f_{dyn}). Stick–slip is normally encountered with machine tools which operate with slow feeds. Stick–slip can cause chatter marks on components (Figure 2.5).

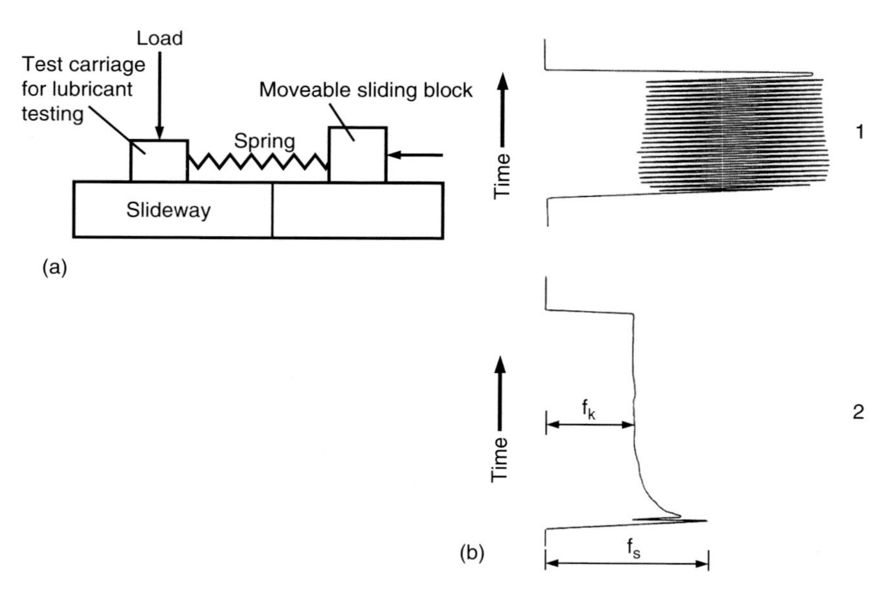

Figure 2.5 Stick–slip. (a) Test equipment for stick–slip. (b) Results of stick–slip behaviour of two oils. 1: oil with bad stick–slip behaviour; 2: oil with good stick–slip behaviour; f_k: relative kinematic coefficient of friction; f_s: relative static coefficient of friction.

2.3.2
Friction and Lubrication Conditions

In tribological systems, different forms of contact can exist between contacting partners.

2.3.2.1 Solid Friction (Dry Friction)

This occurs when two solids have direct contact with each other without a separating non-solid layer. If conventional materials are involved, the coefficients of friction and wear rates are high. Lubrication technology attempts to eliminate this condition.

2.3.2.2 Boundary Friction

The contacting surfaces are covered with a molecular layer of a substance whose specific properties can significantly influence the friction and wear characteristics. One of the most important objectives of lubricant development is the creation of such boundary friction layers in a variety of dynamic, geometric and thermal conditions. Such layers are of great importance in practical applications when thick, long-lasting lubricant films to separate two surfaces are technically impossible. Boundary lubricating films are created from surface-active substances and their chemically reaction products. Adsorption, chemisorption and tribochemical reactions also play significant roles.

Although boundary friction is often allocated to solid friction, the difference is of great significance to lubricant development and the understanding of lubrication and wear processes, especially when the boundary friction layers are formed by the lubricants.

2.3.2.3 Fluid Friction

In this form of friction, both surfaces are fully separated by a fluid lubricant film (full-film lubrication). This film is either formed hydrostatically or, more commonly, hydrodynamically. From a lubricants point of view, this is known as hydrodynamic or hydrostatic lubrication (Figure 2.6). Liquid or fluid friction is caused by the frictional resistance because of the rheological properties of fluids.

If both surfaces are separated by a gas film, this is known as gas lubrication.

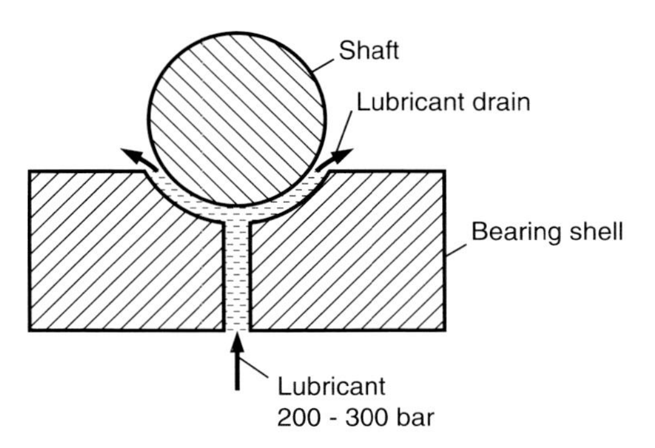

Figure 2.6 Hydrostatic lubrication as a form of fluid friction.

2.3.2.4 Mixed Friction

This occurs when boundary friction combines with fluid friction. From a lubricants technology standpoint, this form of friction requires sufficient load-bearing boundary layers to form. Machine elements which are normally hydrodynamically lubricated experience mixed friction when starting and stopping.

For roller bearings, one of the most important machine elements, it has been shown that the reference viscosity either of lubricating oils or of the base oils of greases is not sufficient to ensure the formation of protecting lubricant layers and the required minimum lifetime. Under mixed friction conditions, it is important to choose the appropriate lubricant, that is that which enables the formation of tribolayers by antiwear and extreme pressure additives [4].

In 2004, Wiersch and Schwarze described a means of calculating mixed lubrication contacts over a wide range of operating conditions and applications. The performance of the mixed friction model was demonstrated using the example of a cam tappet contact [5,6].

2.3.2.5 Solid Lubricant Friction

This special form of friction occurs when solid lubricants are used (Chapter 17). It cannot be allocated to the previously mentioned forms of friction because particle shape, size, mobility and, in particular, crystallographic characteristics of the particles justify a separate classification.

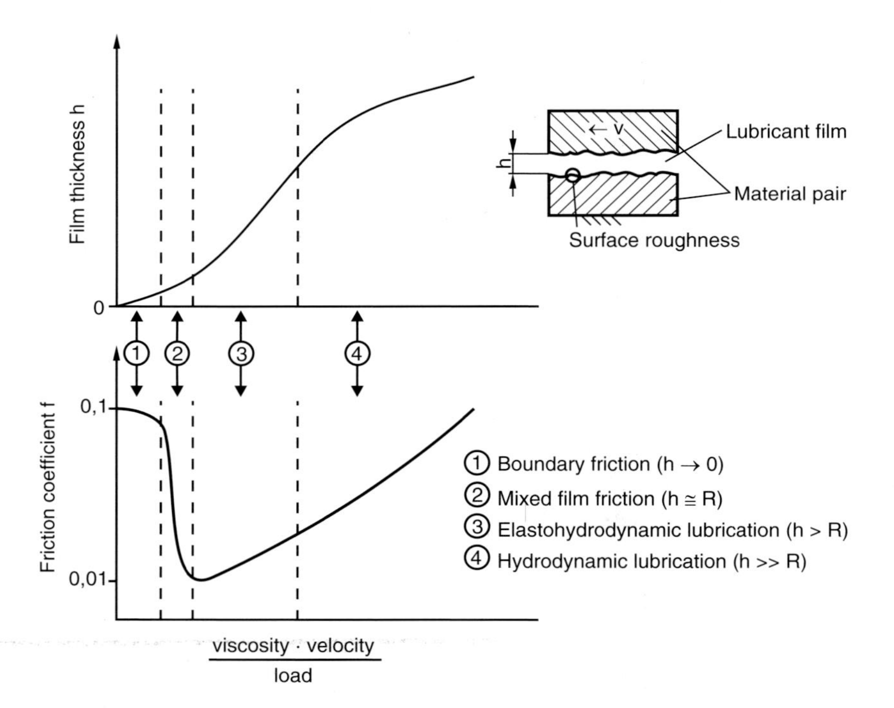

Figure 2.7 Stribeck graph according to Ref. [7].

2.3.2.6 Stribeck Diagram

The friction or lubrication conditions between boundary and fluid friction are graphically illustrated by use of Stribeck diagram (Figure 2.7) [7]. These are based on the starting-up of a plain bearing whose shaft and bearing shells are, when stationary, separated only by a molecular lubricant layer. As the speed of revolution of the shaft increases (peripheral speed), a thicker hydrodynamic lubricant film is created that initially causes sporadic mixed friction but which, nevertheless, significantly reduces the coefficient of friction. As the speed continues to increase, a full, uninterrupted film is formed over the entire bearing faces; this sharply reduces the coefficient of friction. As speed increases, internal friction in the lubricating film adds to external friction. The curve passes a minimum coefficient of friction value and then increases, solely as a result of internal friction.

The lubricant film thickness shown in Figure 2.7 depends on the friction and lubrication conditions, including the surface roughness R.

2.3.2.7 Hydrodynamic Lubrication

Figure 2.8 demonstrates the formation of a hydrodynamic liquid film. The lubricant is pulled into the conical converging clearance by the rotation of the shaft. The created dynamic pressure carries the shaft.

On the basis of the Navier–Stokes theory of fluid mechanics, Reynolds created the basic formula for hydrodynamic lubrication in 1886. Several criteria remained excluded, however, especially the influence of pressure and

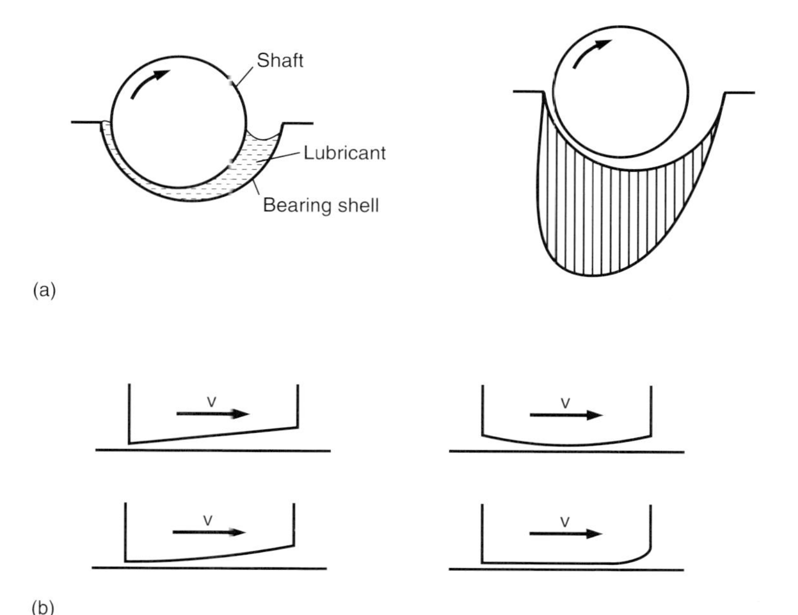

Figure 2.8 Formation of a hydrodynamic liquid lubricant film. (a) Rolling, development of pressure in the hydrodynamic film. (b) Sliding, preferred geometry.

temperature on viscosity. The application of the Reynolds' formula led to theoretical calculations on plain bearings. The only lubricant value was viscosity.

2.3.2.8 Elastohydrodynamic Lubrication (EHD Regime)

Hydrodynamic calculation on lubricant films was extended to include the elastic deformation of contact faces (Hertzian contacts, Hertz's equations of elastic deformation) and the influence of pressure on viscosity (Chapter 3). This enables application of these elastohydrodynamic calculations to contact geometries other than that of plain bearings, for example those of roller bearings and gear teeth. Compared with the 25 μm films found in hydrodynamic bearings, films of generally less than 2 μm characterize EHD lubrication. The geometry of EHD is classified by Figures 2.9 and 2.10 indicating that the contact area between the mating elements is counterformal or slightly conformal.

Figure 2.9 shows the elastic deformation of the ball and raceway of a ball bearing and Figure 2.10 shows an example of Hertzian contacts for various pairs with non-converging lubricant clearance.

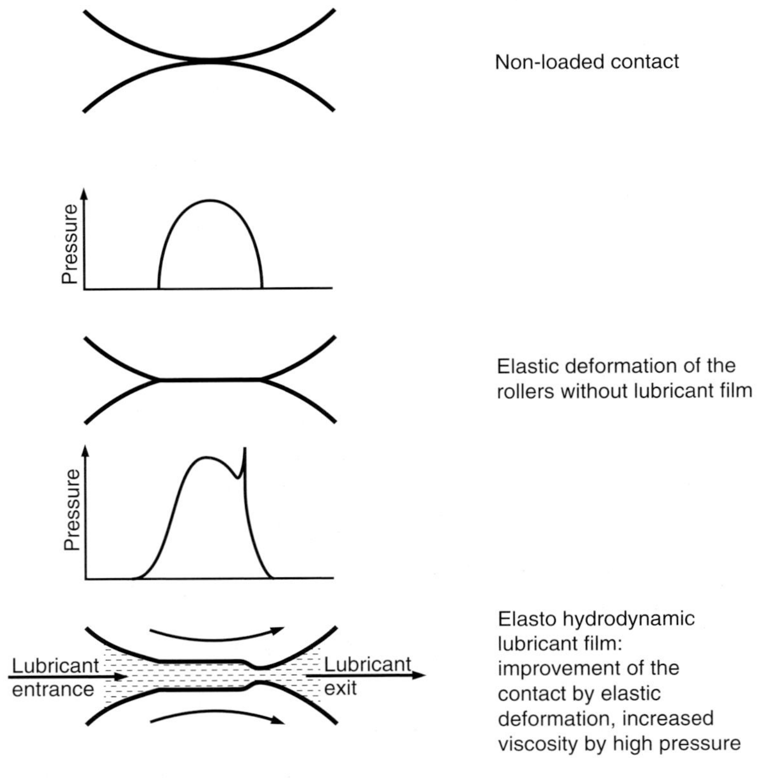

Figure 2.9 Improvement of hydrodynamic lubrication clearance between two rollers by Hertzian deformation (elastohydrodynamic (EHD) contact), pressure distribution in the Hertzian contact.

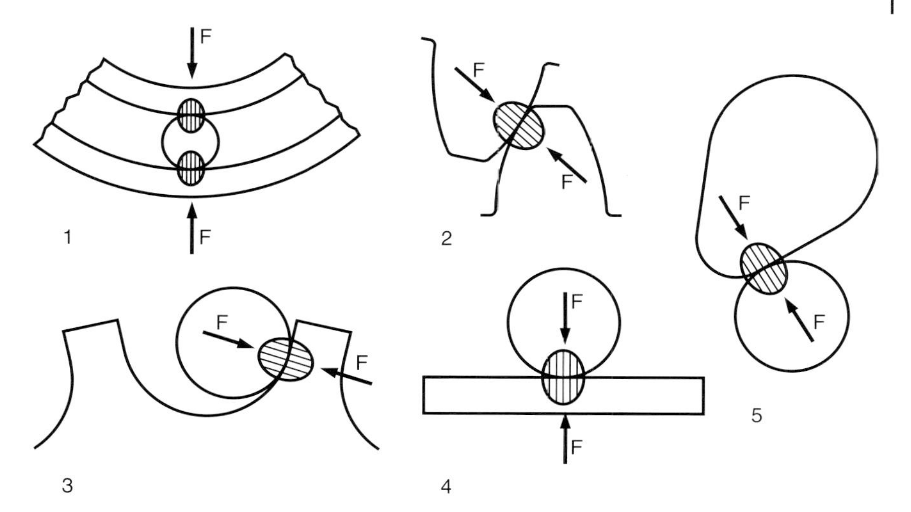

Figure 2.10 Hertzian contacts for different pairs with non-converging lubricant clearance [8]. 1: roller bearing; 2: gear wheels; 3: chain wheels; 4: roller on flat path; 5: cam lifter.

2.3.2.9 Thermo-Elasto-Hydrodynamic Lubrication (TEHD)

TEHD lubrication theory solves the Reynolds equation, including the energy equation of the lubricant film. Calculation of the energy takes into consideration heat convection in all directions, heat conduction in the radial direction, compression and heating caused by viscous and asperity friction.

TEHD lubrication theory has been applied, for example in important areas of automotive engines, using a model including shear rate-dependent viscosity (Chapter 3) and simulation of the lubrication conditions for the main crankshaft bearing of commercial automotive engines [9].

2.4 Wear

According to the German DIN standard 50320, wear is defined as the progressive loss of material from the surface of a contacting body as a result of mechanical causes, that is the contact and relative movement of a contacting solid, liquid or gas to the body.

2.4.1 Wear Mechanisms

Wear is created by the processes of abrasion, adhesion, erosion, tribochemical reactions and metal fatigue which are important to lubrication technology.

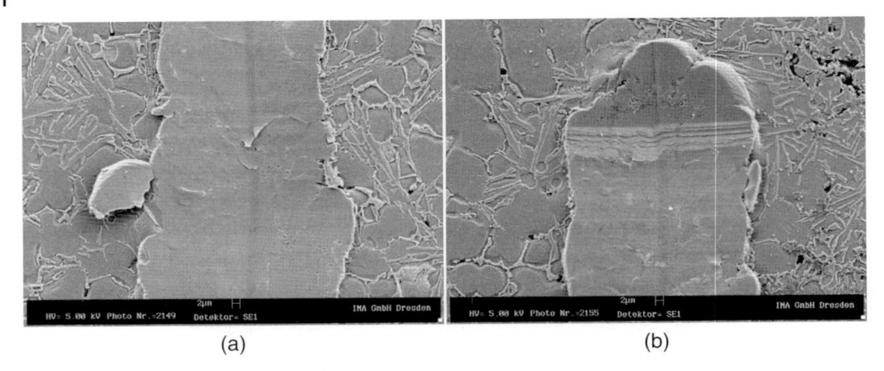

(a) (b)

Figure 2.11 Abrasive wear. (a) Plowing: material is displaced to the side but not necessarily removed. (b) Plowing: material is displaced to the side and to the end of the groove tribological system: AlSi9Cu3 – 100Cr6.

2.4.1.1 **Abrasion**

Abrasive wear or ploughing wear occurs principally when one body is penetrated by harder surface features of a mating body (two-body abrasion) or by third bodies which can be debris (three-body abrasion). Three-body abrasion can also occur by the contamination of lubricants with abrasive media (such as sharp quartz particles). Abrasion causes the surface peaks to break-off or it gouges further abrasive particles from the surface. This form of wear can combine with other wear mechanisms. Abrasive wear is demonstrated in Figure 2.11.

2.4.1.2 **Adhesion**

This is the most complex of wear mechanisms. The basic theory of adhesive wear follows logically from the adhesive theory of friction which was first developed by Bowden and Tabor. Adhesion is one of the two main factors that contribute to friction between dry contacting surfaces in impending or sliding motion. The other factor is deformation. A simplified version of the theory says that contact between two rough surfaces is initially only made with some local asperity tips. Under normal load, some of the asperity tips on the softer material start to deform elastically and finally the plastic contacting asperities are cold-welded onto those on the harder surface, forming strong adhesive bonds. To break these bonds, either at the contact interface or within the softer asperities, a tangential force is needed which becomes the source of the adhesive component of friction. The tangential force causes a shear stress in the junction area. Shearing will take place at that interface, with the resulting wear. Molecular and atomic forces between the two friction partners can tear material particles out of the surfaces. The volume of worn material is proportional to the load and the sliding distance and is inversely proportional to the softer material hardness. The fragments being plucked out will be either carried away attached to the harder material or become detached as debris.

Adhesive wear is shown in Figure 2.12.

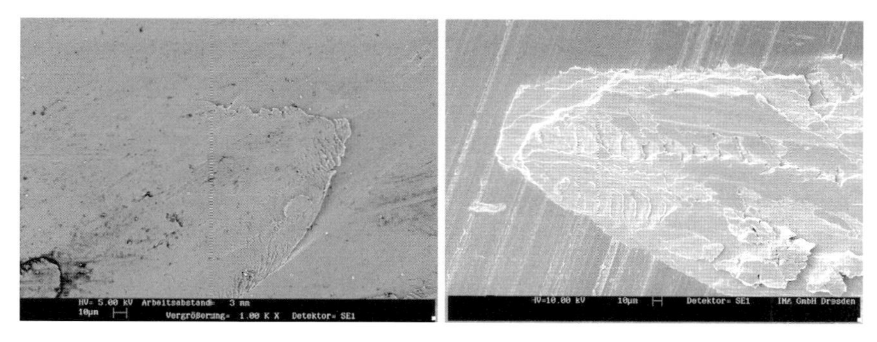

Figure 2.12 Adhesive wear. Tribological system: TiAl6V4 – 100Cr6.

2.4.1.3 **Tribochemical Reactions**

The chemical reactions which occur under tribological conditions are referred to as tribochemistry. It is sometimes wrongly assumed that these reactions are governed by some special laws. Nevertheless, for better understanding of the specific reaction conditions, the laws applying to tribological contact should be observed. This applies in particular to thermal effects (flash temperatures) which are not readily macroscopically recognizable and to frictional effects which lead to chemically reactive surfaces where the valences of the metal's structural matrices play a role in wear and deformation. The removal of chemically reacted surface layers thus formed constitutes tribochemical wear (Figure 2.13).

2.4.1.4 **Surface Fatigue**

Surface fatigue is the result of periodic loads in the contact zones. Frequent spot loading leads to surface fatigue which is a result of material fatigue. Micro-pits are one visible sign of this type of wear. A thick, separating lubricant film either minimizes or eliminates this problem. Machine elements which are subject to periodically severe loads in the contact zone are particularly prone to this type of damage. Figure 2.14 demonstrates wear by tribochemical reactions.

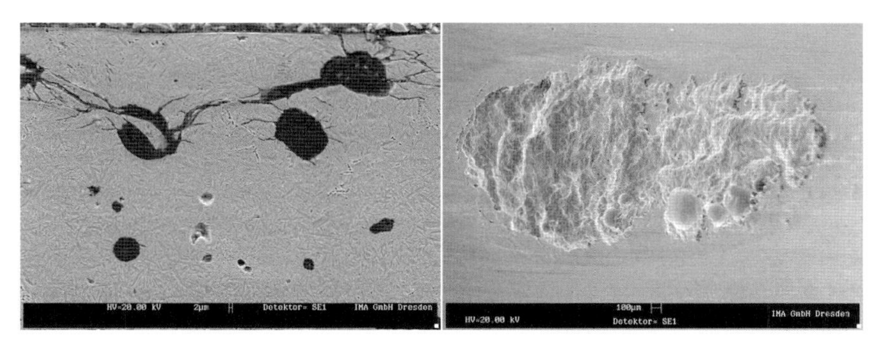

Figure 2.13 Tribochemical wear. Tribological system: GJS-600 – 100Cr6.

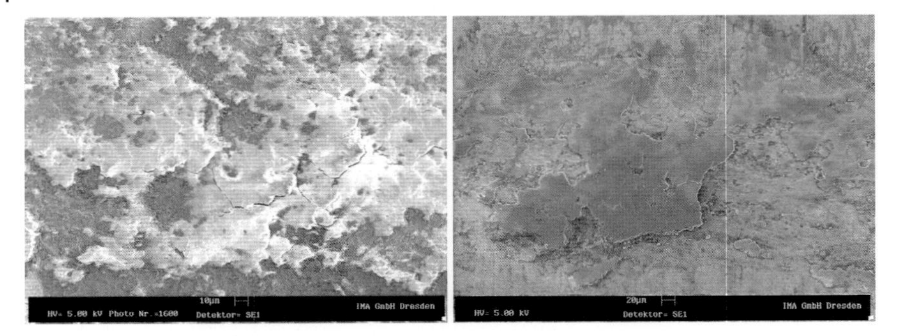

Figure 2.14 Tribochemical reaction layer. Formation of a tribochemical reaction layer system materials: AlSi12CuNi – 100Cr8.

2.4.1.5 Erosion

Erosive wear which is a loss of material from a solid surface occurs during the impingement on a surface by a fluid containing solid particles.

2.4.1.6 Fretting

Fretting is fatigue wear between two nominally static contacting surfaces. It is caused by cyclic relative tangential motion of very small amplitude. Most of the previously mentioned types of wear are included in fretting wear and this explains why fretting wear is seen as a separate form. Besides mechanical components, tribochemical reactions play an important role. Oxidation creates layers and fretting wear is sometimes known as fretting corrosion. Fretting corrosion creates trapped wear debris between the contact faces. Fretting wear can be avoided if suitable surface-active additives are added to lubricants (especially as pastes for threaded fasteners).

2.4.1.7 Cavitation

Materials can be damaged by imploding gas and vapour bubbles entrained in lubricating oils or hydraulic fluids. In systems which carry lubricants, the elimination of dragged-in air, low boiling point substances or the use of surface active components scales down the gas bubbles and thus reduces cavitation.

The condition of surfaces effected by cavitation can be further damaged by corrosion. This process can be controlled by the use of specific inhibitors.

2.4.1.8 Corrosive Wear

Corrosive wear is the removal of chemically reacted surface layers by friction processes. Air is the most common corrosive medium. A corrosive environment can be dangerous especially for steel structures. The best solution to avoid corrosive wear is by applying the correct type and amount of lubricant including speciality additives.

2.4.2
Types of Wear

There are several criteria which could be used to classify the different types of wear, for example according to the types of (kinematic) friction which lead to wear (sliding wear, rolling wear, fretting wear), wear mechanisms (adhesive wear, abrasive wear, tribochemical wear) or the shape of the wear particle.

There is also highly specialized wear terminology for different lubricant applications; this is oriented to the geometry of the various bearing faces (e.g. clearance and crater wear in the field of chip-forming tools).

2.4.3
The Wear Process

Wear can be measured gravimetrically, volumetrically or in terms of area over a period of time or against increasing load. Uniformly decreasing wear which stabilizes at a very low level can be described as running-in wear. This can be controlled by the tribochemical reactions of the additives in the lubricants. Wear can occur at relatively constant speed and ultimately lead to the functional failure of the bearing. Wear at an increasing wear rate can lead to progressive wear. In time, the material damage caused by wear will also lead to the failure of a component. To facilitate projection of possible functionality, the concept of failure analysis was introduced.

Analysis of machine element failure as a result of wear and lubrication can be performed by use of practically oriented tests. Determination of the effect of different greases on the projected failure of roller bearings has become increasingly important.

2.4.4
Tribomutation

Gervé [10] introduced this term to describe the processes of friction and wear from the tribosystem. His aim, in particular, was to separate the interpretation of wear from the confusing association with material characteristics and to highlight much more complex processes in the tribosystem in which lubricants are also included. Accordingly, friction and wear are purely system-related values. Gervé created the basis for his friction and wear concepts by conducting numerous sensitive wear measurements in the nanometre range using the established radionuclide wear measurement technique [11,12] and by successful ion implantation such as that used in metal-cutting tools and wear parts in combustion engines. He described the two most important tribological properties T_{pi} (once for wear and once for friction) in the equation:

$$T_{\mathrm{pi}} = f_i(r, P(\mathrm{M}, G, L)e, \tau,$$

where r are running conditions, for example speed, load, temperature and others, P are parameters such as M materials, G micro and macrogeometry and

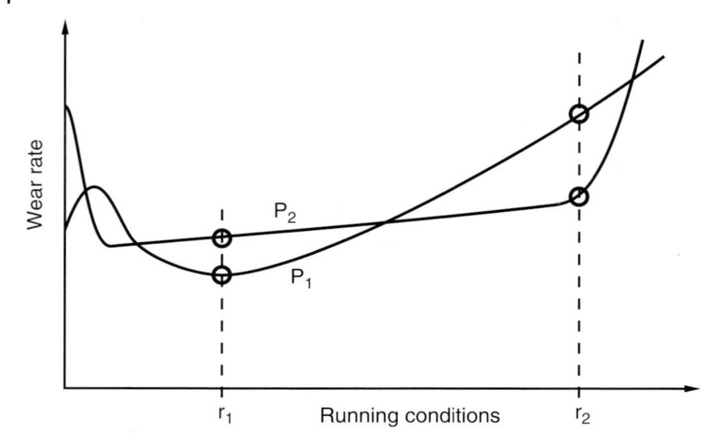

Figure 2.15 Tribomutation. System dependence of wear rate for changing running conditions for two sets of parameters. (According to Ref. [10].)

L the lubricant and the lubrication conditions, e are environmental conditions such as humidity, dust and temperature and τ is the life of the tribosystem.

Figure 2.15 demonstrates the system dependence of wear rate for changing running conditions for two parameters. For the running conditions r_1, the wear rate for parameter p_1 is significantly lower than that for parameter p_2; this is reversed for r_1.

Figure 2.16 shows the alteration of the material caused by tribomutation. The materials M_1 and M_2 at differing running times (t_1 and t_2) show wear occurring in the reverse order. Tribomutated materials have changed their tribological characteristics.

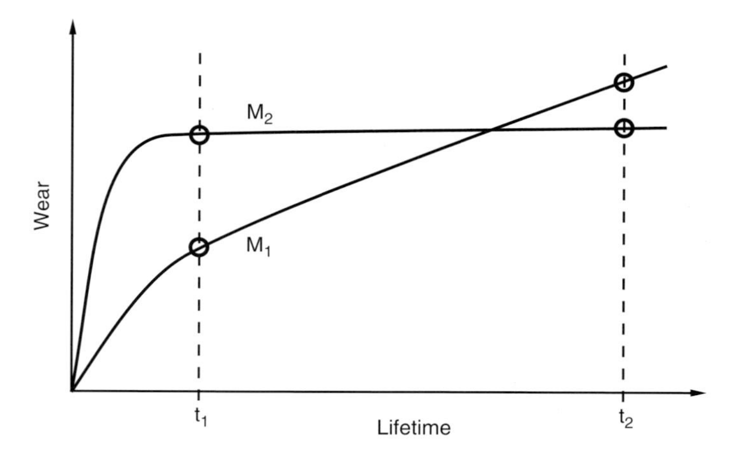

Figure 2.16 Tribomutation. Wear of tribosystem under constant running conditions with different materials. (According to Ref. [10].)

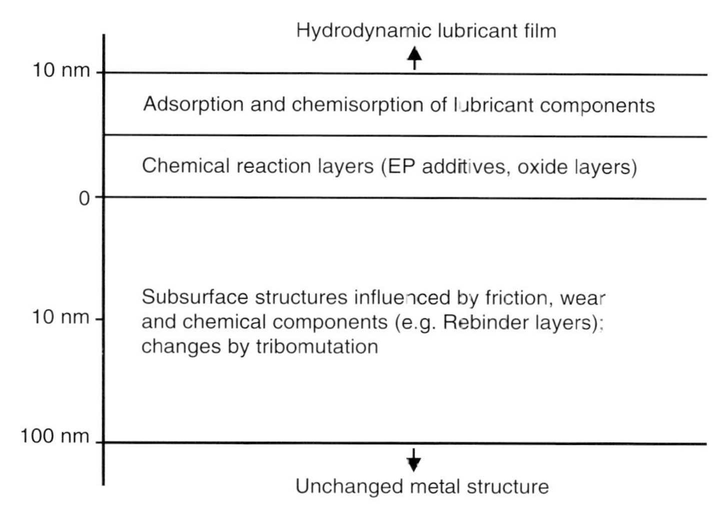

Figure 2.17 Lubrication conditions and possible changes by tribomutation.

Changes to the material caused by tribomutation, which often occur under the surface of the material down to a depth of 150 nm, and which are reinforced by surface wear, can be quantified by sensitive methods of analysis of material composition. Gervé explained the material change caused by tribomutation by the part of the friction energy which is not consumed by heat generation and the binding energy of the wear particles.

Tribological system analysis and the associated tribomutation question the practical application of a series of friction and wear tests. Apart from what happens during running-in, other friction and wear phenomenon can be explained while the metalworking process along with the cutting fluid must be seen as an important influence on the tribological properties of a component.

Figure 2.17 shows the possible changes under tribologically stressed surfaces.

Friction and wear development is highly influenced by the changed conditions created by tribomutation.

2.4.5 Nanotribology

The preceding chapter on tribomutation and Figure 2.17 reveal that deeper knowledge of frictional and wear processes on the micro- and nanoscale can enable explanation of different tribological phenomena. This implies the need for novel measurement methods or adaptation of current measurement procedures to the new tasks, the development and application of which constitute an essential part of nanotechnology in general and of nanotribology in particular. This enables visualization and measurement of different criteria up to the molecular scale. In the last 10 years, public interest in nanotechnology and nanotribology has grown substantially, because they are expected not only to enable theoretical explanation of different phenomena but also to improve economical realization [13]. Gervé [14] assigns three functions to micro- and nanotribology:

1) Analysis of the movement of single atomic groups in relation to the basic material and measurement of unequally distributed friction forces with nanometre resolution. In addition, the new definition of topographical roughness is relevant to understanding of the interrelationship between roughness and friction. This also includes experimental and theoretical analysis of frictional interactions at the atomic level, excitation of lattice oscillations and related reduction of the bond strength of the lattice elements, finally leading to wear.
2) Nano- and microtribology imply measurement and calculation of classical and tribological characteristics.
3) Examination of the effects of frictional phenomena on macroscopic tribosystems in the millimetre and micrometre range. This includes tribomutation. A recently started interesting project detected the effects of tribomutation on the piston–liner system of automotive vehicles. It was proved that production-related material changes (different cutting, grinding and honing processes, special metalworking fluids, dry cutting) had an effect on the friction and wear properties of engine components [15].

One of the most important aspects of nanotribology is the rapid development of microscale and nanoscale test equipment [15–18], for example the microtribometer for detection of forces in the micronewton range on springs, interferometer-based tribometers or fibre-optics-based microtribometers, microanalysis for hardness measurement and scratch testing and nanoscale probe techniques, in which the atomic force microscope (AFM, RKM) has become extremely important. This enables nanoscale scanning of surfaces with a tip passing over the surface with an accuracy of 10^{-} m, corresponding to the diameter of an atom. In combination with the aforementioned spring and optical or capacitive instruments, this creates an image of the surface. The instrument also enables friction force measurement (FFM) on the nanoscale [19].

Also of interest are biological, micro- and nanotribology [20], the objective of which is to gather information about friction, adhesion and wear of biological systems and to apply this new knowledge to the design of micro-electromechanical systems and the development of new types of monolayer lubrication and other technology [20]. Nanotribology is the study of nanomaterials used as friction and wear-reducing lubricant additives. Carbon nanotubes and fullerene-like materials have entered lubricant research laboratories [21]. Submicrometre graphite particles, on the other hand, have been used as lubricant additives for many years.

2.4.6 Tribosystems of Tomorrow

To develop tribosystems and, especially, lubricants for tomorrow, collaborative research centres must be formed to integrate all the necessary competence. This could be done by organizing mutual research of the lubricant industry with universities and with the companies which use lubricants. Creation of a collaborative research centre at the university level must involve all those institutes which can contribute to the complex area of lubrication, wear, friction materials and environmental and toxicological problems. The most fascinating project in this field, 'Environmentally Friendly Tribosystems for the Machine Tool by Use of Suitable Coatings and Fluids', was started in the mid-1990s in Germany at the Technical University of Aachen (RWTH Aachen). The institutes involved are as follows:

- A technical chemistry institute for development of appropriate lubricants
- An institute for characterizing the environmental and ecotoxicological behaviour of the new lubricants
- A surface engineering institute for developing the appropriate coatings (low-temperature PVD coatings for gears, bearings and pistons of hydrostatic units, high-temperature PVD coatings for tools used in metal cutting and forming);
- A manufacturing technology institute
- An institute for machine tools and production engineering
- An institute for fluid power
- An institute for machine elements and machine design [22].

References

1 Czichos, H. (1992) *Basic Tribological Parameters, Friction, Lubrication and Wear Technology*, ASTM Handbook, vol. **18**, p. 474.

2 Bowden, F.P. and Tabor, D. (1954) *The Friction and Lubrication of Solids: Part 1*, Clarendon Press, Alderly, Gloucestershire, UK.

3 Archard, J.F. (1953) Contact and rubbing of flat surfaces. *J. Appl. Phys.*, **24** (8), 981–988.

4 Karbacher, R. (2006) Mixed film lubrications of rolling bearings, 15th International Colloquium Tribology, Automotive and Industrial Lubrication (17–19 January), Technische Akademie Esslingen.

5 Wiersch, P. and Schwarze, H. (2006) EHL simulation under mixed friction conditions using the example of a cam tappet contact, 15th International Colloquium Tribology, Automotive and Industrial Lubrication, Technische Akademie Esslingen.

6 Wiersch, Petra. (2004) Berechnung thermoelastohydrodynamischer Kontakte bei Mischreibung. Dissertation, TU Clausthal.

7 Czichos, H. and Habig, K.-H. (1992) *Tribologie Handbuch*, Vieweg, Wiesbaden.

8 Klamann, D. (1984) *Lubricants and Related Products*, Wiley-VCH Verlag GmbH, Weinheim.

9 Bukovnik, S. *et al.* (2006) Thermo-elasto-hydrodynamic lubrication model for journal bearing including shear rate dependent viscosity. 15th International Colloquium Tribology, Automotive and

Industrial Lubrication, Technische Akademie Esslingen.

10 Gervé, A. (1993) Improvement of tribological properties by ion implantation. *Surf. Coat. Technol.*, **60**, 521–524.

11 Gerve, A. (1971) Radioisotopes in mechanical engineering, Fourth United Nations International Conference on the Peaceful Uses of Atomic Energy, Geneva, Switzerland (6 September), Technical Report, AEC Conference 71-100-55.

12 Gerve, A., Kehrwald, B., Wiesner, L., Conlon, T.W. and Dearnaly, G. (1985) Continuous determination of the wear-reducing effect of ion implantation on gears by double labelling radionuclide technique. *Mater. Sci. Eng.*, **69**, 221–225.

13 Bhusan, B. (1997) *Micro/Nanotribology and Its Application*, Kluwer Academie Publishers.

14 Gerve, A. (2000) Mikro- und Nanotribologie, eine neue Sicht der Tribologie. 12th International Colloquium Tribology 2000-Plus, Esslingen (January 11–13).

15 Santner, E. and Stegemann, B. (2006) Tribological measurements at the nanoscale. 15th International Colloquium Tribology, Automotive and Industrial Lubrication, Technische Akademie Esslingen.

16 Gerbig, Y.B. (2006) Influence of the nanoscale topography on the microfriction of hydrophobic and hydrophilic surfaces. 15th International Colloquium Tribology, Automotive and Industrial Lubrication, Technische Akademie Esslingen (17–19 January).

17 Gold, P.W., Wolf, Th., Loos, J., Reichelt, M., Weirich, Th., Richter, S. and Mayer, J. (2006) Nanomechanical and analytical investigations on the influence of lubricant variation on tribological layers in slow running roller bearings, 15th International Colloquium Tribology, Automotive and Industrial Lubrication, Technische Akademie Esslingen (17–19 January).

18 Myshkin, N.K. and Grigoriev, A.Y. (2006) Measurement of tribological characteristics in the nano range, 15th International Colloquium Tribology, Automotive and Industrial Lubrication, Technische Akademie Esslingen (17–19 January).

19 Luther, R. and Seyfert, C. (2003) *Können Motoren langer leben? Einfluss von Fertigungsmethoden auf die Motorenlebensdauer*, Jahrestagung der Gesellschaft für Tribologie (GfT), Gottingen.

20 Scherge, M. and Gorb, S.N. (2001) *Biological Micro- and Nanotribology*, Springer, Berlin.

21 Mieno, T. and Ohmae, N. (2006) Carbon nanotubes as lubricant additives. 15th International Colloquium Tribology, Automotive and Industrial Lubrication, Technische Akademie Esslingen (17–19 January).

22 Murrenhoff, H. (2006) Introduction into the Collaborative Research Center 442: environmentally friendly tribosystems by suitable coatings and fluids with respect to the machine tool. 15th International Colloquium Tribology, Automotive and Industrial Lubrication, Technische Akademie Esslingen (17–19 January)

3
Rheology of Lubricants

Theo Mang

Consistency, flow properties or viscosity in the case of oils, are key parameters to create lubrication efficiency and the application of lubricants. These are terms which appear in nearly all lubricant specifications. Viscosity is also the only lubricant value which is adopted into the design process for hydrodynamic and elastohydrodynamic (EHD) lubrication.

3.1
Viscosity

Friction generated by a fluid surrounding contacting partners, that is without contact of the partners, is the internal friction of the fluid. In the right-hand branch of the Stribeck graph (Figure 2.7), internal friction increases with bearing speed. The measure of internal friction in a fluid is viscosity. Viscosity and its dimensions are best explained with a model of parallel layers of fluid which could be viewed molecularly (Figure 3.1). If this packet of fluid layers is sheared (τ), the individual fluid layers are displaced in the direction of the shearing force. The upper layers move more rapidly than the lower layers because molecular forces act to resist movement between the layers. These forces create resistance to shearing and this resistance is given the term dynamic viscosity. The difference in velocity between two given fluid layers, related to their linear displacement, is referred to as shear rate S. This velocity gradient is proportional to the shear stress (τ). The proportionality constant η is called dynamic viscosity and has the units Pa·s. Analysis of the dimensions uses the following equations:

$$S = \frac{\mathrm{d}v}{\mathrm{d}y}\left[\frac{\mathrm{m}}{\mathrm{s}\cdot\mathrm{m}}\right] \text{ or } \left[\mathrm{s}^{-1}\right], \tag{3.1}$$

$$\eta = \frac{\tau}{S}\left[\frac{\mathrm{N}}{\mathrm{m}^2\cdot\mathrm{s}^{-1}}\right] \text{ or } \left[\frac{\mathrm{Pa}}{\mathrm{s}}\right] \text{ or } [\mathrm{mPa}\cdot\mathrm{s}] \text{ or (Centipoise).} \tag{3.2}$$

Lubricants and Lubrication, Third Edition. Edited by Theo Mang and Wilfried Dresel.

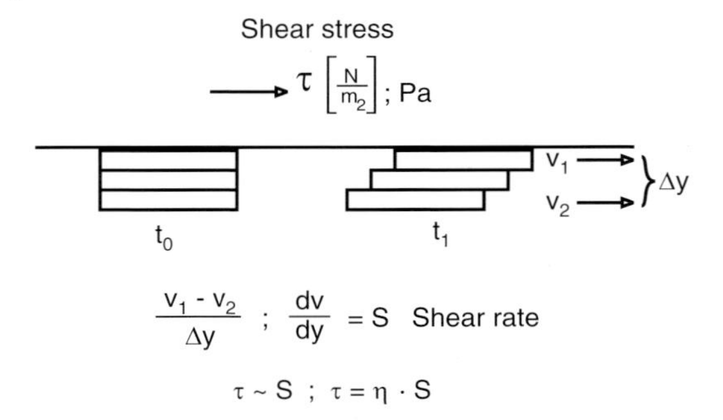

Figure 3.1 Explanation of viscosity.

The laboratory determination of viscosity in run-out or capillary tubes is influenced by the weight of the fluid. The relationship between dynamic viscosity and specific gravity is referred to as kinematic viscosity ν. The following unit analysis applies:

$$\nu = \frac{\eta}{\rho}\left[\frac{\mathrm{Pa}\cdot\mathrm{s}}{\mathrm{kg}\cdot\mathrm{m}^{-3}}\right] \text{ or } \left[\frac{\mathrm{N}\cdot\mathrm{s}\cdot\mathrm{m}^{3}}{\mathrm{m}^{2}\cdot\mathrm{kg}}\right] \text{ or } \left[\frac{\mathrm{kg}\cdot\mathrm{m}\cdot\mathrm{s}^{-2}\cdot\mathrm{s}\cdot\mathrm{m}^{3}}{\mathrm{m}^{2}\cdot\mathrm{kg}}\right] \text{ or } \left[\frac{\mathrm{m}^{2}}{\mathrm{s}}\right]\left[\frac{\mathrm{mm}^{2}}{\mathrm{s}}\right] \text{ or (Centistoke).} \tag{3.3}$$

Fluids which display the above proportionality constant between shear stress and shear rate are referred to as Newtonian fluids, that is the viscosity of Newtonian fluids is independent of shear rate (Figure 3.2). Deviations from this Newtonian behaviour are sometimes referred to as structural viscosity. Those viscosities are named as apparent viscosities.

As described above, kinematic and dynamic viscosity differ by way of density. Taking the example of a medium viscosity paraffinic mineral oil cut (ISO VG 32, Section 3.6), the difference is between 12 and 25% at temperatures between 0 and 100 °C (the higher values apply to kinematic viscosity).

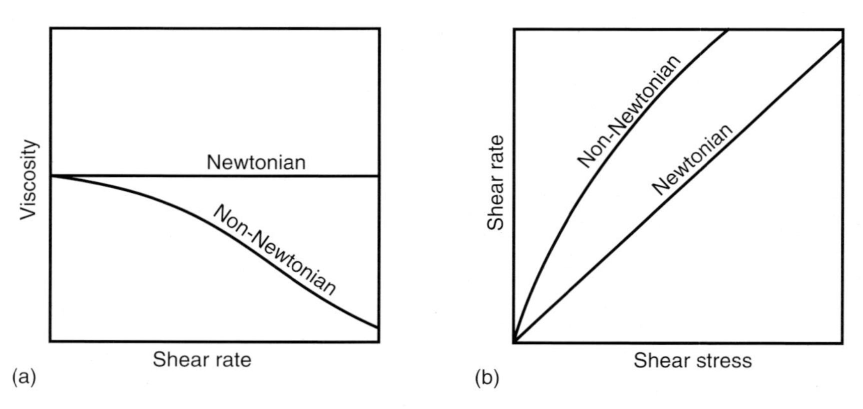

Figure 3.2 Flow characteristics of Newtonian and non-Newtonian lubricants. (a) Viscosity as a function of shear stress. (b) Shear rate as a function of shear stress.

3.2
Influence of Temperature on Viscosity (*V–T* Behaviour)

The viscosity of all oils used for lubrication purposes drops significantly when their temperature increases. In linear systems, this $V-T$ behaviour is hyperbolic and the practical differentiation necessary in practice is difficult to replicate and the interpolation between two measured viscosities is also problematic. For these reasons, $V-T$ behaviour has been allocated to a function which results in a straight-line graph if suitable co-ordinates are selected.

The Ubbelohde–Walter equation has become generally accepted and also forms the basis of American Society for Testing and Materials (ASTM), ISO and DIN calculation guidelines.

$$\lg \lg(\nu + C) = K - m \cdot \lg T. \tag{3.4}$$

In this double-logarithmic formula, C and K are constants, T is temperature in Kelvin and m is the $V-T$ line slope. Figure 3.3 shows the linear and Ubbelohde–Walter double-logarithmic $V-T$ curves of three oils with significantly differing $V-T$ lines (naphthenic oil, paraffinic HC-II/Group III-oil/Chapter 4) and a natural vegetable (rapeseed) oil.

The constant C for mineral oils in the $V-T$ equation is between 0.6 and 0.9.

The constant C only plays a minor role in the viscosity calculation, larger differences are only apparent at very low temperatures.

The use of kinematic viscosity dominates in the lubricants industry, even though dynamic viscosity is a much more important parameter in lubrication technology.

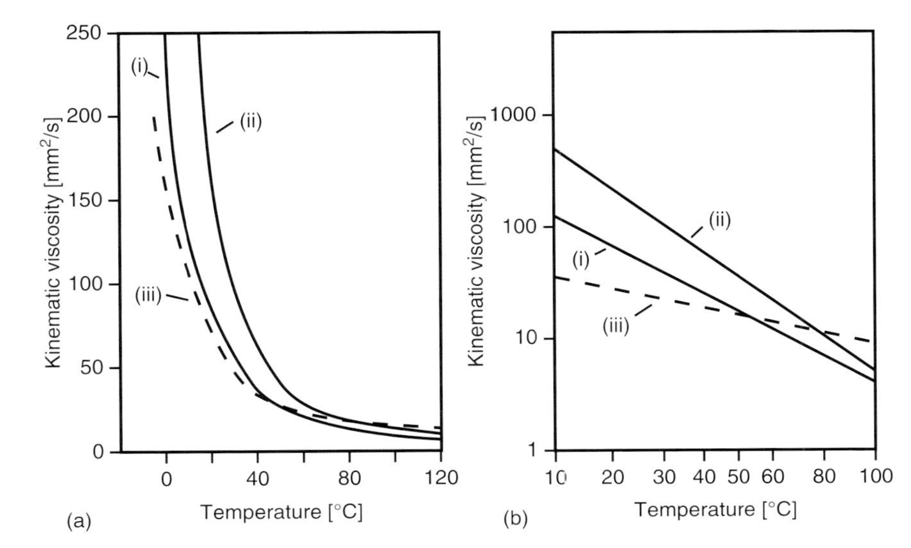

Figure 3.3 V-T behavior of various oils. A linear; B double-logarithmic; (i) paraffinic base oil; (ii) naphthenic base oil; (iii) rapeseed oil

The Vogel–Cameron equation is used for the rapid, computer-based calculation of dynamic viscosity. The Vogel–Cameron viscosity–temperature equation is

$$\eta = \mathrm{A} \cdot \exp\left(\frac{B}{T + C}\right), \tag{3.5}$$

where A, B and C are constants and T is again temperature in Kelvin.

The m value in the Ubbelohde–Walter equation which represents the double-logarithmic V–T graph slope is sometimes used to characterize the V–T behaviour of oils. The m value for lubricant base oils is between 4.5 and 1.1 [1]. The smaller values apply to oils which are less affected by temperature.

$$m = \frac{\lg \lg(\nu_1 + 0.8) - \lg \lg(\nu_2 + 0.8)}{\lg T_2 - \lg T_1} \tag{3.6}$$

The viscosity–temperature constant (VTC) was introduced to better differentiate V–T behaviour when the influence of temperature is low.

$$\mathrm{VTC} = \frac{\nu_{40} - \nu_{100}}{\nu_{40}}. \tag{3.7}$$

3.2.1 Viscosity Index

Neither the m value nor the viscosity–temperature constant have found particular acceptance in lubrication technology. This is not the case for the viscosity index VI, which today defines the international description of viscosity–temperature behaviour. This value was first introduced in the United States in 1928. It was based on the then greatest (VI = 100) and the smallest (VI = 0) temperature dependence of US base oils.

VI is graphically illustrated in Figure 3.4. Values over 100 are calculated as VIE.

Viscosity index is also defined in the most international standards.

The evaluation of V–T behaviour at low temperatures according to Ubbelohde–Walter lines or other straight-line V–T graphs often leads to inaccuracies. The previously described dependency effects do not apply to base oils which can suffer thickening caused by crystallization of some components (e.g. paraffins) at low temperatures or those whose polymer molecules simulate viscosity effects at low temperatures.

Table 3.1 shows various V–T characteristics for a number of oils.

3.3 Viscosity–Pressure Dependency

The significance of viscosity–pressure dependency (V–p behaviour) was, and still is, underestimated for numerous lubrication applications. V–p behaviour has

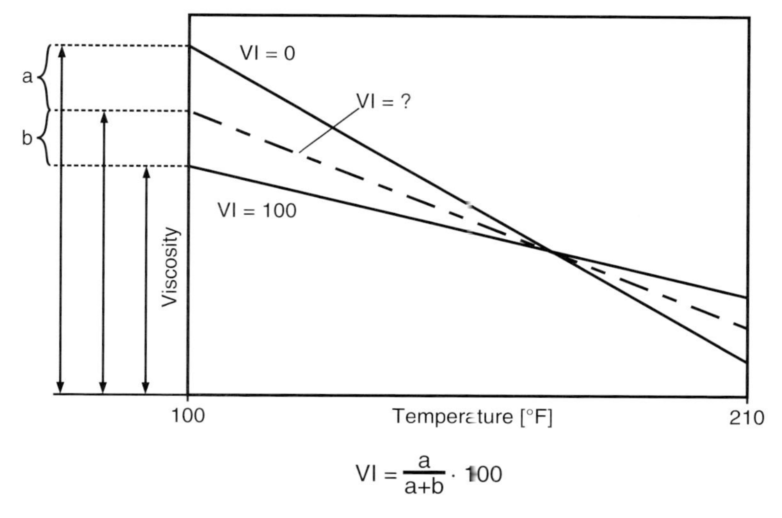

$$VI = \frac{a}{a+b} \cdot 100$$

Figure 3.4 Graphical illustration of viscosity index (VI).

become a part of the calculation of elastohydrodynamic lubricant films The exponential dependence of viscosity on pressure means that viscosity increases very rapidly with pressure. Metal-forming lubricants can be subject to such pressures that the viscosity of such oils can increase by a number powers of 10.

V–p behaviour can be described by the formula

$$\eta_p = \eta_1 \cdot e^{\alpha(p-p_1)}, \tag{3.8}$$

where η_p is the dynamic viscosity at a pressure p, η_0 is the dynamic viscosity at 1 bar and α is the viscosity–pressure coefficient. The following function

Table 3.1 Various *V–T* characteristics for several oils [1].

	Kinematic viscosity ($mm^2\,s^{-1}$)		Viscosity index (VI)	*m* Constant	Viscosity temperature constant (VTC)
	40 °C	100 °C			
Naphthenic spindle oil	30	4.24	40	4.05	0.847
Paraffinic spindle oil	30	5.23	105	3.68	0.819
Medium solvent extract	120	8.0	−50	4.51	0.939
Medium polyglycol	120	20.9	200	2.53	0.826
Medium silicone oil	120	50.0	424	1.14	0.583
Multigrade motor oil (SAE 10W-30)	70	11.1	165	2.82	0.841
Ester oil	30	5.81	140	3.40	0.806

generates α:

$$\alpha = \frac{1}{\eta_p} \cdot \left(\frac{d\eta_p}{dp}\right)_T \tag{3.9}$$

V–p dependence is defined by the chemical structure of the substances with the steric geometry of the molecules being of particular significance.

Figure 3.5 shows the development of viscosity against pressure for a number of oils with different chemical structures.

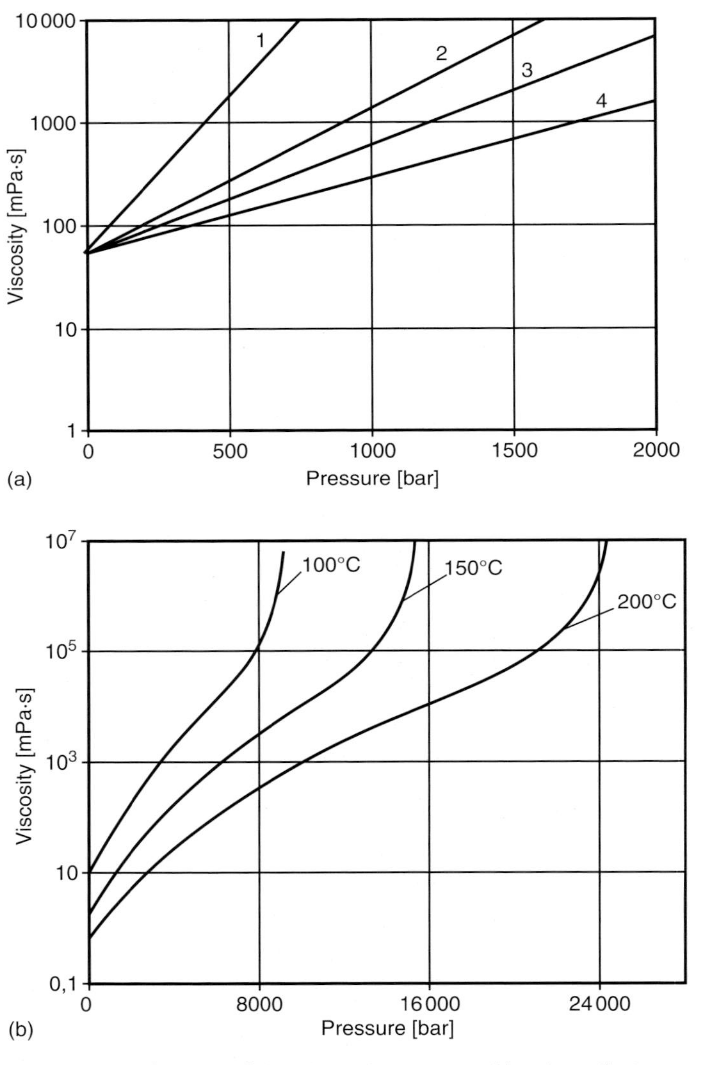

Figure 3.5 Development of viscosity against pressure. (a) various oils: 1: aromatic oil; 2: naphthenic oil; 3: paraffinic oil; 4: biodegradable polyolester. (b) Increase of the viscosity–pressure coefficient with falling temperature according to Ref. [2].

α can increase significantly with falling temperature which has an exponential effect on viscosity. Practical lubrication technology therefore necessitates consideration of pressure and temperature to make a reasonable evaluation of viscosity.

It has recently been shown [3] that VI improvers have an important effect on pressure–viscosity behaviour. This affects the lubricant film thickness between lubricated highly stressed contacts under elastohydrodynamic conditions. Depending on the molecular weight and concentration of the VI improvers (Chapters 9 and 6), lubricants behave differently when such films are formed. At low temperatures, a thin lubricant film is formed. At higher temperatures, VI-improved lubricants form a thicker film as a consequence of a higher viscosity pressure coefficient α.

3.4
The Effect of Shear Rate on Viscosity

The definition of the viscosity of Newtonian fluids (Section 3.1), is a constant (proportionality factor) between the shear force τ and the shear rate S. This means that viscosity does not change (with the exception of temperature and pressure dependence) even when subject to greater shear forces in a friction contact zone or in other words, in isothermal and isobaric conditions. Lubricants which display dependence on shear rate are known as non-Newtonian or fluids with structural viscosity.

Oils containing polymers with specific additives or thickeners and mineral oils at low temperatures (long-chain paraffin effects) display such structure-viscous behaviour. At normal application temperatures, most major lubricant base oils such as hydrocarbon oils (mineral oil raffinates or synthetic hydrocarbons), synthetic esters and natural fatty oils can withstand very high shearing forces (e.g. $10^9\ s^{-1}$) as found in highly loaded machine elements (e.g. gearboxes) and are independent of shear rate.

Engine oils containing polymer VI improvers (Chapter 9) or polymer ash-free dispersants display structural-viscosity effects at low and high temperatures. As a rule, the dependence of viscosity on shear rate is undesirable. However, one can use this effect in fuel-efficiency oils. At high sliding speeds when hydrodynamic lubrications is given, lower viscosity generates lower friction and lower energy consumption. To keep this process under control, HTHS viscosity (high temperature high shear) was introduced. This measures viscosity at higher temperatures (corresponding to the oil temperature at friction points) and at high shear rates. This figure appears in engine oil specifications as a threshold value (Chapter 9).

While the reduction in viscosity caused by the shearing of structure-viscous fluids is reversible, that is after the shearing stops, the original viscosity returns, polymer-based oils can suffer a permanent reduction in viscosity. In these cases, the shearing forces lead to a mechanical change or reduction in the size of the polymer molecules so that their desired effects are minimized. These effects have been observed, in particular, with multigrade engine oils and high-VI hydraulic oils. The shear stability of these polymers is therefore an important quality parameter.

3.5 Special Rheological Effects

Apart from the above-described structure-viscosity and the permanent reduction in viscosity caused by shearing, lubricants are subject to further rheological effects and in particular, colloidal systems consisting of solid or fluid dispersions (solid dispersions or emulsions).

Even small mechanical loading such as vigorous stirring can cause a system to change completely, for example pasty systems can breakdown into low-viscosity systems. Figure 3.6 shows such systems for different structures. If this process is time-based, that is continuous mechanical load causes apparent viscosity to fall over time and the original viscosity is restored after a certain rest period, such fluids are called thixotropic, Figure 3.7. This effect is used for sheet forming lubricants (Chapter 15). In this case, a low-viscosity fluid is preferred for the application of the product (e.g. spraying) but a high-viscosity lubricant film which resists run-off is required on the panels. The rheopexy effect, in which continuous shearing causes viscosity to increase, is not applicable to lubrication technology.

Time-based viscosity changes can also be caused by the separation of colloidal particles such as paraffins from mineral base oils or additives during the storage of lubricants. This undesirable effect can occur during the cooling phase following production or by long-term storage at low temperatures. This separation process can also be independent of temperature and be caused by the solubility status of additives in the base oils.

3.5.1 Greases

The rheological description of greases is also complex because of their complex system structure (fluids, soap thickeners, solid thickeners, additives, Chapter 16).

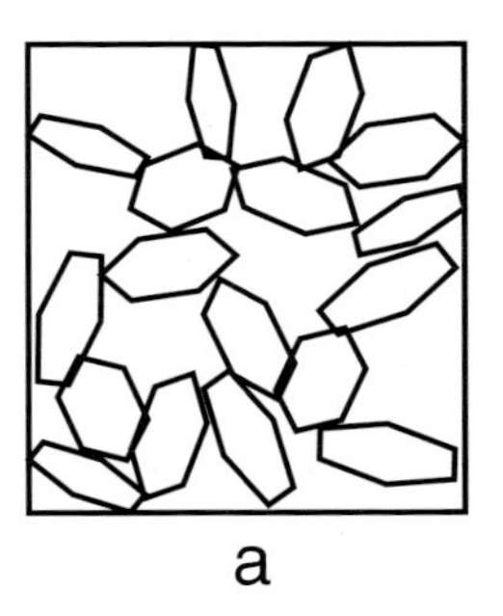

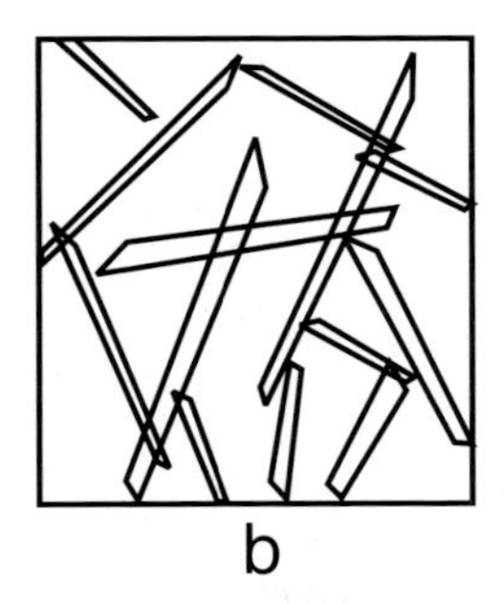

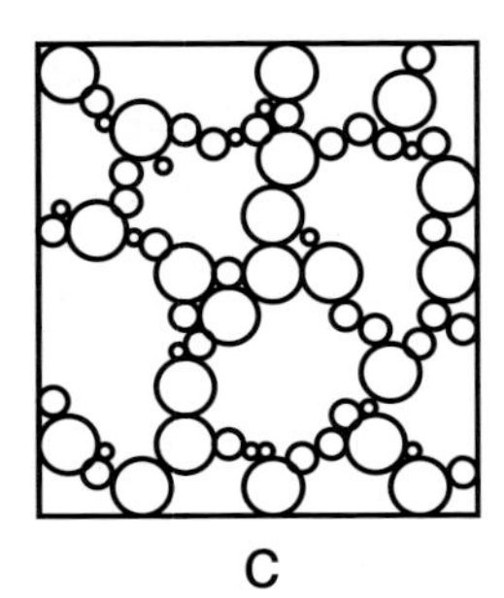

Figure 3.6 Various lubricant structures (a, b, c) with high sensitivity against shear stress.

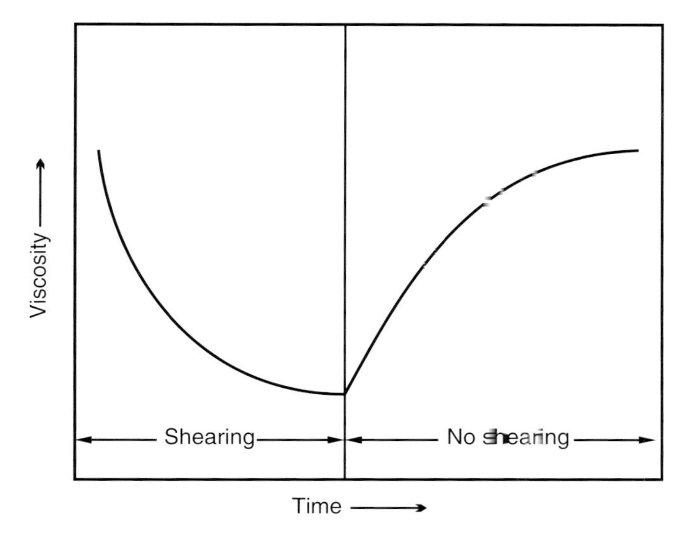

Figure 3.7 Flow characteristics of thixotropic lubricants.

The mathematical description of apparent viscosity has not gained general acceptance in practice. The Sisko equation has developed some significance for shear rates between 10^{-2} and 10^{-4} s^{-1} [1].

$$\eta_s = a + b\gamma^{n-1}, \tag{3.10}$$

where a is dynamic base oil viscosity, γ the shear rate and the constants b and n describe pseudoplastic behaviour.

At lower shear loads, greases behave similarly to structure-viscous substances but at high shear loads, like Newtonian base oils.

3.6 Viscosity Grades

To simplify the classification of lubricants according to their application, viscosity grades were introduced which are now internationally accepted. ISO viscosity grades apply to industrial lubricants while Society of Automotive Engineers (SAE) classifications apply to automotive engine and gear oils.

3.6.1 ISO Viscosity Grades

A total of 18 viscosity grades are laid down in the ISO standard 3448. Over the range from 2 to 2500 mm^2 s^{-1}, these are the international standard number series E6 rounded to whole numbers when the 6 numerals correspond to one power of 10 (the first and fourth power of 10 are reduced). The viscosity grades were also adopted into or added to national standards such as ASTM or DIN.

Viscosity grades are not used for all industrial lubricants. Particularly, oils for chip-forming and chip-less metalworking processes are not classified in this way.

Apart from the viscosity grades, ISO 3448 defines tolerances as well as median viscosities.

3.6.2
Other Viscosity Grades

3.6.2.1 Engine Oils

To define the viscosity of engine oils, two or three viscosity thresholds were selected to define flow properties at low temperatures and to define a minimum viscosity at high temperatures. Maximum viscosity at low temperature should ensure the rapid oil circulation to all lubrication points and permit a sufficiently higher cranking speed for starting and the minimum viscosity at 100 °C should ensure that adequate lubrication of the bearings occurs at high temperatures. Although the classification system was introduced by the SAE together with the ASTM, it is used throughout the world and has been adopted into all national standards. Low temperature viscosity is measured as dynamic viscosity with a specially constructed rotational viscosimeter (cold cranking simulator) at low shear rates (Chapters 9 and 13).

Oils which only fit into one viscosity grade are known as monograde oils. The V–T behaviour of such oils corresponds to that of conventional mineral oils without VI improvers. Oils that cover two or more viscosity grades are multigrade oils and are based on oils containing VI improvers or base oils with high natural VI.

3.6.2.2 Automotive Gear Oils

Specific SAE viscosity grades have been created for automotive gearbox, axle and differential oils. Compared with engine oils and the low temperature behaviour of these oils is more heavily weighted in that a single maximum dynamic viscosity figure and the corresponding maximum temperatures are determined for a number of viscosity grades.

3.6.2.3 Industrial Gear Oils

AGMA (American Gear Manufacturers Association) defines nine viscosity ranges for industrial gear oils.

3.6.2.4 Viscosity Grades for Base Oils

Mineral base oils are traditionally classified according to Saybolt Universal Seconds (SUS). A 150 N base oil shows a viscosity of 150 SUS at 100 °F.

3.6.2.5 Comparison of Viscosity Grades

Figure 3.8 compares the most important viscosity ranges for various applications.

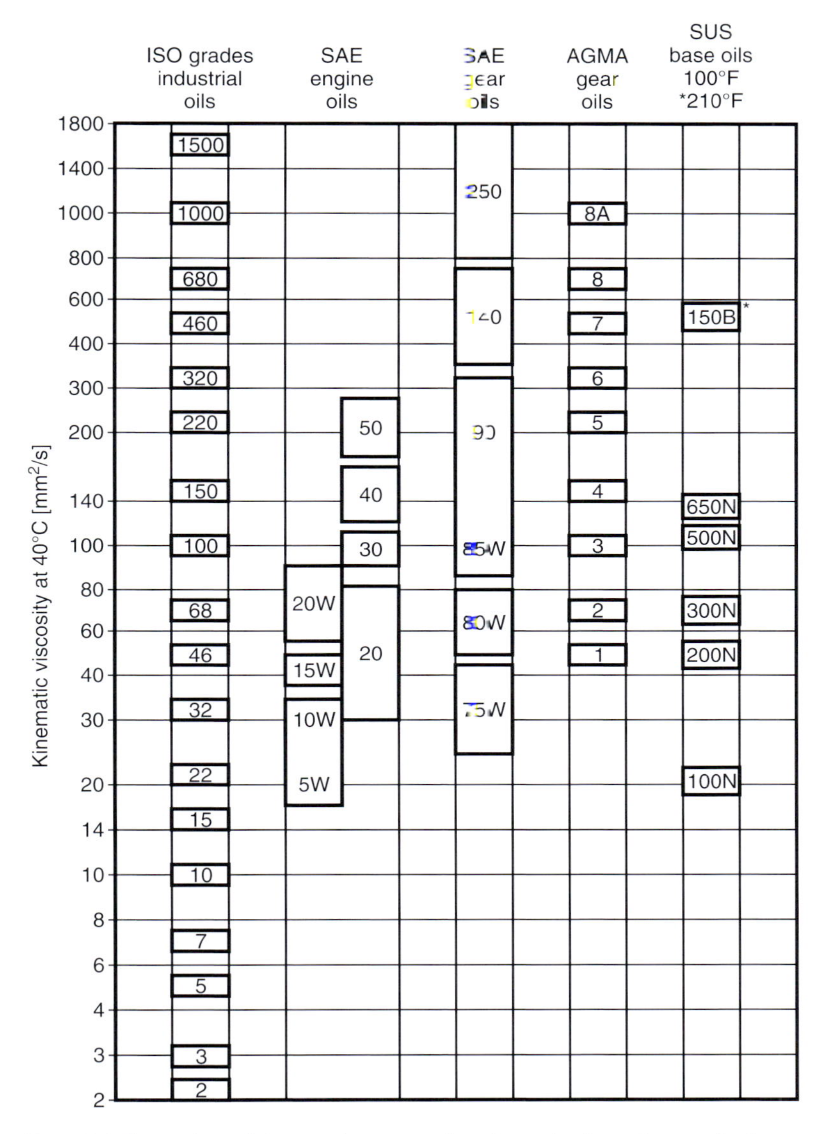

Figure 3.8 Comparison of the most important viscosity grades for various applications.

3.7
Viscosity Measurements

Nael Zaki

3.7.1
Measurement of Viscosity by Capillary Tubes

3.7.1.1 Newtonian Fluids

Viscosity is usually measured from the time of flow of a given quantity of liquid in a capillary under the effect of gravity. Poisueuille's equation shows that, in these conditions, the time of flow is proportional to the viscosity and inversely proportional to the density of the liquid:

$$\eta/\rho = \nu. \tag{3.11}$$

Where ν is called the kinematic viscosity. Its dimensional formula is as follows:

$$L^{-1}M\,T^{-1}/ML^{-3} = L^2T^{-1}. \tag{3.12}$$

In the SI system, it corresponds to $m^2\ s^{-1}$. In the CGS system, it corresponds to $cm^2\ s^{-1}$ or stokes. Considerable use is made of the centistoke (cSt) or square millimetre per second, which is very close to the viscosity of water at 20 °C. The British units are square inch per second.

To measure the viscosity of a fluid in practice, the viscometer is calibrated with standard oils and the flow time of a given quantity of fluid is measured in precise conditions generally set by a standard. Many types of tubes exist for these measurements (Cannon-Fenske, Ubbelohde, Ostwald, etc.) (Figures 3.9 and 3.10). Other types of viscometers (Engler, Saybolt, Redwood) are slightly modified 'funnels'. They give a conventional kinematic viscosity that is more or less, but not exactly proportional to the real viscosity because the flow is not

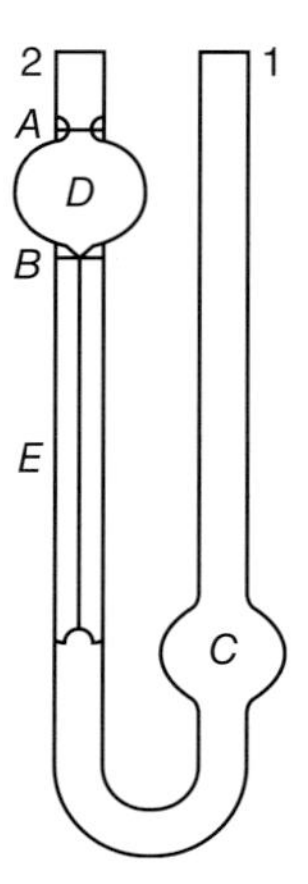

Figure 3.9 Ostwald viscometer. (From Ref. [4])

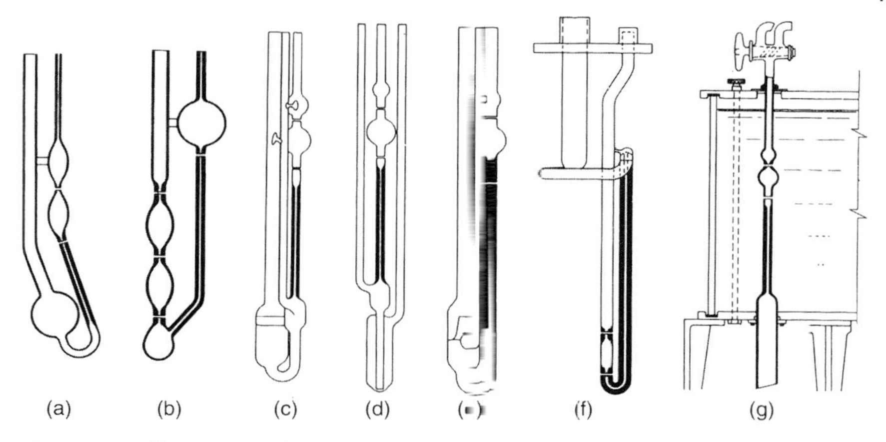

Figure 3.10 Different types of capillary viscometers. (From Ref. [4])

strictly isothermal and tables are used to fix the corresponding values between different methods.

3.7.2
2 Non-Newtonian Fluids

Non-Newtonian fluids have non-linear shear stress–shear rate relationship The flow velocity does not obey a parabolic law. For a Bingham fluid, for example flow occurs only above a given shear stress, and takes place with a central plug (Figure 3.11a) [5].

The shear stress $Pr/2L$ is proportional to the distance from the tube centerline. It is zero at the centre and no fluid shear occurs within the cylinder of radius r as long as $Pr/2L$ is lower than τ_c, the yield value (Figure 3.11b). No flow occurs at all if the radius of the capillary is less than

$$r_o = \tau_c 2L/P, \tag{3.13}$$

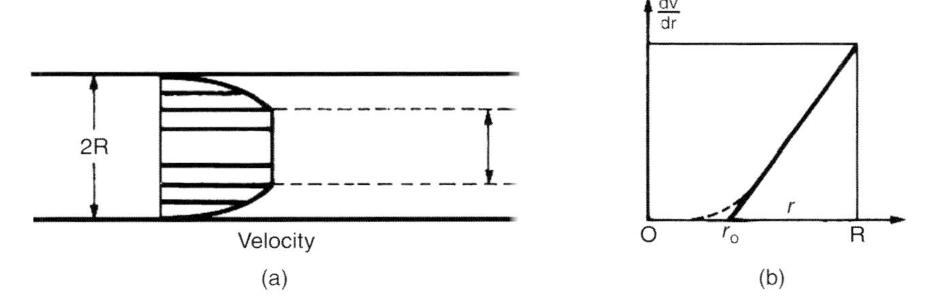

Figure 3.11 Plastic fluid flow in a pipe. (a) Velocity profile. (b) Shear rate as a function of the distance from the central axis.

for a given pressure difference. If R is greater than r_o, plug flow occurs at the centre. If the drive pressure increases, r_o decreases and the diameter of the plug is reduced. Hence, in this case, the shear rate is not proportional to the shear stress. The shear rate varies differently from the shear stress depending on the distance from the centerline.

In case of plastic or pseudoplastic fluids, the apparent viscosity η_a can be defined which, for a given strain rate and shear stress, is equal to

$$\eta_a = \tau/\mathrm{d}\nu/\mathrm{d}r. \tag{3.14}$$

In this case, there is one to one correspondence between the shear stress and the shear rate:

$$\eta_a = f(\mathrm{d}\nu/\mathrm{d}r) \text{ or } \eta_a = f(\tau). \tag{3.15}$$

The characterization of the rheological properties of such a fluid at a given temperature is thus reduced to defining η_a as a function of τ or of $\mathrm{d}\nu/\mathrm{d}r$, or defining of τ as a function of $\mathrm{d}\nu/\mathrm{d}r$. This is done based on the calculations of Saal and Koens [5] and of Mooney [6].

3.7.2.1 Lubricating Greases Viscosity Measurements by the Standard Oil Development (SOD) Viscometer

The SOD is a capillary viscometer that is used to measure the viscosity of lubricating greases. As seen in Figure 3.7, the grease is enclosed in a cylinder and forced by a hydraulically operated piston to flow through a calibrated capillary tube. The flow rate and the shear stresses can be altered. A measurement is taken by noting the fluid flow rate determined from the pump characteristics and its speed of rotation, and the pressure of the hydraulic fluid at the capillary inlet. A plot of the value of $4Q/\pi R^3$ as a function of shear stress $PR/2L$ and, using Mooney's formula, to determine the curve of the strain rate $\mathrm{d}\nu/\mathrm{d}r$ as a function of shear stress. Figure 3.8 gives an example of the application of these formulas to experimental results obtained with a grease on the SOD viscometer [7] (Figures 3.12 and 3.13).

The solid line represents the following expression:

$$4Q/\pi R^3 = f(\tau_R), \tag{3.16}$$

and the dotted line obtained from it by the expression:

$$\mathrm{d}\nu/\mathrm{d}r = f(\tau_R). \tag{3.17}$$

The apparent viscosity $\eta_a = \tau/(\mathrm{d}\nu/\mathrm{d}r)$ corresponding to a given shear rate can be determined from this curve. The SOD viscometer can measure shear rates in the range 0–15 000 s^{-1}.

3.7.3 High Shear Rate Capillary Viscometers

A capillary viscometer can be used to determine the apparent viscosity of non-Newtonian fluids. As Newtonian fluids, the flow rate and pressure loss are

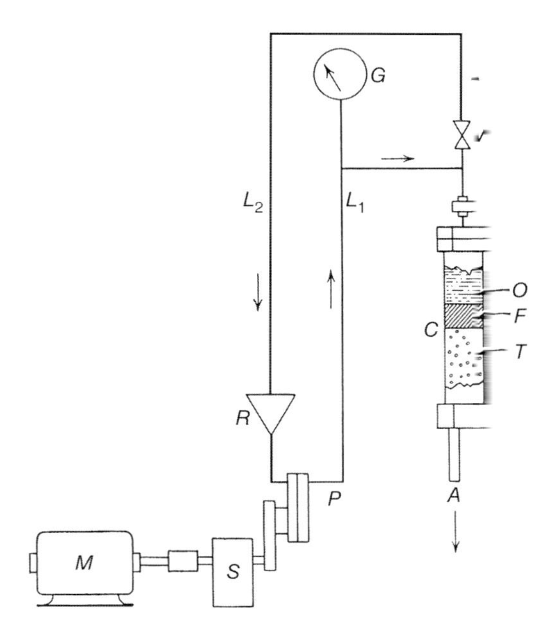

Figure 3.12 Schematic depiction of SOD viscometer (From Ref. [4]) A: Capillary; C: cylinder; F: piston; G: pressure gauge; L_1: oil inlet line; L_2: oil return line; M: motor; O: hydraulic fluid; P: positive displacement pump; R: reservoir; S: reduction gearbox; T: test; V: valve.

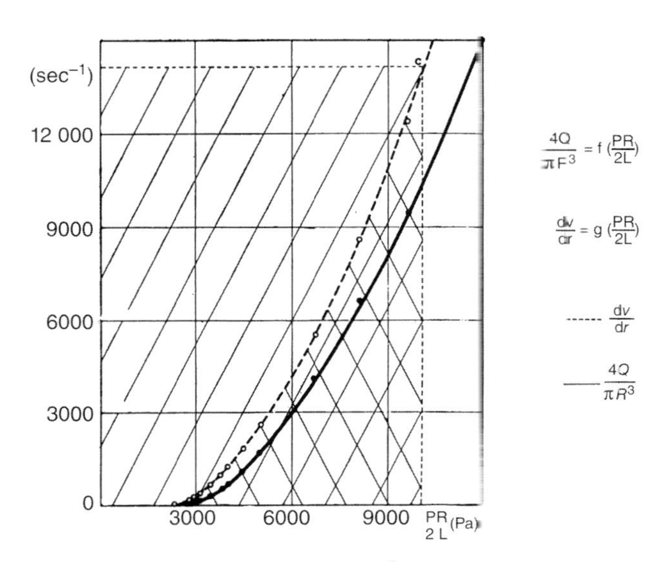

Figure 3.13 Shear rate and value $4Q/\pi R^3$ at the wall as a function of shear stress. (From Ref. [7]).

measured. However, for non-Newtonian fluids several measurements at different wall shear rates must be used to plot a curve, because of the variability of the apparent viscosity with shear rate. ASTM standardized these instruments by the ASTM D 4624.

3.7.4
Rotational Viscometers

3.7.4.1 Coaxial Cylinder Viscometer

The coaxial viscometer has two cylinders, one is static and the other rotates where the material to be tested is placed between the two cylinders (Figure 3.14).

For Newtonian fluids the Margules equation is used:

$$\eta = M/4\pi h\Omega \cdot (1/R_i^2 - 1/R_e^2). \tag{3.18}$$

Where M is the Torque moment, R_i is the radius of the inner cylinder and R_e is the radius of the outer cylinder, h is the height of the cylinder and Ω is the angular rotation rate of the mobile cylinder in rad s^{-1}.

For Non-Newtonian Fluids

Reiner and Rivlin [8] have established an equation for a Bingham fluid in coaxial cylinders, and other equations have been established using power law for different fluids:

$$T = R_o(\mathrm{d}\nu/\mathrm{d}r)^n. \tag{3.19}$$

3.7.4.2 Cone/Plate Viscometer

The cone and plate viscometer is depicted in (Figure 3.15).

The viscosity is given by the relationship

$$\eta = 3\alpha M/2R^3\,\Omega, \tag{3.20}$$

where Ω is the angular velocity, α is the cone angle and R is the rotor radius.

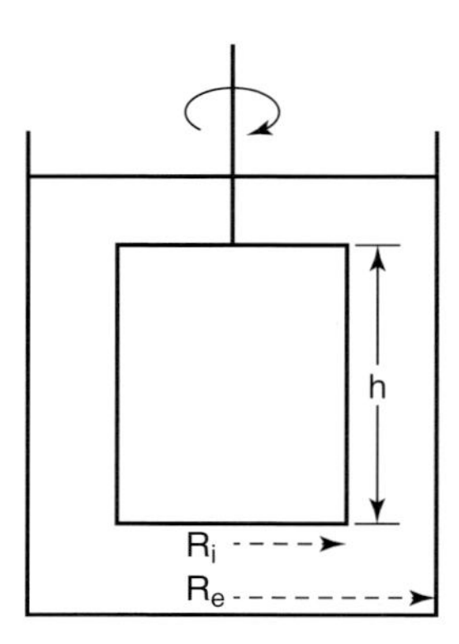

Figure 3.14 Illustration of the coaxial rotational rheometer.

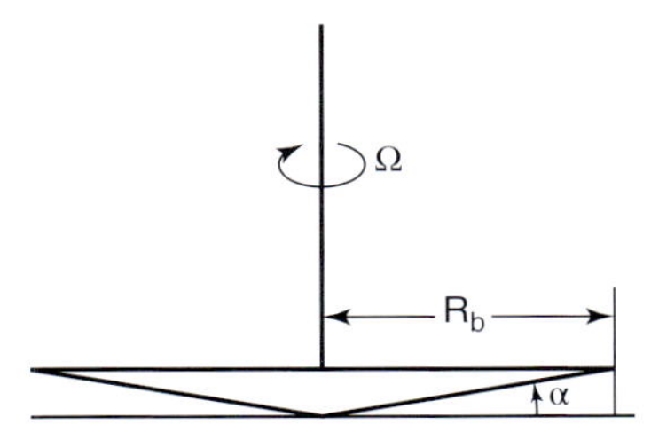

Figure 3.15 Cone and plate viscometer.

Figure 3.16 A photograph showing a viscous fluid exhibiting the Weissenberg's effect. (From Ref. [8]).

For viscoelastic fluids, such as oils containing a viscosity index improver and for greases, the Weissenberg effect (normal forces) may be measured (Figure 3.16).

3.8
Viscosity Measurements at High Pressure

The viscosity of fluids can be determined by applying a known force and measuring the resultant rate of deformation. While there are numerous types of rheometer and viscometer available which operate at atmospheric pressure over a wide range of shear rates and shear stresses, the measurement of viscosity at elevated pressures is not quite so straightforward due to the need for high-pressure containment. Several designs of high-pressure viscometer exist and each is challenged with the need for adequate sealing and remote sensors [9]. A high-pressure viscometer based on a falling sinker design in which gravity was used

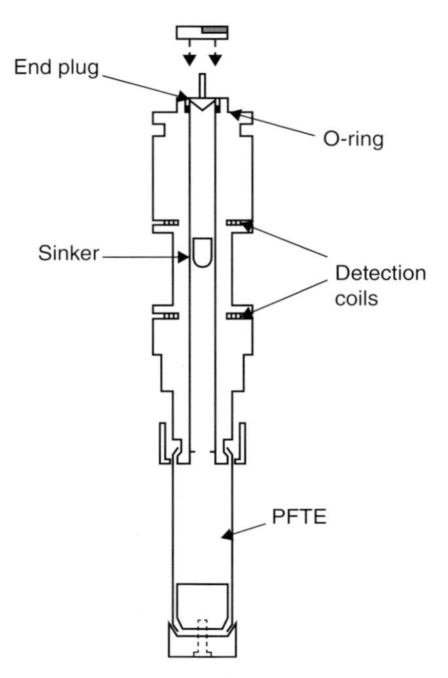

Figure 3.17 High pressure falling sinker viscometer. (From Ref. [9]).

as the applied force is described here as an example. The viscometer consists of a vertical tube through which a sinker falls. Samples of oils are sealed into the tube using a shrinkable PTFE expansion sheath. It is essential that no air remains entrapped within the sample. The viscometer assembly is shown in Figure 3.17.

The tube and associated assembly is housed in a high-pressure vessel rated to 1 GPa and maintained at a constant temperature by air circulation. The pressure is transmitted using an oil mixture as the hydraulic medium by pressure intensification from air pressure at 7 bar through to a maximum operating pressure of 500 MPa. A 1 : 1 mixture offers good lubricating properties with a high freezing point. The whole device can be rotated 180° to return the sinker to its original starting position between each measurement. A piezoresistive pressure gauge is used to measure the high pressure within the viscometer tube. Pressure is related to the voltage output by the polynomial

$$P = a_0 + a_1\, V + a_2\, V^2. \tag{3.21}$$

Where P is the pressure, V is the voltage and a_o and a_1 are the polynomial coefficients.

The viscosity is determined directly from the time taken for the sinker to descend a fixed length. This is the distance between two detection coils wrapped around the outside of the tube.

The duration of the sinker fall between peaks depends on the viscosity but is typically between 2800 and 8200 ms for n-octane with a 95% confidence level of

+125 ms found experimentally. For high viscosity liquids such as lubricating oils, the fall time is notably longer. It is between 393 and 2800 s with a 95% confidence level of +38.056 s (i.e. +2%). For a fluid which exhibits Newtonian behaviour such as mineral oil, the viscosity can be determined from the shear stress profile across the annular gap. From the time taken for the sinker to pass the detection coils, the viscosity is determined from

$$\mu = t(1 - \rho_1/\rho_s)/A[1 + 2\alpha(T - T_o)][1 - 2\beta(P - P_o)]. \qquad (3.22)$$

Where μ is the viscosity of the measured fluid, ρ_1 is the density of the measured fluid and ρ_s is the density of the sinker, α is the coefficient of thermal expansion, β is the bulk compressibility of the sinker, T is the measured temperature and T_o is the standard temperature and A is a coefficient calculated from the dimensions of the tube and the sinker.

References

1 Klamann, D. (1984) *Lubricants and Related Products*, VCH Verlagsgesellschaft, Weinheim.

2 Holland, H. *Information of the Institut für Reibungstechnik und Maschinenkinetik*, Technische Universitat Clausthal, Clausthal-Zellerfeld.

3 Gold, PeterW., Loos, Jorg, Kretschmer, Torsten and Wincierz, Christoph (2006) Influence of VI improvers on pressure-viscosity behaviour. Proceedings of the 15th International Colloquium Tribology, Automotive and Industrial Lubrication, 17–19 January 2006, Technische Akademie Esslingen.

4 Van Waser, J.R., Lyons, J.W., Kim, K.Y. and Colwell, R.F. (1963) *Viscosity and Flow Measurmenets*, Interscience Publishers, John Wiley and Sons, New York.

5 Saal, R.N.J. and Koens, G. (1933) Plastic properties of asphaltic bitumen. *J. Inst. Petroleum Tech.*, **19**, 176.

6 Mooney, M. (1931) Explicit formulas for slip and fluidity. *Rheology*, **2**, 210–222.

7 Briant, J. (1956) Study of rheological properties of oils using the SOD viscometer. *Rev. Inst. Franc. du Petrole*, **11**, 113–133 and 247–287.

8 Weissenberg, K. and Freeman, S.M. (1948/49). Proceedings of the First International Rheology Congress, Holland, pp. 11–12.

9 Isdale, J.D. (1976) Viscosity of simple liquids including measurements and prediction at elevated pressure. PhD thesis. University of Strathclyde, Glasgow, UK.

4
Base Oils

Theo Mang and Georg Lingg

In terms of volume, base oils are the most important components of lubricants. As a weighted average of all lubricants, they account for more than 95% of lubricant formulations. There are lubricant families (e.g. some hydraulic and compressor oils) in which chemical additives only account for 1% while the remaining 99% are base oils. On the other hand, other lubricants (e.g. some metalworking fluids, greases or gear lubricants) can contain up to 30% additives.

The origin of the overwhelming quantity of mineral lubricant base oils has led to lubricants being viewed as a part of the petroleum industry and this is underlined by their inclusion in petroleum statistics. Over the last few years, lubricants have increasingly become a separate discipline with clear differences from petroleum mass products. This was caused by the high added value which is generated in this product segment along with the fact that many high-performance lubricants no longer contain petroleum base oils.

4.1
Base Oils: A Historical Review and Outlook

Although the most important requirement of base oils in the 1950s was the correct viscosity and the absence of acidic components, base oils in the 1960s were downgraded to solvents or carriers for additives in the euphoria surrounding chemical additives. In the 1970s, there was a realization that some synthetic fluids with uniform basic chemical structures offered performance superior to that of mineral base oils. At that time, the considerably higher price of these products hindered their market acceptance. In the 1980s however, lower-price, quasi-synthetic hydrocracked oils were introduced in Western Europe that closely matched the properties of synthetic hydrocarbons (Shell, BP, Fuchs). In the 1990s, base oil developments were influenced by the ever-increasing demands on lubricant performance, and by environmental and health and safety criteria.

Lubricants and Lubrication, Third Edition. Edited by Theo Mang and Wilfried Dresel.

This led to chemically more pure oils such as hydrocracked products, polyalphaolefins and esters gaining acceptance. Natural fatty oils, particularly their oleochemical derivatives, have experienced a renaissance because of their technical characteristics but, above all, because of their rapid biodegradability.

The trend towards ever-greater performance and even better environmental compatibility continues in the first decade of the new millennium. The significantly higher price of the new lubricants, which will be increasingly characterized by their base oils and less so by their chemical additives, will probably be accepted by users who will benefit from long product life and lower overall system costs. In 2004, throughout the world, approximately 7% (w/w) of base oils were synthetic products (including Group III hydrocracked oils; Section 4.6). This segment is forecast to grow to 10% in 2015 [1]. In Germany, furthermore, approximately 5% of lubricant base oils were rapidly biodegradable (natural and synthetic) esters in 2005. In addition, gasification of carbon-containing raw materials, for example natural gas, biomass and carbon, will enable the production of synthetic lubricant base oils of high quality. Significant industrial-scale production of such base oils will start at the beginning of the second decade of the new millennium (Section 4.4.8).

4.2 Chemical Characterization of Mineral Base Oils

The characterization of mineral oil fractions, whether crude oil or lubricant base oil fractions, by use of normal chemical practices to determine their exact structure is not possible without great expense. Crude oil generally consists of many thousands of single components and these are reflected in the processing of each fraction. It was, therefore, always an objective to describe mineral oil fractions by the comparatively simple expedient of defining their technical properties or to identify and quantitatively determine groups of components with similar chemical character. Advanced physicochemical methods are, however, increasingly being used in routine testing.

4.2.1 Rough Chemical Characterization

4.2.1.1 Viscosity–Gravity Constant (VGC)

This value enables only rough chemical characterization of oils. Values near 0.800 indicate paraffinic character whereas values near to 1.000 point to a majority of aromatic structure (ASTM D 2501-91).

4.2.1.2 Aniline Point

The aniline point is also a help when characterizing the hydrocarbon structure of mineral oils. When aniline and oil are mixed and then cooled to a certain temperature (aniline point), two phases form. Because of their good solubility, aromatic structures give the lowest values.

4.2.2
Carbon Distribution

The most important means of analysis for characterization of mineral oil hydrocarbons was, and remains, the determination of carbon in terms of its three categories of chemical bond–aromatic (C_A), naphthenic (C_N) and paraffinic (C_P).

N–D–M analysis uses physicochemical data that are easy to obtain. These include refractive index, density and molecular weight. Molecular weight can be determined by measuring the viscosity at different temperatures (e.g. ASTM D 2502-92). Carbon distribution is given in $\%C_A$, $\%C_N$ and $\%C_P$ (100% in total). N–D–M analysis also determines the average total number of rings per molecule (R_T) and the breakdown into aromatic and naphthenic rings (R_N) per molecule ($R_N = R_T - R_A$).

Brandes [2] created a method of determining carbon distribution according to specific bands in the infrared spectrum. The method has proven itself for lube base oils and can be performed at acceptable expense. Exact determination of aromatic carbon content can be performed by high-resolution nuclear magnetic resonance (NMR) (ASTM D 5292-91).

4.2.3
Hydrocarbon Composition

A further refinement in the characterization of lubricant base oils is the determination of molecular families. Chromatography is used first to separate components and the fractions are then subjected to advanced analytical procedures.

To differentiate mineral oils including lube base oils in the boiling range from 200 to 550 °C, high ionizing voltage mass spectrometry (ASTM D 2786-91) is used for saturated fractions and ASTM D 3239-91 for aromatic fractions. Saturated fractions are separated into alkanes (0-ring), 1-ring, 2-ring, 3-ring, 4-ring and 5-ring naphthenes. The aromatic fractions are subdivided into seven classes: monaromatics, diaromatics, triaromatics, tetraaromatics, pentaaromatics, thiopheno aromatics and unidentified aromatics.

4.2.4
Polycyclic Aromatics in Base Oils

Polycyclic aromatics (PAH (polycyclic aromatic hydrocarbons), or, in general, PCA (polycyclic aromatics)) are carcinogenic, environmentally harmful substances, which are found in crude oils. In general, they are not created when a lubricant is used. PAH formed by the combustion of gasoline can gather in engine oils. They can also gather in quenching oils after long periods of heavy-duty use.

In traditional solvent refining processes, PAH largely remain in the extract. Non-solvent extracted distillates contain PAH in line with their boiling point.

The carcinogenic characteristics of non-severely treated distillates in the petroleum industry were established by the IARC (International Agency for Research on Cancer) in 1983 [3]. This led to considerable limitations in the manufacture and application of naphthenic base oils and the ending of the use of aromatic extracts as lube base oils. Previously, Grimmer and Jakobs [4–6], in particular, confirmed the presence of PAH in lubricant base oils analytically and highlighted their carcinogenic characteristics.

The standardization of analytical methods of determining PAH and the setting of threshold values was the subject of lively discussions between 1985 and 1995. Whereas, one group was primarily interested in establishing exact analyses and others wanted benzo(*a*)pyrene as a reference, another group was pressing ahead with the IP 346 method. This was later adopted into national legislation in several countries. PAH can be determined by HPLC with anthracenes or other aromatics as markers or by GC–MS after appropriate sample preparation.

The IP 346 method does not analyze PAH directly but an extract is obtained in DMSO (dimethyl sulfoxide), in which PAH accumulate. As a rule, the extract, which is often wrongly referred to as a PAH concentrate, largely contains naphthenes or mono-aromatics. According to IP 346, DMSO extracts contain only 0.1% PAH. After numerous skin-painting tests on mice, the carcinogenicity of petroleum products corresponds to the percentage of DMSO extract.

Criticism of the IP 346 threshold value continues. The weakness of the value is clear when used gasoline engine oils are evaluated. The PCA enrichment caused by gasoline combustion is not considered by the IP method. For example, a 1000 ppm increase in PCA only increases the DMSO extract of the 'used' base oil from 1.0 (fresh oil) to 1.1%.

Figure 4.1 illustrates the increasing carcinogenicity of petroleum products corresponding to the percentage of DMSO extract and various analytical methods for the determination of PAH content [7–10].

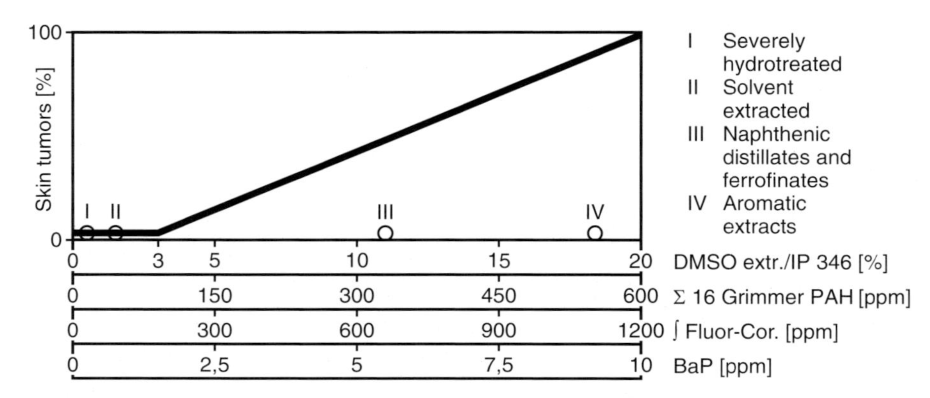

Figure 4.1 Dependence of the carcinogenicity of petroleum products on the PAH content [8]. Skin painting tests and rough statistical classification of different PAH analyses.

4.2.4.1 Aromatics in White Mineral Oils

White oils which are used for pharmaceutical, medical or food processing applications should be free from aromatics, or at least low in aromatics, and preferably contain no traces of polycyclic aromatics. Ultraviolet absorption in the range 260–350 nm is used. Enrichment by extraction is worthwhile. ASTM D 2296-99 describes such a process including pre-enrichment with dimethyl sulfoxide extraction.

4.3 Refining

Since the beginning of the petroleum industry, mineral oils have been used for lubricant base oils. The process of converting crude oil into a finished base oil is referred to as refining. As far as base oil manufacturing is concerned, the actual refining process begins only after the distillation stages. Refining is thus the term often used to describe all the manufacturing stages after vacuum distillation.

Lubricant refineries are divided into integrated and non-integrated plants (Figure 4.2). Integrated refineries are linked to primary crude oil refineries

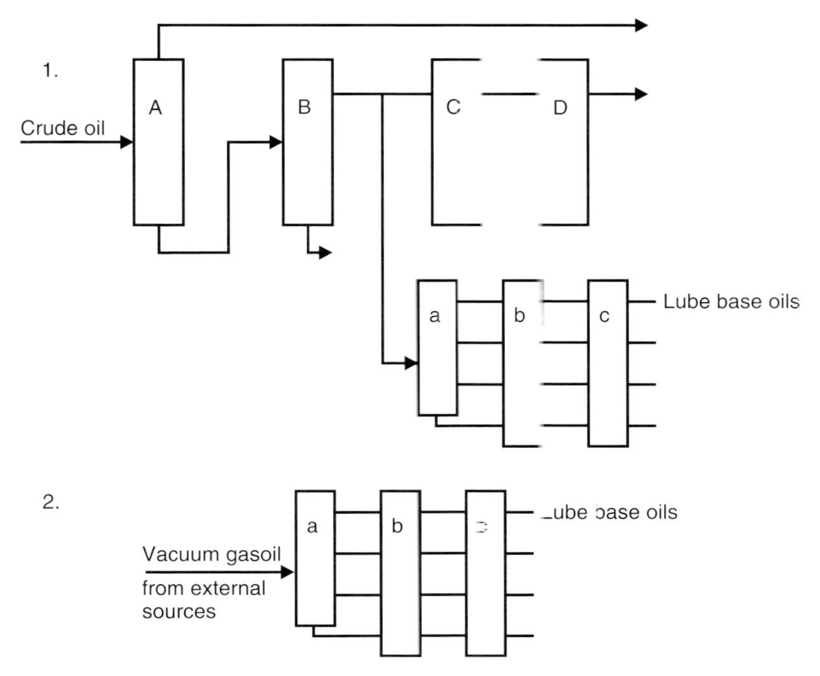

Figure 4.2 Integrated (1) and not integrated (2) lube refinery. A: Atmospheric distillation; B: vacuum distillation; C and D: processing of vacuum distillates for non-lube production. a: Fractionating vacuum distillation; b and c: lube refining processes.

and are fed with vacuum distillate by pipeline. Non-integrated refineries purchase vacuum distillate on the open market or buy atmospheric residues and perform their own vacuum distillation. Occasionally, they perform vacuum distillation on crude oil.

4.3.1
Distillation

By way of fractional distillation, products are removed from crude oil that approximately meet the viscosity grades ultimately required. Often, only four or five cuts suffice to fulfil lubricant requirements. As described later, the viscosity of the primary vacuum distillate is independent of the finished base oils in hydrocracking processes because the hydrocracking process creates new molecule dimensions.

After the corresponding separation of the lighter components from the crude oil by atmospheric distillation, the lubricant components are in the atmospheric residue. Figure 4.3 shows the yields of the various cuts in conventional lube oil refining with the corresponding boiling ranges of a typical lube crude. The atmospheric residue is now subjected to vacuum distillation to remove the components required for lubricants. In a vacuum, the boiling points of the heavier cuts fall so that distillation without thermal destruction (cracking) is possible.

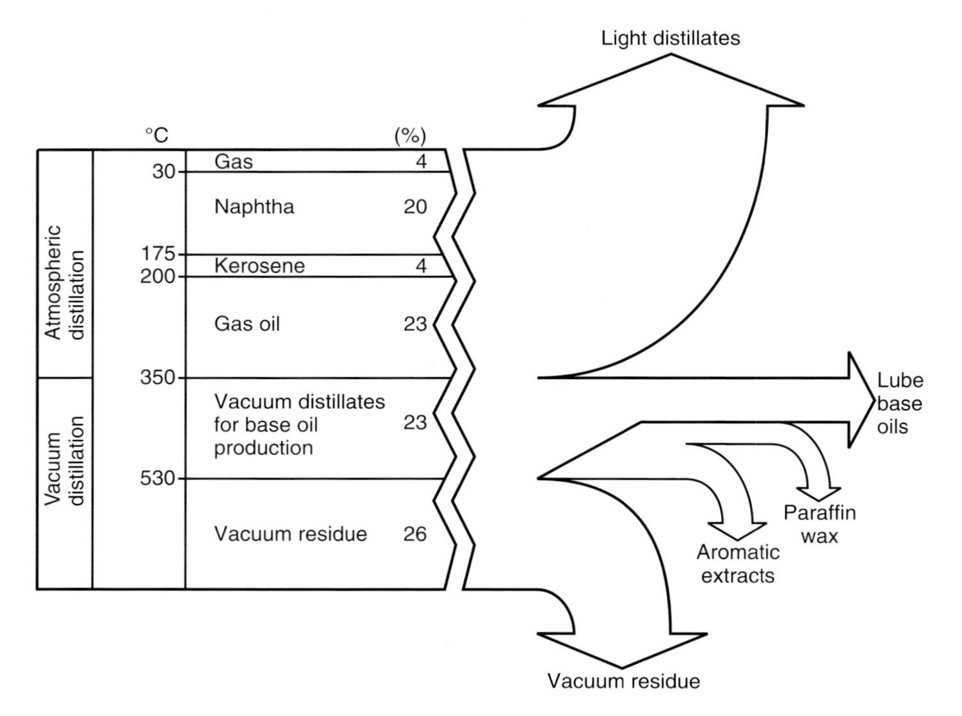

Figure 4.3 Yield of the various cuts in conventional lube oil refining of a typical lube crude.

4.3.2
De-Asphalting

Although the vacuum residue still contains highly viscous hydrocarbons which can supply valuable components for lube base oils, distillation cannot separate these from the asphalt which is also present and extraction processes must be used to separate these highly viscous base oils, commonly known as brightstocks. Brightstocks are produced in lube oil refineries when the use of the asphaltene by-product (hard asphalt) is worthwhile. The quality of the hard asphalt for the manufacture of high quality bitumen depends on the crude oil. Extractive separation uses light hydrocarbons (propane to heptane), of which propane is the leading product for de-asphalting. Brightstocks can be manufactured with viscosities of more than 45 $mm^2 s^{-1}$ at 100 °C.

Figure 4.4 shows a flow chart for the manufacture of suitable feeds for lube oil refining by distillation and propane de-asphalting. Figure 4.5 shows the de-asphalting process.

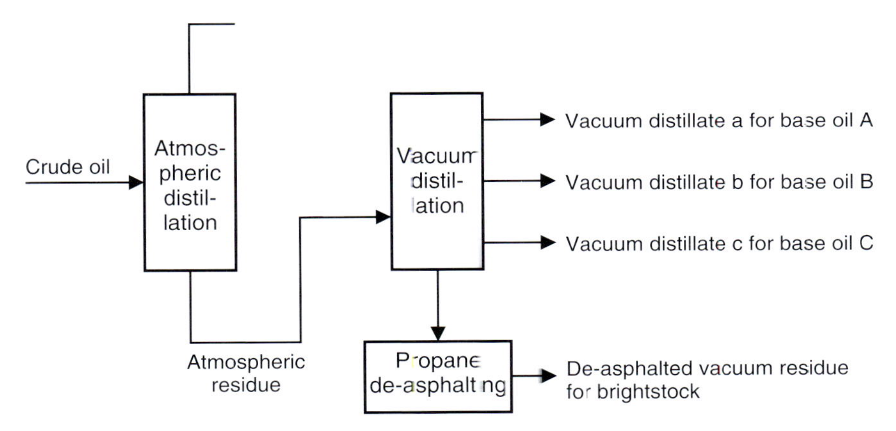

Figure 4.4 Flow chart for the manufacture of suitable feeds for lube oil refining by distillation and propane de-asphalting.

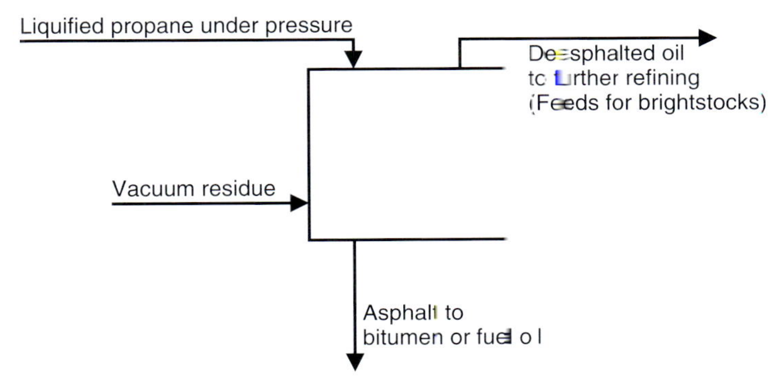

Figure 4.5 Flow chart of the de-asphalting process.

4.3.3 Traditional Refining Processes

Vacuum distillation cuts principally determine the viscosity and flashpoint of later base oils. The precision of the fraction at the upper and lower boiling limits of a cut are of great importance.

The distillates still contain components that can detrimentally affect ageing, viscosity–temperature behaviour and flowing characteristics, and components which are hazardous to health. To eliminate these disadvantages several refining methods were developed, of which solvent refining has become the most accepted method over the past few decades. Today, however, new plants increasingly use hydrotreatment.

4.3.3.1 Acid Refining

Acid refining has become less popular because the acid sludge waste produced is difficult to dispose of and this method has been replaced by solvent extraction. Acid refining is still used to some extent for the re-refining of used lubricating oils (Chapter 8) and for the production of very light-coloured technical or pharmaceutical white oils and petroleum sulfonates as by-products.

When the distillates are treated with concentrated sulfuric acid or fuming sulfuric acid (oleum), substances which accelerate oil ageing are removed. Oleum treatment (wet refining) has a greater chemical effect on the structure of aromatics and not only readily removes reactive oil components such as olefins but also reduces the aromatic content, which in turn increases the viscosity index of the product. Reactions with saturated paraffinic structures lead to refining losses. Acid-refined oils require complex neutralization and absorption follow-up treatment to remove all traces of acid and undesirable by-products. Some lubricant specifications still require base oils to be free from acid even though modern base oils no longer come into contact with acids.

4.3.3.2 Solvent Extraction

Whereas acid refining uses chemical reaction to reduce aromatic content and to eliminate reactive, oil-ageing accelerators, solvent extraction is based on physicochemical separation (Figure 4.6). Solvent extraction creates base oils which are known as solvent raffinates or solvent neutral (SN). Extraction processes using solvents create both a base oil and, after evaporation of the solvent, an aromatic-rich extract. The selectivity of the extraction media for aromatics is an important selection parameter.

In particular, the selectivity towards polycyclic aromatics with three or more aromatic rings has attracted attention because of the carcinogenicity of these compounds. Numerous extraction media have been developed, of which furfural, NMP (*N*-methyl-2-pyrrolidone) and phenol have become economically significant. Sulfur dioxide (SO_2) should also be mentioned because of its historical importance as B. Edeleanu introduced extraction to petroleum technology in

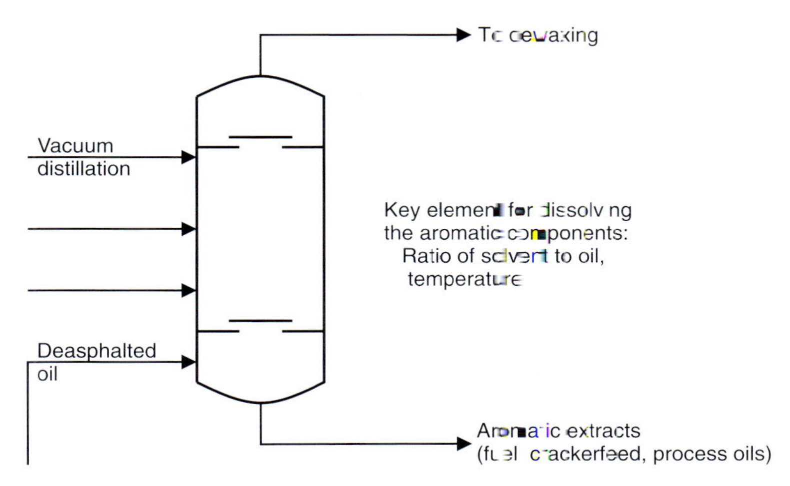

Figure 4.6 Solvent extraction.

1912. In 1999, sulfur dioxide was only used in very few refineries (one in Germany until 1999) for the refining of naphthenic distillates.

In recent years, several furfural extraction plants have been converted into NMP plants, and even phenol plants have taken a back seat. NMP is a non-toxic solvent and can be used in a low solvent-to-oil ratio with high selectivity. This generates significant energy savings. NMP in new plants results in physically smaller units and thus lower capital expenditure.

As a rule, extraction plants are the first refining step for vacuum distillates because subsequent solvent dewaxing is the more complex refining process in terms of capital expenditure and operating costs. This route means that the extract part does not undergo unnecessary dewaxing.

Depending on the crude, the extract part of paraffinic oils can be 30–50%. As the standard requirement of solvent neutral oils is a viscosity index (VI) of at least 95, the extraction severity is matched to this demand.

A higher proportion of aromatic and naphthenic hydrocarbons in the distillates requires greater extraction severity and thus a larger quantity of extract. The percentage share of the extract is a major economic factor in conventional lube refining. In general, extracts can only be used as products which are of lower value than the finished base oils. The large quantities are used as cracker feed and only a comparatively small proportion can be used as process oil. Their use as plasticizers has been severely limited in recent years because their carcinogenicity. The same applies to their use as lubricant base oils. Aromatic extracts have a very high viscosity–pressure coefficient (Chapter 3) and in the past were often used in metal-forming operations such as cold extrusion.

A one-point increase in viscosity index as a result of greater extraction severity creates, on average, 1% more extract. Some refineries these days produce so-called semi-raffinates with VIs of between 70 and 80. It is cheaper to adjust the

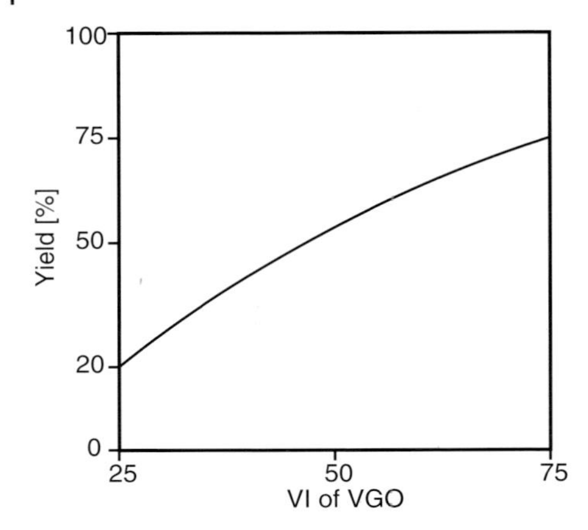

Figure 4.7 Average yield of waxy base oils in solvent extraction to get 100 VI.

VI later with VI improvers than to produce 15 or 20% more extract. These half-raffinates, however, have disadvantages resulting from the lower extraction severity, for example lower oxidation stability, possibly higher amounts of sulfur or higher amounts of polycyclic aromatics.

Solvent extraction is generally only economic to a minimum VI of 50 of the vacuum gas oil [11]. Figure 4.7 shows the average yield of waxy base oils by solvent extraction to achieve 100 VI. It must, however, be remembered that subsequent solvent dewaxing further cuts the yield by approximately 20%.

4.3.4 Solvent Dewaxing

In traditional refining processes, solvent extraction is followed by solvent dewaxing. Long-chain, high melting point paraffins negatively affect the cold flow properties of lube oil distillates and lead to a high pour point. This is caused by the crystallization of waxy substances at low temperatures, which results in turbidity and an increase in viscosity. Their removal has, therefore, been an important consideration since the beginning of crude oil refining.

Dewaxing by crystallization of paraffins at low temperatures and separation by filtration are the principal processes in traditional refining. Compared with catalytic dewaxing with hydrogen, urea-dewaxing to separate *n*-paraffins is of relatively minor importance in lube oil refining. Crystallization methods involve mixing the solvent with the oil; this improves filtration, as a result of dilution, and promotes the growth of large crystal formations.

The important solvents are ketones and chlorinated hydrocarbons. Dewaxing with ketones (dimethyl ketone, methyl ethyl ketone, MEK) is normally used for pour points down to −12 °C. For lower pour points, the Di–Me (dichloroethane–dichloromethane mixtures) method is used. This also enables the manufacture of hard and soft waxes. As for the extracts created by solvent extraction,

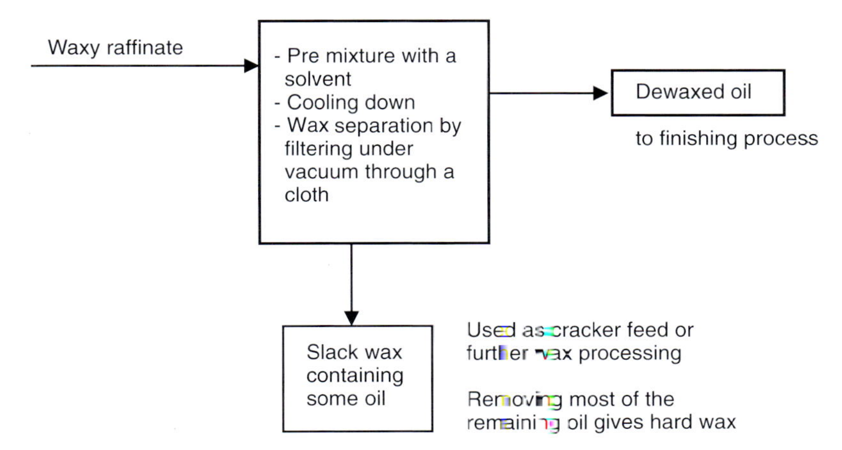

Figure 4.8 Illustration of the conventional solvent dewaxing process.

the by-product situation is again problematic. The paraffins are only worthwhile if their quality and processing is good and they can then be sold as candle wax, for coatings, and for other purposes. They are often used as cracker feed in fuel catalytic crackers (FCC) and the cracker yield is then assessed by use of special formulas.

Figure 4.8 illustrates the conventional solvent dewaxing process.

4.3.5 Finishing

A finishing stage often follows the two refining stages of extraction and dewaxing. In the past, methods which employed absorbents were often used but these days the processes almost all use hydrogen (hydrofinishing). Finishing should improve the colour of the product and remove surface-active substances which can negatively affect the air-release or demulsifying properties of a lubricating oil.

Depending on the temperature, pressure, catalyst, and space velocity of the hydrofinishing, a more or less severe hydrogenating process can be initiated. In general, the hydrofinishing process is referred to as mild hydrotreating and focuses on improving colour, odour, and ultra-violet stability. Ferrofining (BP) has achieved major economic importance as a finishing process. The process parameters do not generally lead to de-sulfurization. More severe processes with pressures in excess of 100 bar can bring about significant de-sulfurization and some de-aromatization.

Additional finishing with absorbents (bleaching clays, bauxite) is sometimes used for the manufacture of refrigerator, transformer, or turbine oils. The subsequent filtration process in filter presses or other filtering equipment represents additional complexity and the disposal of the filter residue is an increasing problem.

Table 4.1 VI and wax content of vacuum gas oil (VGO) from various crudes.

Crude	VI after solvent dewaxing	Wax in VGO (%, w/w)
Pennsylvania	100	0
Ordovician	85	13
Brent	65	19
Arab light	60	9
Arab heavy	55	9
Iranian light	55	16
Lagomedio	45	10
Urals	40	11
Iranian heavy	40	16
ANS	15	8

4.3.5.1 Lube Crudes

Crude oils, which yield high-grade base oils, are preferred for the production of lube base oils. Of importance to the vacuum gas oils (VGO) that are derived from the crude oil and which are the direct feeds for lube refining are, particularly, VI, wax content and sulfur content. A high VGO VI leads to low extraction losses in solvent extraction processes and low hydrogen consumption during hydrogenating processes. High wax contents increase production costs, as a result of the high operating costs of the dewaxing plant, as well as reducing base oil yield. In a base oil production unit using a wax isomerization process, high wax content can also contribute to a higher VI. The finished base oils have a high-sulfur content, especially after conventional solvent refining. Hydrogenating processes almost fully eliminate sulfur but desulfurization consumes hydrogen.

Table 4.1 shows the VI and wax content of VGO from various lube crudes.

4.4 Base Oil Manufacturing by Hydrogenation and Hydrocracking

Traditional solvent refining is the separation of unwanted components from vacuum distillates. Hydrogenation and hydrocracking in the manufacture of lubricant base oils significantly influence the chemical structures of mineral oil molecules. On one hand, unstable molecules are chemically stabilized by the removal of the heteroatoms (sulfur, oxygen, nitrogen) and on the other, severe hydrogenation can convert aromatics into saturated naphthenic or paraffinic structures. In addition to the hydrogenation process, hydrocracking breaks-down or cracks larger molecules into smaller ones. Larger molecular structures can re-form from small fragments. The principal process criteria are temperature, pressure, the catalyst and space velocity. If special conditions are met, a focal point of the process is the isomerization of paraffinic structures. Besides the saturation of aromatics, opening of the naphthenes rings can occur.

It is clear that lubricant base oils can be much more easily tailored using these processes than is possible with simple solvent refining separation. The future of lube base oil production thus lies with hydrogenation and hydrocracking. An additional advantage of advanced hydrocracking is the lower dependence on the quality of the crude oil. Although the economic boundaries of solvent refining are set by yield (extract and paraffin quantities), altering hydrocracking process parameters can compensate for varying crude oil qualities.

The roots of manufacturing lubricant base oils by hydrogenation lie in previous attempts to liquefy coal by high-pressure hydrogenation. On the basis of the results of Bergius (1913, Hanover, Germany), the first technical plant began operations in 1921 in Mannheim–Rheinau, Germany. Lubricating oil cuts were, however, inferior to those obtained from petroleum. The combination of results with the above and experience from ammonia synthesis of the high-pressure behaviour of hydrogen (1910) and methanol synthesis of carbon dioxide and hydrogen (1923, 200 bar, BASF Ludwigshafen, Germany) formed the basis of the catalytic high-pressure hydrogenation of coal, tars and petroleum products. Key contracts between the German IG and the American Standard Oil Co. of New Jersey for the hydrogenation of petroleum in 1927 and 1929 led to the construction of two large-scale hydrogenation plants in Baton Rouge, Louisiana and Baybay, NY in 1930 and 1931. Apart from light products, 'Essolube' lubricant base oils were thus created [12].

Disregarding finishing processes and white oil production, high-pressure hydrogenation processes to create lubricant base oils from petroleum became fashionable again only in the early seventies. In the German Democratic Republic, lubricant base oils were manufactured by the high-pressure hydrogenation of lignite tars up to the early 1990s.

In addition to hydrocracking and hydroisomerization, the following typical chemical processes occur during hydrogenation (Table 4.2).

While the processes mentioned are generally replacing solvent extraction for the manufacture of base oils, there are hydrogenation processes which either crack long-chain paraffins with specific catalysts into light products and remove them from the base oil or convert them into isoparaffins with good low-temperature characteristics without a significant loss in yield. If one combines the previously mentioned hydrogenation processes with this type of catalytic hydrogenating dewaxing, this is called the 'all-hydrogen route'.

4.4.1
Manufacturing Naphthenic Base Oils by Hydrogenation

Only about 10% of the petroleum base oils used in lubricants are naphthenic. Before 1980, naphthenic oils were significantly more important (United States 25%). In 1983, a publication by the IARC< defined a number of mineral oil

Table 4.2 Typical chemical processing during hydrogenation.

Desulfurization	Mercaptans R – SH → RH + H_2S; Disulfides R – S – S – R → RH + H_2S; Sulfides R – S – R → RH + H_2S; Thiophenes (dibenzothiophene → biphenyl + H_2S)		
Denitrogenation	R,R-substituted quinoline (N) → R,R,R-substituted benzene + NH_3		
Polyaromatic saturation	R–anthracene–R → R–perhydroanthracene–R		
	Viscosity index:	−60	20
	Pour point:	+50 °C	+20 °C
Naphthenic ring opening	R–tricyclic naphthene–R → tetra-R-cyclohexane → tetra-R-cyclopentane → Alcanes		
	Viscosity index:	20	~130
	Pour point:	+20 °C	−10 °C
Isomerization	$C_{10} - C(C_5) - C_{10}$ → $C_{10} - C(C(C_2)(C_2)) - C_{10}$		
	Viscosity index:	125	125
	Pour point:	+20 °C	−40 °C

products as carcinogenic. None of these was a severely treated distillate. Naphthenic oil, which normally has good natural low-temperature properties, did not need to be dewaxed and it was possible to convert the vacuum distillates into low viscosity index (LVI) and moderately good ageing behaviour lubricant base oils by simple refining (acid refining or hydrofinishing). In the past these base oils were much cheaper than paraffinic solvates and this also explained their popularity. Although the IARC publication and subsequent legislation and

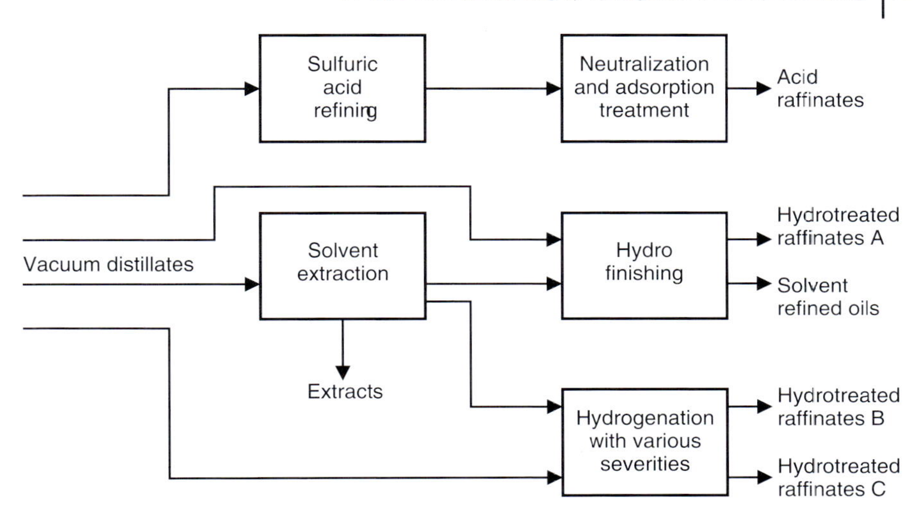

Figure 4.9 Manufacture of naphthenic base oils.

classification requirements led to the disappearance of this type of oil from important markets, naphthenic base oils were still required for some lubricant applications (greases, metalworking oils and refrigerator oils) because of their outstanding solubility and these are now manufactured with severe hydrogenation which produces non-carcinogenic base oils. Solvent extraction (e.g. the Edeleanu process) with its high proportion of extracts and low yields is now less advantageous. Figure 4.9 shows a flow chart for the manufacture of naphthenic base oils. Severely hydrotreated Group C distillates are the most accepted naphthenic base oils in western markets [13].

4.4.2 Production of White Oils

The large-scale production of white oils by high-pressure hydrogenation are part of the first step in the manufacture of lubricant base oils by hydrogenation. White oils–light-coloured, odourless, low-aromatic or aromatic-free mineral oil raffinates – are used in a number of applications as medical and pharmaceutical white oils or technical white oils in, above all, the food and beverage industries as food-grade lubricants. BASF Germany was the first to develop white oil manufacturing from the high-pressure hydrogenation technology used to liquefy coal. Up to that point, white oils were manufactured exclusively by acid treatment.

The feeds for hydrogenated white oil manufacturing are generally naphthenic or paraffinic solvent-treated vacuum distillates or untreated vacuum distillates. In the BASF process hydrogenation takes place in two stages: the first at 300–380 °C and 80–200 bar with a NiMo catalyst and the second at 200–300 °C

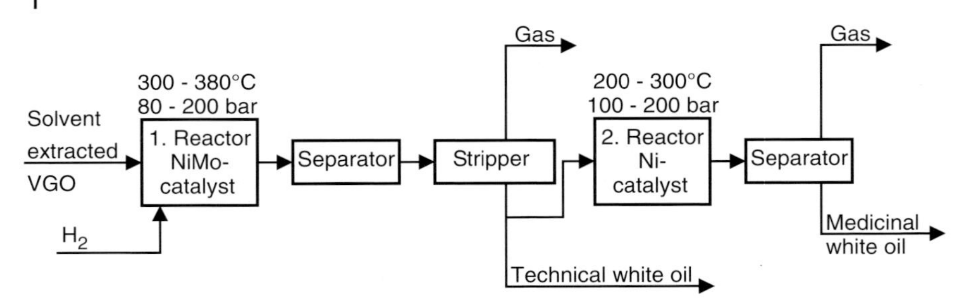

Figure 4.10 Flow chart for white oil manufacture [BASF].

and 100–200 bar with an Ni catalyst. The first stage produces the technical qualities (according to the US Food and Drug Administration) while the second stage produces the medical grades (e.g. German Pharmacopoeia). Figure 4.10 shows a flow chart for white oil manufacture [11].

4.4.3
Lube Hydrocracking

The principal elements of the lube hydrocracking process are the cracking of low-VI components and the saturation of aromatics.

Hydrocracking as a major method of future mineral base oil manufacture can be performed in two types of refinery. The first is a lube oil refinery that operates a hydrocracker with the principal objective of producing lubricating oils and the second is a refinery that operates a fuel hydrocracker to convert vacuum gas oil into high-grade fuels. Fuel hydrocracker residues are excellent feeds for the manufacture of lubricant base oils and hydrocrackers used to make olefin feeds for steam crackers can supply a premium quality feed for lube base oils. The severity has the most important effect on the quality of the base oil produced. High severity (e.g. 80% light products) generates a high-VI and low-evaporation oil. While hydrocracking refineries for base oils with normal VIs and normal Noack evaporation (HVI oils, Group II oils, Section 4.6) at moderate severity are being built in the United States and the Far East, based mainly on Chevron Technology, Western European hydrocracking plants were designed for low-evaporation base oils (Shell, Petit Couronne, France, 1972; Union FUCHS, Wesseling, Germany, 1986 and BP, Lavera, France). These produce Group III oils, on one hand from waxes and on the other from vacuum gas oils with hydrocracker residues.

Hydrocracked base oils differ from solvent-extracted oils by their extremely low aromatic content and their chemical purity, that is only traces of heteroatoms such as sulfur, nitrogen or oxygen. Furthermore, Group III oils which were manufactured at high severity or from waxes have Noack evaporation characteristics, which are about 50% down on equiviscous, solvent-extracted vacuum distillates. If catalytic dewaxing is performed, even lower pour points can be achieved than with SN oils.

Since 1995, more than half of all new base oil manufacturing plants have been built or are planned using hydrocracking technology.

Apart from high-value base oils, sufficiently large hydrocracking plants offer favourable operating costs and greater crude oil flexibility, despite high initial investment costs. A lube oil hydrocracking plant which operates at approximately 50% vacuum distillate severity generates just as many high-quality light products as lube base oils while, at present, just as many low-value extracts and paraffin waxes are generated by solvent extraction.

To increase VI from 100 to 125, the hydrocracking severity has to be increased so much that up to one half of the base oil yield is lost [11].

The increasing number of fuel hydrocrackers, which are a response to increasing demands for low-sulfur oils (diesel fuels with <50 ppm sulfur), represents a growing source of hydrocracked feeds. Fuel hydrocracker residue can be processed into high-quality, hydrocracked base oils by distillation and subsequent dewaxing (wax isomerization) at acceptable cost.

During the hydrocracking process, especially if vacuum gas oils are severely treated to create very high viscosity index (VHVI) oils, polycyclic aromatics can be formed along with saturated structures (aromatics to naphthenes or paraffins and isoparaffins) under some process conditions. These have to be removed by subsequent high-pressure hydrogenation or by extraction if hybrid methods are used. This subsequent hydrogenation can also be manipulated so that it also produces a significant increase in VI.

4.4.4 Catalytic Dewaxing

The most complex stage of traditional base oil refining is solvent dewaxing. In addition to the high capital expenditure and, above all, operating cost, the solvent-related limits to the achievable pour point are a disadvantage.

In recent years, various methods have been developed which can remove unbranched, long-chain or short-chain paraffins, less branched paraffins and some other petroleum components by catalytic and hydrogenation reactions or convert them into such components which improve the low-temperature characteristics of the base oil.

The first technologies were based on the catalytic cracking of these above-mentioned substances. In 1979, Mobil introduced the 'Mobil Lube Dewaxing Process' (MLDW). The development of catalysts since then has led to strong hydroisomerization activity processes (Mobil selective dewaxing, MSDW).

While the MLDW process mainly involves the cracking of long-chain paraffins and the production of larger quantities of light by-products and poor base oil yields, MSDW processes create high-yield, non-waxy isoparaffins from the waxy parts of the feed.

At the beginning of the 1990s, Chevron successfully introduced its isodewaxing process. In 1999, it is the most important technology for catalytic dewaxing [11]. The process combines relatively high yields with low pour points and high VIs. The method uses a hydrogenation component as an intermediate pore-silicoaluminumphosphate molecular sieve (e.g. Pt on SAPO-11). In terms of process economy, high base oil yields and also the creation of mainly high-value C_{5+} liquids as by-products are of significance. In particular, the formation of low-value by-products such as propane is avoided.

In the development of new dewaxing catalysts, it is important to achieve the right balance between hydrogenation activity and acid activity. Increasing hydrogenating properties usually lead to a reduction in isomerization and thus a worsening of the pour point. Higher acid activity increases cracking and thus yield losses. Sulfur and nitrogen in the feed play important roles. Nitrogen is detrimental to acid activity and sulfur is poison for catalysts for metal hydrogenation components.

Although isodewaxing enables manufacture of lubricant base oils with pour points below −45 °C, with catalytic dewaxing processes one must never forget that dewaxing to very low pour points (depending on process conditions and catalysts) leads to VI losses.

A major success of this new dewaxing process are the molecular sieve catalysts–zeolites with exactly defined mesh sizes. Figure 4.11 illustrates the procedure of catalytic dewaxing with zeolite catalysts.

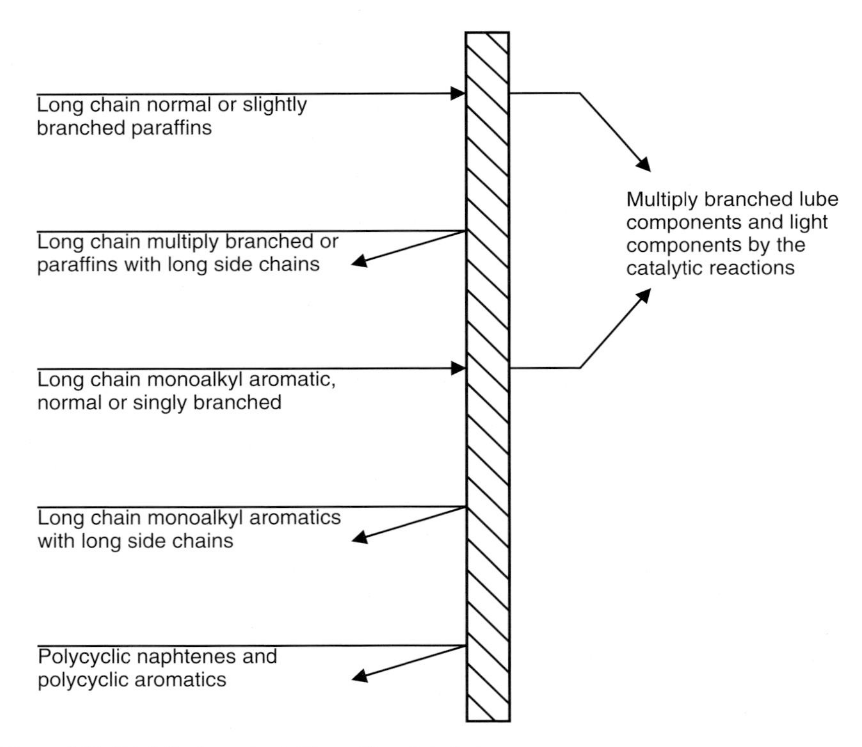

Figure 4.11 Catalytic dewaxing with zeolite catalysts.

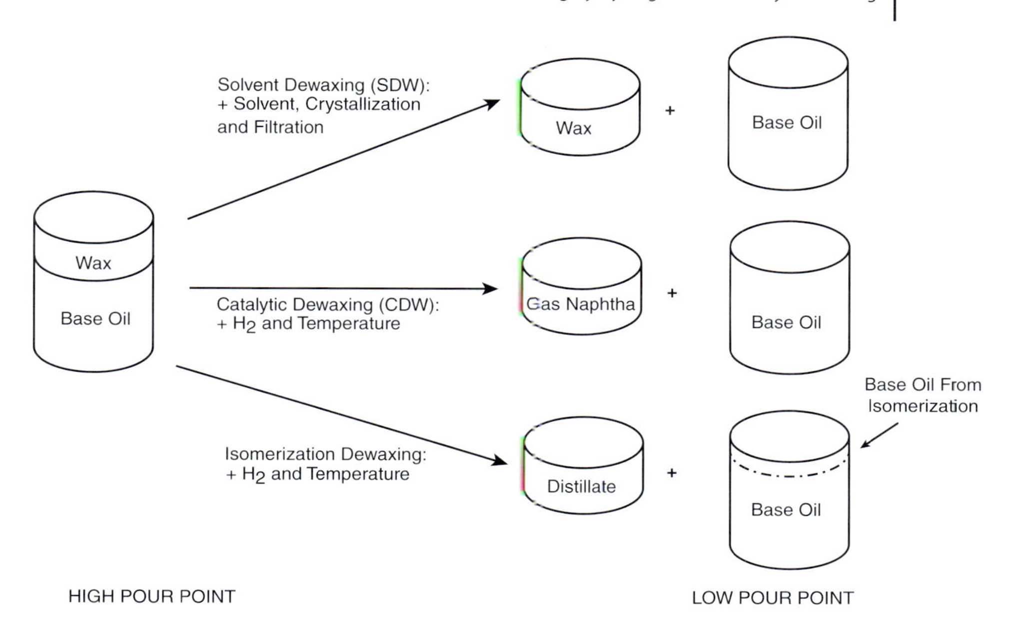

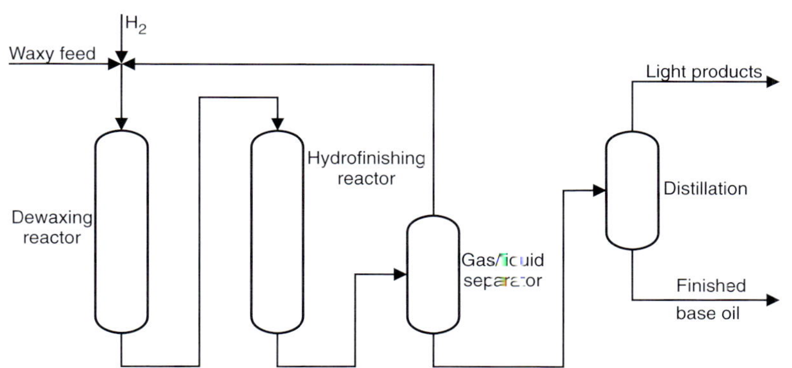

Figure 4.12 (a) Illustration of the differences of the various dewaxing methods: conventional solvent dewaxing, catalytic dewaxing (cracking long chain paraffins) and lube isomerization dewaxing. (b) Flow chart for catalytic dewaxing (hydrodewaxing) including hydrofinishing.

Although highly branched paraffins or polycyclic naphthenes or aromatics are not affected by the zeolites, the long-chain less or unbranched molecules are acted upon by the catalysts [14–16].

Figure 4.12 shows the flow chart for catalytic dewaxing (hydrodewaxing) including a hydrofinishing reactor.

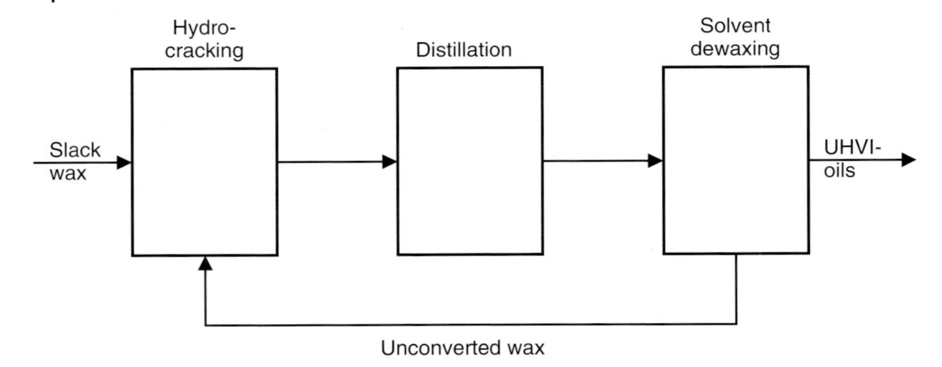

Figure 4.13 Process for the production of ultra-high-viscosity index oils by wax hydrocracking and isomerization.

4.4.5
Wax Isomerization

VHVI oils have been manufactured from waxes since the early 1970s. Feeds were wax cuts from solvent extraction processes. The conversion of long-chain normal paraffins or less branched paraffinic components into isomerized products with good low-temperature characteristics and high VIs succeeded, with suitable catalysts, in isomerizing hydrocracking processes. Because the input feed is already aromatic- and naphthene-free, the catalysts can be fully optimized to the conversion of the paraffinic material. At relatively high yield losses, the process creates high quality but comparatively expensive hydrocracked oils (Shell XHVI, Exxon Exxyn, Mobil MWI-2 catalyst). Figure 4.13 shows the manufacture of XHVI oils by wax isomerization and hydrocracking, including solvent dewaxing.

4.4.6
Hybrid Lube Oil Processing

The combination of traditional solvent refining with severe hydrotreating and hydrocracking processes is known as hybrid processing. On one hand, such processes should represent an extension of existing refineries operations, and on the other, exploit the favourable properties of hydrogen processes. To create higher VIs (>105) and to reduce sulfur content, hydrotreating can follow solvent extraction (e.g. in the manufacture of turbine oils).

The combination of a mild furfural extraction with hydrocracking can produce hydrocracked oils by low-hydrocracking conversion in small hydrocracking reactors. The low hydrogen consumption adds to the economy of such processes given that the extraction plant exists and the hydrocracker can be integrated into the infrastructure of a lube oil refinery. The introduction of a hydrocracking stage into a conventional solvent refinery offers the attractive possibilities of debottlenecking if the corresponding dewaxing capacity is available.

Catalytic dewaxing can, on the other hand, follow solvent refining. This presupposes that the catalytic dewaxing catalysts can withstand, and are not poisoned by, the sulfur and nitrogen components in the solvent raffinate. In the Mobil processes, this is more likely with MLDW catalysts than with MSDW catalysts.

In connection with the isomerization of slack waxes from petroleum distillates, several other high-paraffin components should be considered as future base oil feeds. These could include high-wax (> 70%) natural distillates and Fischer–Tropsch waxes or synthetic fluids from natural gas. Significant future importance is given to this latter group of products.

Hystart

The quality of low-quality lube feeds (VGO) can be improved by hydrogenation before solvent extraction. This process is called Hy-starting or Hystart.

4.4.7 All-Hydrogen Route

The production of base oils by hydrocracking and catalytic, solvent-free dewaxing is called the all-hydrogen route. Figure 4.14 shows a flow chart for such a refinery used to manufacture VHVI oils. Depending on the severity of the hydrocracker, Group II HVI or Group III VHVI oils can be produced (HC-I and HC-2 oils, Table 4.3). Leading examples for HC-I oils are the Chevron refinery in Richmond, California and the Conoco–Pennzoil refinery in Lake Charles (Exel Paralubes) [17]. Some HC-I refineries have been prepared to produce HC-2 oils by the all-hydrogen route by increasing the severity of the hydrocracker. Another example of a hydrocracker refinery is the Petro-Canadas plant, Figure 4.15 [18].

The use of a fuel hydrocracker for the production of base oils using the all-hydrogen route was first realized by the SK Corporation (Ulsan, Korea) in 1995. By recycling the hydrocracker bottom and the special integration of the fuel hydrocracker in the lube oil process, SK also developed a specific method (UCO process, Figure 4.16) [19].

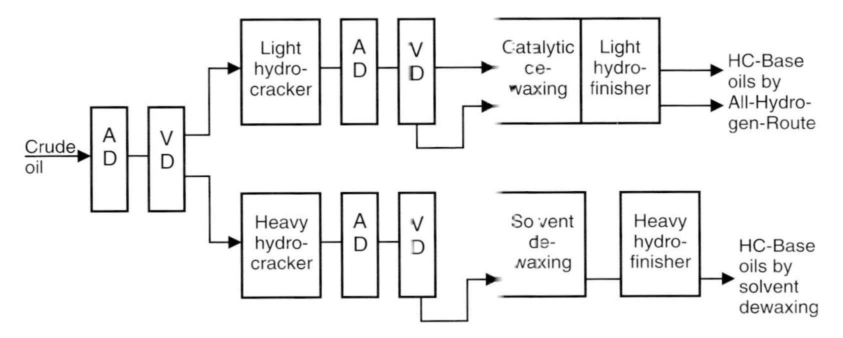

Figure 4.14 All-hydrogen route in Chevron's Richmond (CA, USA) refinery.

Table 4.3 API/ATIEL classification of base oils.

Group	Sulfur (% w/w)		Saturates (% w/w)	Viscosity index
I	>0.03	and/or	<90	80–120
II	≤0.03	and	≥90	80–120
III	≤0.03	and	≥90	>120
IV	All polyalphaolefins (PAO)			
V	All base oils not included in Groups I–IV or VI			
VI	All polyinternalolefins (PiO) (Europe only)			

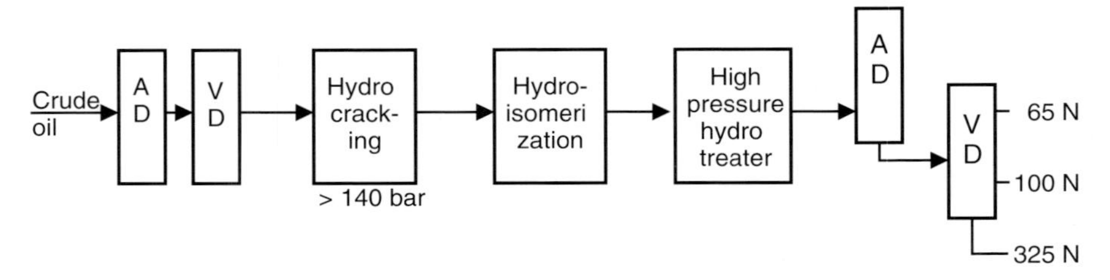

Figure 4.15 Lube base oil production with severe hydrocracking and hydroisomerization. (All-hydrogen route, Petro-Canada)

4.4.8
Gas-to-Liquids Conversion Technology

As a result of efforts to increase the value of natural gas in logistically favourable locations, the chemical liquefaction of natural gas (also the chemical reaction route) was developed (on the basis of the Fischer–Tropsch process).

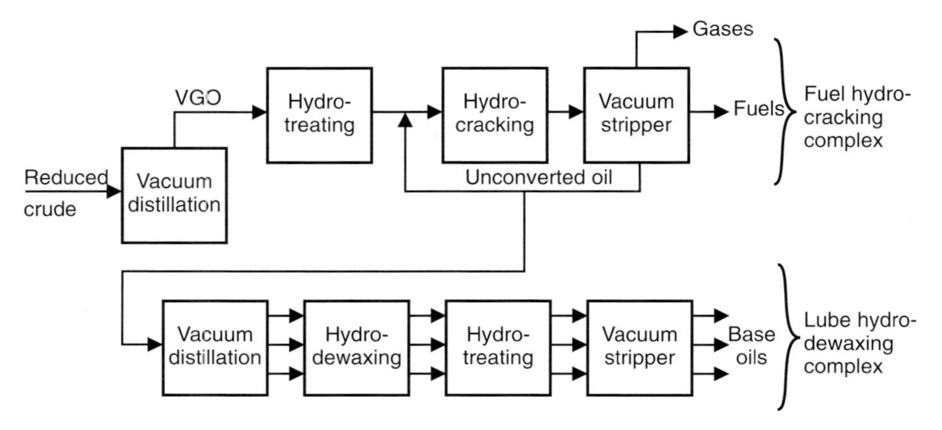

Figure 4.16 VHVI base oil production based on fuel hydrocracker residue (SK Corporation, Korea).

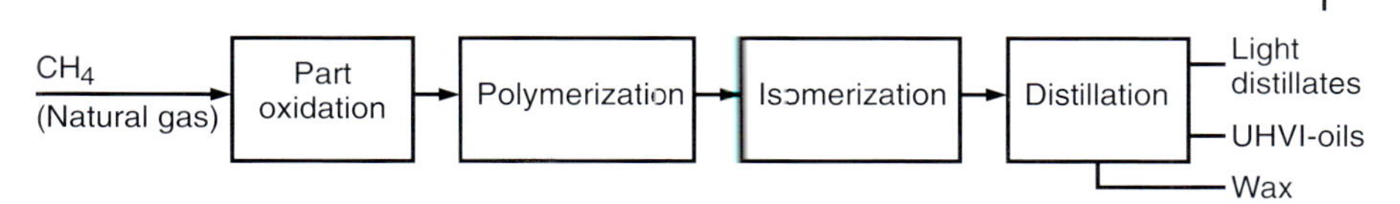

Figure 4.17 UHVI lube base oil production with gas-to-liquids conversion technology.

German Chemists Franz Fischer and Hans Tropsch invented the Fischer–Tropsch Synthesis in 1923. The first realization was the production of fuels from Coal in the World War II.

This process creates high-quality liquid products and paraffin wax. High-quality ultra high viscosity index (UHVI) oils can then be obtained from natural gas by part oxidation, polymerization, and isomerization, Figure 4.17 [20,21].

The base oil market could undergo dramatic changes if Fischer-Tropsch waxes, which are part of gas-to-liquids technology, become more generally available. The 90-year-old Fischer-Tropsch technology has attracted considerable attention in the last few years. The focus of this attention is the better utilization of natural gas. Syngas (CO and H_2) is made from methane, oxygen (air) and water vapour and this, in turn, is made into fluid and solid hydrocarbons in the Fischer-Tropsch reactor. The solid hydrocarbon waxes (>99% paraffins) are hydrocracked, hydro-isomerizated and iso-dewaxed into super-clear base oils. Shell has used such waxes from its Bintulu plant in Malaysia for its XHVIs. New technologies for smaller, efficient plants have been developed by Rentech and Syntroleum in the last years.

It is, nevertheless, most likely that major oil companies will be the first to operate large-scale gas-to-liquid (GTL) plants. ExxonMobil, Shell, and SasolChevron have each announced GTL projects including base-oil-production units in Qatar [22]. The most important activity is Pearl GTL, a partnership between Shell and Qatar Petroleum. The plant is located in Ras Laffan Industrial City (on stream since 2011).

GTL base oils will have premium characteristics, including very high viscosity indices, essentially no sulfur and nitrogen, very low evaporative losses and almost no aromatic content [23]. They will most probably be classified as Group III + base oils because their characteristics will vary between UHVI base oils (Group III) and polyalphaolefins (Group IV).

Besides natural gas, all carbon containing materials can, mainly, be used for production of liquid products and paraffin wax by Fischer–Tropsch technology. Because of the limited availability of crude oil, gasification and liquefaction of carbon, biomass and even oil sands [24] are of increasing interest.

4.5 Boiling and Evaporation Behaviour of Base Oils

In traditional solvent refining, the boiling point distribution of base oils is largely determined during vacuum distillation. Three, four or five (including bright stock) cuts are taken whose boiling point distribution is reflected in the finished base oil. Lighter products can only be created during hydrofinishing as the final refining stage. Components, which negatively affect the evaporation behaviour of the base oil can remain if these light components are not fully stripped.

In hydrocracking processes, the decisive distillation stages take place after cracking or after catalytic dewaxing. The function of the stripper or fractional distillation columns play an important role whenever light cracked products are created.

The target of base oil distillation is the viscosity desired at 40 and 100 °C. The same distillation cut (same boiling point distribution) with different chemical structures leads to different viscosities. A highly naphthenic cut produces a higher viscosity than a paraffinic cut. In other words, equiviscous cuts of different chemical structures have different boiling-point distributions.

In practice, ISO viscosity grades or other required viscosities are created by blending different cuts. If the boiling points of both cuts are too far apart, the flashpoint (Chapter 18) drops significantly and evaporation increases. Flashpoint and evaporation are generally determined by the base oil components with the lowest boiling points. Figure 4.18 shows the distillation curves of three paraffinic solvent refined cuts. One line is a blend of cut A with cut C that creates an equiviscous oil with cut B. The share with a boiling range up to 380 °C increases from 10 (equiviscous cut) to 20% (equiviscous blend).

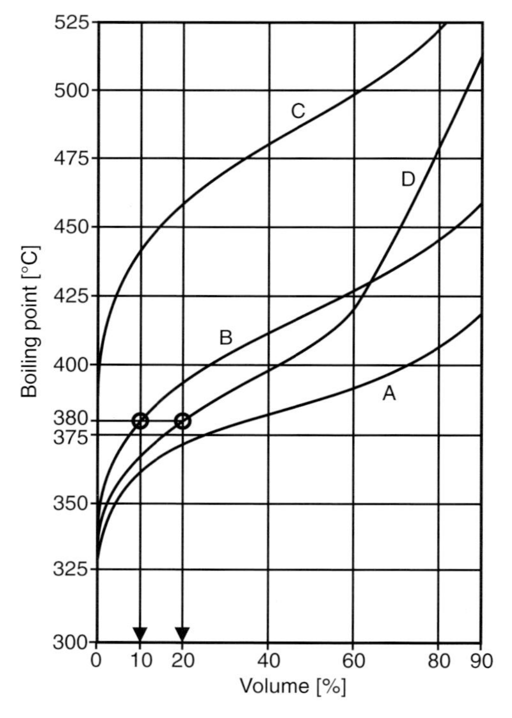

Figure 4.18 Distillation curves of three paraffinic solvent refined cuts. A: Light cut (low boiling); B: medium light cut; C: heavy cut (high boiling); (D) blend of A and C, equiviscous with B.

Apart from the use of a variety of laboratory distillation apparatus to determine the boiling point distribution of a base oil, gas chromatographic methods of determining boiling point have gained popularity over the last few years. The use of gas chromatography to determine the boiling range has developed continuously, especially as a result of improvements in columns and software, and has found its way into various standards. Both non-polar packed and capillary gas chromatographic columns can be used. The injection port temperature is between 360 and 390 °C depending on the length of the column. The initial column temperature can be as low as −50 °C, the final column temperature is between 360 and 390 °C and the programming rate is usually below $10^{\circ}\,min^{-1}$. The maximum temperatures stated are the decomposition thresholds of the fractions being tested or of the column material.

Figure 4.19 shows a typical calibration curve (ASTM 2887-93). A mixture of normal paraffins including n-C_5 to n-C_{44} is used for calibration. The boiling points of n-paraffins are: n-C_{16} 287 °C, n-C_{30} 449 °C and n-C_{44} 555 °C. ASTM D 5307-92 describes another method of determining boiling-range distribution by gas chromatography.

Figure 4.20 shows the first application of high-temperature gas chromatography to determine the boiling point distribution of high-boiling-point coal tars [25]. Normal paraffins in a non-polar column were used for reference. Figure 4.21 shows the distillation curve of a mixture of two solvent-refined cuts determined by gas chromatography and the latest software.

In recent years, the evaporation characteristics of lubricants have become increasingly important quality criteria. The reasons for this are the emissions created when a lubricant evaporates, and the accompanying change to the lubricant's composition. The topic of evaporation behaviour has become increasingly

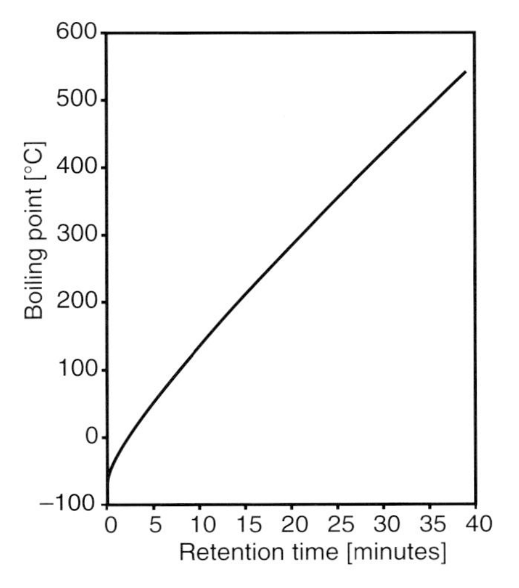

Figure 4.19 Typical calibration curve for determination of the boiling-point distribution of lube base oils by gas chromatography.

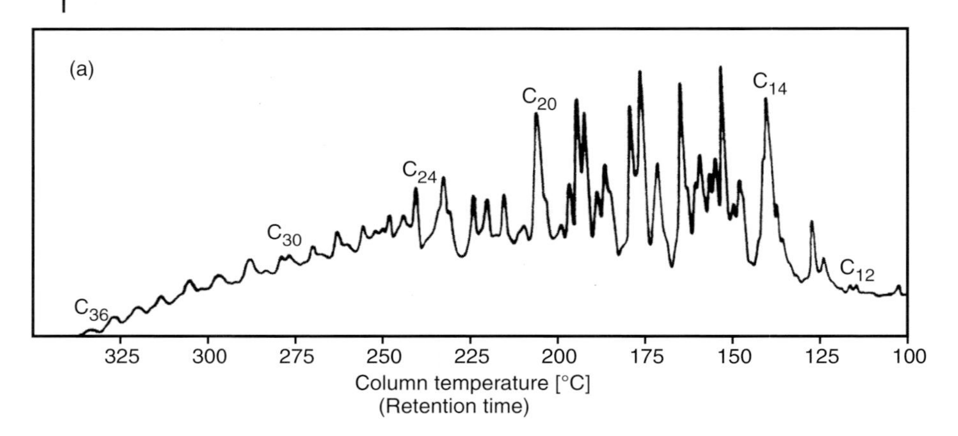

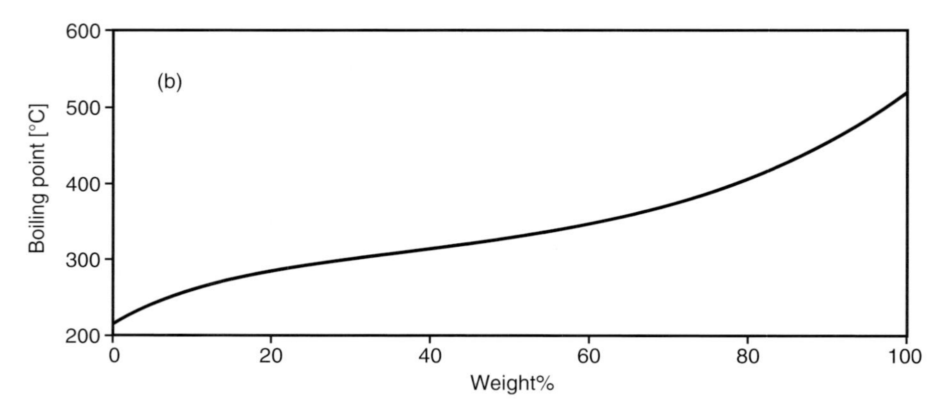

Figure 4.20 Use of high-temperature gas chromatography to determine boiling point distribution [25]. (a) Gas chromatogram, (b) boiling point distribution.

important in line with the trend towards lower viscosity oils for most applications (energy-saving oils).

Evaporation is dependent on the vapour pressure of the base oil components at a given temperature and the ambient conditions (such as atmospheric pressure, turbulence, etc.). Vapour pressure is dependent on temperature and follows the Antoine equation for base oil components. Mathematical models have been created for complex hydrocarbon mixtures. As a simple laboratory method, Noack evaporation (1 h at 250 °C) has become the established method of characterizing evaporation behaviour in lubricant specifications with the evaporation losses being given in % w/w. Gas chromatographic methods are also used but these can produce somewhat deviating results.

This led to a simulated Noack evaporation method using gas chromatography. In the United States, a gas chromatographic process was developed to determine engine oil volatility (ASTM D5480). In this case, evaporation takes place at

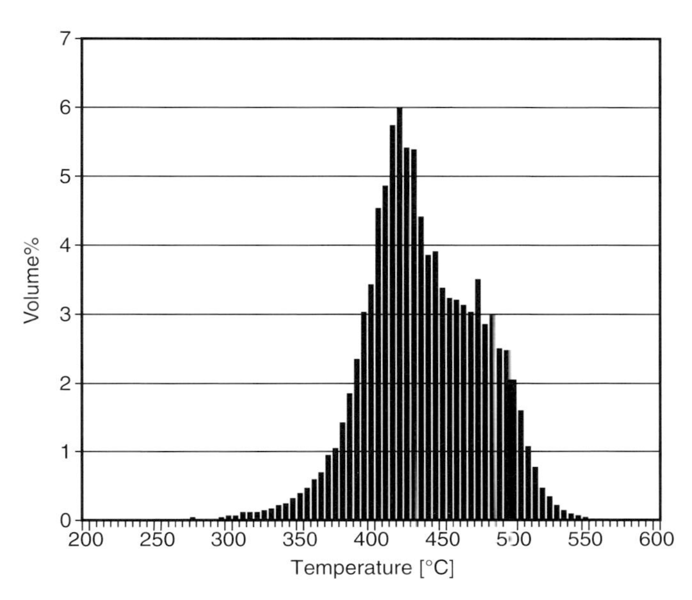

Figure 4.21 Distillation curve of a blend of two solvent-refined base oils obtained by gas chromatography and modern software.

371 °C (700 °F) and the evaporating components are measured to about n-C_{22}. Although this process is easily reproducible, the values obtained are not comparable with those from the Noack method.

Figure 4.22 shows typical evaporation losses for well-cut paraffinic solvates, values for VHVI oils (HC-2 oils, API Group III oils) and typical ester base oils (Chapter 5) are shown for comparison.

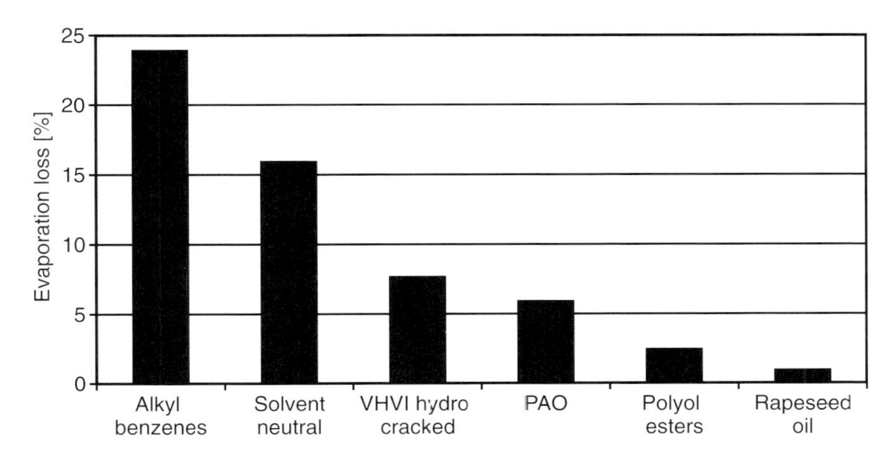

Figure 4.22 Typical evaporation losses (Noack evaporation) of various lube base oils (250 °C, 1 h).

4.6
Base Oil Categories and Evaluation of Various Petroleum Base Oils

The American Petroleum Institute (API) and the Association Technique de l'Industrie Européenne des Lubrifiants (ATIEL) have classified base oils according to their chemical composition. Initially there were four groups; after the introduction of VHVI oils in Europe this was increased to five. The most important reason for these groups was the necessity to regulate base oil interchangeability for engine oils (Chapter 9). The classification of petroleum base oils (Groups I–III) considers three parameters: saturates content, sulfur content and viscosity index. Table 4.3 shows this classification. Accordingly, Group I oils are solvent extracted HVI oils (SN oils), Group II oils are hydrogenated or hydrocracked oils (as the sulfur content of >300 ppm shows) and Group III products are VHVI oils manufactured by severe hydrocracking and or wax isomerization (VI > 120, sulfur < 300 ppm).

Table 4.4 shows typical data of various hydrocracked base oils (HC-oils), in comparison with solvent refined oils and polyalphaolefins. Although there are several technical intermediate possibilities, severely hydrocracked Group III oils have become well-established on the market.

Figure 4.23 illustrates the reduction of aromatics (as indicated by infrared spectroscopy) by different refining processes including production of HC-2 oils [26].

HC-2 oils are classified as VHVI oils. HC-3 oils are also known as extra or UHVI oils. They are generally made by hydrocracking and isomerization slack waxes followed by solvent dewaxing.

Table 4.5 shows typical hydrocarbon compositions of HC-oils in comparison with conventional solvent refined oils and polyalphaolefins [18].

The first characteristics of hydrocracked oils which attracts attention is their low sulfur content, or even the absence of sulfur, and the high VI and low evaporation characteristics of HC-2 and HC-3 oils. The expected, poorer additive response of antioxidants (Chapter 6), pour point depressors or other additive groups were not confirmed when hydrocracked oils were introduced.

Table 4.4 Typical data for 4 $mm^2 s^{-1}$ base oils – HC oils – in comparison with conventional solvent refined oils and polyalphaolefins.

	Solvent refined	HC-1	HC-2	HC-3	Polyalphaolefins
Viscosity ($mm^2 s^{-1}$) at 100 °C	4	4	4	4	4
Viscosity index	100	105	125	130	125
Volatility, Noack evaporation loss (% w/w).	23	18	14	13	12
Pour point (°C)	−15	−15	−18	−20	−65
API group	I	II	III		IV

Table 4.5 Typical hydrocarbon composition of HC-oils in comparison with conventional ($4\,mm^2\,s^{-1}$) solvent refined oils and polyalphaolefins.

	Solvent refined SN 100	HC-1	HC-2	HC-3	Polyalphaolefins
n- and iso-paraffins	25	30	55	75	96
Monocycloparaffins	20	35	24	15	4
Polycycloparaffins	30	34	20	10	—
Aromatics	24	0.5	0.3	0.1	—
Thiophenes	0.5	—	—	—	—

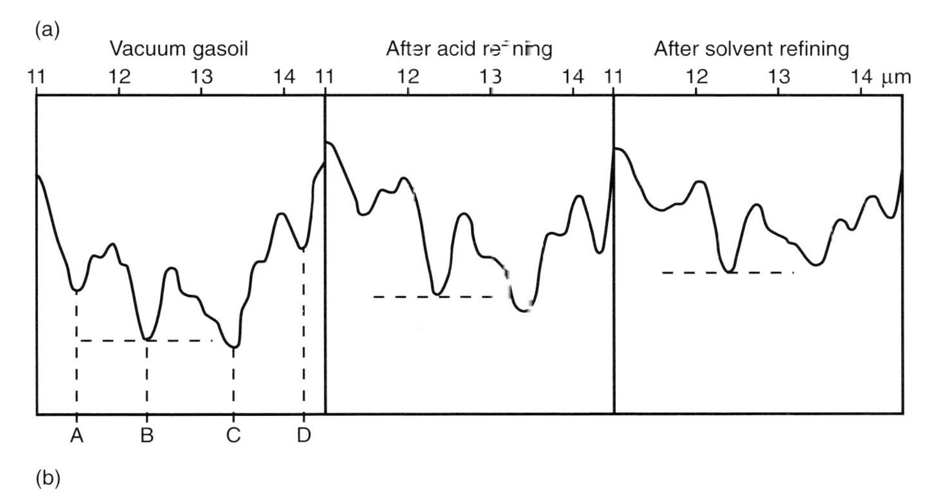

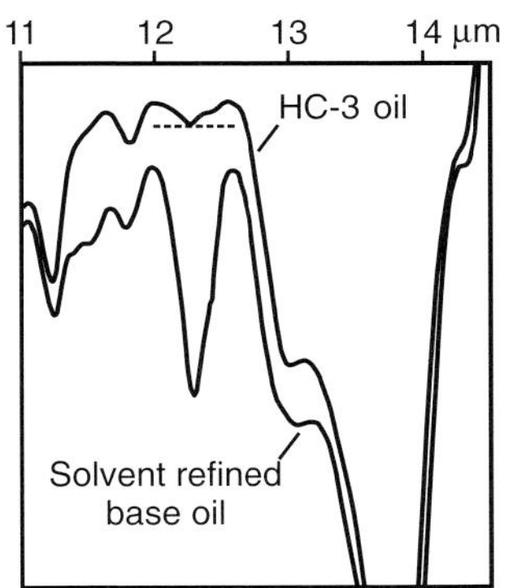

Figure 4.23 Reduction of aromatics by different refining processes. (a) Infrared spectra of the aromatic fraction (out-of-plane oscillations). (b) Infrared spectra of non-pretreated base oils (solvent refined and hydrocracked).

When inhibitor compositions were optimized, all hydrocracked oils were more stable to oxidation than solvent refined oils (if the right additives were used). This is primarily a result of their high chemical purity. The missing 'natural' sulfur inhibitors are more than compensated for by the addition of suitable antioxidants.

HC-2 and HC-3 oils are being increasingly compared with synthetic polyalphaolefins (PAO); the close similarity of the performances of some hydrocracked oils and PAO, and their significantly lower cost, make them increasingly attractive for lubricant formulations. The term 'synthetic', which has been used in Europe and recently also in the United States, has led to heated legal arguments. As considerable synthesis takes place during severe hydrocracking and catalytic dewaxing, the term seems justified, although this then eliminates the terminological differentiation into HC-2 and HC-3 oils. In Germany, the term HC-synthesis has gained acceptance for HC-2 oils. The synthesis terminology applies even more to oils which originate from liquefied natural gas and which can be classified as HC-3.

Figure 4.24 Illustrates the development of VHVI and UHVI oils (HC-2 and HC-3 oils) and PAO in the different regions of the world. Whereas Shell and BP in France and Union Fuchs in Germany were the only manufacturers of VHVI oil in the 1980s, by the mid-1990s similar activities had spread from Western Europe around the world [20].

Because of the high quality and performance of UCBOs, these products have a higher market value than Group I or Group II base oils. The market price for future GTL base fluids will be determined by product quality, economies of scale and the marketing strategies of the GTL players.

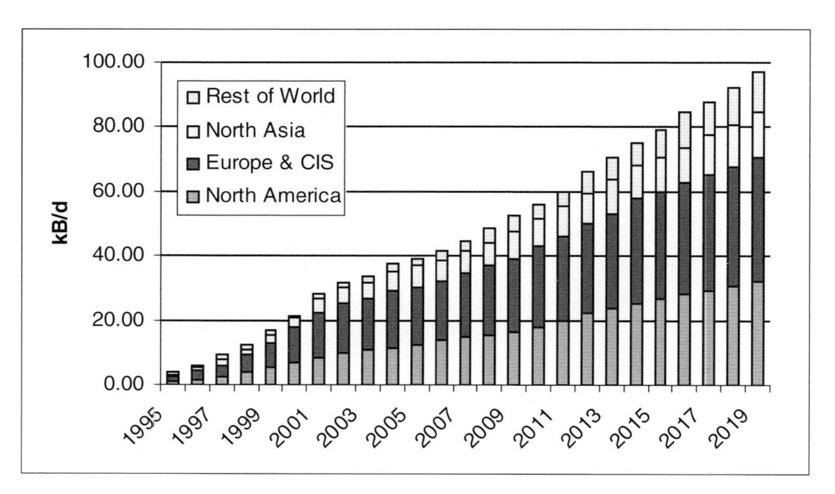

Figure 4.24 Development of VHVI and UHVI oils (HC-2 and HC-3) and PAO in the different regions of the world.

References

1 Downey, W.R. Jr. (2005) *Impact of GTL Technology on the Future of the Lubes Business*, UEIL World Congress, Rome.

2 Brandes, G. (1956) Lubricant composition with low deposition tendency, *Brennstoff-Chem.*, **37**, 263–267.

3 IARC (1984) Classification of Mineral oils According to their Carcinogenicity, vol. 33.

4 Grimmer, G., Jacob, J. and Naujack, K.-W. (1981) Profile of the polycyclic aromatic hydrocarbons from lubricating oils, inventory by GCGC–MS – PAH in environmental materials, Part 1. *Fresenius' Z. Anal. Chem.*, **306**, 347–355.

5 Mang, T. (1989) Lubricants and legislation in the Federal Republic of Germany. *Erdol Kohle*, **42**, 400–407.

6 Grimmer, G., Jacob, J. and Naujack, K.-W. (1983) Profile of the polycyclic aromatic compounds from crude oils, Part 3 inventory by GCGC–MS – PAH in environmental materials. *Fresenius' Z. Anal. Chem.*, **314**, 29–36.

7 Mang, T. (1997) A new generation of non hazardous and environmentally safe metalworking oils. International Tribology Conference, Melbourne.

8 Mang, T. (1989) Legislative influences on the development, manufacture, sale and application of lubricants in the Federal Republic of Germany. CEC Symposium, Paris.

9 Mang, T. (1991) Recent European Trends in metalworking lubricants. IX National Conference on Industrial Tribology, Bangalore.

10 Mang, T. (1992) Umweltbedingte Alternativen zu Mineralölen als Grundkomponenten in Hydraulikflüssigkeiten. 10. Aachener Fluidtechnisches Kolloquium.

11 Miller, S.J., Shippey, M.A. and Masada, G.M. (1992) Advances in lube base oil manufacture by catalytic hydroprocessing. 1992 NPRA National Fuels and Lubricants Meeting, Houston.

12 Annonymous (1969) *Ullmann's Enzyklopadie der technischen Chemie*, Urban und Schwarzenberg.

13 Okamoto, T. (1996) *Naphthenic Baseoil*, UEIL General Assembly, Lyon.

14 Jacob, S.M. (1998) Lube base oil processing for the 21st Century. 4th Annual Fuels and Lubes Asia Conference, Singapore.

15 Weitkamp, J., Karge, H.G., Pfeifer, H. and Holderich, W. (1994) Zeolites and related microporous materials: state of the art 1994 10th International Zeolite Conference, Garmisch–Partenkirchen.

16 Miller, S.J. (1994) New molecular sieve process for lube dewaxing by wax isomerization. *Micropor. Mater.*, **2**, 439–449.

17 Johnson, E.S. and Thomas, K. (1995) Excel paralubes – base oils for tomorrow. National Fuels and Lubricants Meeting, Houston.

18 Cohen, S.C. and Mack, P.D. (1996) HVI and VHVI base stocks, The World Base Oils Conference, London.

19 Moon, W.S., Cho, Y.R., Yoon, C.B. and Park, Y.M. (1998) VHVI base oils from fuels hydrocracker bottoms. China Lube Oil Conference, Beijing, China.

20 Min, P.Y. (1998) VHVI base oils: supply and demand. 4th Annual Fuels and Lubes Asia Conference, Singapore.

21 Mang, T. (2000) Future importance of base oils in lubricants. 12th International Colloquium Tribology, Stuttgart/Ostfildern.

22 Skledar, G. and Rosenbaum, J. (2006) GTL base oils – countdown to delivery. World Base Oil Conference, London.

23 Snyder, P.V. (1999) GTL lubricants: the next step. NPRA Lubricants and Waxes Meeting, Houston.

24 (2005) *Gasification News, Vol. IX No. 2*, Hart Energy Publishing, Houston.

25 Mang, T. (1967) Gravimetrische Verfolgung der Vakuumpyrolyse von Steinkohlen und Analyze der Teere mit Hilfe physikalisch-chemischer Meßmethoden unter besonderer Berücksichtigung von Transporterscheinungen, Dissertation.

26 Kagler, S.H. (ed.) (1987) *Neue Mineralölanalyse, Band 1: Spectroskopie (2.)*, Hüthig, Heidelberg.

5
Synthetic Base Oils

Wilfried Dresel

According to Zisman [1] the rise of synthetic lubricating oils on an industrial scale began in 1931, when Sullivan [2] and his co-workers published the results of their attempts to make tailor-made saturated lubricating oils with low pour points by catalytic polymerization of olefins. Gunderson and Hart [3] in 1962 edited an outstanding book on synthetic lubricants. It covered nine classes of lubricating oils and also contained Zisman's contribution to their history. In 1993, Shubkin and in 1999 Rudnik and Shubkin [4] edited books on the subject that omitted some of the older classes and added some new ones instead (Table 5.1). In 2005 and in 2013, Rudnik [5] updated them.

Although many of the synthetic base oils available today had been developed decades ago, their use on a large technical scale has increased only slowly because of their considerably higher cost (Heckenkamp, Jutta, Fuchs Petrolub SE, Private Communication.) [6–8] (Table 5.2).

Although the consumption of synthesized lubricating oils is, on the whole, responding only reluctantly to benefits which have long been recognized [9], and to the needs of machinery that has to work under increasingly extreme conditions, in terrestrial applications mainly at higher temperatures and pressures [10–12], in space applications mainly at lower temperatures and very low pressures [13,14], the consumption of polyalphaolefins, the most common synthetic lubricating oils, has increased enormously in the last two decades [15–17].

In contrast with mineral oil-based oils, which contain many different hydrocarbons and nitrogen-, oxygen- and sulfur-containing chemical derivatives of these hydrocarbons, which must be purified and refined and distilled (Chapter 4), synthetic base oils usually are prepared by reaction of a few defined chemical compounds – although in many cases based on petroleum also – and tailored to their application by the right choice of reaction conditions. This comparably simple chemistry not only has advantages, of course, but, mainly with regard to additive response and elastomer compatibility, some disadvantages also [18,19].

Synthetic base oils have been classified according to both the production process and their chemical composition [7,20]. From a chemical point of view, the latter method is preferable.

Lubricants and Lubrication, Third Edition. Edited by Theo Mang and Wilfried Dresel.

Table 5.1 Classes of synthetic lubricant.

Gunderson and Hart (1962) [3]	Shubkin *et al.* (1993) [4]
	Alkylated aromatics
Chlorofluorocarbon polymers	Chlorotrifluoroethylene
	Cycloaliphatics
	Dialkylcarbonates
Dibasic acid esters	Esters
Fluoro esters	
Neopentyl polyol esters	
	Perfluoroalkylpolyethers
Phosphate esters	Phosphate esters
	Phosphazenes
	Polyalphaolefins
	Polybutenes
Polyglycols	Polyalkylene glycols
Polyphenyl ethers	
	Silahydrocarbons
Silicate esters	
Siloxanes	Siloxanes

Table 5.2 Relative costs of synthetic base oils.

	2011	2004
Mineral oils	1	1
Alkylated aromatic compounds	1–6	2–3
Polybutenes	1.5–10	3–5
Dibasic acid esters	3–10	4–15
Polyalphaolefins	1.5–8	4–15
Poly(alkylene glycol)s	2–9	6–15
Neopentyl polyol esters	3–7	10–20
Silicones	3–50	25–200
Perfluoralkyl polyethers	30–200	350–800

5.1
Synthetic Hydrocarbons

Synthetic hydrocarbons were developed simultaneously in Germany and in the United States. In Germany, low-temperature performance and the need to overcome the general shortage of petroleum base stocks were the driving force behind the work of Zorn [21]. It is known today that all synthetic hydrocarbons – and the other economically important synthetic lubricating oils – can be synthesized

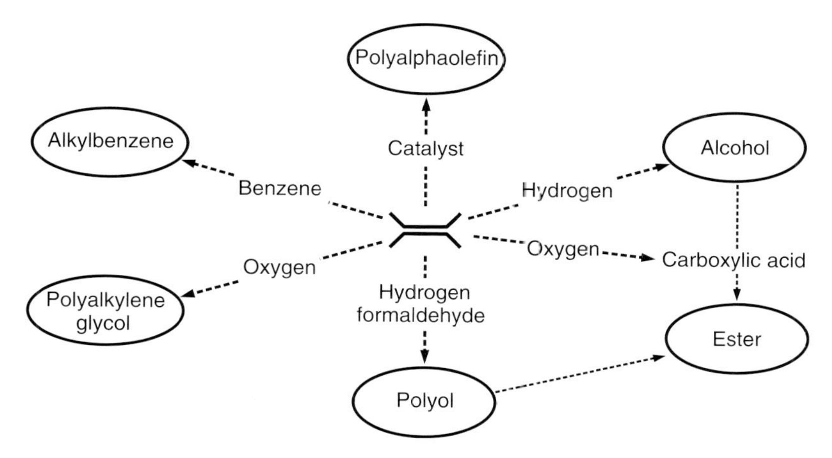

Figure 5.1 Ethylene (a source of many synthetic base oils).

starting with ethylene (Figure 5.1). Ethylene itself is one of the most important petrochemicals [22] and today mainly produced in steam crackers.

5.1.1 Polyalphaolefins

The term polyalphaolefin (PAO) is derived from the source of this class of base oil, usually α-decene or a mixture of α-olefins containing, in general, a minimum of six and a maximum of twelve carbon atoms. The oligomers are saturated, that is hydrogenated, and therefore belong to the aliphatic or branched paraffinic hydrocarbons. Linear α-olefins were first used for the synthesis of lubricating oil by Montgomery, Gilbert and Kline [23]. Polyalphaolefins are described in more detail by Rudnik [17].

Free radical – or thermal – oligomerization of α-olefins is possible but is no longer important because of high activation energy, low yield and quality, even when peroxides are used as catalysts. Oligomerization with Ziegler–Natta catalysts of the aluminium triethyltitanium tetrachloride type tends to yield a wide range of oligomers that can be controlled more easily when a catalyst of the alkylaluminium halide–alkoxide–zirconium halide type is used. Boron trifluoride-based Friedel–Crafts oligomerization with alcohols as co-catalysts has, nevertheless, proved superior, although even today the mechanism is not fully understood. Possible mechanisms have been described in detail by Mortier and Orszulik *et al.* [24]. The boron trifluoride method has become predominant for the production of grades with lower viscosities, 2–10 $mm^2 s^{-1}$. For higher viscosities, 40 and 100 $mm^2 s^{-1}$, other catalysts must be used, and for the production of grades with viscosities of 10–25 $mm^2 s^{-1}$, intended to replace disadvantageous blends and not yet available in comparable amounts, C-12 and C-14 olefins can be used; these yield mainly dimers [25]. The

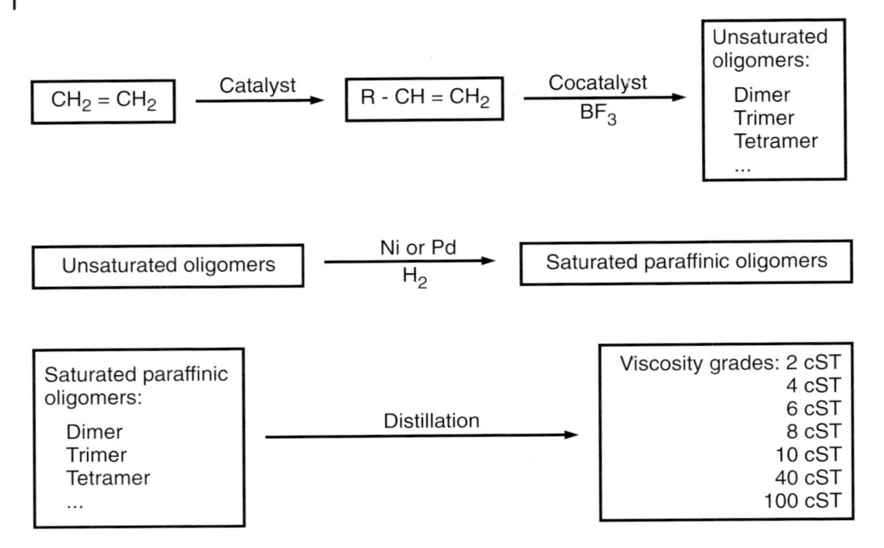

Figure 5.2 The three steps of polyalphaolefin production.

next step of the manufacturing process comprises the catalytic hydrogenation of the unsaturated olefins. This is achieved with the aid of classical catalysts, for example nickel on kieselguhr or palladium on alumina. In a third and final step, the saturated oligomers are distilled (Figure 5.2). The oligomerization of α-decene or α-olefin mixtures always results in complex mixtures of isomers with more branching than expected. Because the rearrangements are intramolecular, the molecular weights of the products can be kept within a narrow range. A typical saturated α-decene trimer for example looks like a three-pointed star [26] (Figure 5.3).

Polyalphaolefins satisfy some of the requirements of ideal hydrocarbon lubricants that can be predicted from chemical structure considerations – ideal hydrocarbon lubricants should have straight chains, be completely saturated and crystallize at low temperatures. The viscosities of straight-chain alkanes do indeed increase with chain length; the same is true for their pour points and viscosity indexes (VI). At constant molecular mass branching leads to increase viscosity and a decrease in pour point and VI. The length and position of the side chains influence all three properties. When branching occurs in the middle of the main chain, pour points are lower. Long side chains improve viscosity–temperature (*V*–*T*) characteristics.

Polyalphaolefins therefore have several advantages: narrow boiling ranges, very low pour points, VI > 135 for all grades with kinematic viscosity > 4 $mm^2\,s^{-1}$ at 100 °C. Their volatility is lower than that of all possible and equiviscous mineral oil grades and they contain only small quantities of unsaturated and polycyclic aromatic compounds and only traces of nitrogen, sulfur or other impurities. Although in some oxidation tests without additives some mineral oils seem

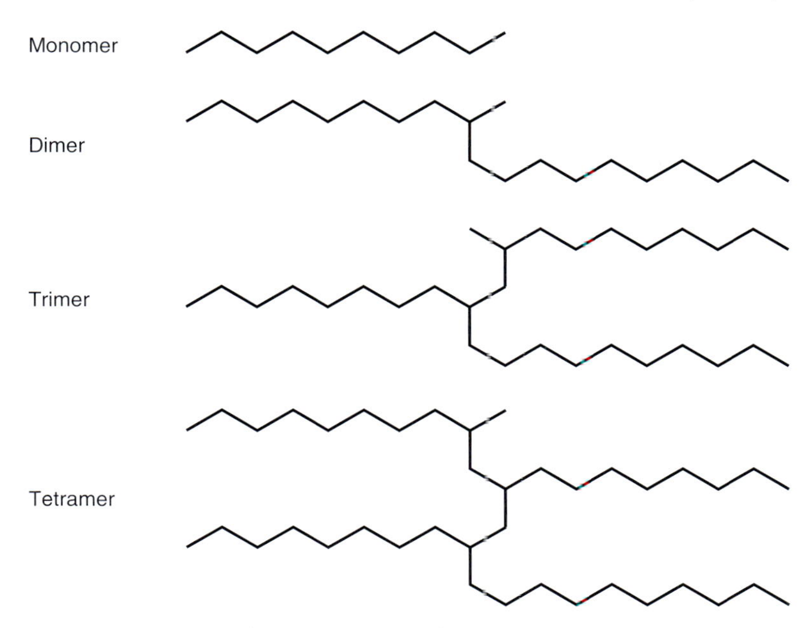

Figure 5.3 The shapes of typical polyalphaolefin oligomers.

superior to polyalphaolefins, this is because of the presence of natural antioxidants in the mineral oils that have survived the refining process – the oxidation of mineral oils has been described by Rasberger [27] in detail. The response of the synthetic products to antioxidants and their EP/AW synergists is better [28–30] than that of mineral oils. Their low polarity, on the other hand, leads to poorer solvency for very polar additives and can cause problems with seals. Therefore, they tend to be used in combination with smaller amounts of (dicarboxylic acid) esters or solvent-refined mineral oils.

Polyalphaolefins have been used traditionally in aerospace and lifetime applications, but today they are used in a wider variety of applications and have gained additional importance because of the increasing need for them in automotive lubricants [31].

5.1.2
Polyinternalolefins

Polyinternalolefins (PIO) are rather similar to polyalphaolefins. Both kinds of hydrocarbon are prepared by the oligomerization of linear olefins. The difference is that polyinternalolefins are made from cracked paraffinic base stocks. Plants for the manufacture of polyinternalolefins and polyalphaolefins therefore look similar also (Figure 5.4). Internal olefins are more difficult to oligomerize and the resulting products have VIs 10–20 units lower than the VIs of equiviscous

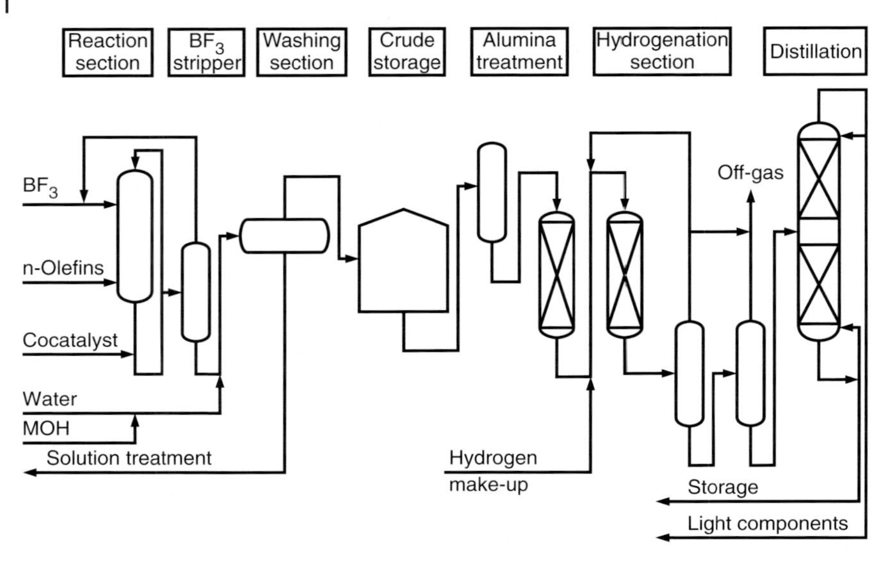

Figure 5.4 Diagram of a polyinternalolefin (polyalphaolefin) plant.

polyalphaolefins [32]. Polyinternalolefins are described in more detail by Navarrini, Ciali and Cooley [33].

Mixtures of polyalphaolefins with polyethylenes with kinematic viscosities from 100 to 2000 $mm^2 s^{-1}$ at 100 °C are available commercially. It can be expected that polyinternalolefins having these viscosities genuinely will be available in the future.

While the cationic polymerization of ethylene with aluminium chloride yields oils with VIs up to 120 and molecular masses from 400 to 2000, the polymerization of propylene gives oils with lower VIs and poor thermal stability. Both kinds of lubricating oil have lost importance, but copolymers of both base materials are said to have the potential for gaining importance again.

5.1.3
Polybutenes

Polyisobutylene rubbers were manufactured as early as the 1930s. It was one or two decades before their liquid analogues became available. Polybutenes (PBs) as synthetic base oils are described in more detail by Casserino and Corthouts [34]. They consist mainly of isobutene and are, therefore, often also known as polyisobutenes (PIBs).

They are produced by the polymerization of a hydrocarbon stream that besides isobutene **1** contains the other two butenes (**2**, **3**) and the butanes. The main sources of the C-4 feedstock are naphtha steam crackers and refinery catalytic crackers. The Lewis acid-catalyzed process yields a copolymer with a backbone built mainly from isobutene units. The higher the molecular weight, the lower the content of other butenes, that is the lower the molecular weight the more complex the structure. At the end of the carbon chain there remains one double bond. Therefore, polybutenes are less resistant to oxidation than polyalphaolefins, polyinternalolefins and alkylated aromatics – above 200 °C they begin to depolymerize and form gaseous products. Normally the end-group is *cis*- and *trans*-trisubstituted. It is possible to replace it by a disubstituted vinylidene group, that is more reactive. With maleic anhydride it yields polybutenylsuccinic anhydride derivatives that are used as corrosion inhibitors and detergents.

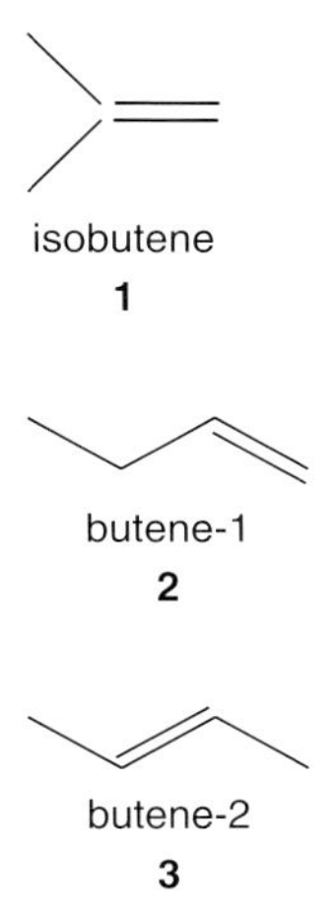

Polybutenes with molecular masses from about 300 to 6000 are important as VI improvers. They are used as components in two-stroke oils, gear and hydraulic oils, metal working lubricants, greases, compressor lubricants and wire-rope protectives [35]. One of the options for future polybutene applications is the increased use as synergistic components in binary or ternary base oil systems [36].

5.1.4 Alkylated Aromatics

Alkylnaphthalenes as lubricants have been available in Germany since ca. 1930 [37]. Alkylated aromatics are described in more detail by Wu and Ho [38]. Low-cost dialkylbenzenes are produced as by-products in the manufacture of linear and branched monoalkyl benzenes that themselves are feedstocks for the manufacture of detergents. The Friedel–Crafts alkylation of benzene with olefins tends to yield polyalkylated products, because of the reactivity of the primary product. With excess benzene, a suitable olefin, for example propene, and the right choice of a catalyst the reaction can be controlled and specialized alkyl benzenes can be synthesized. The structure of a typical alkylbenzene comprises six propene units **4**. The properties of synthetic alkyl benzenes vary widely and are difficult to generalize.

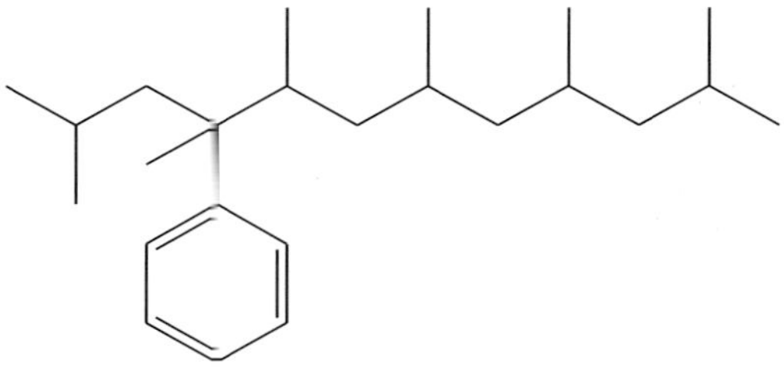

4

Some of the properties of alkylated benzene can be explained by the chemical structure considerations – ring formation leads to a larger increase in viscosity and decrease in VI values than alkyl substitution; side chains on a naphthenic ring have the opposite effect. A shift of cyclohexyl substituents from the end toward the middle of an alkyl chain has only a small effect on viscosity, but a strong negative effect on *V–T*; it also lowers the pour point; cyclopentane rings have a similar effect and alkylated aromatics have effects similar to those of equimolecular alkylnaphthenes but the products have lower viscosities and less favourable *V–T* dependence.

Low-cost dialkylbenzenes are used in a wide variety of industrial areas, for example in transformer oils, where high resistance to gas evolution is important. Tailored grades, especially the linear products, although having poorer properties than polyalphaolefins, because of their excellent solvency, their suitability for low temperatures and their compatibility with elastomers still are an option [39]. The most widespread use of these products is, however, as refrigerator oils.

5.1.5 Other Hydrocarbons

Alicyclic saturated hydrocarbons, cyclohydrocarbons, occur naturally as components of mineral oils and here are known under the name naphthenes. While paraffins have their analogues in polyalphaolefins, and aromatics in alkylbenzenes, synthetic analogues of naphthenes, they are described in more detail by Hamid [40], have rather differing properties. Some polyalkylated oligomers have been found suitable as high-temperature lubricants [41]. Some synthetic cycloaliphatic hydrocarbons have coefficients of friction up to 1.2. One reason is that when bulky cyclopentyl or cyclohexyl groups are introduced into branched short-chain molecules, intramolecular rotation is hindered. The high molecular packing state and the stiffness of some polycyclic hydrocarbon molecules also make them suitable as components of traction fluids for friction gears [42], where these properties of the lubricant are important design elements [43,44].

The basic material for some of the best described traction fluids is 2,4-dicyclohexyl-2-methylpentane **5**, the hydrogenated dimer of α-methylstyrene.

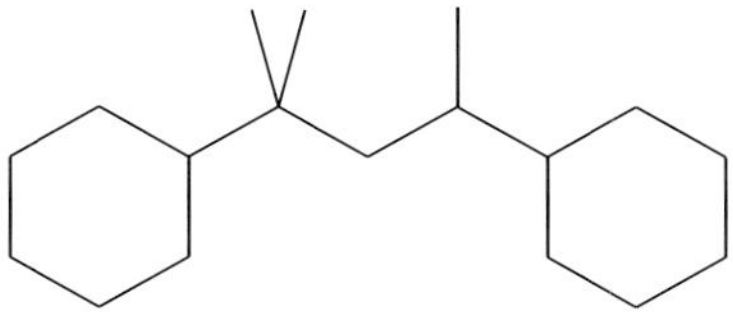

5

Alkylcyclopentanes that have potential as lubricating oils are described in detail by Venier [45]. Of these compounds tris(2-octyldodecyl)cyclopentane **6** has found limited use in high-vacuum applications, for example in greases, because of its extremely low volatility [46].

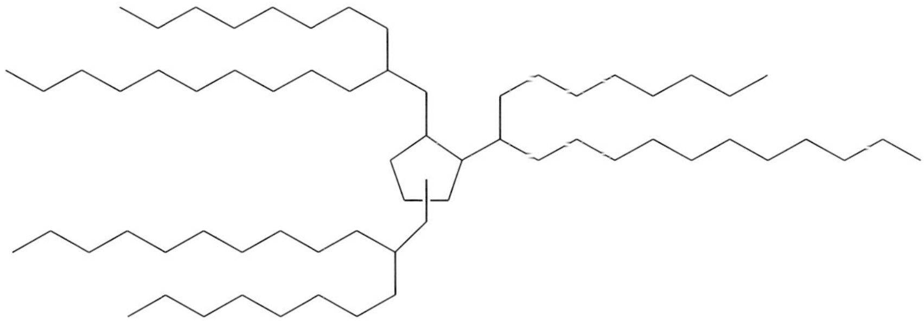

6

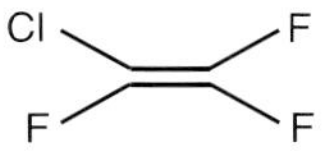

7

Tri(*n*-alkyl)methylmethanes and tetra(*n*-alkyl)methanes have been synthesized and have been compared with polyalphaolefins of similar viscosity. The products seem to be useful in applications that demand extreme thermo-oxidative stability [47]. Alkylated biphenyls and diphenylmethanes have also been proposed as high-temperature lubricants, but are not yet available. Adamantane-based synthetic lubricants have been synthesized and seem to have some potential as electrically insulating fluids [48].

5.2
Halogenated Hydrocarbons

Chlorinated hydrocarbons are very stable and not combustible. They were once widely used as insulating oils, heat-transfer fluids and hydraulic fluids, but this use has been discontinued because of the environmental problems they cause.

Commercial liquid chlorofluorocarbons, all oligomers of chlorotrifluoroethylene **7** contain between 2 and 12 units. The first products were developed in the 1940s and have already been described by Ashton and Strack [49]. Polychlortrifluorethylenes are described in detail by Epstein and Ferstandig [50]. They are suitable for use as lubricants mainly because of their extraordinary chemical stability, including against oxygen.

Their low corrosivity, the good low-temperature characteristics of the low-viscosity grades and their lubricating properties are also advantageous. Disadvantages are their high volatility and their not ideal V–T dependence. The properties of the chlorofluorocarbons are a consequence of the larger volume of the chlorine and fluorine atoms, which hinders the flexibility of the molecule, of the reduced intermolecular cohesive forces, and of the different bond lengths. Their density naturally is relatively high (between 1.7 and $2\,g\,ml^{-1}$ at 40 °C). They are used as lubricants in oxygen compressors, in pumps for mineral acids, halogens and oxygen and in mills or mixers for strong oxidizing agents. They are also suitable lubricants for turbine pumps in rocket engines and as base oils for non-flammable hydraulic fluids [51]. Although the halogenated hydrocarbons are comparable with perfluorinated polyethers (see Section 5.5) with regard to their performance, but not as expensive, their future is unclear.

5.3
Synthetic Esters

5.3.1
Esters of carboxylic acids

Synthetic ester lubricants for the jet age were developed in Germany during World War II [52].

Hydrocarbon oils could not satisfy the demands of aircraft engine oils and alkyl esters of aliphatic carboxylic acids had favourable properties. Carboxylic groups, because of their strong dipole moments, reduce the volatility and increase the flash point of lubricating oils, and at the same time positively affect thermal stability, (the bonds of the COO group are thermally more stable than the C—C bond), solvency, lubricity and biodegradability; on the other hand, however, they negatively affect the hydrolytic stability of a lubricant and the reactivity with metals or alloys that contain copper or lead [53]. The general properties of synthetic esters of carboxylic acids have been described in detail in Ullmann's Encyclopedia of Industrial Chemistry [54,55] and by Boyde and Randles [56].

In principle all esters of carboxylic acids can be prepared in the same manner:

1) The carboxylic acid is reacted with excess alcohol in the presence of a catalyst (mineral acids, ideally adsorbed on a solid, ion exchangers, Lewis acids, e.g. boron trifluoride, ideally as the etherate, and amphoteric hydroxides such as aluminium hydroxide can be used as catalysts).

2) To shift the equilibrium to the product side, the reaction water is removed during the process.
3) The unreacted acid is neutralized with sodium carbonate or calcium hydroxide and then removed by filtration and the ester is distilled.

5.3.1.1 Dicarboxylic Acid Esters

Dicarboxylic acid esters have been described in detail by Dukek and Popkin [57]. Two types of ester molecule or their combinations have been found to be most suited as lubricant components:

1) Esters of branched primary alcohols with straight dicarboxylic acids 5.8
2) Esters of straight primary alcohols with branched dicarboxylic acids 5.9

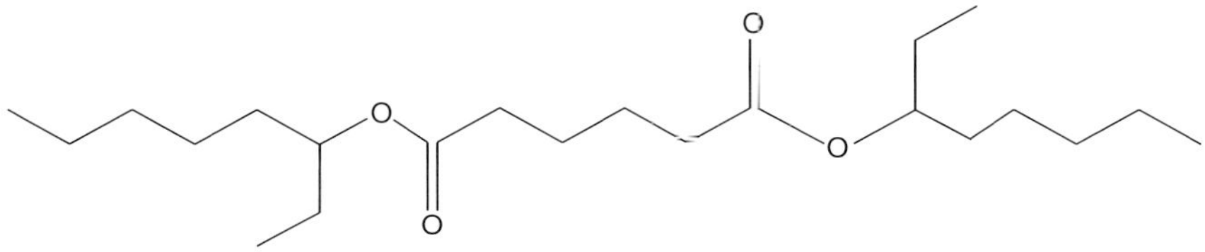

di(2-ethylhexyl)adipate

8

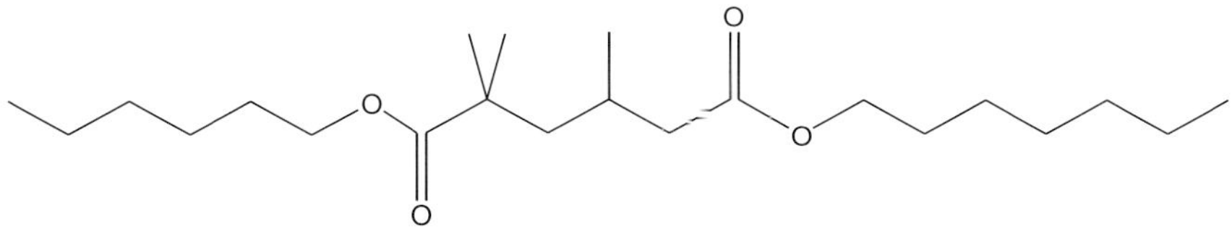

diundecyl(2,2,4-trimethyl)adipate

9

The alcohols that are needed for the production of dicarboxylic acid ester oils can be obtained synthetically by the hydroformylation of olefins with carbon monoxide and hydrogen (Eq. (5.1)). The dicarboxylic acids can be obtained either by oxidation of vegetable oils, for example castor oil, as for azelaic and sebacic acid [58], or by cleavage of corresponding alicyclic hydrocarbons with oxygen, as for adipic acid (Eq. (5.2)) [59].

$$R{-}CH = CH_2 \xrightarrow{CO+H_2} R{-}CH_2{-}CH_2{-}CHO \xrightarrow{H_2} R{-}CH_2{-}CH_2{-}CH_2OH \quad (5.1)$$

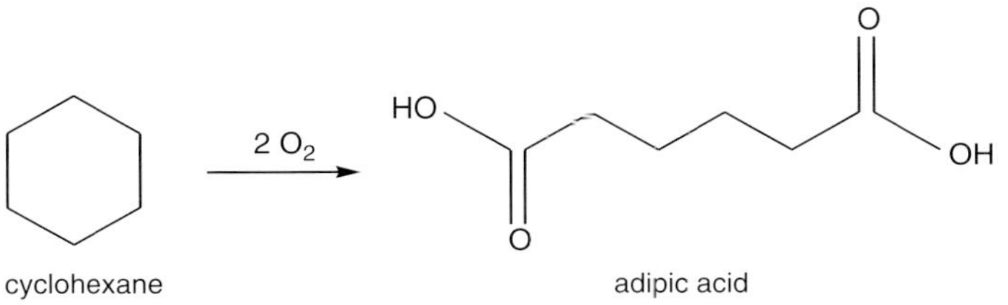

$$\text{cyclohexane} \xrightarrow{2\,O_2} \text{adipic acid} \quad (5.2)$$

Esters of straight-chain dicarboxylic acids have better *V–T* characteristics than mineral oils, and higher VI values. The values decrease with increasing branching, but branching improves the low-temperature properties. Esters with little branching have the best properties, especially esters with methyl groups adjacent to the carboxyl group. When the esters contain tertiary hydrogen atoms, however, their stability toward oxidation is not improved (Eq. (5.3)). The sterically hindered esters have favourable thermal and hydrolytic stability. The hydrolytic stability has been further improved with the aid of carbodiimides [59]. At low temperatures, the viscosity of esters follows a time function and additives can have a negative effect on the viscosity and the pour point. With methacrylates as VI improvers values between 170 and 180 are reached. With complex esters as VI improvers degradation as a result of shear stress is less pronounced. Additives are mainly added to improve the EP performance, although the load-carrying capacity of esters in general is twice that of mineral oils.

$$R_2-C^{\beta}H_2-C^{\alpha}H_2-O-C(=O)-R_1 \longrightarrow \text{[cyclic transition state: } R_2HC^{\beta}\text{–H}\cdots O{=}C(R_1)\text{–O–}C^{\alpha}H_2\text{]} \longrightarrow R_2HC{=}CH_2 + R_1-C(=O)OH \quad (5.3)$$

Dicarboxylic acid ester oils can improve the *V–T* characteristics of automotive engine oils without having a negative effect on the low-temperature viscosity or on the Newtonian flow behaviour. Such oils are suitable for diesel engines also. Esters have also gained importance as lubricity additives in greases and in polyalphaolefins as additives for the improvement of the compatibility with elastomers.

5.3.1.2 **Polyol Esters**

Since the 1940s neopentyl glycol ((2,2-dimethyl)-1,3-propandiol) **10**, trimethylolethane and trimethylolpropane (1,1,1-tris(hydroxymethyl)ethane and 1,1,1-tris (hydroxymethyl)propane) **11** and pentaerythritol (2,2-bis(hydroxymethyl)-1,3-propanediole) **12** have been known as the alcoholic components of esters. They have been described in detail by Smith [60]. Beside excellent thermal and oxidation stability, the esters have good *V–T* and lubricating properties and excellent viscosity characteristics at low temperatures. The reason is that the thermal stability of a primary OH group is greater than that of a secondary group, and the hydroxy derivatives of neopentane contain only primary hydroxy groups, and short side chains in a hydrocarbon not only reduce its pour point, but tertiary C or H atoms at the same time also facilitate the attack of oxygen; the polyols, however, have no tertiary H atoms and no H atoms in the β-position that could facilitate thermal cleavage.

$$HO-CH_2-C(CH_3)_2-CH_2-OH$$

10

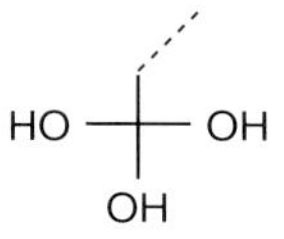

HO OH
OH

11

OH
HO OH
OH

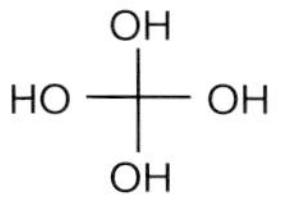

12

The mechanism of the polymerization of polyol esters during ageing has been studied [61] and benzotriazole derivatives have been proposed as multifunctional additives [62]. Since the 1960s the polyol esters have gained importance as lubricants for high-temperature applications, for example in aircraft engines that are meant for speeds beyond Mach 2, and they have become even more important since the rise of the biodegradable esters.

5.3.1.3 **Other Carboxylic Esters**

The structural counterparts of the polyol esters are the esters of some aromatic polycarboxylic acids, mainly phthalic acid **13**, trimellitic acid **14** and pyromellitic acid **15**, with monofunctional alcohols. The latter, also, are suited as high-temperature lubricants. The time dependence of the low-temperature viscosity of some grades is rather pronounced. Therefore, their use in low-temperature applications is problematic.

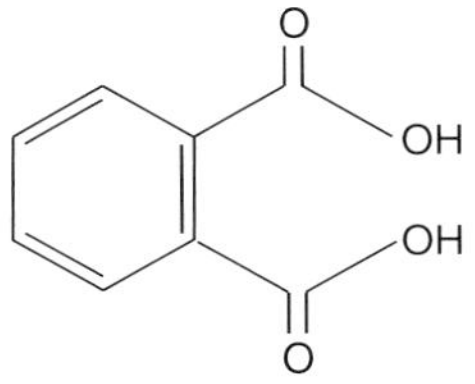

O
OH
OH
O

13

O O
HO OH
OH
O

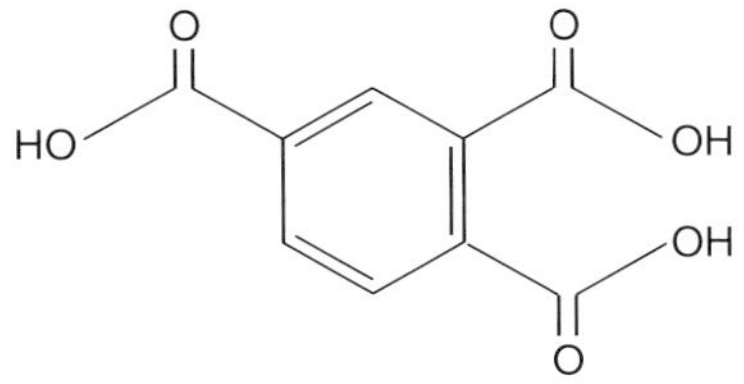

14

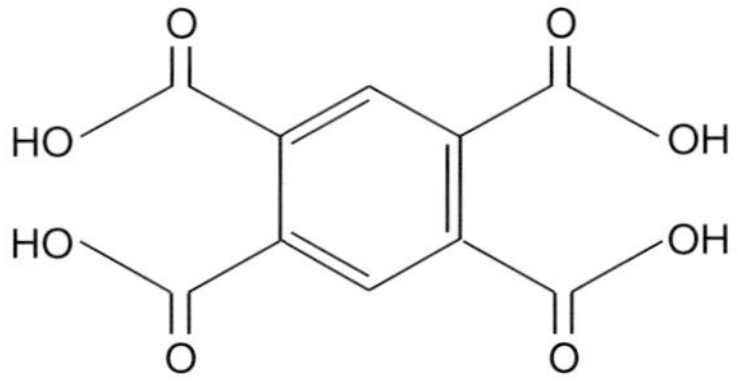

15

Monoesters, the reaction products of fatty acids and monofunctional alcohols, are used in metalworking. The sulfurized esters are excellent EP additives for all kinds of lubricant [63].

Dimeric acid esters **16**, the acids are made from oleic or tallow fat acid, are used as engine oil and thickener components.

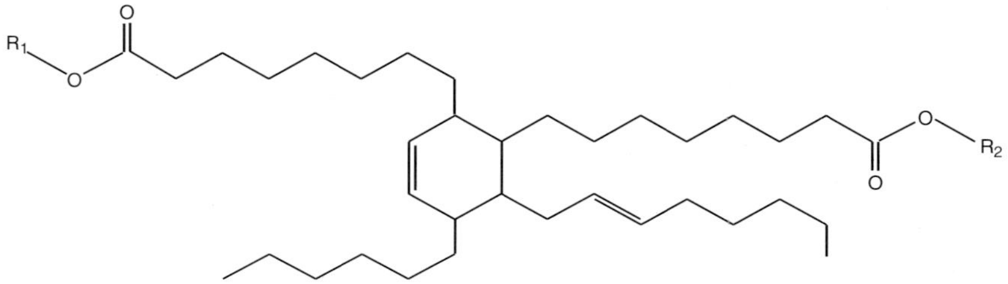

16

5.3.1.4 Complex Esters, Complex Polymer Esters

Esters that contain straight or branched diols or polyalkylene glycols and straight or branched dicarboxylic acids as well as (mono)carboxylic acids and monofunctional alcohols have attracted interest. Usually, first the diol is esterified with the dicarboxylic acid, then, depending on the desired product, the intermediate product is reacted with either a carboxylic acid or a monoalcohol. Two types of complex ester have become important:

ROH...HOOCrCOOH...HORrROH...HOOCrCOOH...HOR

17

RCOOH...HORrROH...HOOCrCOOH...HORrROH...HOOCR

18

The same scheme is valid when neopentane-derived polyols [64] or glycerol [65] are involved.

That complex esters have higher molecular weights and higher viscosities than the common esters was one of the main reasons for their introduction. Esters of type **18** are superior to type **17** with regard to flash point, pour point and low-temperature viscosity. In type **17** polyalkylene glycols lead to lower pour points than aliphatic glycols. Medium molecular-weight oligomeric esters starting from (mainly vegetable) fatty acid triglycerides [66] are of interest for eco-friendly and high-performance applications, as are esters starting from adipic acid or sebacic acid and neopentyl glycol (type 5.17), because their biodegradability and because their pour points are approximately the same as those of their monomers, although their viscosities are higher [67].

High-molecular-weight complex esters of the type

R... (HOrOH...HOOCrCOOH)$_n$...R

19

have also been named polymer esters and find use mainly in applications where mineral oil and solvent resistance are required. Polymer esters are described in more detail by Wallfahrer [68].

When alphaolefins and alkylmethacrylates are co-oligomerized another kind of polymer ester results [69] and a third kind is based on polymers that have been functionalized [70]. With such products attempts have been made to combine the advantages of synthetic hydrocarbons and synthetic esters at a high viscosity level.

5.3.1.5 Fluorinated Carboxylic Acid Esters

Esters of fluorinated acids are readily hydrolyzed and the free acids have a strong oxidizing effect, therefore only esters with fluorinated alcohols can be used as lubricants. The thermal stability of fluorinated esters is better than that of the non-fluorinated analogues. The oxidation stability is also higher. Some types of rubber shrink in their presence and some antioxidants have a negative effect. Fluorinated esters were in the first place intended for high-temperature applications. They have been described in detail by Murphy [71]. It can be assumed that their use will be limited because of competition from the fluorinated polyethers. Esters containing sulfur as a part of the acidic or alcoholic component are known but have not gained any importance.

In the future, mixtures of all kinds of synthetic hydrocarbon and synthetic ester promise much improvement in all fields of application [72–74].

5.3.2 Phosphate Esters

Tertiary esters of phosphoric acid with alcohols or phenols were prepared as long as 150 years ago but were not introduced as antiwear additives before 1930. They have also gained importance as plasticizers, fire-resistant hydraulic fluids, compressor oils, and synthetic lubricants. Phosphate esters have been described by Hatton [75] and by Marino [76] (neutral phosphate esters are described in detail by Phillips, Placek and Marino [77]). They are usually divided into triaryl, trialkyl and alkylaryl phosphates and are made by reaction of phosphoryl chloride with phenols or alcohols (Eq. (5.[illegible])). Their properties range from low-viscosity fluids to high-melting solids. With increasing molecular weight trialkyl phosphates change from water-soluble to water-insoluble liquids. The triaryl phosphates have higher viscosities and are insoluble in water. An aryl side chain reduces the melting point. The properties of alkylaryl phosphates are between those of alkyl and aryl phosphates.

$$3\,ROH + POCl_3 \rightarrow OP(OR)_3 + 3\,HCl \tag{5.4}$$

The hydrolytic stability varies between good and poor. Length and branching of the alkyl chain increase hydrolytic stability; tolyl substitution is superior to a phenyl. Alkylaryl phosphates are more susceptible to hydrolysis than trialkyl or triaryl phosphates. Phosphoric acid esters are less stable than carboxylic acid esters, but more stable than silicic or boric acid esters. Their hydrolytic stability can be increased with, for example epoxides [78]. With ion-exchange treatment and vacuum dehydration, hydrolysis of the phosphoric acid esters used as hydraulic fluids can be virtually eliminated [79].

The thermal stability of triaryl phosphates is better than that of the alkyl compounds. With aminic antioxidants and rust inhibitors, for example some amine salts of primary or secondary phosphates, triaryl phosphates can be used up to 175 °C, but alkyldiaryl phosphates only up to 120 °C. Branching of the alkyl radicals reduces the thermal stability, the effect becomes stronger with decreasing length. Trialkyl phosphates and alkyldiaryl compounds behave similarly. In general phosphoric acid esters are not corrosive. But thermal decomposition leads to the formation of phosphoric acid and this is a corrosion hazard. The high spontaneous ignition temperatures of up to 600 °C underline the good fire resistance of some products. Trialkyl and alkylaryl products have pour points down to −65 °C, when they contain VI improvers.

The lubricating properties of the phosphoric acid esters are excellent, particularly on steel. They can be blended with almost all lubricants and additives. Their dissolving power is, on the other hand, responsible for their incompatibility with rubbers, varnishes and plastics. Nylon-, epoxy- and phenol-formaldehyde resins are stable.

5.4 Polyalkylene Glycols

The first polyalkylene glycols suitable as lubricants were developed during World War II. The first patent was probably published in 1947 by Roberts and Fife [80]. Polyalkylene glycols have been described in detail by Gunderson and Millet [81], Kussi [82], by Matlock and Clinton [83] and by Greaves [84].

Polyalkylene glycols are prepared by the reaction of epoxides, usually ethylene and propylene oxide, with compounds that contain active hydrogen, usually alcohols or water, in the presence of a basic catalyst, for example sodium or potassium hydroxide (Figure 5.5). Variation of the ratio of the epoxides and the end groups leads to different products. Polymers with statistically distributed

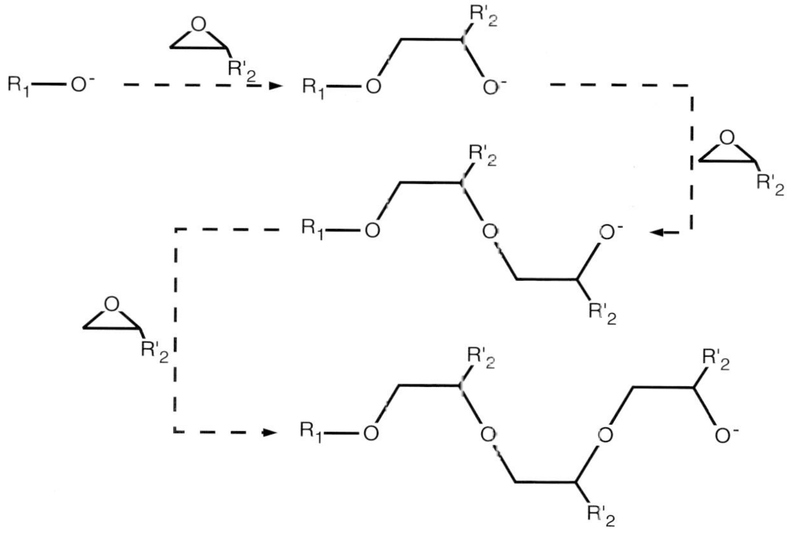

Figure 5.5 Preparation of polyalkylene glycols.

alkylene groups are made by use of a mixture of alkylene oxides. Separate addition leads to block copolymers. Because ethylene oxide is more reactive than propylene oxide, random copolymers tend to have the propylene oxide units at the chain ends.

Terpolymers with, for example tetrahydrofuran have also been prepared. Pure tetrahydrofuran polymers can be obtained by polymerization of tetrahydrofuran in the presence of Friedel–Crafts catalysts. They are colourless oily or waxy substances of very low toxicity.

Polyalkylene glycols have at least one hydroxyl group at one end of the molecule, and can therefore be regarded as alcohols. The number of hydroxyl groups is increased by use of water or multifunctional starters. The reaction of the alcohols with acids leads to esters; reaction with, for example strong acids and olefins leads to ethers.

Because the carbon–oxygen bond is stronger than the carbon–carbon bond, polyalkylene glycols have solvent properties somewhat different from those of hydrocarbons. Miscibility with water increases with the number of ethylene oxide units. The solubility is a result of hydrogen bonding of water with the free electron pairs of the oxygen. In solution, the water-soluble grades are practically non-flammable. The hygroscopic character of the polyalkylene glycols is dependent on their hydroxyl group content – it decreases with increasing molecular weight and the number of ether bonds. The solubility in water decreases in the same way. That it also decreases with increasing temperature can be explained by a loss of hydrogen bonding. Solubility in hydrocarbons increases with the molecular weight. Polyalkylene glycols in general are soluble in aromatic hydrocarbons.

The molecular weight and viscosity of polyalkylene glycols can be significantly influenced during production and be adjusted within narrow limits. The possibility of engineering products in this way distinguishes them from many other lubricants.

Low-molecular-weight polyalkylene glycols containing more than 50% propylene oxide have pour points that go down to −65 °C. The lateral methyl groups are responsible for the disruption of crystallization. On the other hand pure high-molecular-weight polyethylene glycols are wax-like solids with pour points near +4 °C.

Kinematic viscosities range from 8 to 100 000 $mm^2 s^{-1}$ at 40 °C. With the change from diols to monoethers, ester ethers and diethers, at the same molecular weight, viscosities are reduced, particularly at low temperatures. In comparison with mineral oils, which give straight lines, the V–T diagrams of polyglycols show their viscosities are too high at both low and high temperatures. The VI values of polyalkylene glycols usually lie around 200. High-molecular-weight polyethylene glycols have VI values of up to 400.

Prolonged heating of polyalkylene glycols to above 150 °C leads to depolymerization. The resulting aldehydes react further to give acids. It is an advantage that only soluble or volatile products are formed. Traces of alkali or alkaline earth metals promote the degradation. The decomposition can be prevented by addition of aminic antioxidants, to the extent that the oils can be used as heat-transfer fluids up to 250 °C. Antirust and EP additives for use in polyalkylene glycols must have pronounced water tolerance.

The polar nature of polyalkylene glycols gives the products strong affinity for metals, and so the lubricating film remains intact even at high surface pressure, a property useful in lubricants and metal cutting fluids. Because swelling of elastomers decreases with increasing viscosity, polyalkylene glycols can be used with natural and with synthetic rubbers in hydraulic oils and brake fluids. Easy removal by washing with water makes these suitable for applications that are difficult for other products. The toxicity of polyalkylene glycols is similar to that of glycerol for low-viscosity products and similar to that of isopropanol in more viscous products. This is of advantage in the food, pharmaceutical, tobacco and cosmetics industries. Products with high ethylene oxide content are up to 80% biodegradable. Polyalkylene glycols lower the freezing point of water. High viscosity water-soluble products are shear-stable liquid thickeners. Hydrophilic and hydrophobic fractions in block copolymers give them surfactant properties. The lubricating properties in mixed-friction areas and in rolling friction contacts can be improved by use of polyalkylene glycols with a broad molecular-weight range, for example; these can bridge the differences between water and the polyethers in terms of adhesion, viscosity and volatility.

5.5
Other Polyethers

Alkylated aryl ethers, such as alkyl ethers, have lower viscosities, lower VI values, lower pour points and lower boiling points than the corresponding alkanes.

Asymmetric substitution also reduces the pour point, but has an adverse effect on the $V–T$. Although such products have low viscosities, some have found use as components of lubricants used in the life sciences.

5.5.1
Perfluorinated Polyethers

Perfluoroalkylpolyethers (PFPE) as lubricants were probably first mentioned by Gumprecht in 1965 [85]. Schwickerath, Del Pesco, and Walther, Bell and Howell have described them in detail [86–88].

Photochemical polymerization in the presence of oxygen, followed by fluorination with elemental fluorine leads to products of type A (Eq. (5.5)), with tetrafluoroethylene. With hexafluoropropylene, products of type B (Eq. (5.6)) result. Anionic polymerization of hexafluoropropylene epoxide leads to products of type C (Eq. (5.7)), and Lewis acid-catalyzed polymerization of 2,2,3,3-tetrafluorooxetane leads to products of type D (Eq. (5.8)).

$$F_2C=CF_2 \xrightarrow{O_2,F} F_3C-O-\left[\underset{}{\overset{CF_3}{\overset{|}{C}}}F-CF_2-O\right]_m-\left[CF_2-O\right]_n-CF_3 \tag{5.5}$$

$$F_2C=CF-CF_3 \xrightarrow{O_2,F} F_3C-O-\left[CF_3-CF_2-O\right]_m-\left[CF_2-O\right]_n-CF_3 \tag{5.6}$$

$$\underset{\diagdown\;\diagup \atop O}{F_2C-CF}-CF_3 \xrightarrow[(CsF)]{F} F_3C-CF_2-CF_2-O-\left[\overset{CF_3}{\overset{|}{C}}F-CF_2-O\right]_n-CF_2-CF_3 \tag{5.7}$$

$$\begin{matrix}O-CF_2\\ | \quad\; | \\ H_2C-CF_2\end{matrix} \xrightarrow[(BF_3)]{F} F_3C-CF_2-CF_2-O-\left[CF_2-CF_2-CF_2-O\right]_n-CF_2-CF_3 \tag{5.8}$$

The density of the PFPEs is nearly twice that of hydrocarbons. They are immiscible with most of the other base oils and non-flammable under nearly all practical conditions. The more common types A and B have a good to very good $V–T$ and $V–p$ dependence, and low pour points [89]. The viscosity of linear PFPEs changes less with temperature and pressure than that of the non-linear variety, but has a non-negligible deviation from linearity, as has recently been reported [90]. In air, they are stable up to 400 °C. Trifluoromethyl groups that are adjacent to the ether bonds shield them from acid-catalyzed cleavage, but difluoroformyl groups contribute to a decrease in stability at higher temperatures. PFPEs are remarkably inert chemically, this is including elastomers. Their hydrolytic stability is excellent. Good radiation stability, that was stated earlier, has been doubted [91].

The shear stability of PFPEs is better than that of other polymeric lubricants, but in the presence of steel and under boundary lubrication conditions PFPEs do not perform very well [92]. It has been reported that impurities adversely affect wetting properties [93]. The effect of humidity [94] and their thermo-oxidative behaviour have been investigated [95]. It has been found that the performance can be improved with α,β-diketones [96].

Perfluoropolyalkyl ethers have all the properties required by modern spacecraft: as lubricants and hydraulic fluids they resist to thermal and oxidative attack above 260 °C, possess good low-temperature flow characteristics and are fire-resistant. They can also be used as power transmission and inert fluids and in transformers and generators as dielectrics with outstanding properties.

5.5.2
Polyphenyl Ethers

Polyphenyl ethers **20** are the reaction products of phenols and halogenated aromatic compounds (Eq. (5.9)). They have been described in detail by Mahoney and Barnum [97], and are described by Hamid and Burian [98]. Abbreviated formulas are in use. They contain the substitution position and the number of phenyl rings (ϕ) and ether bonds:

ppp5P4E stands for ϕ—O—ϕ—O—ϕ—O—ϕ—O—ϕ (p—p—p)

20

O O H Br O → O O O

(5.9)

The aromatic groups increase the stability, but negatively affect the *V–T* dependence of these polyethers. Alkyl groups lower the high melting points. *para*-Derivatives have lower volatilities; for *ortho*-products, the volatilities are higher. Spontaneous ignition occurs between 550 and 600 °C and alkyl substitution reduces this by about 50 °C. With the usual elastomers swelling occurs.

The oxidation stability of polyphenyl ethers is only slightly lower than that of polyphenyls or tetraarylsilanes. Alkyl substituents reduce it. The thermal decomposition temperatures are up to 465 °C. Short-chain substituents reduce it to below 380 °C and with higher alkyl groups it drops to below 350 °C, that is to temperatures typical for aliphatic hydrocarbons. Trifluoromethyl groups have a worse effect, the thermal decomposition temperature drops below 270 °C. Coke formation is low. It increases with alkyl substitution, particularly in the presence of methyl groups.

Polyphenyl ethers are the most radiation-resistant lubricants. At low temperatures, radiation affects the viscosity more pronouncedly than at high temperatures. It increases viscosity, acidity, evaporation loss, corrosivity and coke formation, but reduces flash and ignition point.

Between 200 and 300 °C, the lubricating properties of polyphenyl ethers are reported to be comparable with those of mineral and ester oils and better than those of polysiloxanes and aromatic hydrocarbons. Alkylated polyphenyl ethers have better properties than the unsubstituted ethers. Polyphenyl ethers are suited as high-temperature or radiation-resistant hydraulic fluids and lubricants [99], and as lubricants for optical switches [100].

5.5.3 Polysiloxanes (Silicone Oils, Silicones)

Polysiloxanes can be liquids or solids. A well-known book about their chemistry and technology has been written by Noll [101]. Silicone oils suitable as lubricants are generally straight-chain polymers of the dimethylsiloxane **21** and phenylmethylsiloxane series **22**. They are described in detail by Awe and Schiefer, by Demby, Stoklosa and Gross and by Perry, Quinn, Traver and Murthy [102–104].

$$CH_3-\overset{CH_3}{\underset{CH_3}{Si}}-O-\left[\overset{CH_3}{\underset{CH_3}{Si}}-O\right]_n-\overset{CH_3}{\underset{CH_3}{Si}}-CH_3$$

21

$$\cdots\overset{CH_3}{\underset{C_6H_5}{Si}}-O\cdots\left[\overset{CH_3}{\underset{CH_3}{Si}}-O\right]_m\cdots\overset{C_6H_5}{\underset{C_6H_5}{Si}}-O\cdots$$

22

Methyl silicone oils are simply made from quartz and methanol in the end (Figure 5.6), but commercial technology employs sophisticated steps, for

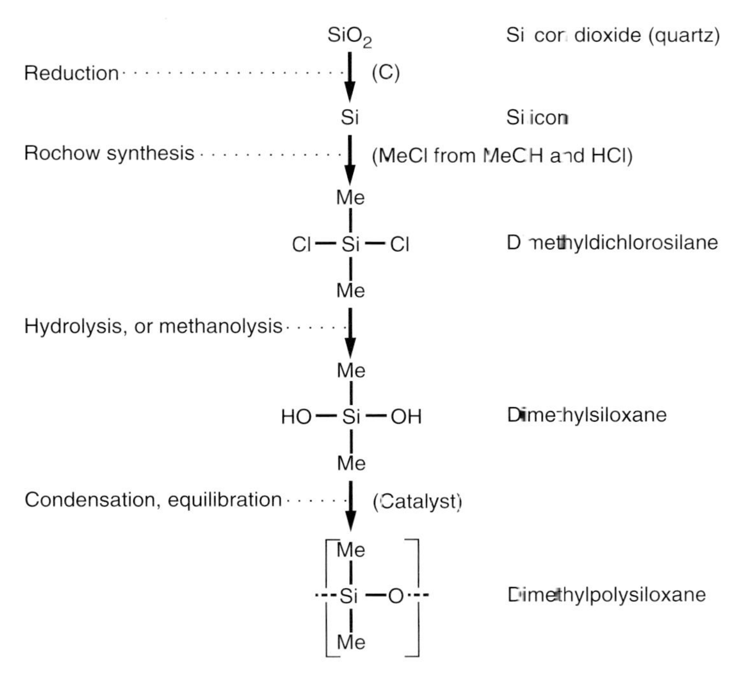

Figure 5.6 Preparation of dimethylsilicone oils.

example the so-called equilibration – treatment of a mixture of siloxanes of different molecular weight with strong acid or basic catalysts with the aim of achieving a narrow Gaussian molecular weight distribution [105].

Among the unique properties of silicone lubricating oils is their immiscibility with many organic fluids, the low-temperature dependence of their physical properties and their physiological inertness. It is possible to make them with both very low pour points and high viscosity by increasing the asymmetry of the molecules, usually by partial replacement of the dimethylsilyl groups by phenylmethylsilyl groups. The density of silicone oils is in the region of that of water, those of dimethylsilicone oils are slightly below, those of phenylmethyl silicone oils slightly above.

The extremely low viscosity–temperature coefficients (VTC) [106], lower than 0.6 for low-viscosity dimethylsilicone oils are mainly due to the extraordinary flexibility of their Si–O chains. Low-to-medium viscosity silicone oils have Newtonian behaviour up to high shear rates, but with increasing viscosity the apparent viscosity decreases with increasing shear rate, that is pseudoplastic behaviour is observed. Compressibility and viscosity changes at high pressure depend strongly on the methyl-to-phenyl ratio of the silicone oils and are relatively large, again mainly because of the flexibility of their Si–O chains.

The thermal decomposition of silicone oils begins at approx. 300 °C. In general, the decomposition products are not corrosive, but they lower the viscosity of the oils. Up to 200 °C the oxidation stability of silicone oils is superior to that of hydrocarbons, esters and polyalkylene glycols. The stability of phenylmethylsilicone grades, because of the resonance stabilization of the phenyl radicals, has been reported to exceed that of dimethylsilicone grades by up to 50 °C. Correlation of thermoanalytical and viscometric data of aged silicone oils has indicated an even larger figure [107]. At higher temperature, the formation of siloxyl and silyl radicals via cross-linking leads to the formation of polymer molecules and thus results in gels.

Some iron and cerium compounds are especially suitable as oxidation inhibitors. At elevated temperatures, gel formation nevertheless occurs readily in the presence of selenium or tellurium. In the presence of chlorine even explosions are possible. Si—O bonds can also be broken by hydrolytic attack, but in contrast with silicic ester oils formation of silica or silicic acid gel does not occur.

The surface tension and the foaming tendency of silicone oils are much lower than those of mineral oils, but the thermal expansion of the same order of magnitude; silicone oils are, therefore, ideal defoamers for the hydrocarbons.

The lubricating properties of phenylmethylsilicone oils are better than those of the dimethylsiloxanes; trifluoropropylmethylsilicone oils perform even better [108]. Despite this they have been widely superseded by the perfluorinated polyethers.

Silicone oils can be found in all kinds of industrial and military installations. Dimethylsilicone oils are used as lubricants for bearings and gears with rolling friction. In sliding friction, their performance depends on the metal pairs, they are used, for example as lubricants for bronze or brass on aluminium, copper or zinc at low surface pressures and as lubricants for porous bronze bearings. Silicone oils are among the best lubricants for plastic bearings, but in precision instrument applications spreading should be prevented by use of an epilamization agent. They are suitable lubricants for rubber parts, and also serve as switch and transformer oils. Silicone oils with higher phenyl substitution are mainly used for the lubrication of turbines, ball bearings and all kinds of instrument, especially at high temperatures. Their radiation stability is also remarkable.

Silicone oils can be used as base oils for all kinds of lubricating greases, be it for sealing and damping applications, for reasons of chemical inertness, or for high-performance low- and high-temperature applications [109].

5.6 Other Synthetic Base Oils

There have always been two reasons for the development of new synthetic lubricating oils. The first is to achieve a lower temperature-dependence of the physical properties, that is to make either higher or lower temperatures accessible; the second is to obtain better inertness, that is to prolong performance time. More recently a third reason must be added, better ecological properties combined, if possible, with physiological inertness.

Below −75 °C, the performance of commercial dicarboxylic acid esters and siloxane oils ends and only some branched hydrocarbons, for example 3,3-dimethylhexane, seem to have suitable properties (melting points < −125 °C) but at the same time have all the disadvantages of their kind. Above 300 °C, the practical performance of commercial perfluorinated polyethers ends and fluids comprising a stable element – carbon nucleus, for example C–N in some triazine derivatives and alkylated phenyl or silyl groups, must be considered. Above 400 °C, the performance of perfluorinated polyethers ends, even under ideal conditions, and in the triazine derivatives the alkylated phenyl or silyl groups must be replaced by phenoxyphenyl groups. Tailored molecules with no C—H bonds at all can be imagined that will be suitable for temperatures above 500 °C [110].

Of the following group of synthetic fluids that have been proposed as base oils for lubricants or that have been developed especially for that use, none has gained yet real economic or commercial importance in that area during the last 30 years, all still wait for their niche application, even though some have really astonishing properties.

Acetals **23** have been investigated because of their high stability in alkaline media; they are, however, sensitive to acidic hydrolysis.

$$R_1-\underset{\underset{\displaystyle O-R_3}{|}}{\overset{\overset{\displaystyle O-R_2}{|}}{C}}-H$$

23

Adamantane derivatives, mainly the esters of 1,3-dihydroxy-5,7-dimethyladamantane **24**, reminding of the extraordinary oxidative stability of adamantane itself, have been reported as being superior to the common dicarboxylic acid esters [111].

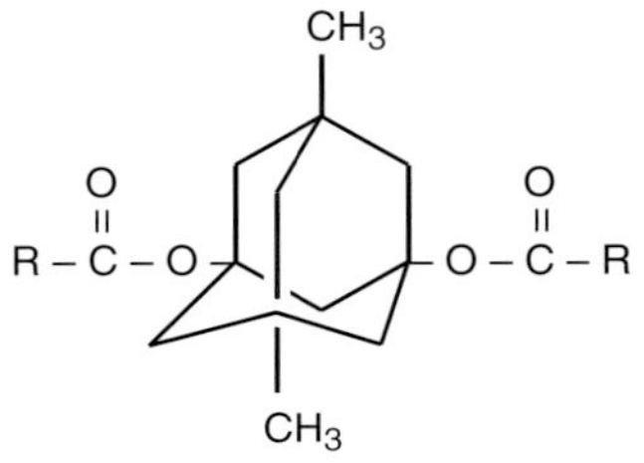

24

Alkylated carbosilanes [112], molecules with alternating Si and C atoms, are somewhat comparable with the tetraalkylsilanes, although of higher viscosity, and are even more stable thermally. Although known for decades, they are still waiting for suitable niches as lubricants, mainly because of problems with their technical scale synthesis.

Aromatic amines, in particular their trialkylsilyl-substituted derivatives **25** have been considered as lubricants because of their good heat resistance [110].

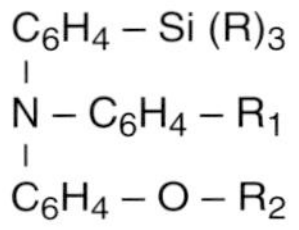

$C_6H_4 - Si(R)_3$
|
$N - C_6H_4 - R_1$
|
$C_6H_4 - O - R_2$

25

The performance of dialkylcarbonates, the diesters of carbonic acid **26**, is similar to that of common dicarboxylic acid esters, and have been reported to be preferable with regard to toxicity and seal compatibility [113]. They are described in more detail by Rudnick and Zecchini [114]. Interesting is that no acid compounds are formed on decomposition.

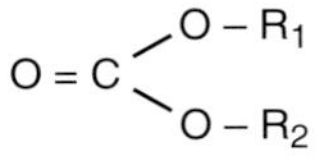

$O = C(O - R_1)(O - R_2)$

26

Heterocyclic boron, nitrogen and phosphorus compounds are ring systems that do not consist only of carbon. It is replaced partially, for example by nitrogen. Triazine derivatives such as melamine **27** and cyanuric acid **28** form the nuclei for fire-resistant and high-temperature lubricants [110,111]. Carbon can be in the minority as for example in polycarboranesiloxanes **29**, that have been reported to be comparable with silicone oils, but can be used at temperatures above 220 °C [115] and carbon can be entirely replaced by nitrogen and boron [116] or by nitrogen and phosphorus, as in the fluorine-containing phosphazene derivatives **30** that have been developed mainly for the use in fire-resistant high-temperature hydraulic fluids [117] and for use as lubricity additives in perfluorinated polyethers [118,119]. Phosphazenes are described in more detail by Singler and Gomba [120].

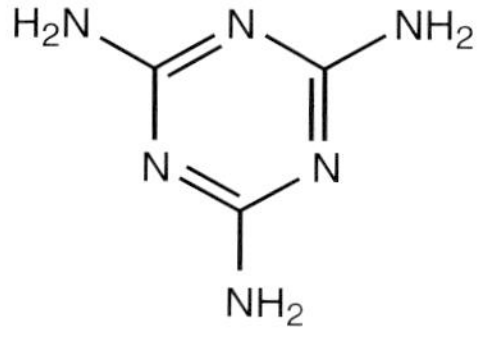

27

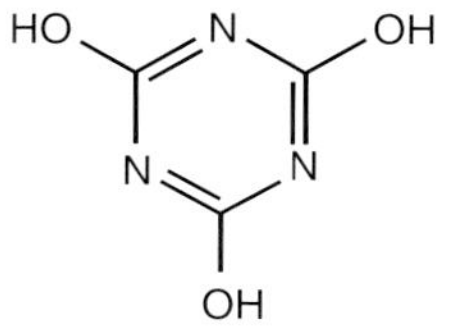

28

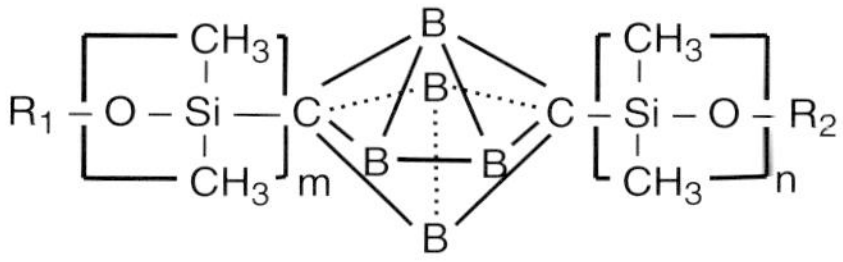

29

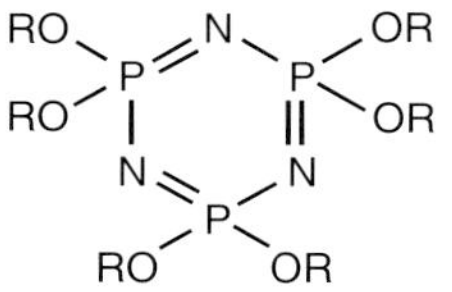

$R = C_6H_4F,\ C_6H_4-CF_3, \cdot\cdot$

30

Room temperature ionic liquids, salts that are liquid at room temperature, attracted interest as lubricants when the preparation of these air- and moisture-stable, neutral products became possible [121]. The vapour pressure of these liquids is too low to measure, and they are non-flammable, thermally stable, have a wide liquid range and have solvating properties for all kinds of material. Depending on the length of the cation side chain and on the choice of the anion, they can be made miscible with water or with organic solvents.

They consist of bulky organic cations, for example 1-alkyl-3-methylimidazolium **31**, and a wide range of anions, from tetrafluoroborate or hexafluorophosphate to organic anions, for example bis-trifluorosulfonimide or tosylate. 1-Butyl-3-methylimidazolium tetrafluoroborate, for example, is a colourless liquid with a viscosity of approximately $100\,mm^2\,s^{-1}$ at room temperature and a freezing point below $-80\,°C$. Ionic liquids can be used – with water as additive – as lubricants for steel–steel contacts [122] or for steel–aluminium contacts [123], with or without water as heat-transfer fluids [124].

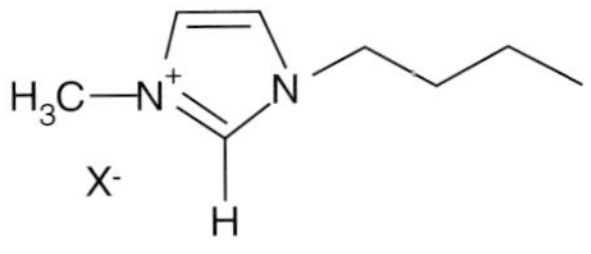

H_3C—N^+ (imidazolium ring) N—butyl; C2—H; X^-

31

Polyalkylenesulfides, or Polythioether Oils are polyalkylene glycols in which all C—O—C bonds are replaced by C—S—C bonds; they have good oxidation stability, and higher viscosities and lower pour points than the corresponding hydrocarbons [125].

Polyphenylsulfides, or polyphenylthioethers, or C-ethers are polyphenylethers in which all C—O—C bonds are replaced by C—S—C bonds. This gives them lower pour points and better boundary lubrication properties, but reduced oxidative and thermal stability [126,127].

Compared with their hydrocarbon analogues, silahydrocarbons, mainly tetraalkylsilanes **32** have lower pour points, lower volatilities, higher VI values up to 155 and superior thermal stability [110,128–130]. They respond to antioxidants and anti-wear additives in the same way as PAOs [131].

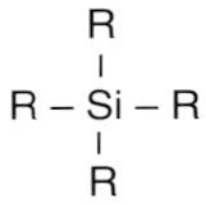

$$SiR_4 \quad R = CH_3,\ C_2H_5 \cdots C_{12}H_{25}$$

32

Silicate or (ortho)silicic acid esters **33** have been known since the middle of the nineteenth century. They have properties that differ strongly from those of siloxanes. Despite their good thermal stability and their low pour points, their use is restricted to hydraulic, heat-transfer and cooling fluids and to oils for automatic weapons because of their poor hydrolytic stability [132]. With the development of polysilicate clusters, the shortcomings of the simple esters can, perhaps, be overcome [133].

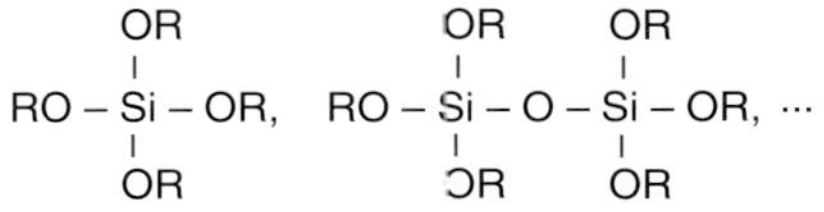

$$Si(OR)_4,\quad (RO)_3Si-O-Si(OR)_3,\ \cdots$$

33

Table 5.3 Performance ranking of base oils.

	A	B	C	D	E	F	Σ
Alkylbenzenes	2.5	1.0	2.5	3.0	2.0	3.0	2.3
Naphthenic mineral oils	2.5	2.0	2.0	2.5	2.0	2.5	2.2
Paraffinic mineral oils	2.0	3.0	2.0	2.5	2.0	2.0	2.3
Hydrocracked mineral oils	1.5	3.0	2.0	2.0	2.0	1.5	2.0
Polyalphaolefines	1.0	1.0	1.0	2.0	1.5	1.5	1.3
Polyolesters	1.5	1.5	2.0	1.0	1.5	1.5	1.5
Poly(alkylene glycol)s	2.0	1.5	2.0	1.5	1.5	1.5	1.7
Silicones	1.0	1.0	1.0	1.0	1.5	1.0	1.1
Perfluorinated polyethers	1.0	1.5	1.0	1.5	1.5	1.5	1.3
Vegetable oils	3.0	2.0	3.0	1.0	1.0	2.0	2.0

1.0 = Excellent, 2.0 = fair, 3.0 = poor.
A = High temperature, B = low temperature, C = ageing.
D = Evaporation loss, E = toxicity, F = VT-behaviour.

5.7 Comparison of Synthetic Base Oils

It is of interest, of course, to compare the performance of synthetic base oils. Because these oils are mostly niche products for very specific applications and their costs are very different, there was some reluctance to publish a general comparison. A performance-related general ranking has, nevertheless, been attempted (Table 5.3) [8].

It is beyond question that – when it comes to detail – there is no simple answer. High-temperature performance is an open field and, fortunately, tribochemistry [134], after two decades of being kind of a sleeping beauty, has again attracted the attention of lubrication specialists, with regard to comparison of engine oils and the role of micelles, for example [135]. Low-temperature performance proves to be equally complex, even with use of low-temperature DSC and NMR as tools [136]. And (non)toxicity, now environmental friendliness, is about to affect the results of comparisons more than in the past [137].

5.8 Mixtures of Synthetic Lubricants

Today, few lubricants contain only one base oil, firstly because mixing two or more base oils with different properties often leads to a lubricant with the desired performance, and secondly because many of the more polar synthetic base oils serve as additives in less polar oils, for example esters in hydrocarbons and vice versa. In thickened systems even immiscible base fluids such as hydrocarbons and complex esters, perfluorinated ethers or polyalkylene glycols can be combined.

References

1 Zisman, W.A. (1962) Historical review, lubricants and lubrication, in *Synthetic Lubricants* (eds R.C. Gunderson and A.W. Hart), Reinhold, New York, pp. 6–60.

2 Neely, A.W., Shankland, R.V., Sullivan, F.W. and Vorhees, N.N. (1931) *Ind. Eng. Chem.*, **23**, 604.

3 Gunderson, R.C. and Hart, A.W. (eds) (1962) *Synthetic Lubricants*, Reinhold, New York.

4 R.L. Shubkin (ed.) (1993) *Synthetic Lubricants and High-Performance Functional Fluids*, Marcel Decker, New York; Rudnik, L.R. and Shubkin, R.L. (eds) (1999) *Synthetic Lubricants and High-Performance Functional Fluids*, Marcel Decker, New York.

5 Rudnick, Leslie R. (ed.) (2013) *Synthetics, Mineral Oils, and Bio-Based Lubricants: Chemistry and Technology*, 2nd edn, CRC Press, Boca Raton.

6 Rao, A. Madhusudhana, Srivastava, S.P., Mehta, K.C. (1987) Synthetic lubricants in India – an overview. *J. Synth. Lubr.*, **4**, 137–145.

7 Rudnik, L.R. and Bartz, W.J. (2013) Comparison of synthetic, mineral oil, and bio-based fluids, in *Synthetics, Mineral Oils, and Bio-Based Lubricants: Chemistry and Technology* (ed. L.R. Rudnik), CRC Press, Boca Raton, pp. 347–366.

8 Lingg, G. (2004) Unconventional Base Oils for Liquid and Semi-Solid Lubricants, 14th International Colloquium, Esslingen, Vol I, pp. 1–4.

9 Beerbower, A. (1984) Environmental Capabilities of Liquid Lubricants, ASLE Special Publication, SP-15, pp. 58–69.

10 Loomis, W.R. (1984) Overview of Liquid Lubricants for Advanced Aircraft, ASLE Special Publication, SP-15, pp. 33–39.

11 Snyder, C.E. Jr. and Gschwender, L.J. (1984) Liquid Lubricants for Extreme Conditions, ASLE Special Publication, SP-15, pp. 70–74.

12 Lansdown, A.R. (1994) *High Temperature Lubrication*, Mechanical Engineering Publications, London, pp. 1–7, 95–142.

13 Sicre, J., Flamand, L., Reynaud, P., Vergne, P. and Godet, M. (1994) Rheological and tribological characterisation of six wet lubricants for space. *J. Synth. Lubr.*, **11**, 35–44.

14 Fusaro, R.L. (1995) Lubrication of space systems. *Lubr. Eng.*, **51**, 182–194.

15 Williamson, E.I. (1985) Commercial developments in synthetic lubricants – a European overview. *J. Synth. Lubr.*, **2**, 329–341.

16 Plagge, A. ((1985) /1987) Gebrauchseigenschaften synthetischer Schmierstoffe und Arbeitsflüssigkeiten. *Tribol. Schmierungstech.*, **32**, 270–278/ 148–156.

17 Rudnik, L.R. (2013) Polyalphaolefins, in *Synthetics, Mineral Oils, and Bio-Based Lubricants: Chemistry and Technology* (ed. L.R. Rudnik), CRC Press, Boca Raton, pp. 3–40.

18 Mühlemeier, J. (1992) Vorteile von synthetischen Schmierstoffen, dargestellt an ausgewählten Beispielen. *Mineralöltechnik*, **37**, 1–4.

19 Nagdi, K. (1987) Wechselbeziehungen zwischen synthetischen Flüssigkeiten und Dichtungsmaterialien. *Tribol. Schmierungstech.*, **34**, 156–169.

20 Bartz, W.J. (1992) Comparison of synthetic fluids. *Lubr. Eng.*, **48**, 765–774; Bartz, W.J. (1993) Vergleich synthetischer Flüssigkeiten. *Mineralöltechnik*, **38**, 9.

21 Jantzen, E. (1995) The origins of synthetic lubricants: the work of Hermann Zorn in Germany Part 1 basic studies of lubricants and the polymerization of olefins. *J. Synth. Lubr.*, **12**, 283–301.

22 Miller, S.A. (1969) *Ethylene and its Industrial Derivatives*, Benn, London.

23 Montgomery, C.W., Gilbert, W.I. and Kline, R.E. (1951) US Patent 2,559,984.

24 Mortier, R.M., Orszulik, S.T. *et al.* (1997) Synthetic base fluids, in *Chemistry and Technology of Lubricants, 2nd* edn (eds R.M. Mortier and S.T. Orszulik), Blackie, London, pp. 34–74.

25 Kumar, G. and Shubkin, R.L. (1993) New polyalphaolefin fluids for specialty applications. *Lubr. Eng.*, **49**, 723–725.

26 Christakudis, D. (1986) Synthetische Schmieröle und ihre Bedeutung für die Herstellung von Motorenölen – Ein Überblick. *Schmierungstechnik*, **17**, 232–237.

27 Rasberger, M. (1997) Oxidative degradation and stabilization of mineral oil based lubricants, in *Chemistry and Technology of Lubricants*, 2nd edn (eds R.M. Mortier and S.T. Orszulik), Blackie, London, pp. 98–143.

28 Kristen, U., Müller, K., Schumacher, R. and Chasan, D. (1984) Aschefreie EP/AW – Additive für PAO, *Ester und Polyalkylenglykole*. International Colloquium, Esslingen, 56, pp. 1–8.

29 van der Waal, G.B. (1989) Properties and applications of ester base fluids and PAO's. *NLGI Spokesman*, **53**, 359–368.

30 Shkolnikov, V.M., Zvetkov, O.N., Chagina, M.A. and Kolessova, G.V. (1990) Improvement of antioxidation and antiwear properties of polyalphaolefin oils. *J. Synth. Lubr.*, **7**, 235–241.

31 Benda, R., Bullen, J. and Plomer, A. (1996) Polyalphaolefins – base fluids for high-performance lubricants. *J. Synth. Lubr.*, **13**, 41–57.

32 Corsico, G., Pacor, P., Ciali, M., Zatta, A. and Gatti, N. *PIO: A New Synthetic Basestock,* Presentation of Mixoil.

33 Navarrini, F., Ciali, M. and Cooley, R. (2013) Polyinternalolefin, in *Synthetics, Mineral Oils, and Bio-Based Lubricants: Chemistry and Technology* (ed. L.R. Rudnick), CRC Press, Boca Raton, pp. 41–50.

34 Casserino, M. and Corthouts, J. (2013) Polybutenes, in *Synthetics, Mineral Oils, and Bio-Based Lubricants: Chemistry and Technology* (ed. L.R. Rudnik), CRC Press, Boca Raton, pp. 273–300.

35 Wilson, B. (1994) Polybutenes – the multipurpose base oil and additive. *Ind. Lubr. Tribol.*, **46**, 3–6.

36 Fotheringham, J. (1995) Hyvis/Napvis polybutene. *Tribol. Schmierungstech.*, **42**, 93–96.

37 Zorn, H., Mueller-Conradi, M. and Rosinski, W. (1930) German Patent 565,249.

38 Wu, M.M. and Ho, S.C. (2013) Alkylated aromatics, in *Synthetics, Mineral Oils, and Bio-Based Lubricants: Chemistry and Technology* (ed. L.R. Rudnik), CRC Press, Boca Raton, pp. 149–168.

39 Sowle, E.D. and Lachocki, T.M. (1996) Linear dialkylbenzenes as synthetic base oils. *Lubr. Eng.*, **52**, 116–120.

40 Hamid, S. (2013) Cyclohydrocarbons, in *Synthetics, Mineral Oils, and Bio-Based Lubricants: Chemistry and Technology* (ed. L.R. Rudnick), CRC Press, Boca Raton, pp. 177–184.

41 Nemes, S., Borbély, J. and Kovacs, M. (1998) Properties of naphthene oligomers polyalkylated with linear 1-alkenes. *J. Synth. Lubr.*, **15**, 285–291.

42 Matzat, N. (1984) Einsatz und Entwicklung von Traktionsflüssigkeiten. International Colloquium, Esslingen, pp. 16.1–16.26.

43 Hata, H. and Tsubouchi, T. (1998) Molecular structures of traction fluids in relation to traction properties. *Tribol. Lett.*, **5**, 69–74.

44 Simon, M. and Vojacek, H. (1998) Entwicklung von Schmierstoffen mit hohen Reibungszahlen unter elastohydrodynamischen Bedingungen, 11th International Colloquium, Esslingen, Vol I, 3.4, pp. 359–372.

45 Venier, C.G. (2013) Alkylcyclopentanes, in *Synthetics, Mineral Oils, and Bio-Based Lubricants: Chemistry and Technology* (ed. L.R. Rudnik), CRC Press, Boca Raton, pp. 259–272.

46 Venier, C.G., Casserly, E.W. and Gunsel, S. (1992) Tris(2-octyldodecyl) cyclopentane, a low volatility, wide liquid range, hydrocarbon fluid. *J. Synth. Lubr.*, **9**, 237–252.

47 Squicciarini, M.P., Heilman, W.J. and Bremmer, M.L. (1996) Evaluation of tetraalkylmethanes as synthetic lubricants. *Lubr. Eng.*, **52**, 111–114.

48 Podehradskà, J., Vodicka, L. and Stepina, V. (1989) Synthesis and properties of adamantane synthetic lubricants. *J. Synth. Lubr.*, **6**, 123–131.

49 Ashton, W E. and Strack, C.A. (1962) Chlorofluorocarbon polymers, in *Synthetic Lubricants* (eds R.C. Gunderson and A.W. Hart), Reinhold, New York, pp. 246–263.

50 Epstein, R.M. and Ferstandig, L.L. (2013) Chlorotrifluoroethylene, in *Synthetics, Mineral Oils, and Bio-Based Lubricants: Chemistry and Technology* (ed. L.R. Rudnik), CRC Press, Boca Raton, pp. 203–212.

51 Gschwender, L.J., Mattie, D., Snyder, C.E. and Warner, W.M. (1992) Chlorotrifluoroethylene oligomer based nonflammable hydraulic fluid. Part 1, fluid, additive, and elastomer development. *J. Synth. Lubr.*, **9**, 187–202.

52 Jantzen, E. (1995) The origins of synthetic lubricants: the work of Hermann Zorn in Germany Part 2 esters and additives for synthetic lubricants. *J. Synth. Lubr.*, **12**, 283–301.

53 Wildersohn, M. (1985) Esteröle – Struktur und chemisch–physikalische Eigenschaften. *Tribol. Schmierungstech.*, **32**, 70–78.

54 Riemenschneider, W. (1987) Organic esters, in *Ullmann's Encyclopedia of Industrial Chemistry*, vol. **A 9**, Wiley-VCH, Weinheim, pp. 565–585.

55 Catogan, D.F. and Howick, C.J. (1992) Plasticizers, in *Ullmann's Encyclopedia of Industrial Chemistry*, vol. **A 20**, Wiley-VCH, Weinheim, pp. 139–458.

56 Boyde, S. and Randles, S.J. (2013) Esters, in *Synthetics, Mineral Oils, and Bio-Based Lubricants: Chemistry and Technology* (ed. L.R. Rudnik), CRC Press, Boca Raton, pp. 51–80.

57 Dudek, W.G. and Popkin, A.H. (1962) Dibasic acid esters, in *Synthetic Lubricants* (eds R.C. Gunderson and A.W. Hart), Reinhold, New York, pp. 151–245.

58 Hurd, P.W. (1996) The chemistry of castor oil and its derivatives. *NLGI Spokesman*, **60**, 14–23.

59 Feßenbecker, A. and Roehrs, I. (1997) A new additive for the hydrolytic and oxidative stability of ester based lubricants and greases. *NLGI Spokesman*, **61**, 310–317.

60 Smith, T.G. (1962) Neopentyl Polyol Esters, in *Synthetic Lubricants* (eds R.C. Gunderson and A.W. Hart), Reinhold, New York, pp. 388–401.

61 Koch, B. and Jantzen, E. (1987) Thermal oxidation behaviour of synthetic oils: basic studies in the polymerization of polyol ester oils. *J. Synth. Lubr.*, **4**, 321–335.

62 Tianhui, Ren, Jian, Xia, Longzhen, Zheng, Junyan, Zhang, Weimin, Liu and Qunji, Xue (1999) A tribological study of an S,P-containing benzotriazole derivative as an additive in pentaerythritol ester. *J. Synth. Lubr.*, **15**, 263–269.

63 Bongardt, F. (1990) Einfluß der chemischen Struktur auf das Luftabscheidevermögen und die hydrolytische Stabilität von Estern. *Fat Sci. Technol.*, **92**, 473–478.

64 Venkataramani, P.S., Kaira, S.L., Raman, S.V. and Srivastava, H.C. (1988) Synthesis evaluation and applications of complex esters as lubricants: a basic study. *J. Synth. Lubr.*, **5**, 271–289.

65 Schmidt, H.-G. (1994) Komplexester aus pflanzlichen Ölen, 9th International Colloquium, Esslingen, Vol II, pp. 2.2-1–2.2-9.

66 Wagner, H., Luther, R. and Mang, T. (2001) Lubricant base fluids based on renewable raw materials: their catalytic manufacture and modification. *Appl. Catal. A-Gen.*, **221**, 429–442.

67 Kolwzan, B. and Gryglewicz, S. (2003) Synthesis and biodegradability of some adipic and sebacic esters. *J. Synth. Lubr.*, **20**, 99–107.

68 Wallfahrer, U. (2013) Polymer esters, in *Synthetics, Mineral Oils, and Bio-Based Lubricants: Chemistry and Technology* (ed. L.R. Rudnik), CRC Press, Boca Raton, pp. 109–122.

69 Jelitte, R. and Beyer, C. (1993) Synthetic fluids between synthetic hydrocarbons and ester fluids: Co-oligomers of alphaolefins and alkylmethacrylates. *J. Synth. Lubr.*, **10**, 285–307.

70 Wallfahrer, U. (1996) High viscosity polymer esters as an alternative to PAO 100. *J. Synth. Lubr.*, **13**, 263–297.

71 Murphy, C.M. (1962) Fluoroesters, in *Synthetic Lubricants* (eds R.C. Gunderson and A.W. Hart), Reinhold, New York, pp. 361–387.

72 Denis, J. (1984) The relationships between structure and rheological properties of hydrocarbons and oxygenated compounds used as base stocks. *J. Synth. Lubr.*, **1**, 201–238.

73 Luksa, A. (1990) Viscosity properties of base lubricating oils containing mineral oil, synthetic ester oil and polymeric additives. *J. Synth. Lubr.*, **7**, 187–192.

74 Vilics, T., Ciorbagiu, F. and Ciucu, B. (1996) Ester base stocks for synthetic lubricants. *J. Synth. Lubr.*, **13**, 289–296.

75 Hatton, R.E. (1962) Phosphate esters, in *Synthetic Lubricants* (eds R.C. Gunderson and A.W. Hart), Reinhold, New York, pp. 103–150.

76 Marino, M.P. (1993) Phosphate esters, in *Synthetic Lubricants and High-Performance Functional Fluids* (ed. R.L. Shubkin), Marcel Decker, New York, pp. 67–100.

77 Phillips, W.D., Placek, D.C. and Marino, M.P. (2013) Neutral phosphate esters, in *Synthetics, Mineral Oils, and Bio-Based Lubricants: Chemistry and Technology* (ed. L.R. Rudnick), CRC Press, Boca Raton, pp. 81–108.

78 Okazaki, M.E. and Abernathy, S.M. (1993) Hydrolysis of phosphate-based aviation hydraulic fluids. *J. Synth. Lubr.*, **10**, 107–118.

79 Phillips, W.D. and Sutton, D.I. (1997) Improved maintenance and life extension of phosphate esters using ion-exchange treatment. *J. Synth. Lubr.*, **13**, 225–261

80 Roberts, F.H. and Fife, H.R. (1947) US Patent 2,425,755.

81 Gunderson, R.C. and Millett, W.H. (1962) Polyglycols, in *Synthetic Lubricants* (eds R.C. Gunderson and A.W. Hart), Reinhold, New York, pp. 61–102.

82 Kussi, S. (1985) Chemical, physical and technological properties of polyethers as synthetic lubricants. *J. Synth. Lubr.*, **2**, 63–84; Kussi, S. (1986) Eigenschaften von Basisflüssigkeiten für synthetische Schmierstoffe. *Tribol. Schmierungstech.*, **33**, 33–39.

83 Matlock, P.L. and Clinton, N.A. (1993) Polyalkylene glycols, in *Synthetic Lubricants and High-Performance Functional Fluids* (ed. R.L. Shubkin), Marcel Decker, New York, pp. 101–123.

84 Greaves, M.R. (2013) Polyalkylene glycols, in *Synthetics, Mineral Oils, and Bio-Based Lubricants: Chemistry and Technology* (ed. L.R. Rudnick), CRC Press, Boca Raton, pp. 123–148.

85 Gumprecht, W.H. (1965) PR-143 – a new class of high-temperature fluids. *ASLE Trans.*, **9**, 24–30.

86 Schwickerath, W. (1987) Perfluorether – Basisflüssigkeit für Hochtemperaturschmierstoffe. *Tribol. Schmierungstech.*, **34**, 22–38.

87 DelPesco, T. W. (1993) Perfluoroalkylpolyethers, in *Synthetic Lubricants and High-Performance Functional Fluids* (ed. R.L. Shubkin), Marcel Decker, New York, pp. 145–172.

88 Walther, H.C., Bell, G.A. and Howell, J.L. (2013) Perfluoroalkylpolyethers, in *Synthetics, Mineral Oils, and Bio-Based Lubricants: Chemistry and Technology* (ed. L.R. Rudnick), CRC Press, Boca Raton, pp. 185–202.

89 Corti, C. and Savelli, P. (1992) Perfluoropolyether lubricants. *J. Synth. Lubr.*, **9**, 311–330.

90 Maccone, P., Di Niccolò, E. and Boccaletti, G. (2006) Viscosity dependence on temperature for PFPE lubricants: An empirical approach. *NLGI Spokesman*, **69**, 28–35.

91 Wunsch, F. (1991) Perfluoralkylether. *Tribol. Schmierungstech.*, **38**, 130–136.

92 Carré, D.J. (1990) Perfluoropolyalkylether lubricants under boundary conditions: Iron catalysis of lubricant degradation. *J. Synth. Lubr.*, **6**, 3–15.

93 Shogrin, B., Jones, W.R. and Herrera-Fierro, P. (1996) Spontaneous dewetting of a perfluoropolyether. *Lubr. Eng.*, **52**, 712–717.

94 Helmick, L.S. and Sharma, S.K. (1996) Effect of humidity on friction and wear for a linear perfluoropolyether fluid under boundary lubrication conditions. *Lubr. Eng.*, **52**, 437–442.

95 Koch, B. and Jantzen, E. (1995) Thermo-oxidative behaviour of

perfluoropolyalkylethers. *J. Synth. Lubr.*, **12**, 191–204.

96 Jones, W.R., Ajayi, O.O. and Wedeven, L.D. (1997) Enhancement of perfluoropolyether boundary lubrication performance. *Lubr. Eng.*, **53**, 24–31.

97 Mahoney, C.L. and Barnum, E.R. (1962) Polyphenyl ethers, in *Synthetic Lubricants* (eds R.C. Gunderson and A.W. Hart), Reinhold, New York, pp. 402–463.

98 Hamid, S. and Burian, S.A. (2013) Polyphenylether lubricants, in *Synthetics, Mineral Oils, and Bio-Based Lubricants: Chemistry and Technology* (ed. L.R. Rudnick), CRC Press, Boca Raton, pp. 169–176.

99 Herber, J.F., Joaquim, M.E. and Adams, T. (2001) Polyphenyl ethers: lubrication in extreme environments. *Lubr. Eng.*, **57**, 9–14.

100 Stone, D.S. and Joaquim, M.E. (2002) The many roles of polyphenyl ethers, *Sensors*, pp. 1–6.

101 Noll, W. (1968) *Chemie und Technologie der Silicone (Chemistry and Technology of Silicones)*, edn. 2, Verlag Chemie, Weinheim.

102 Awe, R.W. and Schiefer, H.M. (1962) Silicones, in *Synthetic Lubricants* (eds R.C. Gunderson and A.W. Hart), Reinhold, New York, pp. 264–322.

103 Demby, D.H., Stoklosa, S.J. and Gross, A. (1993) Silicones, in *Synthetic Lubricants and High-Performance Functional Fluids* (ed. R.L. Shubkin), Marcel Decker, New York, pp. 183–203.

104 Perry, R., Quinn, C., Traver, F. and Murthy, K. (2013) Silicones, in *Synthetics, Mineral Oils, and Bio-Based Lubricants: Chemistry and Technology* (ed. L.R. Rudnick), CRC Press, Boca Raton, pp. 213–226

105 Huber, P. (1985) Siliconöle. *Tribol. Schmierungstech.*, **32**, 64–69; Huber, P. and Kaiser, W. (1986) Silicone fluids: synthesis, properties and applications. *J. Synth. Lubr.*, **3**, 105–120.

106 Moretto, H.-H., Schulze, M. and Wagner, G. (1993) Silicones, in *Ullmann's Encyclopedia of Industrial Chemistry*, vol. **A 24**, Wiley VCH, Weinheim, pp. 57–93.

107 Kopsch, H. and Ströfer, W. (1991) Ermittlung des Alterungsverhaltens thermooxidativ beanspruchter Silikonöle. *Tribol. Schmierungstech.*, **38**, 36–42.

108 Schiefer, H.M., Azzam, H.T. and Miller, J.W. (1969) Industrial fluorosilicone applications predicted by laboratory tests. *Lubr. Eng.*, **25**, 210–220.

109 Jungk, M. and Hesse, D. (1998) Silicone oil-based fluids as a tool to tailor high-performance lubricating greases. *Eurogrease*, 31–35.

110 Baum, G. (1962) Novel synthetic lubricants, in *Synthetic Lubricants* (eds R.C. Gunderson and A.W. Hart), Reinhold, New York, pp. 464–489.

111 Stepina, V. and Vesely, V. (1992) *Lubricants and Special Fluids*, Elsevier, Amsterdam, pp. 186–191.

112 Fritz, G. and Matern, E. (1986) *Carbosilanes, Syntheses and Reactions*, Springer, Berlin.

113 Fisicaro, G. and Gerbaz, G. (1993) Dialkylcarbonates, in *Synthetic Lubricants and High-Performance Functional Fluids* (ed. R.L. Shubkin), Marcel Decker, New York, pp. 229–239.

114 Rudnick, L.R. and Zecchini, C. (2013) Dialkyl carbonates, in *Synthetics, Mineral Oils, and Bio-Based Lubricants: Chemistry and Technology* (ed. L.R. Rudnik), CRC Press, Boca Raton, pp. 245–258.

115 Gmelin Borverbindungen 6 (1974) 4.9.4 110.

116 Klamann, D. (1984) *Lubricants and Related Products*, Verlag Chemie, Weinheim, pp. 151–153.

117 Singler, R.E. and Bieberich, M.J. (1993) Phosphazenes, in *Synthetic Lubricants and High-Performance Functional Fluids* (ed. R.L. Shubkin), Marcel Decker, New York, pp. 215–228.

118 Zhao, Q. *et al.* (1999) Tribological study of phosphazene-type additives in perfluoropolyether lubricant for hard disk applications. *Lubr. Eng.*, **55**, 16–21.

119 Kang, H.J. *et al.* (1999) The use of cyclic phosphazene additives to enhance the performance of the head/disk interface. *Lubr. Eng.*, **55**, 22–27.

120 Singler, R.E. and Gomba, F.J. (2013) Phosphazenes, in *Synthetics, Mineral*

Oils, and Bio-Based Lubricants: Chemistry and Technology (ed. L.R. Rudnik), CRC Press, Boca Raton, pp. 235–244.

121 Wilkes, J.S. and Zaworotko, W.J. (1992) Air and water stable 1-ethyl-3-methylimidazolium based ionic liquids. *Chem. Commun.*, 965–967.

122 Lu, Q., Wang, H., Ye, C., Liu, W. and Xue, Q. (2004) Room temperature ionic liquid 1-ethyl-3-hexylimidazolium-bis (trifluoromethylsulfonyl)-imide as lubricant for steel–steel contact. *Tribol. Int.*, **37**, 547–552.

123 Chen, Y., Ye, C., Wang, H. and Liu, W. (2003) Tribological performance of an ionic liquid as a lubricant for steel/aluminum contacts. *J. Synth. Lubr.*, **20**, 217–225.

124 Van Valkenburg, M.E., Vaughn, R.L., Williams, M. and Wilkes, J.S. (2005) Thermochemistry of ionic liquid heat-transfer fluids. *Thermochim. Acta*, **425**, 181–188.

125 Klamann, D. (1984) *Lubricants and Related Products*, Verlag Chemie, Weinheim, p. 116.

126 Stepina, V. and Vesely, V. (1992) *Lubricants and Special Fluids*, Elsevier, Amsterdam, pp. 161–162.

127 Lansdown, A.R. (1994) *High Temperature Lubrication*, Mechanical Engineering Publications, London, p. 132.

128 Pettigrew, F.A. and Nelson, G.E. (1993) Silahydrocarbons, in *Synthetic Lubricants and High-Performance Functional Fluids* (ed. R.L. Shubkin), Marcel Decker, New York, pp. 205–214.

129 Snyder, C.E. Jr, Gschwender, L.J., Tamborski, C., Chen, G.J. and Anderson, D.R. (1982) Synthesis and characterization of silahydrocarbons – a class of thermally stable wide-liquid-range functional fluids. *ASLE Trans.*, **25**, 299–308.

130 Snyder, C.E. and Pettigrew, F.A. (2013) Silahydrocarbons, in *Synthetics, Mineral Oils, and Bio-Based Lubricants: Chemistry and Technology* (ed. L.R. Rudnik), CRC Press, Boca Raton, pp. 227–234.

131 Thomas, S.G., Campbell, D.G. and Hsu, C.J. (1993) Improvement of silahydrocarbon performance with additives. *J. Synth. Lubr.*, **10**, 195–212.

132 Hatton, R.E. (1962) Silicate esters, in *Synthetic Lubricants* (eds R.C. Gunderson and A.W. Hart), Reinhold, New York, pp. 323–360.

133 Stepina, V. and Vesely, V. (1992) *Lubricants and Special Fluids*, Elsevier, Amsterdam, pp. 178–180.

134 Heinicke, G. (1984) *Tribochemistry*, Oxford University Press, Oxford.

135 Pawlak, Z. (2003) *Tribochemistry of Lubricating Oils*, Elsevier Science, Amsterdam.

136 Boyde, S. (2001) Low-temperature characteristics of synthetic fluids. *Synth. Lubr.*, **18**, 99–114.

137 Hörner, D. (2002) Recent trends in environmentally friendly lubricants. *J. Synth. Lubr.*, **18**, 327–347.

6
Additives

Jürgen Braun

Base fluids – mineral oil and also synthetic products – generally cannot satisfy the requirements of high performance lubricants without using the benefit of modern additive technology. Additives are synthetic chemical substances that can improve lots of different parameters of lubricants. They can boost existing properties, suppress undesirable properties and introduce new properties in the base fluids.

Additives can be classified regarding different aspects. Important and helpful for the understanding of additives is the following differentiation that takes into consideration which part of the tribo system is influenced by the additives. According to this consideration additives can be classified into types that

1) influence the physical and chemical properties of the base fluids
   - physical effects: for example VT characteristics, demulsibility, low temperature properties,
   - chemical effects: for example oxidation stability
2) affect primarily the metal surfaces modifying their physicochemical properties, for example reduction of friction, increase of EP behaviour, wear protection, corrosion inhibition.

Additives are used at treat rates of a few ppm (antifoam agents) up to 20 or even more weight percentages. They can assist each other (synergism) or they can lead to antagonistic effects. For instance, anticorrosion additives with their high surface activity compete with other polar additives like antiwear and extreme pressure additives for the metal surface and can therefore reduce their efficiency. Thus a FZG damage load stage of >12 can be reduced down to a damage load stage of 8 if highly efficient and polar anticorrosion additives are added. Some additives are multifunctional products that decrease the possibility of additives interfering with each other negatively.

Towards the end of a development process of a new lubricant or if an existing lubricant has to be adjusted to new requirements often one specific property remains to be adjusted or optimized. In such cases so-called 'boosters' are readily recommended to improve this specific property. Due to possible negative

Lubricants and Lubrication, Third Edition. Edited by Theo Mang and Wilfried Dresel.

antagonistic effects such boosters have to be scrutinized very carefully to ensure that the other properties of the original formulation would not be influenced negatively.

For the evaluation of additives, many specific tests have been developed in the course of time and stipulated in national, international or in-house test standards. These test procedures in the chemical lab or dynamic test field should be able to simulate their dedicated use in the field. In order to get results in acceptable time, the tests have to be run at more severe conditions compared to the real application to accelerate the investigation and get a kind of fast motion. This is done typically by increasing test parameters like temperature, catalysts, load, and so on or by using unfavourable geometries in tribological tests. Although the necessity of the fast motion effect is well perspicuous and definitely indispensable the test conditions have to consider the real application and should not distance themselves too far from the real conditions of the tribological application. As higher and more exaggerated the stress in a test is as more carefully the results have to be regarded. Finally, new thoroughly tested formulations should be tested in controlled field tests before being introduced in the general market to avoid unpredictable problems in the 'real life'.

Also additives have a very big influence on the performance of lubricants that make it possible to fulfil new performance levels, of course there are some properties that cannot be influenced by additives, for example volatility, air release properties, thermal stability, thermal conductivity, compressibility, boiling point.

Although well balanced and optimized additive systems can improve the performance of lubricants enormously, the formulation of high performance lubricants requires also excellent high-quality base fluids. The present trend to use more and more hydrocracked and severely hydrotreated highly refined mineral oils as well as to synthetic esters and PAOs underlines this statement.

6.1 Antioxidants

Industrial lubricants must fulfil many different functions, for example power transmission and wear protection just to name two of them. Mostly the function of a lubricant is limited by the ageing of the lubricant base stock. Typical characteristics of aged lubricants are discoloration and a characteristic burnt odour. In advanced stages the viscosity will begin to rise significantly, acidic oxidation products are build, which in turn may induce corrosion and lubricant problems. This ageing process can be delayed tremendously by the use of antioxidants.

6.1.1 Mechanism of Oxidation and Antioxidants

The ageing of lubricants can be differentiated into two processes: the oxidation process by reaction of the lubricant molecules with oxygen and the thermal

Initiation

$RH \xrightarrow{+O_2} R\bullet$

Chain propagation

$R\bullet + O_2 \longrightarrow ROO\bullet$

$ROO\bullet + RH \longrightarrow ROOH + R\bullet$

Chain branching

$ROOH \longrightarrow RO\bullet + \bullet OH$

$RO\bullet + RH \longrightarrow ROH + R\bullet$

$\bullet OH + RH \longrightarrow H_2O + R\bullet$

Termination

$2\,R\bullet \longrightarrow R - R$

$R\bullet + ROO\bullet \longrightarrow ROOR$

$2\,ROO\bullet \longrightarrow ROOR + O_2$

Figure 6.1 Mechanism of autoxidation.

decomposition (cracking) at high temperatures. In practice, the oxidative ageing of the lubricant is the dominating process which influences significantly the lifetime of the lubricant. Caused by steadily increased power density and reduced lubricant volumes (higher load-to-oil ratio) as well as extended service life in the last years, the thermal stress on the lubricant molecules grows constantly.

The oxidation of hydrocarbons can be described by the well-known free radical mechanism via alkyl and peroxy radicals [1]. The main reaction steps are shown in Figure 6.1.

The initiation of the so-called autoxidation consists in the hydrogen abstraction of the hydrocarbon by oxygen attack that will lead to the formation of an alkyl radical. The alkyl radical can react with oxygen to form an alkyl peroxy radical. The next step in the chain propagation scheme is the hydrogen abstraction by a peroxy radical from another hydrocarbon that will lead to a hydroperoxide and an alkyl radical which can again react with oxygen as described above. The difference in reactivity of the miscellaneous radicals explains why linear and unbranched hydrocarbons exhibit a much higher oxidation stability compared to branched, aromatic and unsaturated hydrocarbons. Further steps are chain branching (homolytic cleavage of hydroperoxides, increasing number of reactive free radicals, autocatalytic phase of the autoxidation) and the termination of the radical chain reaction by recombination of two radicals to yield unreactive, non-radical species.

The typical oxidation products that will be formed by these oxidation processes are alkylhydroperoxides (ROOH), dialkyl peroxides (ROOR′), alcohols (ROH), aldehydes (RCHO), ketones(RR′C=O), carboxylic acids (RCOOH), esters (RCOOR′) and so forth. By polycondensation processes high molecular weight oxidation products are formed. These products are responsible for the typical viscosity increase of aged oil. Further polycondensation and polymerization of these still oil soluble oxidation products lead finally to oil-insoluble polymers that can be observed as sludge and varnish-like deposits.

Because of the acidic character of most of the oxidation products the danger of corrosion is increased. Also the attack of alkylperoxy radicals on the metal surface may be responsible for corrosive wear. Furthermore such dissolved metals can form salts that also precipitate as sludge.

Non-hydrocarbon-based lubricants may behave totally different. Thus the ageing of polyalkylene glycols will lead to a decreasing of the viscosity caused by the decomposition of the polymeric structure.

Normally, refined mineral base stocks contain traces of nitrogen-, sulfur- and oxygen-containing heterocycles as well as mercaptans (RSH), thioethers(RSR) and disulfides (RSSR) that may act as so-called 'natural antioxidants' or as pro-oxidants that will accelerate the oxidation of the lubricant. Usually, the antioxidants are dominating thus providing a relatively good oxidation stability of the base oil. On the other hand, hydrocracked base stocks (API group II and III) as well as synthetic lubricants like PAOs (API group IV) demand a well-balanced antioxidant combination in order to benefit the advantages of that kind of lubricants [2–4].

Antioxidants can be differentiated as primary antioxidants (radical scavengers) and secondary antioxidants (peroxide decomposers).

Radical scavengers compete successfully with the lubricant molecules in the reaction with reactive radicals of the propagation process. They react preferably with the radical oxidation products forming resonance-stabilized radicals that are so unreactive that they will stop the propagation of the autoxidation. Peroxide decomposers convert hydroperoxides into non-radical products thus also preventing the chain propagation reaction.

In finding the optimum synergistic antioxidant and metal deactivator combination, the maximum delay of oil oxidation is achieved.

6.1.2
Compounds

6.1.2.1 Phenolic Antioxidants

Sterically hindered mono-, di- and polynuclear phenol derivatives belong to the most effective antioxidants acting as radical scavengers and are used in many applications. Typically, the phenols are substituted in the two and six positions with tertiary alkyl groups. The most common substituent is the tertiary butyl group. The most simple derivatives are 2,6-DTB (2,6-di-tert-butylphenol) **1** and BHT (butylated hydroxytoluene, 2,6-di-tert-butyl-4-methylphenol) **2**. The advantage of polynuclear phenols like 4,4′-methylenebis(2,6-di-tert-butylphenol) **3** or types with high-molecular mass substituents in the four position is the reduced volatility due to the higher molecular weight that makes these products suitable for high-temperature applications. Phenolic antioxidants can form degradation by-products (e.g. quinones) with intense yellow or red colours. Hindered phenols with large substituents in the para position are less prone to this effect, because the reaction pathway is more hampered.

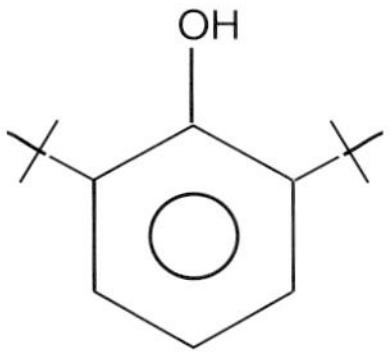

OH

1

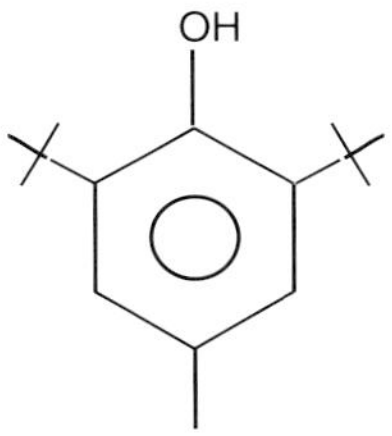

OH

2

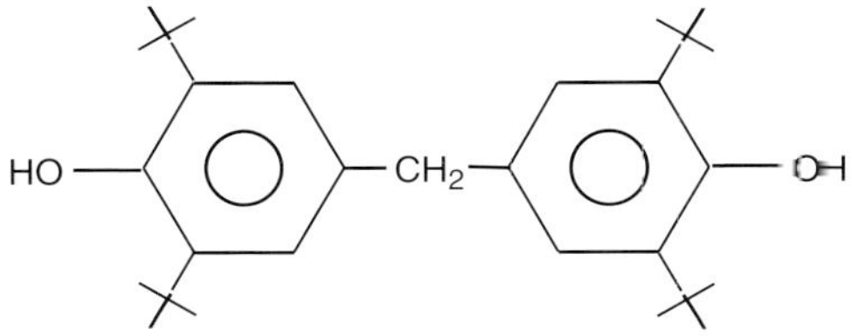

HO — CH$_2$ — OH

3

6.1.2.2 Aromatic Amines

Oil-soluble secondary aromatic amines represent another important class of antioxidants that act as radical scavengers. Typical products are a large number of alkylated diphenylamines **4**, N-phenyl-1-naphthylamine (PANA) **5** and the polymeric 2,2,4-trimethyldihydroquinoline (TMQ) **6**. Because of its poor solubility in mineral oil the latter is commonly used in greases and polar lubricants.

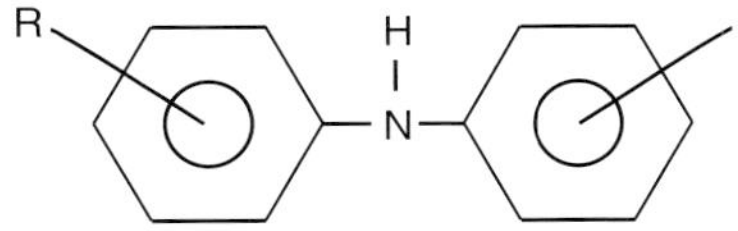

R H N R

4

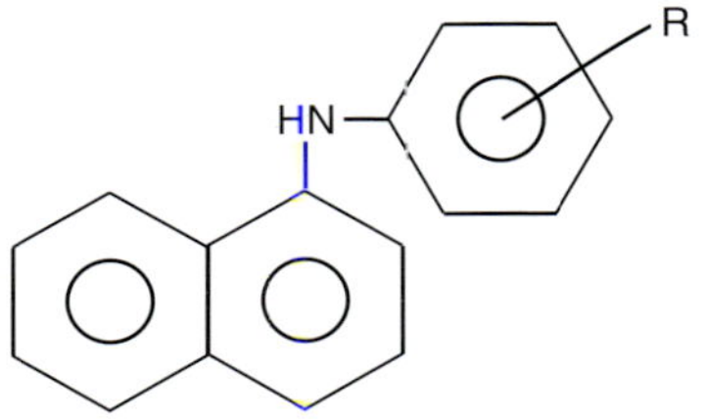

HN R

5

CH_3 H H CH_3 CH_3 N H n

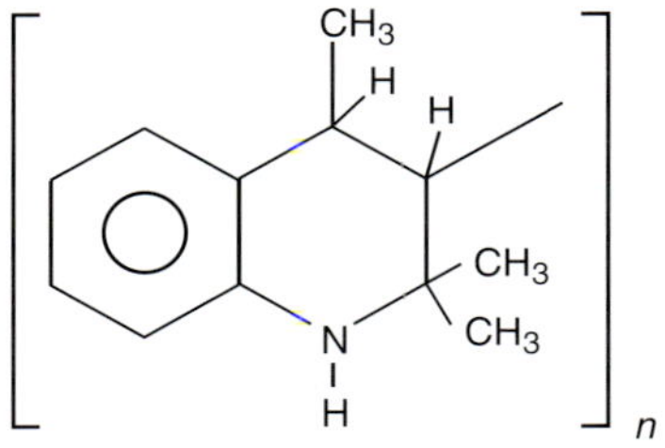

n = 9–10

6

At moderately elevated temperatures ($<120\,^{\circ}$C) one diphenylamine molecule can eliminate four peroxy radicals whereas a monophenol can only scavenge two peroxy radicals. Under high-temperature conditions it could have been shown that aminic antioxidants are even still more superior to their phenolic counterparts [5]. Aminic antioxidants, on the other hand, tend to form strongly coloured, dark brown degradation by-products (polyconjugated systems) finally resulting in dark coloured oils even after relatively short usage. Aminic antioxidants, especially unsubstituted PANA, also have a greater tendency to form sludge than phenolic antioxidants [6].

6.1.2.3 Compounds Containing Sulfur and Phosphorus

The most famous representatives of this group of additives are the zinc dithiophosphates which mainly act as radical scavengers [5]. Because of their multifunctional properties (antioxidant, antiwear additive, extreme pressure additive, corrosion inhibitor) zinc dithiophosphates retain their dominating role as standard additives since the early days of modern tribology in the beginning of this century.

A survey of the antioxidant properties of primary, secondary and aryl zinc dithiophosphates is given in Section 6.8.

Beside the well established group of metal dithiophosphates there are also a large number of ashless dithiophosphoric acid derivatives, the so-called O,O,S-triesters, which are reaction products of dithiophosphoric acid with olefins, cyclopentadiene, norbornadiene, α-pinene, polybutene, unsaturated esters like acrylic acid esters, malenic acid esters and other chemicals with activated double bonds [5]. All these additives show antiwear and also antioxidant properties, although their performance as antioxidant is not really as good as that of the metal dithiophosphates.

6.1.2.4 Organosulfur Compounds

These products are typical peroxide decomposers. Numerous types of organosulfur compounds have been proposed as antioxidants: dialkyl sulfides, diaryl sulfides, polysulfides, modified thiols, thiophene derivatives, xanthates, thioglycols, thioaldehydes, sulfur containing carboxylic acids, heterocyclic sulfur nitrogen compounds like dialkyldimercaptothiadiazoles, 2-mercaptobenzimidazoles and others.

Zinc- and methylenebis(dialkyldithiocarbamates) have been found to be highly efficient [5]. See also Section 6.8.3.

6.1.2.5 Organophosphorus Compounds

Beyond this group triaryl- and trialkylphosphites are the most common types. Moreover they are not only peroxide decomposers, but also they can limit photodegradation. Because of their relatively poor hydrolytic stability their application is restricted to sterically hindered derivatives.

6.1.2.6 Other Compounds

Surprisingly organocopper compounds in combination with peroxide decomposers can act as antioxidants, although catalytic quantities of copper ions usually act as pro-oxidants [5]. Overbased phenates and salicylates of magnesium and calcium behave as antioxidants at higher temperatures. Many compounds have been proposed as antioxidants but none of these have gained real relevance.

6.1.2.7 Synergistic Mixtures

Commonly several types and combinations of antioxidants are used. Well known is the synergistic effect when aminic and phenolic antioxidants are combined. This effect is based on the ability of the phenol to regenerate the more efficient aminic antioxidant [7]. The combination of radical scavengers and peroxide decomposers is called heterosynergism. So the combination of phenolic antioxidants with phosphites is known as highly efficient especially in hydrotreated base stock [8]. As metals may act as catalysts in the oxidation process, typical rust inhibitors and metal passivators are also used as synergistic compounds (see also Section 6.10).

6.1.3 Testing of the Oxidation Stability

For evaluation of antioxidants, several bench tests have been established: turbine oil oxidation stability test (TOST) (ISO 4263 1-3, ASTM-D 943, ASTM-D 4310), rotating pressure vessel oxidation test (RPVOT, formerly known as rotary bomb test, ROBOT) (ASTM-D 2272), IP 48, to name just a few. A detailed list of methods can be found in Chapter 18.

All tests are made under severe conditions by increasing the temperature, using catalysts (iron, copper) and high exposure to oxygen in order to shorten the time required to decompose the lubricant. Depending on the given set of oxidation conditions the antioxidants will exhibit different performance in different tests.

Therefore, a variety of different oxidation tests are necessary to obtain an actual description of the lubricant's oxidation stability and the performance of the antioxidants. On the other hand, a surprisingly good correlation of TOST and RPVOT results could have been shown recently for a given steam turbine formulation in different base stocks [9].

Basically, oxidation tests can be differentiated into two groups: one kind of tests describe the condition of the lubricant after a defined test period by measuring several ageing indicating parameters like acid number, viscosity change and sludge formation. The 1000-h TOST test according to ISO 4263-1-appendix C and IP 280 oxidation test are examples of that kind of test. Another group of oxidation tests measures the so-called induction time (time from the beginning of the oxidation to the autocatalytic phase of the autoxidation) by recording constantly or after defined periods of time the ageing indicating parameters. Typical examples of this group are the lifetime TOST test according to ISO 4263-1 (~ ASTM-D 943), the chemiluminescence method (DIN 51835-1) or the rotating pressure vessel oxidation test (ASTM-D 2272), where the drop of oxygen pressure indicates the ageing of the lubricant. More recently, the high pressure differential scanning calorimetry (HPDSC) and the sealed capsule differential scanning calorimetry (SCDSC) have seen more application measuring the stability of different antioxidants and formulations [10,11].

6.2 Viscosity Modifiers

6.2.1 Physical Description of Viscosity Index

A fundamental characteristic of every fluid is its viscosity (see Chapter 3). The kinematic viscosity of a fluid is dependent on the external parameters of pressure and temperature. A great number of applications specify viscosity at a defined temperature, in the case of hydraulic equipment at 40 and 100 °C, to achieve optimum pump efficiency [12].

The effect of temperature on a fluid can be illustrated double-logarithmically according to Walter (see Section 3.2), whereby the gradient defines the viscosity index. The viscosity index is defined exactly by the gradient between 40 and 100 °C.

This simple-to-calculate characteristic has a significant impact on lubrication technology and is practically a standard feature of every specification. The viscosity index defines molecular structure and has been exactly described for defined molecules [13].

Depending on their source, conventionally refined base oils (API group I) display between 85 and 100 VI points and the most common, central European paraffinic solvent cuts have VIs of around 95 points. Modern API group II and III base oils exhibit higher "natural" VIs between 100 and 135 (see Chapter 4).

6.2.2 VI Improvement Mechanisms

In the simplest of cases, a desired viscosity index can be achieved by mixing fluids with corresponding VIs. Usually, however, the viscosity requirements of modern lubricant specifications can be met only by addition of viscosity modifiers (VMs) also known as viscosity index improvers (VIIs).

As opposed to those present in low-molecular base fluids, viscosity modifiers have a polymer nature [14]. These molecules are described as being chain-like molecules whose solubility depends on chain length, structure and chemical composition [15].

As a rule, the base oil solubility of these polymer chains deteriorates as the temperature falls and improves with increasing temperature so that an increase in viscosity induced by viscosity modifiers also increases the viscosity index. In 1958, Selby published a descriptive explanation of the mechanism of VMs [16]. Because of poor solubility at low temperatures, the chain-like VM molecules form coils of small volume and as the temperature is increased these molecules expand and unravel, resulting in an increasing beneficial effect on high-temperature viscosity.

The absolute increase in viscosity and the VI depends on the type, the molecular weight and the concentration of viscosity modifiers in the formulation [17]. In practice and depending on the projected application, molecular weights of 15 000 to 250 000 $g\,mol^{-1}$ for PAMAs and 2 000 to 5 000 $g\,mol^{-1}$ for PIBs, to name just two, are used. Concentrations are usually between 3 and 30% (w/w). As a result of their high molecular weight, viscosity modifiers are always dissolved in a base fluid.

Apart from their thickening effect which is schematically illustrated in Figure 6.2a as a function of molecular weight, shear stability serves as a second characteristic [18]. According to Figure 6.2b, increasing molecular weight reduces shear stability if the polymer concentration remains constant.

The reason for this effect is mechanically- or thermally-induced chain degradation [19]. As opposed to Newtonian fluids, whose viscosity is independent of the rate of shear or the velocity gradient, long-chain compounds subject to high shearing are mechanically broken. Depending on the type and duration of the load, a number of different molecular sizes are created. The resulting drop in viscosity is described by the permanent shear stability index (PSSI) which describes the percentage loss of the contribution of the polymer to the viscosity (viscosity increase by the VM compared with base oil viscosity) (Figure 6.3).

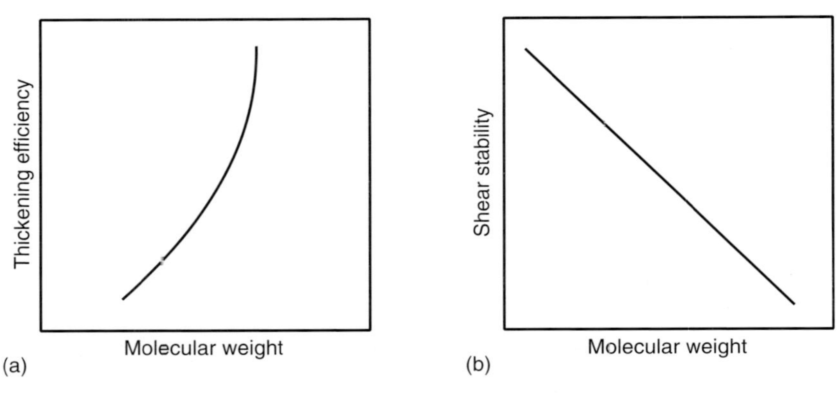

Figure 6.2 (a) Thickening efficiency and (b) shear stability of VMs.

Under conditions of high shear stress, if the relaxation time for the polymer chains is short (10^{-6} s) the high-molecular weight molecules adopt a temporary alignment. In Selby's model, the polymer coils are deformed in the direction of the shear force and lose part of their original contribution to the viscosity. This reversible drop in viscosity is described by the temporary shear stability index (TSSI).

Both values are of great importance to automotive applications, especially to engine, gear and hydraulic oils because the specified characteristics of such an oil do not just apply to the fresh oil but should remain throughout the drain interval. Although the reduction in molecular weight is, in practice,

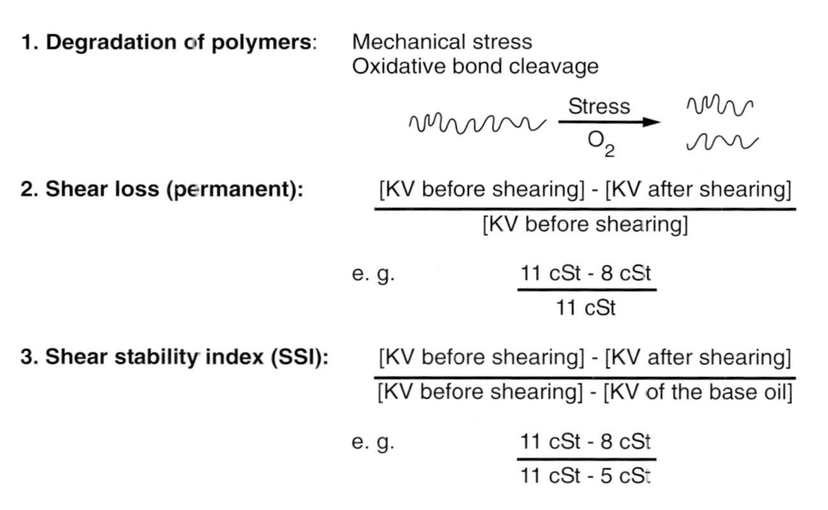

Figure 6.3 Mechanism and description of the shear loss.

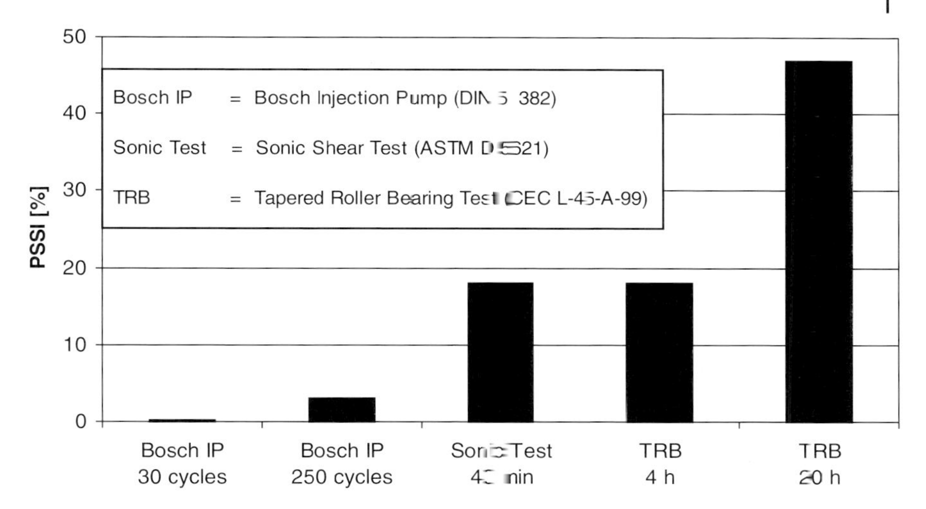

Figure 6.4 Laboratory shear-stability tests.

overshadowed by oxidation and other effects, a series of laboratory tests has been established to characterize shear stability [20]. The tests are shown in Figure 6.4; they are listed in order of test severity which increases in the order Bosch diesel injection pump test, 30 cycles < Bosch diesel injection pump, 250 cycles < tapered roller bearing test, 4 h ~ sonic shear test, 40 min < tapered roller bearing test 20 h.

6.2.3
Structure and Chemistry of Viscosity Modifiers

Apart from the fundamental description of thickening effect and shear stability, the expert differentiates viscosity modifiers by the molecular structure, composition and chemical nature of the individual chain links (monomers).

The most important monomers for viscosity modifiers are shown in Table 6.1 as structural formulas in the typical nomenclature for homo- and copolymers.

As a listing according to monomer units would not be worthwhile due to the vast number of combination possibilities, a series of polymer types have found favour for certain applications (Table 6.1).

It is also possible to sort these polymers according to their supramolecular chain structure, independently of the base monomers used (Figure 6.5).

In this case, a differentiation is made between linear and branched chains. Chain branching can create comb- or star-shaped structures. Depending on the polarity of the monomer, the viscosity modifiers are either dispersing or non-dispersing. Dispersing viscosity modifiers are mainly the link to ashless dispersants which are discussed in Section 6.4.3

Table 6.1 Types of polymeric viscosity modifier.

VM Type	Description	Main applications
OCP $-[(CH_2-CH_2)_A-(CH_2-CH(CH_3))_B]_x-$	Olefin copolymers	Engine and hydraulic oils
PAMA $-[CH_2-C(CH_3)(C(=O)-O-C_NH_{2N+1})]_x-$	Polyalkyl(meth) acrylates	Gear and hydraulic oils
PIB $-[CH_2-CH(CH_3)(CH_3)]_x-$	Polyisobutylene	Gear oils, raw material for ash-less dispersants
SIP $-[(CH_2-CH_2-CH(CH_3)-CH_2)_A-(CH_2-CH(C_6H_5))_B]_x-$	Hydrogenated styrene–isoprene copolymers	Engine oils
SBR $-[(CH_2-CH_2-CH_2-CH_2)_A-(CH_2-CH(CH_2-CH_3))_B-(CH_2-CH(C_6H_5))_C]_x-$	Hydrogenated styrene–butadiene	Engine oils

In 2008, two new PAMA types with different structure have been introduced onto the market: asteric polymers have radial or 'star' architecture versus a linear structure of conventional PAMAs. These PAMAs deliver superior performance regarding thickening efficiency, shear stability and viscosity index that limits conventional PAMA viscosity modifier technology.

Similiar benefits exhibit new comb-type PAMAs providing excellent thickining performance. These polymers have a comb structure that combines the chemistry of linear polyolefin and linear PAMA [21]. Compared to conventional viscosity modifiers comb polymers exhibit significantly improved viscosity/

Polymer type	General structure	Chemical type
Homopolymer	A-A-A-A-A-A-A-A-A-A	PIB, PAO
Random	A-B-A-B-B-A-B-A-A-B	PMA, OCP
Alternating	A-B-A-B-A-B-A-B-A-B	MSC (Maleate Styrene Copolymer)
Random Block	A-A-A-B-B-A-A-B-B-B	SBR, SIP, OCP
Biblock (A-B block)	A-A-A-A-A-B-B-B-B-B 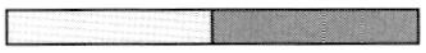	SIP
Tapered Block		SBR
Graft copolymers	A-B-A-B-B-A-B-A-A-B X X	OCP-g-PMA or SBR-g-PMA
Star polymers		SBR, OCP

Figure 6.5 Viscosity modifiers, chain structures and monomer make-up.

temperature properties such as higher VI (XHVI, extra high VI) and lower Brookfield viscosity at −40 °C.

6.3
Pour Point Depressants (PPD)

With the exception of polyalkylated naphthalenes, pour point depressants (PPD) are closely linked to a series of viscosity modifiers. The major difference of these polymers is their application concentration and the selection of monomer building blocks [22]. Molecular weight and thickening efficiency only play a subordinate role in a band from 0.1 to max. 2%.

An additional thickening effect is always welcome but is usually limited by solubility thresholds.

As is generally the case with flow improvers, although paraffin crystallization cannot be suppressed, the crystalline lattice and thus the morphology of the paraffin crystals can be significantly altered. While polyacrylates and ethylene vinyl acetate copolymers are used for crude oils and petroleum, special polyalkyl methylacrylates are normally used for mineral oil-based formulations. The fundamental background for this is the previously-mentioned co-crystallization of paraffinic components in the base oil and the polymer chain. This interaction results in an alteration of the crystal morphology. Instead of the needle-like paraffin crystals which rapidly cause paraffin gelation, densely packed, round crystals are formed which hardly effect flowing properties even at temperatures below the pour point. The corresponding PAMA-PPDs are in principle, comb-like and contain C_{12}—C_{24} paraffinic side chains [23]. By determining the polymer solubility and crystallization of paraffinic components in the base oil, the optimum PPD-mix and concentration can be determined for every base oil mixture. However, as a wide range of base oils are used in practice, standard products are often used which cover almost the entire spectrum. In such cases, efficiency is controlled via PPD concentration.

6.4 Detergents and Dispersants

Detergents and dispersants, often called DD or HD (heavy-duty) additives have been indispensable for the development of modern engine oils for gasoline and diesel combustion motors. These lubricants are especially severely stressed due to the high temperatures that they are exposed to and the additional influence of aggressive blow-by gases of the combustion process. DD additives keep oil-insoluble combustion products in suspension and also prevent resinous and asphalt-like oxidation products from agglomerating into solid particles. The overbased metal-containing compounds additionally are able to neutralize acidic combustion products as well as oxidation products by their alkaline reserve. Thus, DD additives prevent oil thickening, sludge and varnish deposition on metal surfaces and corrosive wear.

The original definition of detergents refers to their cleaning properties similar to the detergents in washing agents, although their function appears to be more the dispersing of particulate matter like abrasive wear and soot particles rather than cleaning up existing dirt. Historically, these kind of additives have been metal-containing compounds, often with high alkaline reserve.

To meet the dramatically increased requirements of modern high-performance engine oils, new ashless dispersants with improved dispersing properties have been developed. As these ashless compounds possess also cleansing properties, in fact there is no real difference between detergents and dispersants and it is more an arbitrary definition to call the metal-containing compounds as detergents and the ashless types as dispersants. Thus, it seems to be more appropriate to speak of metal-containing and ashless DD or HD additives [24,25].

6.4.1 Mechanism of DD Additives

Detergents and dispersants are generally molecules having a large oleophilic hydrocarbon 'tail' and a polar hydrophilic head group. The tail section serves as a solubilizer in the base fluid, while the polar group is attracted to contaminants in the lubricant. A multitude of dispersant molecules are able to envelope solid contaminants forming micelles whereby the non-polar tails prevent the adhesion of polar soot particles on metal surfaces as well as the agglomeration into larger particles. Figure 6.6 illustrates this process that is generally known as peptidization.

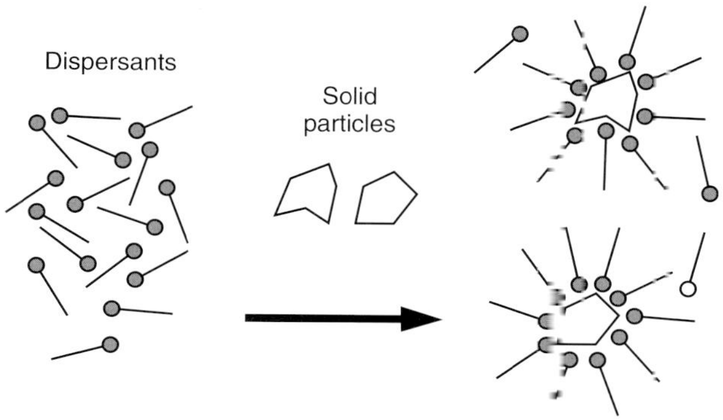

Figure 6.6 Function of dispersants – peptidization process.

6.4.2 Metal-Containing Compounds (Detergents)

6.4.2.1 Phenates

Phenates represent an important class of detergents which are synthesized by reaction of alkylated phenols with elemental sulfur or sulfur chloride followed by the neutralization with metal (calcium, magnesium, barium) oxides or hydroxides. Calcium phenates **7** are currently the most widely used types. Basic calcium phenates **8** can be produced by using an excess of the metal base. Beside their good dispersant properties they also possess greater acid neutralization potential.

6.4.2.2 Salicylates

Salicylates **9** are generally prepared by carboxylation of alkylated phenols with subsequent metathesis into divalent metal salts. Typically also these products are overbased by an excess of metal carbonate (calcium and magnesium) to form highly basic detergents that are stabilized by micelle formation. Salicylates exhibit additional antioxidant properties and have proven effective in diesel engine oil formulations.

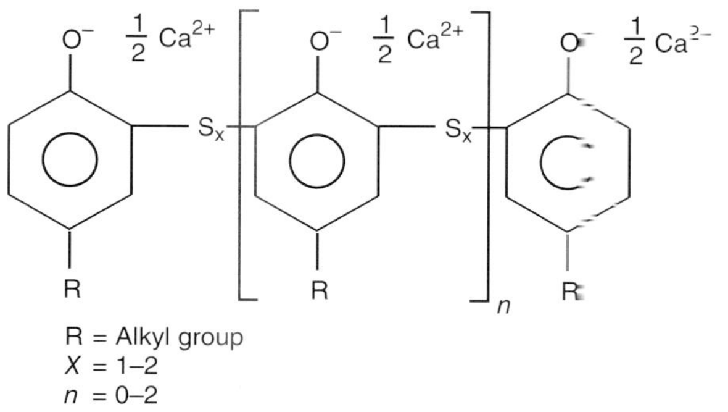

7

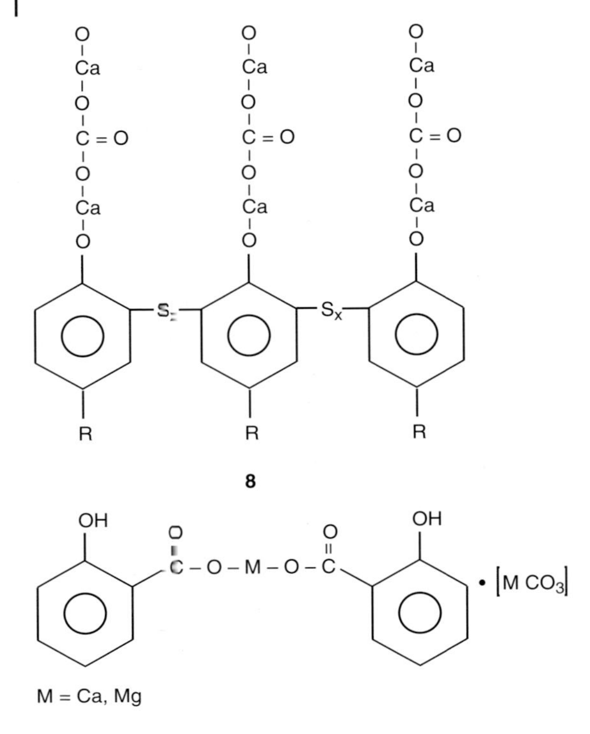

8

9

6.4.2.3 **Thiophosphonates**

These representatives of the detergents are produced by the reaction of polybutene (molecular weight from 500 to 1000) with phosphoruspentasulfide followed by hydrolysis and formation of metal (calcium, formerly also barium) salts.

The reaction products consist mainly in thiopyrophosphonates **10** combined with thiophosphonates **11** and phosphonates **12** [26]. Overbased products have almost vanished from use.

Thiopyrophosphonate

10

Thiophosphonates

11

```
     O
     ||
R – P – O
     |   |
     O – M
```

Phosphonate

12

6.4.2.4 Sulfonates

Sulfonates are metal salts of long-chain alkylarylsulfonic acids which can be divided into the petroleum and synthetic types. For more details see Section 6.10.1. Beside their excellent anticorrosion properties neutral and especially so-called overbased sulfonates with colloidally dispersed metal oxides or hydroxides (Figure 6.7) additionally exhibit an excellent detergent and neutralization potential that makes them very cost-effective multifunctional DD additives for engine oils.

Calcium sulfonates are relatively cheap products with good general performance. Magnesium compounds distinguish by excellent anticorrosion properties but tend to form hard ash after thermal decomposition. Deposits of hard ash can lead to bore polishing. Barium sulfonates are hardly used anymore due to toxicological concerns.

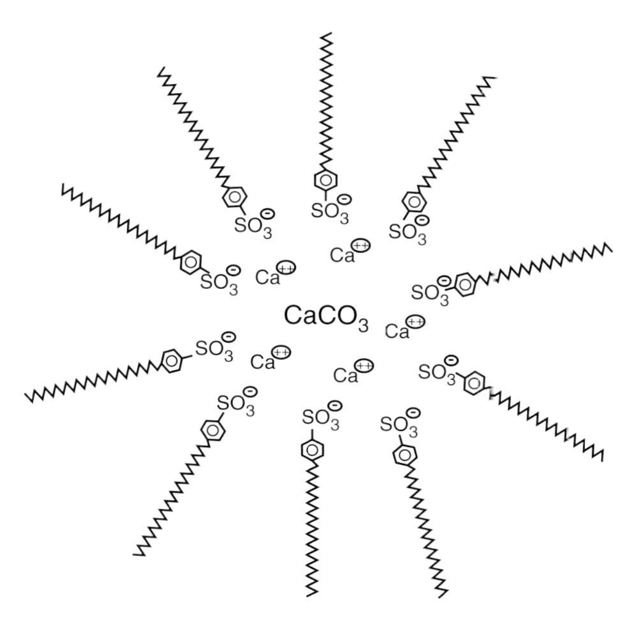

Figure 6.7 Model of the structure of overbased sulfonates.

6.4.3
Ashless Dispersants (AD)

As opposed to detergents containing metals, ADs are by definition free of metals. They are generally derived from hydrocarbon polymers. The best known and economically most interesting raw material group are polybutenes with molecular weights of 500–3000 g mol^{-1} [27].

As a result of the cationic polymerization of a C4 cut, polybutenes (PIB) with double bonds are formed, which can be thermally coupled with maleic anhydride (MA), to give PIBSA (polyisobutene succinic acid anhydride). In addition to the thermal synthesis of PIBSA, a chlorine-catalyzed production process is also still in use. In a further reaction step oligomeric aminoalkylenes are added to the anhydride to form thermally stable imides. By using suitable raw materials and reaction control, this two-stage, single vessel reaction can produce a variety of products. The reaction schematic in Figure 6.8 shows just one possible reaction and, for illustration purposes only, just one PIBSI (polyisobutene succinimide) as the final product. In general, there are mono- and bis-succinimides.

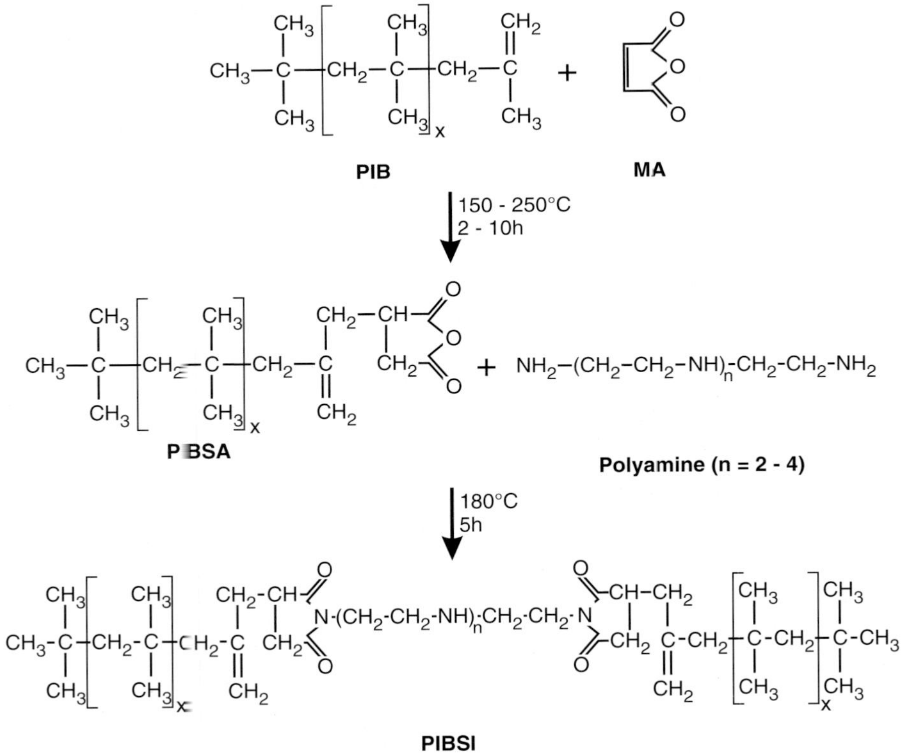

PIB + MA

150 - 250°C
2 - 10h

PIBSA + $NH_2-(CH_2-CH_2-NH)_n-CH_2-CH_2-NH_2$

Polyamine (n = 2 - 4)

180°C
5h

PIBSI

Figure 6.8 Reaction mechanism for the manufacture of ashless dispersants.

The amine functions of PIBSI and free-unreacted, low-molecular weight amines in particular, participate in undesirable reactions with fluorocarbon elastomers. Different methods are used to reduce the aggressiveness of amine groups, for example blocking by chemical reaction with boric acid. In addition to bifunctional amines (triethylene tetramines, tetraethylene pentamines), alcohols such as pentaerythritols are used as polar groups to improve thermal stability and elastomer compatibility. These polyisobutene succinic esters (PIBSE) **13** are often used in combination with PIBSIs, because they are less efficient as dispersants than PIBSIs. Intramolecular combinations (polyisobutene succinic acid ester amides) are also known.

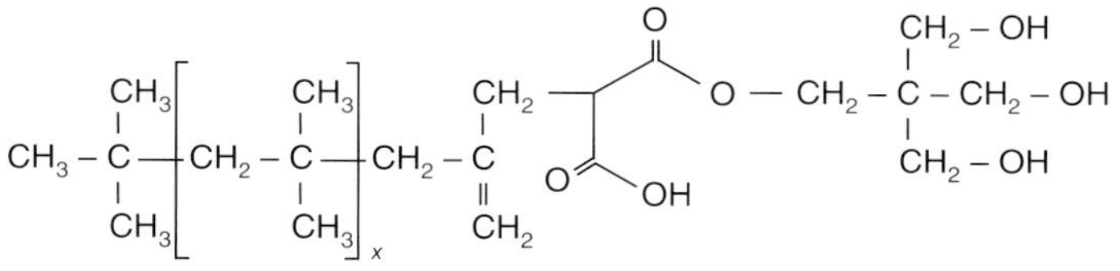

X = 8–45, typically 18

13

Alkylphenolamines (Mannich bases) **14** have lost importance, because they are very aggressive towards fluorocarbon elastomers.

$NH_2 - (CH_2 - CH_2 - NH)_n - CH_2 - CH_2 - NH_2$

R = long-chain alkyl

n = 1–4

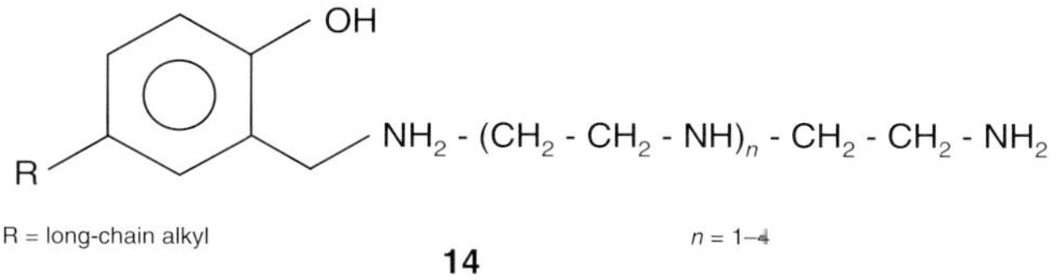

14

Because advanced PCMO formulations contain almost twice the amount of ADs as in the past, the search is on for highly efficient alternatives. Dispersion of oxidation, nitration, and soot particles can be controlled by selection of the appropriate product mixture and by use of components of adequate purity.

The disadvantage of conventionally manufactured PIB-based oligomers is their limited functionality and their wide molecular weight distribution. Interesting developments [28,29] have lead to hydrocarbon oligomers with a high percentage of double bonds of the end of the chain. This can be achieved by the cationic polymerization of isobutylenes or by the metallocene oligomerization of α-olefins [30]. The raw materials can be further altered using gentle, chlorine-free processes.

Equally chlorine-free alternatives are low-molecular weight, dispersant PAMAs. Contrary to hydrocarbon-based oligomers, the dispersing units are spread along the chains, whereby the type of dispersing groups can vary greatly [31]. Current research reports advantages with regard to soot dispersing and improved fuel efficiency characteristics [32].

6.5 Antifoam Agents

The foaming of lubricants is a very undesirable effect that can cause enhanced oxidation by the intensive mixture with air, cavitation damage as well as insufficient oil transport in circulation systems that can even lead to lack of lubrication. Beside negative mechanical influences the foaming tendency depends very much on the lubricant itself and is influenced by the surface tension of the base oil and especially by the presence of surface-active substances such as detergents, corrosion inhibitors and other ionic compounds.

Very important for the understanding of foaming effects is the difference between the so-called surface foam and the inner foam.

Surface foam can be controlled by antifoam agents. Effective defoamers possess a lower surface tension compared to the lubricant base oil, are usually not soluble in the base oil and therefore have to be finely dispersed in order to be sufficiently stable even after long-term storage or use. The particle size of the dispersed defoamers should be smaller than 100 μm or even smaller than 10 μm [33].

The inner foam refers to finely dispersed air bubbles in the lubricant that can form very stable dispersions. Unfortunately, the common defoamers dedicated to control the surface foam tend to stabilize the inner foam. Generally the air release properties of lubricants cannot be improved by additives. In the contrary, lots of additives have a negative influence. Lubricants which need excellent air release properties, for example turbine oils have to be formulated using specially selected base oils and additives.

Air is not only present in the lubricants in form of dispersed air bubbles (surface and inner foam) but also can be truly physically dissolved up to 9% (v/v) in mineral oil. Also this air can cause severe problems like cavitation ('diesel effect') but this effect cannot be controlled by additives.

6.5.1 Silicon Defoamers

Liquid silicones, especially linear **15** and cyclic polydimethylsiloxanes **16**, are the most efficient antifoam agents at very low concentrations of 1 to max. 100 mg kg^{-1}. To guarantee a stable dispersion, silicones usually are predissolved in aromatic solvents.

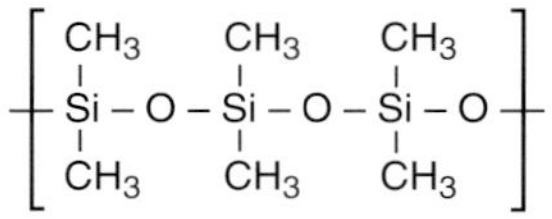

$$\left[-\underset{CH_3}{\overset{CH_3}{Si}}-O-\underset{CH_3}{\overset{CH_3}{Si}}-O-\underset{CH_3}{\overset{CH_3}{Si}}-O- \right]$$

15

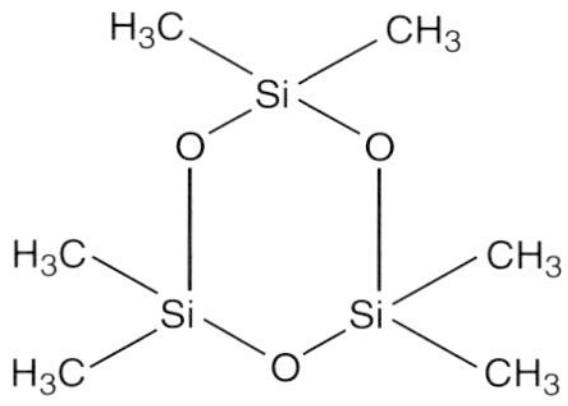

16

Compared to other additives, silicone defoamers have the disadvantage of being particularly easily carried out of the lubricant due to their insolubility and their strong affinity to polar metal surfaces.

6.5.2 Silicone-Free Defoamers

Nowadays, silicone-free defoamers are used more and more in many applications. Especially in metalworking processes, cutting fluids as well as hydraulic fluids used close by have to be silicone-free to guarantee the subsequent application of paints or lacquers on the workpieces. Silicones have caused lots of problems in this application.

The main representatives of silicone-free defoamers are special polyethylene glycols (polyethers), polymethacrylates and miscellaneous organic copolymers. Also tributylphosphate has been proposed as antifoam agent.

6.6 Demulsifiers and Emulsifiers

6.6.1 Demulsifiers

Most of the industrial oils in circulation systems (hydraulic, gear, turbine and compressor oils) require good or excellent demulsification properties to separate water contamination from the lubricating system. Without demulsifiers lubricating oils can form relatively stable water-in-oil emulsions.

In principle, all surface-active substances are suitable demulsifiers. One of the first known types have been alkaline-earth metal salts of organic sulfonic acids, particularly barium and calcium dinonylnaphthenesulfonates. Nowadays special poly-ethylene glycols and other ethoxylated substances have proved to be highly efficient demulsifiers which are often part of many additive packages.

Surprisingly, the same class of chemical substances is used as emulsifiers. Here the molecular weight, the degree of ethoxylation and the treat rate are very important to guarantee the demulsifying properties.

6.6.2
Emulsifiers

Because of their importance for the formulation of water-based metal working fluids, types and mechanism of emulsifiers are described in Chapter 14.

6.7
Dyes

For marketing, identification or leak detection purposes some lubricants contain dyes which are classified according to the International Color Index [34]. Most of these substances are solids which often are suspended in mineral oil or dissolved in aromatic solvents to make their handling easier.

Generally oil-soluble azo dyes are used. Some of these products may be removed in the near future because they have to be labelled as potentially carcinogenic.

Fluorescent dyes are typically used to detect leaks under UV light when the coloration of lubricants is not appreciated.

6.8
Antiwear (AW) and Extreme Pressure (EP) Additives

When two contacting parts of a machinery start to move and the hydrodynamic lubrication has not yet build up or in the case of severe stress and strong forces the lubricating system runs in the area of mixed friction. In this case, antiwear (AW) and extreme pressure (EP) additives are necessary in any metalworking fluid, engine oil, hydraulic fluid or lubricating grease to prevent welding of the moving parts to reduce wear.

6.8.1
Function of AW/EP Additives

Because of their polar structure these additives form layers on the metal surface by adsorption or chemisorption that guarantees their immediate availability in the case of mixed friction conditions. When the hydrodynamic lubricating film is not yet or no longer present, temperature will increase and the AW and EP additives can react with the metal surface forming tribochemical reaction layers (iron phosphides, sulfides, sulfates, oxides and carbides – depending on the chemistry of the additive) that will prevent direct contact between the sliding metals. These friction-reducing, slideable reaction layers can smooth the asperity of the metal surface by plastic deformation and reduce wear that would occur due to microwelding processes, respectively, avoiding real welding of the moving parts under extreme pressure conditions.

Layers formed by only physically adsorbed polar substances like fatty oils, fatty acids and others exhibit only poor or moderate high pressure properties. These kind of additives are called friction modifiers (see Section 6.9). More effective and more stable are chemically reactive products (AW and EP additives) that can form tribochemical reaction layers.

AW additives are mainly designed to reduce wear when the running system is exposed to moderate stress, whereas EP additives are much more reactive and are used when the stress of the system is very high in order to prevent the welding of the moving parts that otherwise would lead to severe damage. Typically, EP additives increase wear effects due to their high reactivity.

This differentiation cannot be precise and there are many additives that can be related to both groups.

6.8.2
Compounds

6.8.2.1 Phosphorus Compounds

Under condition of medium stress organic phosphorus compounds work excellent as antiwear additives. Some of them are highly efficient FZG boosters.

Most of these additives are neutral and acidic phosphoric acid ester derivatives, their metal or amine salts or amides. As the acidic form of these compounds is the most reactive one, the reactivity decreases with the degree of neutralization.

Trialkyl- and triarylphosphates represent the neutral phosphoric acid triesters, where the tricresylphosphate (TCP) **17** is the most known species. For toxicological reasons, TCP should be free of *o*-cresol.

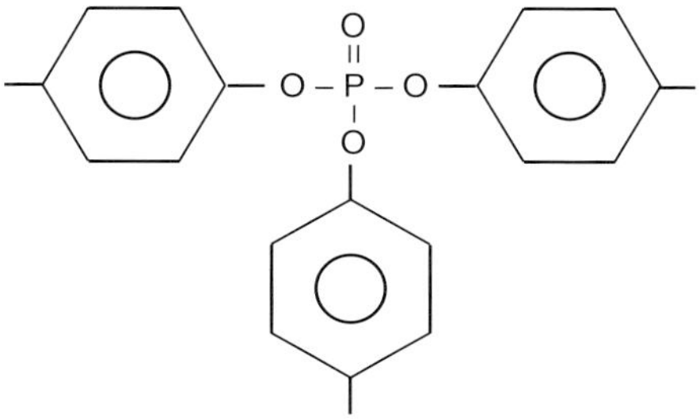

O
||
O – P – O
|
O

17

Amine neutralized mixtures of mono- and dialkyl phosphoric acid partial esters **18** are highly efficient FZG boosters and widely used in ashless hydraulic oil formulations. Some types exhibit additional anticorrosion properties.

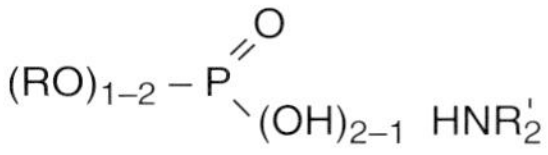

$(RO)_{1-2} - P(=O)(OH)_{2-1}\ HNR'_2$

18

Ethoxylated mono- and dialkylphosphoric acids are even more polar by their hydrophilic structure what makes them more efficient.

Also phosphites have generated much interest. Because of the inherent hydrolytic instability of this chemical group in practice only sterically hindered derivatives like triarylphosphites and long chain trialkylphosphites are used for some applications. Beside their antiwear properties, they are able to scavenge free sulfur by forming thiophosphates by which their use in gear oils is based.

Phosphonates and phosphines, the derivatives of the phosphonic and phosphinic acid, have also been proposed as antiwear additives but are not commonly used [35].

6.8.2.2 Compounds Containing Sulfur and Phosphorus

The most important and well known additive group of the sulfur–phosphorus compounds are the zinc dialkyldithiophosphates (ZnDTP) [36]. They are synthesized by reaction of primary and secondary alcohols (C_3—C_{12}) as well as alkylated phenols with phosphorus(V)sulfide followed by neutralization of the resulting dialkyldithiophosphoric acids with zinc oxide. Usually, this last reaction step is done in mineral oil solution but it is also possible to make mineral oil free compounds by using solvents that can be removed afterwards by distillation. Beside the neutral species also basic ZnDTPs can be obtained when the neutralization is done with an excess of zinc oxide.

ZnDTPs based on isopropanol or *n*-butanol are solids, whereas mixtures of short and long chain alcohols are liquid. The thermal and hydrolytic stability of ZnDTPs and thus their reactivity (AW/EP-performance) can be influenced by the structure of the alkyl groups. So the thermal stability increases with the chain length of the alkyl groups and their structure in the sequence secondary, primary and aromatic. By carefully directed alcohol composition, the specific requirements of different applications can selectively be adjusted. Beside excellent antiwear and extreme pressure properties ZnDTPs are also efficient antioxidants and even metal passivators. These multifunctional properties makes them the widest spread cost effective additive group that is used nowadays in huge quantities in engine oils, shock absorber oils and hydraulic fluids. The influence of alcohol composition and structure on the properties of zinc dithiophosphates is shown in Figure 6.9.

Beyond ZnDTPs also ammonium, antimony, molybdenum and lead dialkyldithiophosphates are known, which the latter no longer has real significance because of toxicological and ecotoxicological concerns. MoDTPs are effective antiwear additives with remarkable friction reducing properties and excellent antioxidant behaviour.

Ashless dialkyldithiophosphoric acid-O,O,S-triesters **19** distinguish by improved hydrolytic stability in comparison to metal salts of the dithiophosphoric acid. On the other side, their antioxidant properties are reduced. Similar to the ZnDTPs, the reactivity can be influenced by variation of the organic substituents.

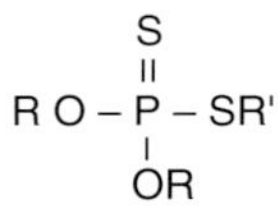

```
       S
       ‖
R O – P – SR'
       |
       OR
```

19

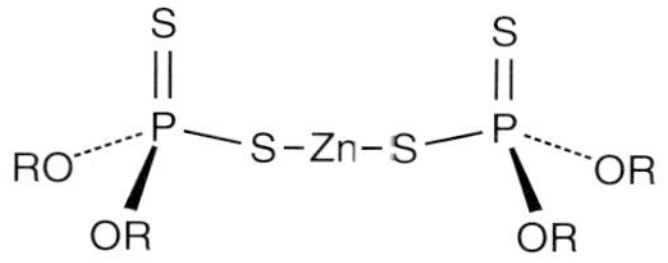

Zinc-dialkyldithiophosphate/
Zinc-diaryldithiophosphate

Alkylgroup R	Reactivity (antiwear properties)	Thermal stability	Hydrolytic stability	Antioxidant properties
$R = C_3$ → $R = C_8$	+ (top) to - (bottom)	- (top) to + (bottom)	- (top) to + (bottom)	No systematic influence
Structure of R				
Primary alkyl	+/-	+	+/-	+/-
Secondary alkyl	++	-	+	+
Aryl	- -	++	- -	-

Primary alkyl group = $-CH_2-R$

Secondary alkyl group = $-CH(R')-R$

Aryl = $-C_6H_4-R$

Figure 6.9 Influence of alcohol composition and structure on the properties of ZnDTPs.

The triphenylphosphorothionate (TPPT) **20** belongs to the neutral thiophosphoric acid esters and is distinguished by its extraordinary thermal stability what makes it suited for high temperature lubrication.

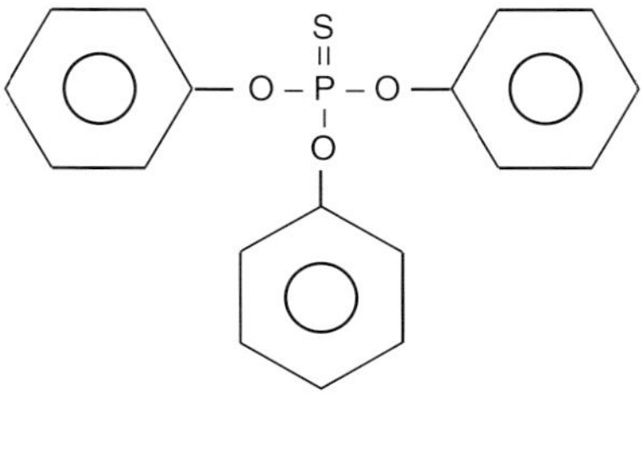

20

6.8.2.3 **Compounds Containing Sulfur and Nitrogen**

Zinc-bis(diamyldithiocarbamate) and the ashless methylene-bis(di-n-butyl dithiocarbamate) **21** are highly effective EP additives and excellent antioxidants. Beside these main species also antimony and tungsten derivatives are known. Dithiocarbamates are predominantly used in lubricating greases and to some extent in gear oil formulations.

$$(H_9C_4)_2N-\overset{\displaystyle S}{\overset{\|}{C}}-S-CH_2-S-\overset{\displaystyle S}{\overset{\|}{C}}-N(C_4H_9)_2$$

21

Dialkyl-2,5-dimercapto-1,3,4-thiadiazole (DMTD) derivatives are usually known as metal passivators and sulfur scavengers, but they also exhibit excellent EP properties. The chemical structure of DMTD is given in Section 6.10.

6.8.2.4 **Sulfur Compounds**

From the early days of lubrication until now elemental sulfur has been added directly to mineral oil (up to 1.5 %) to improve the EP properties of metal working fluids. Oil-soluble organic sulfur compounds, the so-called sulfur carriers, of the general formula $R-S_x-R$ offer improved solubility and better control over the reactivity of the sulfur.

Fundamentally, there can be differentiated between inactive and active sulfur carriers. The inactive types with predominantly disulfide bridges ($x = 2$) possess relatively stable C—S bonds which will react at elevated temperatures only. The active forms with x between 3 and 5 (so-called pentasulfides) are much more reactive as the sulfur of the relatively labile polysulfide bridges can easily be made available even at low temperatures. Moreover numerous sulfur carriers with specific distribution of the different polysulfide bridges ($x = 1$–5) are used to cover the whole field of application with its varying stress requirements. The mechanism of sulfur carriers under EP conditions can be described as to begin with physical adsorption followed by chemisorption and finally cleavage of the sulfur and its reaction with the metal surface (Figure 6.10). Generally, this reaction takes place at temperatures over 600 °C [35].

Active sulfur carriers are excellent EP additives that will prevent welding by a kind of controlled wear when the slideable reaction layers are removed continuously under severe loads. Because of the high reactivity with non-ferrous metals

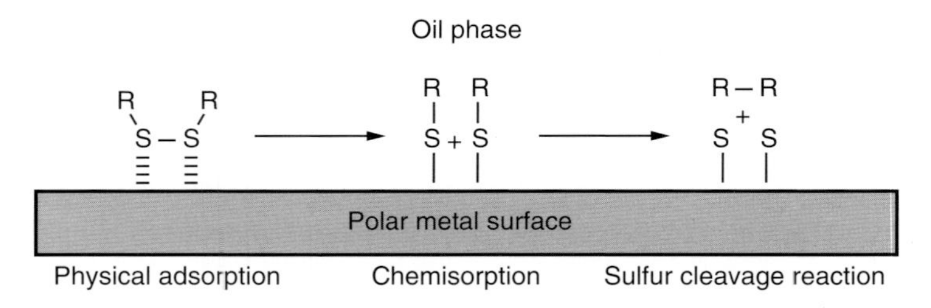

Figure 6.10 Mechanism of sulfur carriers under extreme pressure conditions.

active sulfur carriers cannot be used in the machining of non-ferrous metals, when non-ferrous metals are incorporated in an engine or other aggregates. Inactive sulfur carriers need higher temperatures to set the sulfur free. Therefore, they are much more compatible with non-ferrous metals and show to some extent antiwear properties.

Contrary to this well-known behaviour, active sulfur carriers exhibit surprisingly good antiwear properties in unsaturated esters or fatty oils like trimethylolpropanetrioleate or rape seed oil [37].

The polarity and thus the affinity to the metal surface is determined by the organic substituents. The polarity increases in the sequence-sulfurized hydrocarbons, esters, alcohols and fatty acids whereas the solubility in mineral oil decreases in the same order.

Sulfurized hydrocarbons **22** are obtained by direct sulfurization of olefins (e.g. diisobutylene, terpenes) with elementary sulfur in the presence of H_2S, their reaction with dichlorodisulfide or oxidation of mercaptans to disulfides. The first method will lead to products with high sulfur contents up to 45%. The other will form inactive sulfur carriers that are predominantly used in gear oils and mild metalworking fluids (cutting fluids). Active pentasulfides are used for severe machining processes, for example broaching

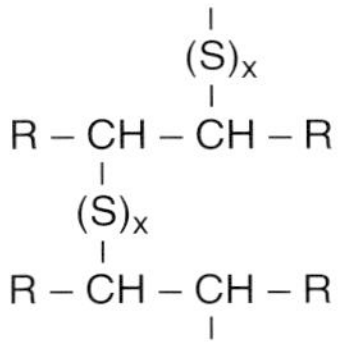

$$\begin{array}{l} \qquad\quad | \\ \qquad\;(S)_x \\ \qquad\quad | \\ R - CH - CH - R \\ \quad\;\; | \\ \quad (S)_x \\ \quad\;\; | \\ R - CH - CH - R \\ \qquad\quad | \end{array}$$

22

Sulfurized fatty oils and fatty acid esters **23** are produced by sulfurization of the unsaturated raw materials with sulfur (dark coloured products) or with sulfur in the presence of H_2S to obtain light coloured products that are used predominantly in modern lubricants instead of the formerly used sperm oil products. The combination of sulfur with the friction reducing properties of the fatty raw materials leads to excellent EP additives with high-load carrying capacity in the four ball testing machine. Sulfurized fatty acid methyl esters will lead to products with low viscosities that are used, for example for deep-hole drilling.

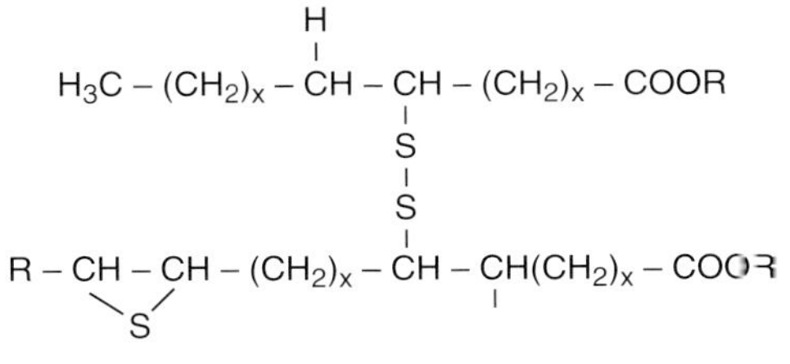

$$\begin{array}{l} H_3C - (CH_2)_x - \overset{H}{\overset{|}{C}}H - CH - (CH_2)_x - COOR \\ \qquad\qquad\qquad\qquad\quad\;\; | \\ \qquad\qquad\qquad\qquad\quad\;\; S \\ \qquad\qquad\qquad\qquad\quad\;\; | \\ \qquad\qquad\qquad\qquad\quad\;\; S \\ \qquad\qquad\qquad\qquad\quad\;\; | \\ R - CH - CH - (CH_2)_x - CH - CH(CH_2)_x - COOR \\ \quad\;\; \diagdown\, S \,\diagup \qquad\qquad\qquad\quad\; | \end{array}$$

23

6.8.2.5 PEP Additives

Overbased sulfonates, especially calcium and sodium salts, can be used as highly efficient boosters in combination with active sulfur carriers to formulate metal working fluids with extremely high load carrying behaviour. These overbased products are called PEPs (passive EP).

6.8.2.6 Chlorine Compounds

The excellent AW/EP properties of chlorine compounds have been conventionally explained by their ability to coat the metal surface with a slideable metal chloride film under the influence of high pressure and in the presence of traces of moisture. Because of the formation of hydrogen chloride alkaline buffers have to be added to avoid severe corrosion. Another approach refers to the formation of high pressure stable adsorption layers due to the affinity of the additive's heteroatom (Cl) showing outstanding lubricity at low machining speed and moderate temperature but reduced efficiency with increasing temperature (speed) [38].

For environmental and toxicological reasons concerning the disposal of the chlorinated used fluids, chlorine compounds are increasingly replaced although this additive technology has been extremely successful in metalworking fluids. Beside the corrosivity of their decomposition products, primarily hydrogen chloride, the biggest problem is the possible formation of highly toxic dioxins at the incineration of used oil especially when the incineration temperatures are not high enough.

Typical additives have been chlorinated paraffins with a chlorine content of 35–70% and a hydrocarbon base of 10–20 carbon atoms as well as chlorinated sulfur carriers.

6.8.2.7 Solid Lubricating Compounds

Mainly, finely ground powders of graphite and molybdenum disulfide as well as their dispersions are used as solid additives. They are distinguished by excellent emergency running properties when the oil supply is breaking down. Also for extreme high temperature applications solid lubricating compounds are used due to their high thermal stability. Other compounds that are used especially in lubricating greases are polytetrafluoroethene, calcium hydroxide and zinc sulfide. For more detailed information see Chapter 17.

6.9
Friction Modifiers (FM)

In the case of liquid (hydrodynamic) lubrication, friction can be reduced only by the use of base oils with lower friction coefficients and lower viscosity. In the area of low slide velocities, moderately increased loads and low viscosities at higher temperatures, the liquid lubrication can easily proceed to mixed friction conditions. In this case so-called friction, modifiers have to be used to prevent stick–slip oscillations and noises by reducing frictional forces. They work at temperatures where AW and EP additives are not yet reactive by forming thin monomolecular layers of physically adsorbed polar oil-soluble products or tribochemical friction reducing reaction layers that exhibit a significantly lower friction behaviour compared to typical AW and EP additives [39,40]. Therefore, friction modifiers can be regarded as mild AW or EP additives working at moderate temperatures and loads in the area of beginning mixed friction.

Friction modifiers can be classified into different groups regarding their function [41]: mechanically working FMs (solid lubricating compounds, e.g. molybdenum disulfide, graphite, PTFE, polyamide, polyimide, fluorinated graphite), adsorption layers forming FMs (e.g. long chain carboxylic acids, fatty acid esters, ethers, alcohols, amines, amides, imides), tribochemical reaction layers forming FMs (saturated fatty acids, phosphoric and thiophosphoric acid esters, xanthogenates, sulfurized fatty acids), friction polymer-forming FMs (glycol dicarboxylic acid partial esters, dialkyl phthalic acid esters, methacrylates, unsaturated fatty acids, sulfurized olefins) and organometallic compounds (molybdenum compounds like molybdenum dithiophosphates, molybdenum dithiocarbamates, their synergistic combination with ZnDTPs [42], copper containing organic compounds).

A widely spread group of friction modifiers are the adsorption forming agents whose effect increases with increasing molecular mass in the order alcohol < ester < unsaturated acid < saturated acid. The main products are carboxylic acids with 12–18 carbon atoms, fatty alcohols and synthetic (methyl, butyl) or natural esters of fatty acids (glycerides).

Typical applications of friction modifiers are modern fuel economy oils, slide way oils, automatic transmission fluids (ATF) that contain so-called anti-squawk additives and lubricants for limited-slip axles that contain so-called anti-chatter additives.

6.10
Corrosion Inhibitors

Corrosion inhibitors are used in nearly every lubricant to protect the metal surface of any machinery, metalworking tool or work piece from the attack of oxygen, moisture and aggressive products. These mostly acidic products may be formed by the thermal and oxidative decomposition of the lubricant (base oil and additives), brought in directly from the environment (acid atmosphere) or caused by the specific application (aggressive blow-by gases in internal combustion engines). The base oil itself forms a kind of protective layer on the metal surface. But in general this will not be sufficient, especially when highly refined oils without natural inhibitors are used. Then highly efficient anticorrosion additives are necessary.

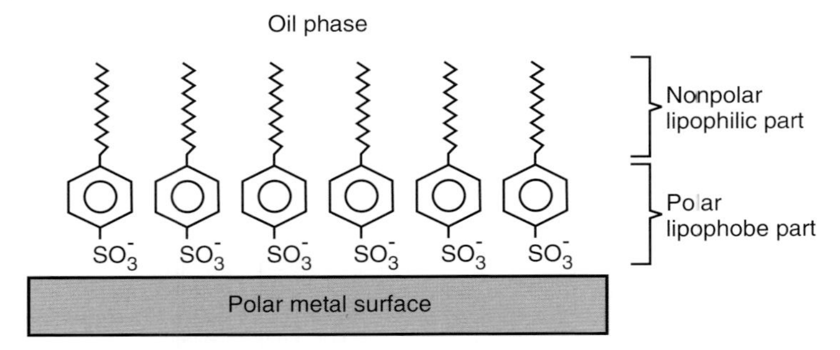

Figure 6.11 Function of rust inhibitors.

6.10.1
Mechanism of Corrosion Inhibitors

The mechanism of anticorrosion additives is relatively simple. Anticorrosion additives are molecules with long alkyl chains and polar groups that can be adsorbed on the metal surface forming densely packed, hydrophobic layers as shown in Figure 6.11. The adsorption mechanism can base on a physical or chemical interaction of the polar anticorrosion additive with the metal surface [43].

Because of this high surface activity, anticorrosion additives compete with other polar additives like antiwear and extreme pressure additives for the metal surface and can therefore reduce their efficiency. Thus a FZG damage load stage of >12 can be reduced down to a damage load stage of 8 if highly efficient and polar anticorrosion additives are added.

Corrosion inhibitors can be divided into two main groups: antirust additives for the protection of ferrous metals and metal passivators for non-ferrous metals.

6.10.2
Antirust Additives (Ferrous Metals)

6.10.2.1 Sulfonates

Petroleum sulfonates (mahogany sulfonates) are by-products at the production of white oils by treatment with oleum. The resulting acid tar contains long chain alkylarylsulfonic acids that can be neutralized with lyes. Sodium sulfonates with low molecular weights (below approximately 450) are typically used as low-priced emulsifiers and detergents with additional anticorrosion properties in water based metal working fluids, engine oils and rust preventatives. Sulfonates with higher molecular weights distinguish as highly efficient corrosion inhibitors especially when based on divalent cations like calcium, magnesium and barium. The importance of the barium compounds is going to decrease constantly due to toxicological and ecotoxicological concerns.

Nowadays synthetic alkylbenzene sulfonates are used preferably in spite of a higher price level due to their higher and more constant quality. They are reaction products of specifically designed monoalkylbenzenesulfonic acids (typically C_{24} alkyl groups) and dialkylbenzenesulfonic acids (typically two C_{12} alkyl groups) with alkaline and earth alkaline metal hydroxides.

A special group of synthetic sulfonates are the dinonylnaphthenesulfonates **24** of which the neutral calcium and barium salts distinguish by additional demulsifying properties and a good compatibility with EP additives.

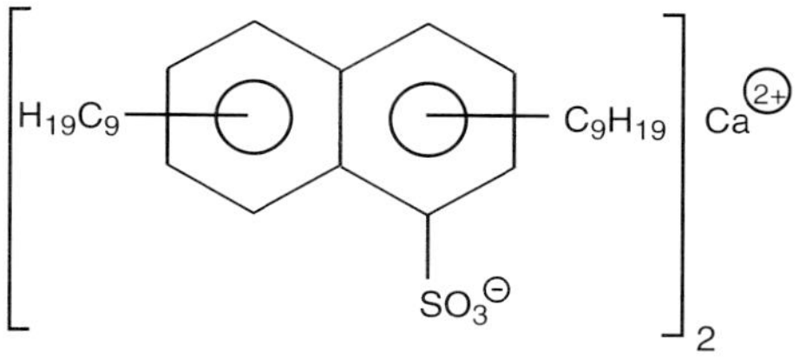

$$\left[H_{19}C_9\text{–}C_{10}H_5(SO_3^{\ominus})\text{–}C_9H_{19} \right]_2 \; Ca^{2+}$$

24

Beside the neutral or only slightly basic sulfonates overbased sulfonates with high alkaline reserve (TBN 100–400 mg KOH g^{-1}) play an important role especially in the formulation of engine oils. There they exhibit detergent properties and can neutralize acidic oxidation products. Moreover, in metal working fluids they act as so called passive EP additives (see Section 6.8) to prevent welding under high pressure conditions.

6.10.2.2 **Carboxylic Acid Derivatives**

Many different long-chain carboxylic acid derivatives have been proposed as corrosion inhibitors at which the carboxylic group acts as polar part that can easily be adsorbed on the metal surface.

Lanolin (wool fat) and salts of the lanolin fatty acid mostly in combination with sulfonates have long been known as corrosion inhibitors in rust preventatives.

Oxidized paraffins with their high polarity because of the high content of hydroxy and oxo carboxylic acids are still used for that purpose.

Zinc naphthenates are especially used in lubricating greases, whereas lead naphthenates are not used anymore for toxicological reasons.

Alkylated succinic acids, their partial esters **25** and half amides are known as highly efficient, not emulsifying antirust additives even at very low treat rates of 0.01–0.05%. Therefore, these additives are used preferably in turbine oils and hydraulic fluids.

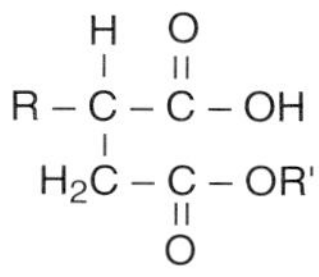

$$R\text{–}CH(\text{–}C(=O)\text{–}OH)\text{–}H_2C\text{–}C(=O)\text{–}OR'$$

25

4-Nonylphenoxyacetic acid **26** and derivatives have a similar performance.

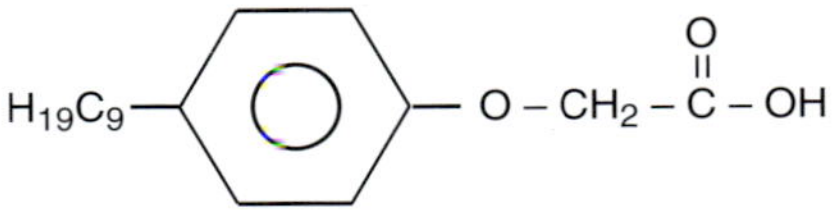

$$H_{19}C_9-C_6H_4-O-CH_2-C(=O)-OH$$

26

Another wide spread group are amides and imides as reaction products of saturated and unsaturated fatty acids with alkylamines and alkanolamines. The most known product of this type is *N*-acylsarcosine **27** that shows a strong synergistic effect with imidazoline derivatives **28**. In addition, these additives have good water-displacing properties.

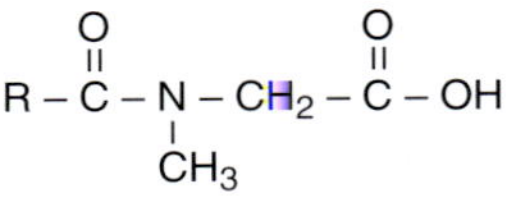

$$R-C(=O)-N(CH_3)-CH_2-C(=O)-OH$$

27

$$\text{2-}C_{17}H_{33}\text{-imidazoline}-N-CH_2\,CH_2\,OH$$

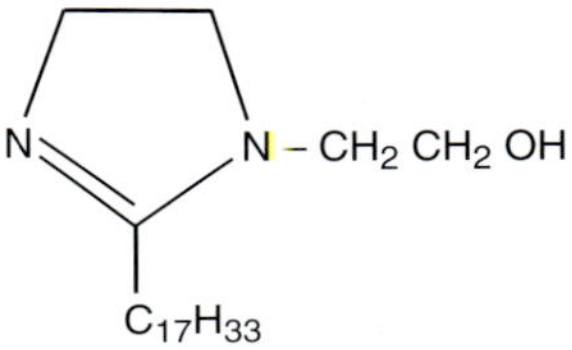

28

Aspartic acid (2-aminobutanedioic) esters are discussed as highly efficient corrosion inhibitors providing good demulsibility and filterability in ashless lubricating oil compositions. They do not have detrimental effects on FZG gear test performance as it is typical for most of the succinic acid-based ashless rust inhibitors [44].

6.10.2.3 Amine-Neutralized Alkylphosphoric Acid Partial Esters

Some special amine salts of mono- or dialkylphosphoric acid partial esters exhibit excellent anticorrosion properties in addition to their highly efficient antiwear properties [45]. Because of the well-known antagonism of anticorrosion and antiwear additives this behaviour makes them one of the mostly used components in ashless industrial oils.

6.10.2.4 Vapour Phase Corrosion Inhibitors

Vapour phase corrosion inhibitors (VCIs) for closed systems are substances with high affinity to metal surfaces and relatively high vapour pressure to guarantee their availability on parts that are not steadily in direct contact with the corrosion inhibited lubricant. The mostly used product group for this application are amines. Morpholine, dicyclohexylamine and diethanolamine have proved to be highly efficient for that purpose. Because of toxicological concerns that refer mainly to the nitrosamine forming potential of secondary amines, these products are going to be partly substituted by tertiary amines like diethanolmethylamine and similar products.

Another group of oil soluble VCIs are low molecular weight carbonic acids (n-C_8 to n-C_{10}).

6.10.3
Metal Passivators (Non-Ferrous Metals)

The metal passivators can be classified into three groups: film forming compounds, complex forming chelating agents and sulfur scavengers.

The fundamental function of the film forming types consists in the building of passivating protective layers on the non-ferrous metal surface thus preventing the solubilization of metal ions that would work as pro-oxidants. The complex forming agents are able to build oil-soluble complexes with significantly reduced catalytic activity regarding the influence of non-ferrous metal ions on the oxidative ageing process of lubricants. Sulfur scavengers are even able to catch corrosive sulfur by integrating sulfur into their molecular structure. Mechanism and function of these three types of metal passivator are shown in Figures 6.12–6.14.

Metal passivators in combination with antioxidants show strong synergistic effects as they prevent the formation of copper ions and suppress their behaviour as pro-oxidants. Thus, these additives are used in nearly every formulation of modern lubricants.

The mostly used metal passivators are benzotriazole and tolyltriazole as well as their alkylated liquid derivatives **29** with improved solubility. Typical treat rates of these film forming passivators are between 0.005 and 0.03%.

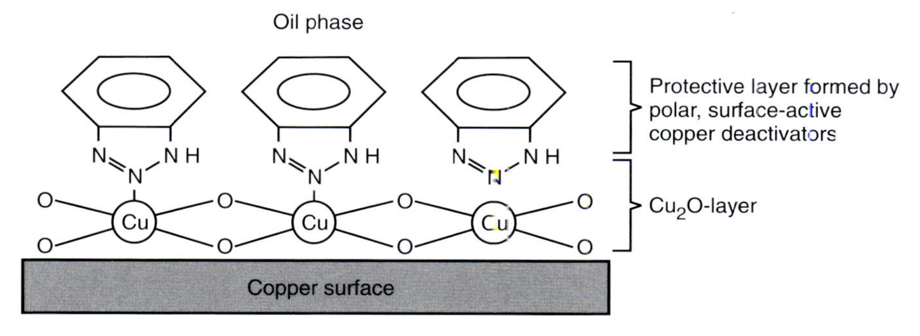

Figure 6.12 Function of benzotriazole as film forming agent.

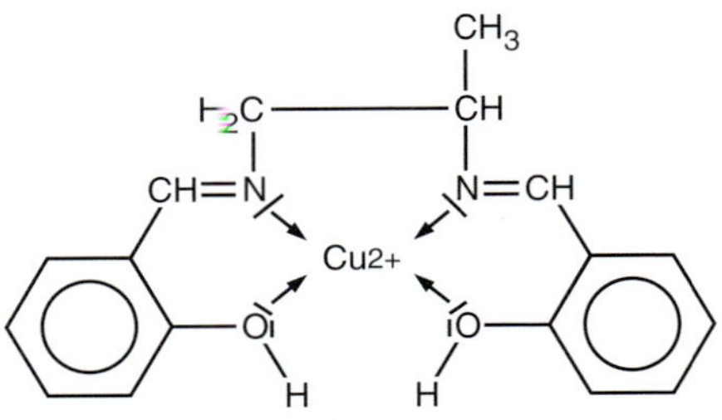

Figure 6.13 Function of N,N'-disalicylidene-propylenediamine as chelating agent.

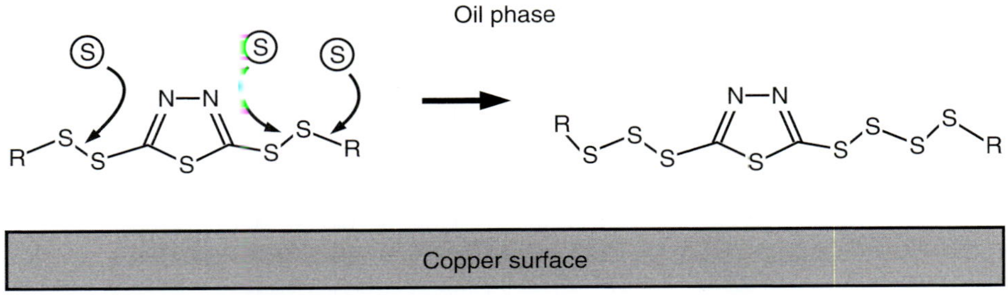

Figure 6.14 Function of dialkyl-2,5-dimercapto-1,3,4-thiadiazole as sulfur scavenger.

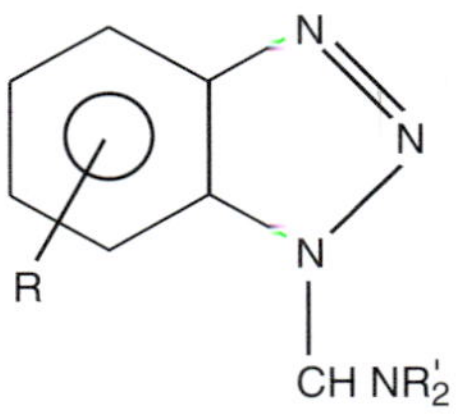

29

Furthermore 2-mercaptobenzothiazole (reduced importance) and especially 2,5-dimercapto-1,3,4-thiadiazole derivatives are used as highly efficient film forming passivators. The later can also act as sulfur scavenger by building in sulfur into the alkyl–sulfur–thiadiazol bonds (see Figure 6.14).

N,N'-disalicylidenealkylenediamines (where 'alkylene' is ethylene or propylene) belong to the group of chelating agents. Beside their EP and antiwear properties zinc dialkyldithiophosphates and dialkyldithiocarbamates have some metal-passivating properties (see Section 6.8).

References

1 Rasberger, M. (1992) Oxidative degradation and stabilisation of mineral oil based lubricants, in *Chemistry and Technology of Lubricants* (eds R.M. Mortier and S.T. Orszulik), Blackie Academic and Professional, pp. 83–93.

2 Rayn, H. (1999) Use of group II, group III base stocks in hydraulic and industrial applications. Proceedings of the 5th Annual Fuels and Lubes Asia Conference, Fuels and Lubes Asia Publication, Manila.

3 Migdal, C.A (1999) The effect of antioxidants on the stability of base oils. Proceedings of the 5th Annual Fuels and Lubes Asia Conference, Fuels and Lubes Asia Publication, Manila.

4 Gatto, V.J. and Grina, M.A. (1999) Effects of base oil type, oxidation test conditions and phenolic antioxidant structure on the detection and magnitude of hindered phenol/diphenylamine synergism. *Lubr. Eng.*, **55** (1), 11–20.

5 Rasberger, M. (1992) Oxidative degradation and stabilisation of mineral oil based lubricants, in *Chemistry and Technology of Lubricants* (eds R.M. Mortier and S.T. Orszulik), Blackie Academic and Professional, pp. 97–98, 104–107, 108, 102–103, and 99–100.

6 Yano, A., Watanabe, S., Miyazaki, Y., Tsuchiya, M. and Yamamoto, Y. (2004) Study on sludge formation during the oxidation process of turbine oils. *Tribol. T.*, **47**, 111–1228.

7 Jensen, R.K., Korcek, S., Zinbo, M. and Gerlock, J.L. (1995) Regeneration of amines in catalytic inhibition and oxidation. *J. Org. Chem.*, **60** (19), 5396–5400.

8 Cohen, S.C. (1991) A new synthetic screw compressor fluid based on two-stage hydrotreated base oils. *Synth. Lubr.*, 7 (4), 267–279.

9 Mookken, R.T., Saxena, D., Basu, B., Satapathy, S., Srivastava, S.P. and Bhatnagar, A.K. (1997) Dependence of oxidation stability of steam turbine oil on base oil composition. *Lubr. Eng.*, **53** (10), 19–24.

10 Bowmann, W.F. and Stachowiak, G.W. (1998) Application of sealed capsule differential scanning calorimetry – Part I. *Lubr. Eng.*, **54** (10), 19–24.

11 Bowmann, W.F. and Stachowiak, G.W. (1999) Application of sealed capsule differential scanning calorimetry – Part II. *Lubr. Eng.*, **55** (5), 22–29.

12 Stambough, R.L., Kopko, R.J. and Roland, T.F. (1990) SAE Technical Paper 901633.

13 Briant, J., Denis, J. and Parc, G. (eds) (1989) *Rheological Properties of Lubricants*, Editions Technip, Paris, Chapter 7.

14 Bartz, W.J. (ed.) (1986) *Additive für Schmierstoffe*, Bd. 2, Vincent Verlag, Hannover, p. S.161ff.

15 Omeis, J. and Pennewiß, H. (1994) *ACS Polym. Preprints*, **35** (2), 714.

16 Selby, T.W. (1958) The non-Newtonian characteristics of lubricating oils. *ASLE Trans.*, **1**, 68–81.

17 Schodel, U.F. (1992) Automatic transmission fluids. International Tribology Colloquium, TAE, Esslingen, 3, 22.7–1.

18 Winter, H., Michaelis, K. and O'Conor, B.M. (1986) Mineralöltechnik, p. 8.

19 Bair, S., Queresti, F. and Winer, W.O. (1992) ASME/STLE Tribology Conference, San Diego, Paper No. 92/3.

20 Mortimer, M. (1996) Laboratory shearing tests for viscosity index improvers. *Tribotest J.*, **2–4** (6), 329.

21 Stöhr, T., Eisenberg, B. and Müller, M. (2009) A new generation of high performance viscosity modifiers based on comb polymers, SAE Technical Paper Series, 2008-01-2462.

22 Bottcher, W. and Jost, H. (1991) SAE Technical Paper 912410.

23 (1968) Petroleum additives – pour point depressants. *Encycl. Polym. Sci. Tech.*, **9**, 843–844.

24 Raddatz, J.H. (1994) Detergent-dispersant additive, ihre herstellung, anwendung und wirkungsweise, in *Additive für Schmierstoffe* (ed. W.J. Bartz), Expert Verlag, pp. 163–196.

25 Small, N.J.H. (1978) Dipersants, in *Katalysatoren, Tenside und Mineralöladditive* (eds J. Falbe and

U. Hasserodt), Georg Thieme Verlag, Stuttgart, pp. 235–239.

26 Stewart, W.T. and Stuart, F.A. (1963) Lubricating oil additives, in *Advances in Petroleum Chemistry and Refining*, vol. 7, Interscience, New York, pp. 3–64, Chapter 1.

27 Omeis, J. and Pennewiss, H. (1996) European Patent EP0699694.

28 Rath, P. and Mach, H. (1999) Highly reactive polyisobutene as a component of a new generation of lubricant and fuel additives, *Lubr. Sci.*, **11** (2), 175–185.

29 Verkouw, H.T., Jonkman, L., deJong, F., Sant, P. and Barnes, J. (1990) Mineralöltechnik.

30 Bansal, J.G. and Lefcourt, S. (1998) Proceedings of the 4th Annual Fuels & Lubes Asia Conference, p. 21.

31 Auschra, C., Omeis, J. and Pennewiss, H. (1997) US Patent 5597871.

32 Hedrich, K. *et al.* (2000) International Tribology Conference, Nagasaki.

33 Hellberg, H. (1978) Schaumdampfer, Farbstoffe und Geruchsüberdecker, in *Katalysatoren, Tenside und Mineralöladditive* (eds J. Falbe and U. Hasserodt), Georg Thieme Verlag, Stuttgart, p. 257.

34 Colour Index International (1997) *Pigments and Solvent Dyes*, Edition, The Society of Dyers and Colourists, Bradford, West Yorkshire, UK.

35 Forbes, E.S. (1970) Antiwear an extreme pressure additives for lubricants. *Tribology*, **3**, 145.

36 von Eberan-Eberhorst, C.G.A., Hexter, R.S., Clark, A.C., O'Connor, B. and Walsh, R.H. (1994) Aschegebende extreme-pressure- und verschleissschutz-additive, in *Additive für Schmierstoffe* (ed. W.J. Bartz), Expert Verlag, pp. 53–83.

37 Fessenbecker, A., Rohrs, I. and Pegnoglou, R. (1996) Additives for environmentally acceptable lubricants. *NLGI Spokesman*, **60** (6), 9–25.

38 Rossrucker, T. and Fessenbecker, A. (1999) Performance and mechanism of metalworking additives: new results from practical focused studies, STLE Annual Meeting, Las Vegas.

39 Voltz, M. and Rulfs, H. (1982) Leichtlauf durch schmieröle – ursachen und wirkungen. *Mineralöltech.*, **27** (6), 1–23.

40 Papay, A.G. (1983) Oil-soluble friction reducer. *Lubr. Eng.*, **39** (7), 419–426.

41 Christakudis, D. (1994) Friction modifier und ihre prüfung, in *Additive für Schmierstoffe* (ed. W.J. Bartz), Expert Verlag, pp. 134–162.

42 Grossiord, C., Martin, J.M., Mogne, T. Le. and Inoue, K. (1999) Friction reduction mechanisms of MoDTC/ZnDTP combination: new insights in MoS2 genesis. *J. Vac. Sci. Technol.*, **A17** (3), 884.

43 Ford, J.F. (1968) Lubricating oil additives – a chemist's eye view. *J. Inst. Petrol.*, **54**, 198.

44 Röhrs, I., Hunter, M.E. and Hessell, E.T. (2000) Next generation of ashless rust and corrosion inhibitors for modern hydraulic fluids. Proceedings of the 12th International Colloquium Tribology, 11–13 January 2000, Stuttgart/Ostfildern, Germany.

45 Olszewski, W.F. and Neiswender, D.D. (1976) Development of a rear axle lubricant, SAE Paper 760, p. 326.

7
Lubricants in the Environment

Rolf Luther

7.1
Definition of 'Environment-Compatible Lubricants'

Ecological aspects are gaining importance in our society. Bearing in mind that our environment is being increasingly contaminated with all kinds of pollutants, any reduction is welcome. From an environmental point of view and compared to a number of other chemical products, lubricants are not particularly problematic. A large proportion of the lubricants pollute the environment either during or after use. This can be technically desired (total-loss lubrication) or a result of mishaps such as leaks, emissions, spillages or other problems.

Lubricants and functional fluids are omnipresent due to their widespread use and they thus pollute the environment in small, widely-spread amounts and rarely in large, localized quantities.

The terminology used in connection with 'environmental compatibility' has to be split between the subjective and the objective.

Subjective criteria (non-measurable) are as follows:

- Environmentally friendly.
- Environmentally compatible.

Objective criteria (measurable or provable), for example are as follows:

- Biodegradability.
- Water solubility, water pollution.
- Ecological toxicity and physiological safety.
- Performance, approvals, oil change intervals.
- Efficiency improvements, lower energy consumption.
- Emission reduction in use.
- Compatibility with conventional lubricants and materials.
- The use of renewable raw materials (bio-based content).
- Environmental awards.

Lubricants and Lubrication, Third Edition. Edited by Theo Mang and Wilfried Dresel.

Other wordings for 'environment-compatible lubricant' have got greater promulgation in the last years. Since biodegradability of at least 60% according to OECD 301 is considered as an important objective criterion, the term 'biolubricant' came up more and more over the last decade. However, the 'bio-'prefix itself is not adequately defined, as to be seen in the following overview:

Origin of material	**Biodegradability**	**Example**	**The meaning of the prefix 'bio-'**
Renewable	Rapidly Biodegradable	Rapeseed oil, TMP trioleate (TMP-O)	Biodegradable and bio-based
Non-renewable	Biodegradable	Di-isotridecyl-adipate (DITA)	Biodegradable
Renewable	Slowly or non-biodegradable	Hydrocarbons from process 'biomass-to-liquid' (BtL)	Bio-based
Non-renewable	Non-biodegradable	White oil for foodgrade lubricants	Biocompatible

This unclear concept was the main reason for a standardization process in regard to bio-based products in Europe. In 2007, the European Commission has started 'A Lead Market Initiative for Europe' (LMI; COM (2007) 860 final). Within this program 'bio-based products' were identified as one very promising market for the industrial future of Europe. 'Biolubricants' are explicitly mentioned as product group of special interest, for example 'bioplastics' and 'biosolvents'.

A summarizing report of the 'Ad-hoc Advisory Group for Bio-based Products' in the framework of the European Commission's Lead Market Initiative was published in November 2009: 'Taking Bio-Based from Promise to Market', which recommends different measures to promote the market introduction of innovative bio-based products.

As an important issue, the LMI asked for standards in regard to bio-based products. The EU Mandate M/430 of October 2008 for the development of European Standards for Bio-Polymers and Bio-Lubricants in relation to bio-based product aspects was addressed to CEN, the European Standardization Body. First step was the establishing of a Working Group 'Bio-Lubricants' (CEN TC19/WG33), under the umbrella of the CEN Technical Committee 19 'Gaseous and liquid fuels, lubricants and related products of petroleum, synthetic or biological origin'. WG33 is dedicated to all aspects regarding standardization of biolubricants.

In the first step, a technical report was prepared and published under CEN TR16227 (2011). Within this report a definition of the term 'biolubricant' is proposed. A European Standard based on this technical report is under development (CEN prEN 16807); the criteria will be nearly the same.

7.1.1 CEN Technical Report 16227 – Standard Designation of the Term Biolubricant

Generic terms such as biolubricant frequently appear in the common language, mainly as marketing tools, and lubricants are far from the only offenders – plastic and detergent products are other examples of the widespread use of the bio-prefix. Without reference to clear and agreed definitions, there are many unsubstantiated claims, which can be deceptive.

Hence, any claim using a 'bio' prefix should refer to an internationally agreed standard. A clear distinction should be made between the origin of the raw materials and their functionality. Rules should be standardized to define when such claims are justified. This is the only way to bring transparent and non-misleading information to consumers (as defined in ISO (International Standards Organization): ISO 14021 on environmental self-claims).

The umbrella definition for all types of biolubricants should be properly designed to deliver lubricants reasonably well for the intended application and for the (changing) needs of the identified market outlets, preferably at minimum costs to society. The outcome could be that a clear identification and a separate collection may be advisable to solve this concern.

Definitions and standards forming the basis for certification purposes are in place for the biodegradability and the toxicity of lubricants. A similar system may be useful to avoid misleading claims on bio-based content. The term 'biomass-based', found on the test methods described by ASTM D 6866 or equivalent, could be used to indicate products that are partly or totally made from renewable raw materials (i.e., with a biological carbon content).

For the sake of an unequivocal B2C communication and, therefore, to avoid market-hindering confusion, the two pieces of information (bio-based content and biodegradability/toxicity) should be provided together.

Moreover, in order to utilize the technological potential of renewable resources, it is important to have a maximum freedom for innovations. Transparency in the market without restrictive quotas and limitation in the raw materials applications is a framework required for an efficient and environment-conscious utilization of resources.

For these purposes, the CEN Technical Report 16227 proposes the following:

European-wide Standardized Minimum Requirements for 'Biolubricants'

- Renewability (i.e. in the view of TR 16227 the bio-based carbon content):
  Content of renewable raw material ≥25% according to ASTM D-6866. (Radiocarbon analysis) or equivalent CEN version.
- Biodegradability:
  ≥60% according to OECD 301 for oils.
  ≥50% according to OECD 301 for lubricating greases.

- Toxicity:

  Not to be labelled as 'dangerous to the environment' according to CLP-Directive 1272/2008/EC. This can be proven for the fully formulated product by testing according to OECD 201/202/203: EC50/LC50/IC50>100 mg l^{-1}.
- Performance:

  'Fit for purpose' or 'Fit for use'.

In particular, it is remarkable that according to these criteria every announcement with regard to biodegradability, toxicity and bio-based content can be measured through the final product in hands of the customer.

This is a real distinction to all ecolabels, whose reputation is based on certification processes by authorities; in general, the certified data can not be measured by the end user.

Based on CEN/TR16227 an European Standard came into force in October 2016: EN 16807 "Liquid petroleum products – Bio-lubricants – Criteria and requirements of bio-lubricants and bio-based lubricants", with adequate criteria.

Moreover, for reasons of statistical distinguishability, a new Combined Number (CN) within the EU Common Customs Tariff was created for Bio-Lubricants:

CN code: 3403 19 20

Description: Lubricants having a bio-based carbon content of at least 25% by mass and which are biodegradable at a level of at least 60%.

7.2 Current Situation

Worldwide mineral oil and its derivatives are dominating the lubricants market. But this triumphant progress is limited to the last century. In former days of mankind the friction and wear decreasing properties of natural oils and fat were well known and used absolutely in many different ways. In this respect the development in the last 35 years to formulate biodegradable lubricants based on natural oils followed the tradition – even if the market share today only amount to few percent.

The statistical data and the main legislation in this chapter apply to experiences in Germany, which has – beside some countries in Western Europe (Holland, Austria, Switzerland) and Scandinavia – the highest amount of biodegradable lubricants worldwide.

7.2.1 Statistical Data

Every year, about 40–50% of the approximately 5 million tonnes of used lubricants in Europe end up polluting the environment. This interface with the environment and the resulting pollution of the atmosphere, the ground or water is either technically desired (total-loss lubrication) or follows leaks or other problems.

The environmental damage caused by mineral oil-based lubricants is largely caused by the approximately 40% of lubricants which are not properly disposed

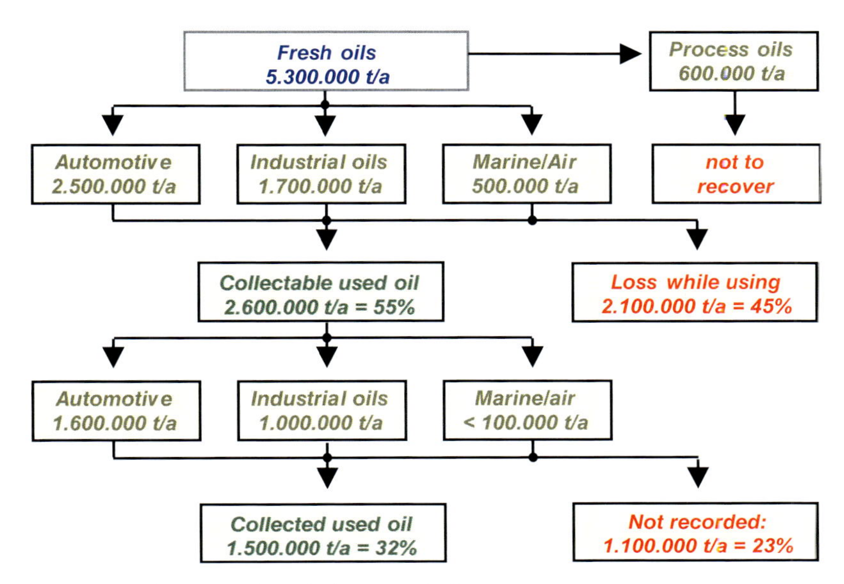

Figure 7.1 Lubricants and residual amounts in Europe.

of. This figures included total-loss applications, the residual oil in about 90 million oil cans and 20 million oil filters, spillages during topping-up, leaks, drips from separated oil-line and hydraulic couplings, accident losses and all manner of emission losses (CONCAWE Report no. 5/96, see Figure 7.1). CONCAWE is the oil companies' European organization for environment, health and safety.

Even though mineral oil products can be relatively rapidly biodegraded by the microorganisms present in nature, these natural degrading systems are overwhelmed by the volume of the losses. In recent years, rapidly biodegradable total-loss lubricants based on esters, polyglycols and natural oils have significantly eased lubricant-based pollution.

In the past, mineral oil-based products were used almost exclusively. Leakages and other causes allowed a part of these lubricants to pollute the environment. Soil and ground water is severely contaminated by lubricants and particularly some of the ingredients in such products. The problem has spurred society and politics into action. This has been reflected by diverse legislation and recommendations, as well as the issuing of environmental awards to products that cause less damage. The aim of this all is to avoid or, at least, reduce environmental pollution. The use of environment-compatible materials and substances will play an increasingly important part in the scope and procedures of the EU legislation. The most effective measure to prevent environmental pollution is and remains the avoidance of physical containment by lubricants. The technical prerequisites for this already exist and should be fully utilized.

Technically, disregarding development and overall costs, it is accepted that more than 90% of all lubricants could be made rapidly biodegradable. However, this change will only take place if the ecological benefits of using harvestable raw materials is made clear and, best of all, quantified in monetary terms.

For Germany, a relatively new statistical evaluation is available (Marktanalyse Nachwachsende Rohstoffe, Schriftenreihe Nachwachsende Rohstoffe, Band 34 (2011)).

This survey has resulted in the following numbers:

Values in mt per annum	RRM amount >50%	RRM amount >25%	Biodegradable according to OECD 301
Engine, compressor and turbine oils	0	1 500	2 000
Gear oils	0	1 500	2 000
Hydraulic fluids	5 000	5 000	10 000
Transformer oils	0	10	10
Machine oils (including chainsaw oils etc.)	2 500	4 500	4 500
Non lubricating oils, for example mould release agents	1 000	2 500	3 800
Process oils	0	0	0
Metal working fluids, including coolants	0	1 500	5 000
Lubricating greases	600	000	2 500
Total	9 100	18 510	29 810

In comparision to former numbers, which have assumed approximately 40 000 mt for Germany, this is a remarkable reduction; however, the explanation seems obvious. The former results included additionally all products which were biodegradable according to the obsolete test method CEC L-33-A-93.

7.2.2
Economic Consequences and Substitution Potential

Now that corresponding laws and guidelines are in place, environmental legislation has economic repercussions. In addition to national laws, European guidelines are gaining relevance in the different European countries. Accordingly, production facilities must comply with EU machinery guidelines. For example, this supercedes previously applicable German Legislation concerning safety and the protection of human health and, along with thorough documentation, includes mandatory measures to reduce operator stress and fatigue. Compared to older plant, this provides the operators of new machinery with greater safety. The implementation of these regulations in practice requires the use of environmentally harmless lubricants.

In Germany the selection of rapidly biodegradable lubricants, especially those classified as 'not water pollutant' reduce the expense of oil spillages or disposal as well as the cost of complying with health and safety regulations to a minimum. Compared to mineral oil-based products, the selection of low-evaporation ester and vegetable oil-based lubricants can allow lower viscosities to be used. This enables faster machining and cuts drag-out losses. This, in turn, increases economy and productivity and generates product optimization.

Active environmental protection is steadily gaining importance. The trend is towards clean and environment-friendly factories in which people like to work and which combine the requirements of advanced industrial technologies with environmental protection considerations. This is a major objective for a number of companies in a variety of industries. In the field of metalworking, for example ecologically friendly cutting fluids can significantly cut the costs of the following:

- Disposal (see Chapter 14 'Water-Miscible Cutting Fluids')
- Complying with water laws
- Maintenance (see Chapter 14)
- Downtime
- Washing operations between machining steps
- Water-reconditioning in splitting plants.

Furthermore, environmentally harmless products are generally non-toxic and skin compatibility is mostly much better than mineral oil products. This creates ecological and economic points of view which promote the use of environment-friendly lubricants.

The vast majority of rapidly biodegradable lubricants currently on the market are based on saturated or unsaturated ester oils. The acids required for esterification are mostly from vegetable oils. Compared to equivalent mineral oils, ester oils are significantly more expensive. This is a major barrier to the wider acceptance of such oils. But the overall cost of a lubricant results from the interaction between the following characteristics (in alphabetical order):

- Ageing and temperature stability.
- Disposal costs.
- Machine compatibility (multifunctional oil).
- Machine investments.
- Maintenance measures.
- Oil change intervals.
- Operating costs.
- Reconditioning possibilities.
- Reduction in storage costs.
- Reduction in emission reduction measures.
- Reduction in the cost of preventative health and safety measures.
- Risk minimization and lower clean-up costs after spillages.
- Simpler plant license procurement.

Completely new mix of economic aspects can result from every new application. In particular, the use of fully saturated synthetic esters offers a number of opportunities. Pure vegetable oil formulations are now only used for lower demands, for example total-loss chain saw or mould release oils and some types of greases.

Users are greatly interested in the potential life of the new generation of environment-friendly lubricants. Rapeseed oil-based hydraulic fluids are normally used in lightly stressed systems and changed at the same intervals as mineral oils. Increasing use is being made of ecologically friendly fluids based on saturated and unsaturated synthetic esters. These satisfy the technical demands which are defined as higher performance, longer life, low oil volume and more rapid oil circulation times. At the same time, the disposal of ester-based lubricants is generally unproblematic.

7.2.3
Agriculture, Economy and Politics

Analysis of the economics of renewable raw materials often includes high European Union agricultural subsidies, of rapeseed oil, for example as a critical factor. This is relatively insignificant for the manufacture of biolubricants. European Union vegetable oil prices depend on world market prices. The world market price of vegetable oils is largely affected by the price of soybeans. Soya accounts for a large fraction of total vegetable oils (Table 7.1). Nearly half of the world's soybean production is in the United States; the volume of soybean oils produced in South America (Brazil, Argentina) is increasing. Vegetable oil prices, and thus also oil acid or oil ester prices are, therefore, heavily dependent on the size of the soya harvest.

In 2011, the worldwide consumption of vegetable oils and fats amounts to 176.5 million mt. During the period between 2006 and 2011 the consumption of vegetable oils is increased by approximately 20%. In particular, the demand for palm oil increased by 36%, rapeseed oil by 34%, palm kernel oil by 30% and soybean oil by 23%.

A significant correlation of the prices of vegetable oils with the price of crude oil must be recognized in some respects, because the use of vegetable oils in combustion engines, both for energy production and by traffic, is increasing. One driving force is the European Directive 2003/30/EC of 8 May 2003 on promotion of the use of biofuels or other renewable fuels for transport.

Minimum proportions of biofuels and other renewable fuels have been laid down as strategic commercial targets. A reference value for these targets was 2% by the end of 2005 and 5.75% by end of 2010, calculated on the basis of the energy content of all petrol and diesel for transport purposes. If these European targets are to be achieved, domestically grown oil crops will not be sufficient to

Table 7.1 Consumption of important natural oils and fats.

Oil seed	World consumption of oil in 2011 (million mt)	Largest producer
Soy	42.4	USA
Palm	47.7	Malaysia
Palm Kernel	4.8	Malaysia
Coconut	4.0	Indonesia
Rapeseed	24.7	China
Sunflower	12.7	Russia

Source: Oil World, 2011.

support the fuel production (mainly biodiesel and ethanol) necessary. Global markets will be able to provide sufficient quantities of oil seed and oils, but with a consequent effect on the prices of vegetable oils.

In Germany, currently the leading country in the use of biodiesel (the methyl ester of rapeseed oil), approximately 2.4 million tonnes of biodiesel were produced in 2011.

Technical crops can replace the reduction in the over-production of food and animal feed products in Europe. Rapeseed oil and derived oleochemical products are especially key products for rapidly biodegradable lubricants. In Europe, rapeseed and sunflower oils as lubricant base oils have been in the forefront, for the following reasons:

- They are abundantly available, growing well in central European climates
- Their thermal oxidation stability is acceptable for some applications.
- Their flowing properties are better than those of other vegetable oils.
- Their oleochemical derivatives combine performance with good environmental properties.

When subjected to severe thermal exposure, especially, pure vegetable oils have application boundaries comparable with similar mineral oil products. The main disadvantage of these oils is that their oxidation, hydrolytic, and thermal stability is not adequate to enable their use in circulating systems.

Apart from pure vegetable oils, oleochemical esters are being used increasingly, because their stability to thermal oxidation compared with that of rapeseed oil opens up a wider field of applications. These products are, however, substantially more expensive, a factor which, in turn, limits their use.

Environmentally harmless lubricants based on renewable material have good potential, even if prices of the raw materials are also rising. In addition to direct environmental advantages, synthetic esters have technical advantages for lubricants in comparison with conventional base oils.

In 2005, approximately 20 000 hectares of rapeseed were cultivated in Germany solely for application in lubricants. This corresponds to approximately 20 000 tonnes rapeseed oil per year, used both as biolubricants and as (lubricity) components of conventional lubricants. It can be assumed that approximately 20 000–25 000 tonnes of oleochemical esters are also used in lubricants every year. In Germany the percentage of biodegradable lubricants amounts to approximately 3% of all lubricants, approximately 30 000 tonnes (2011).

7.2.4 Political Initiatives

In Germany, currently the most important market for biodegradable lubricants, several political initiatives were started in the last decade of these, some were included in two comprehensive reports submitted to the German parliament.

- *Resolution dated 16 June 1994*: Measures to promote the use of rapidly biodegradable lubricants and hydraulic fluids

- *Resolution dated 28 November 1995*: News from the Federal Ministry for Food, Agriculture and Forestry: 'Report on the use of rapidly biodegradable lubricants and hydraulic fluids and measures taken by the Government' (New Edition 1999)

As a consequence, the German Ministry of Agriculture launched the Market Introduction Program for bio-based fuels and lubricants in the year 2000, the focal point of which was lubricants. Using targeted promotional activity, the Ministry hopes that special vegetable oil-based lubricants will gain a stronger footing in the German market.

Technically, rapidly biodegradable alternatives based on harvestable raw materials are available as replacements for most mineral oil lubricants. Bio-based lubricants are also available for particularly demanding applications, for example engine lubrication. Nevertheless, bio-based lubricants have failed to achieve market penetration reflecting their technical performance and ecological potential, mainly because of their high price.

To improve this situation, the Market Introduction Program for bio-based lubricants was started in Germany. This program promotes changing from mineral oil-based lubricants to rapidly biodegradable products made from renewable raw materials. The principal condition is application in the ecologically sensitive areas of agriculture, forestry and water supply. This thus exploits the risk reduction potential of bio-based lubricants without the need for special environmental legislation.

After 5 years of experience with this Market Introduction Program it can be stated that the approach was a success under difficult conditions. The program is ongoing.

Research and development on ecologically harmless lubricants is sponsored by European and national projects. The government offers research sponsorship for projects aimed at improving and demonstrating the performance of biodegradable lubricants, for example producing lubricants which are thermally and hydrolytically stable, and stable to oxidation, without significantly altering their biodegradability.

At a European level the 'European Renewable Resources and Materials Association' (ERRMA), a representative body promoting the use of renewable raw materials in technical applications, focuses its efforts on biolubricants and bioplastics, because of the forecasts for these product groups.

7.3 Tests to Evaluate Biotic Potential

The possibility of measuring the environmental impact of lubricants was often discussed. This issue is to differentiate in tests regarding the biodegradability in the natural environment (water, soil) and in measuring the ecotoxicological potential for damage the health of creature.

7.3.1 Biodegradation

The limit for rapid biodegradation is not exactly defined, but the most legislation and recommendations are working with these numbers for not water soluble lubricants:

Degradation>60% according to OECD 301B.

The most important tests are shown in Table 7.2 –beyond these tests are many ISO, ASTM and national procedures with more or less similar methods and restrictions.

The CEC L-33-A-93 test is obsolete since years; no important regulation refers to this test any longer. It is normally substituted by an OECD 301 procedure.

Meanwhile (2010), a CEC Working Group has developed a new test method, published under CEC-L-103. However, this method again is primarily based on degradation, similar to CEC L-33-A-93. The new method was not considered in subsequent directives or regulations (e.g. eco-labels) as of now.

7.3.2 Ecotoxicity

To develop environmentally acceptable lubricants, toxicological criteria must be considered. The aim is to protect life in various areas, especially in water

Table 7.2 Most important methods for testing the biodegradation of lubricants.

Test	Short description – relation to other tests	Limit for 'rapid degradation'
OECD 301B	'Modified sturm test', aerobic degradation, ultimate biodegradation	≥60% in 28 days
OECD 301C	'Modified MITI test', aerobic degradation, measurement of O_2 consumption, for volatile components	≥60% in 28 days
OECD 301D	'Closed bottle test', aerobic degradation, preferred for water soluble products, but possible for not water soluble substances	≥60% in 28 days
OECD 301F	'Manometric respirometry test'; for water soluble and not water soluble substances	≥60% in 28 days
OECD 302B	'Zahn–Wellens/EMPA test', assesses inherent biodegradability for water soluble substances	—
CEC-L-103	'Inherent degradation', aerobic degradation, water soluble part of degradation not considered, only for bad water soluble substances. Because of primary degradation the values are higher than OECD 301 results – up to 20%.	≥80% in 21 days
BODIS test, ISO 10708	Two phases 'closed bottle' test, similar OECD 301D	—

(aquatic area) and in the non-aquatic area (terrestrial area). The following, especially in the ecolabelling systems used test procedures are of importance:

- Bacteria toxicity, according to DIN 38 412-8; this determines cell-multiplication inhibiting (EC_{10} and EC_{50} values). The Pseudomonas type used for this test is found in wastewater and soil.
- The bacteria test, according to OECD 209 or ISO 8192, determines acute toxicity through the inhibiting of oxygen consumption; results of this test are EC_{50} values.
- The algae toxicity test according to OECD 201 or DIN 38 412-9 is yet another test for aquatic systems (measurement of chlorophyll fluorescence and determination of EC_{10} and EC_{50} values).
- One of the most important test procedures in German legislation concerning the aquatic area is a test on small living organisms called the 'Daphnia test' (daphnia magna STRAUS, water flea, small crustacean) according to DIN 38 412-11 or OECD guideline 202.
- In the aquatic area, fish toxicity, according to OECD 203 or DIN 38 412-15, performed on the Goldorfe (*Leuciscus idus*), is of importance. Test results are the LC_0, LC_{50} and LC_{100} values.
- Fish toxicity, according to OECD guideline 204, is incorporated in the German eco label 'Blue Angel'. Possible pollution of the non-aquatic terrestrial area, that is soil and plants, is evaluated with the plant growth test according to OECD guideline 208 (i.e. wheat, cress and rape seeds).

Naturally, toxicity testing for environmental protection purposes must also include mammal and human toxicity and all the more occupational safety aspects.

7.3.3 Emission Thresholds

Every evaporating lubricant pollutes the atmosphere with its emission.

The evaporation loss of rapidly biodegradable lubricants is generally lower than that of conventional oils. This feature of biodegradable lubricants can help to meet such emission limits as known from cutting oils or engine oils.

7.3.4 Water Pollution

Because the national legislation in Germany is very ambitious, the principles of environmental relevant laws will be shown exemplary for this European country.

7.3.4.1 The German Water Hazardous Classes

The German Water Law serves to protect waters and, indirectly, the ground. A law specifically aimed at protecting the ground, a German Ground Law, is currently being draughted.

Amongst other points, the Water Law defines the water pollution potential of substances and mixtures with water hazardous categories (WHCs).

The measurements and calculations should be performed by a certified laboratory. In Germany, the figures thus gained are submitted to a 'Commission for the Evaluation of Water Polluting Substances'. A substances evaluation and classification in the catalogue of water-polluting substances is then performed by this Government commission. In 1996, new provisions concerning water-polluting substances were implemented and the substances classified in one of four 'water pollution categories' or 'water hazardous classes' (WHC 1–4). In addition, these provisions contained a legally binding directive on how the water pollution category of mixtures should be calculated.

These 1996 provisions were revised on 1 June 1999. The most important change was the linking of water pollution categories to the R-phrases in European Hazardous Substances Legislation with the ultimate objective of harmonizing the evaluation of substances. The principal weakness of the previous system, the poor transferability to international regulations, has thus been eliminated. Water pollution categories neither evaluate the correct use of substances nor what is finally disposed of into sewers or waterways.

The procedure is as follows:

- Classification of the points to each R-phrases (in correlation of the meaning for the safety aim and the previous classification procedure).
- Definition of basis data (acute oral toxicity, one aqueous toxicity, biological degradation, bioaccumulation); distribution of default-values, in case these characteristics have not been determined.
- Derivation of a WHC from the total amount of points.

In the latest version of the WHC system, the former WHC 0 no longer exists; it was replaced by the new category 'not water hazardous' ('nwh') – on the one hand with more restrictions but on the other hand with more advantages in use. The WHCs are defined as follows.

WHC 1 = Slightly water polluting (e.g. mineral oils without additives or simple oils)
WHC 2 = Water polluting (e.g. mineral oils with additives)
WHC 3 = Highly water polluting (e.g. water-miscible cutting fluids, lubricants with emulsifiers, special lubricants).

The determination of WHC of formulated lubricants, that means blends, is described in the 'mixture regulation' (Table 7.3). For WHC 1–3, both the classification of the overall formulation and the classification according to the mixture regulation are possible. For 'not water hazardous' blends, only the mixture regulation is provided.

The principal means of generating the WHC is shown in Figure 7.2.

Table 7.3 Mixture regulation.

Containing materials				
WHC 3 (%)	≥3	0.2–3	<0.2, as supplement	<0.2, no supplement allowed
WHC 2 (%)	—	≥5	0.2 bis 5	<0.2
WHC 1 (%)	—	—	≥3	<3
Not water hazardous				
R 45 phrase (%)	≥0.1	≥0.1, but WHC 2	<0.1, assupplement	<0.1, no supplement allowed
(⟹ Result:	WHC 3	WHC 2	WHC 1	Not water hazardous

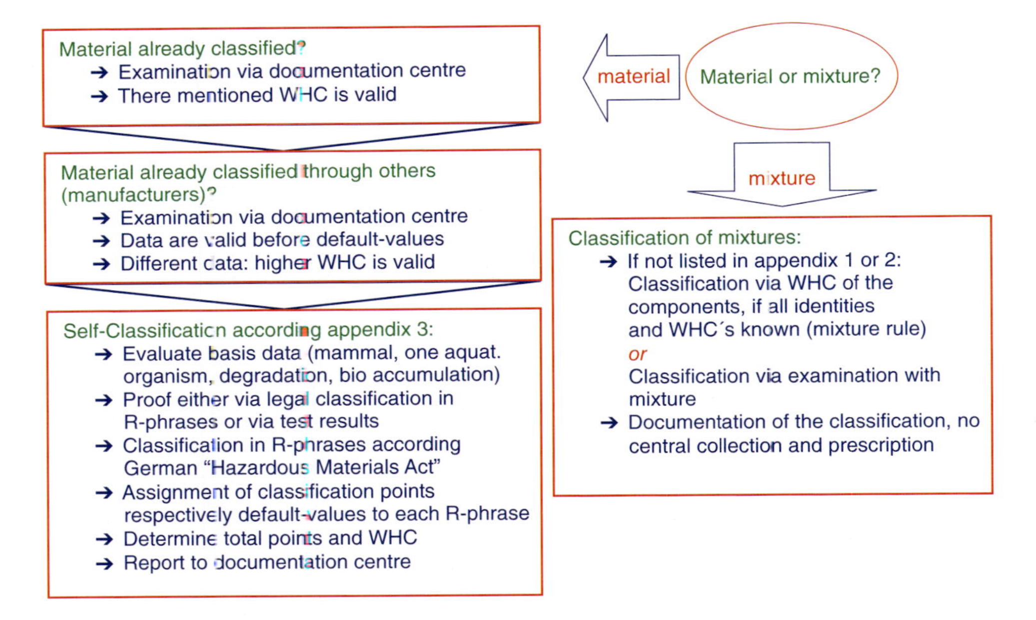

Figure 7.2 Scheme for determination of the WHC.

7.3.4.2 German Regulations for Using Water-Endangering Lubricants (VAwS)

Water hazard classes have a major influence on the storage and handling of these substances. The German regulation for using water-endangering lubricants (VAwS) considerably will take greater care with highly water polluting substances than for non-water-hazardous substances (Table 7.4). Hazard potential based on water hazard class and quantity is divided into three groups when manufacturing, processing, storing or application of water polluting substances is

Table 7.4 German regulations for use of water-endangering lubricants.

Volume of the stock plant (m^3)	Non-water hazardous materials and preparations are not subject to the VAwS	WHC 1	WHC 2	WHC 3
≤1		F0 + R0 + I0	F0 + R0 + I0	F1 + R2 + I0
>1–≤10		F1 + R0 + I1	F1 + R1 + I1	F2 + R2 + I0/ F1 + R3 + I0
>10–≤100		F1 + R1 + I1	F1 + R1 + I2/ F2 + R1 + I1	F2 + R2 + I0/ F1 + R3 + I0
>100		F1 + R1 + I2/ F2 + R1 + I1	F2 + R2 + I0/ F1 + R3 + I0	F2 + R2 + I0/ F1 + R3 + I0

planned. The highest level is achieved when only 1 m^3 of water-miscible cutting fluid (WHC 3) is stored. On the other hand, a rapidly biodegradable lubricant with a nwh-certificate must no longer fulfil the requirements of WHC 1–3.

Ground areas:	F0–no special requirements, internal company requirements are valid
	F1–impervious areas
	F2–like F1, but with proof
Storage capability:	R0–no special requirements, internal company requirements are valid
	R1–storage capability is sufficient up to taking effect of the security precautions
	R2–storage capability equals to the volume of the liquid
	R3–double wall with leak detector
Infrastructure:	I0–no special requirements, internal company requirements are valid
	I1–independent signalling unit in connection with permanent occupied production sites
	I2–alarm and measurement plan

Therefore, it is one objective for ester-based products in Germany, especially in metalworking, to reach the classification 'not water hazardous' – the better classification creates an 'added value' for the higher cost of those lubricants.

Up-Coming Regulation in the Area of the German Water Law (WHG)

§19 of the German Water Law regularized in general:

'Water hazardous substances . . . are solid, liquid and gaseous substances . . . , which possibly harmfully change the physical, chemical or biological properties of water'.

Since some years a new regulation is under development, starting with a revised version of the WHG. Derived from this general approach, a draft regulation was presented in regard to handling of water hazardous substances (VUmwS). Particularly, the system with the three WHC and the classification as 'not water hazardous' (mentioned earlier) will be continued.

Subsequently, the derived draft 'General administrative directive for water hazardous classes' (VwVWGK) was presented, according to the following scheme:

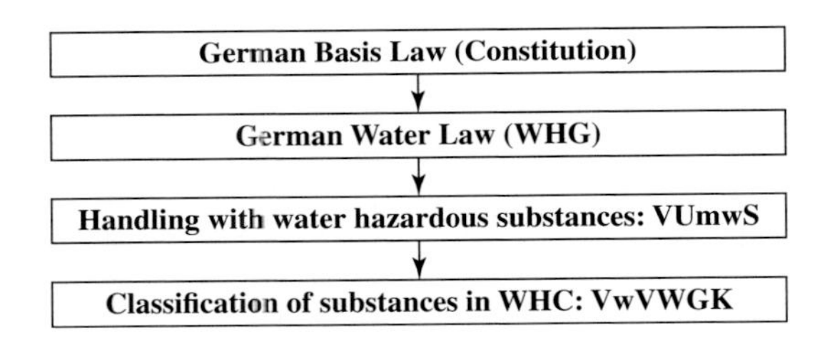

A disruption to the well-accepted regulation of 1999 is proposed for the classification of substances as 'not water hazardous' – with the so-called 'Floater Criterion' the authorties want to consider the floating of not water soluble substances on water:

2.1.1 Substances are not water hazardous, if the following requirements are fulfilled:
- a) The total number of points indicated by the R-phrases has to be zero.
- b) The water solubility of liquid substances is $<10\,mg\,l^{-1}$.
- c) The water solubility of solid substances has to be $<100\,mg\,l^{-1}$.
- d) No test procedure is known, which indicates an acute toxicity for fish (96 h LC50), daphnia (48 h EC50) or the inhibition of algae growth (72 h EC50) below the solubility limit. Valid tests for two of the mentioned organisms are done.
- e) Organic liquids have to be rapidly biodegradable.
- f) An under standard conditions solid organic substance is either rapidly biodegradable or shows no increased bioaccumulation potential (consideration of biodegradability for solid substances, too).

2.1.2 Newly introduced will be the 'Floater Criterion':

The regulation under 2.1.1 cannot be applied for organic liquids, which under standard conditions show a density $\leq 1000\,kg\,m^{-3}$, a vapour pressure $\leq 0.3\,kPa$ and a water solubility $\leq 1\,g\,l^{-1}$.

With this floater criterion, all vegetable oils for using as and in food would be classified no more as 'not water hazardous', but as WHC 1 'slightly water polluting'.

This draft regulation is – due to different reasons and controversial discussions – not entered into force up to now (2014); hence, the well-known old regulation (from 1999) is still valid, including the possibility to formulate 'not water hazardous' lubricants.

7.4 Environmental Legislation 1: Registration, Evaluation and Authorization of Chemicals (REACh)

On 13 February 2001, the European Commission adopted the White Paper 'Strategy for a Future Chemicals Policy'. This proposes a wide-ranging fundamental overhaul of the European Union chemical control legislation, that is the Dangerous Substances Directive (DSD), including the notification scheme for new substances, the Dangerous Preparations Directive (DPD), the Existing Substances Regulation, and the Marketing and Use Directive. In essence, legislation for new and existing substances would be merged. The current European Union chemical control measures result in too great disparity between new and existing substances, with the high cost of notification of new substance stifling innovation. Furthermore, existing substances account for >99% by volume of chemicals in commerce, but are poorly assessed and controlled in comparison.

The European Commission published the first draft of legislation intended to implement the White Paper on 7 May 2003. This is the Registration, Evaluation, and Authorization for Chemicals (REACh). The next step was an eight-week Internet consultation on the 'workability' of this legislation; approximately 6400 contributions were received. A revised version emerged in September 2003, jointly from the Enterprise and Environment Directorates of the European Commission. After consultation within the Commission, the final proposed Regulation was presented on 29 October 2003. These formal legislative proposals had to be discussed by the Council of Ministers and the European Parliament, under the Co-decision Procedure. The European Parliament approved a compromise text on 17 November 2005 and the Competitiveness Council reached political agreement on 13 December 2005. This paves the way for REACh to become law in the summer of 2007, with the European Chemicals Agency (ECA) as administration body, and to become fully operational a year later in Helsinki, Finland. Meanwhile, work on the REACh Implementation Projects (RIPs) continues, with the EU Joint Research Centre (JRC) getting ready to take a greater role in preparing the technical guidance documents. The new and existing substances regimes will continue until REACh starts to become operational.

REACh will place a duty on companies that produce, import and use chemicals to assess the risks arising from their use, and to take the necessary measures to manage any risks identified. Hence, the burden of proof that commercial chemicals are safe will be transferred from the regulators to industry. Testing results will have to be shared, to reduce animal testing, and registration of information on the properties, uses, and safe use of chemical substances will be an integral part of the new system. These registration requirements will vary, depending on the volume of a substance produced and on the likelihood of exposure of humans or the environment. A phase-in system lasting up to 11 years is planned. High-tonnage substances would require the most data, and would have to be registered first; lower-tonnage substances would require less data and be registered later.

Tighter controls will be introduced for the chemicals of highest concern; thus carcinogens, mutagens and reproductive toxicants (CMRs); persistent, bioaccumulative and toxic substances (PBTs) and very persistent and very bioaccumulative substances (vPvBs) will require authorization and hence will be registered early. Other substances of concern, for example endocrine disrupters, will be included on a case-by-case basis within the authorization system. Substances subject to authorization will have to be approved for a specific use, with decisions based on a risk assessment and consideration of socioeconomic factors.

The first reading of the proposed REACh Regulation by the European Parliament occurred on 17 November 2005, and considered over 1000 tabled amendments. A compromise text was approved. Pre-registration of all existing 'phase-in' substances will be in a single phase 18 months after the regulation comes into force, but with a further six months allowed for SMEs and downstream users. Earlier registration will be required for PBT and vPvB substances. Full safety data will only be required for registration of substances at 1–10 tonnes per annum if they are suspected of being carcinogenic, mutagenic or toxic for reproduction (CMR), or are assessed as classified as dangerous to human health or the environment and are for diffusive use, particularly if used by consumers. Reproduction toxicology data will normally be needed at 100 tonnes per annum and above instead of at 10 tonnes per annum. Waiving of studies on the grounds of low exposure is introduced for specific tests at all tonnages. The 'one substance one registration' (OSOR) requirement was agreed, but with the possibility of opting out if the cost would be disproportionate, where there would be a breach of confidentiality, or if there is disagreement about the hazardous properties, but sharing of animal testing results would still be mandatory, as also would sharing of results from non-animal testing if requested by one potential registrant. Authorization of very-high-concern substances will be subject to periodic review.

The European Commission considers that work on existing substances under the Existing Substances Regulation will be completed. New substance notifications made in the run up to REACh will be handled as their equivalents under REACh. Similarly, classification and labelling work will continue, but with a shift towards using the REACh documentation from 2006, and the 31st Adaptation to Technical Progress (ATP) of the Dangerous Substances Directive is planned for end of 2006.

The European Commission plans to bring into force a new regulation implementing the Global Harmonized Classification Scheme (GHS) at the same time as REACh becomes operational; this would be in accord with the stated political intention to implement GHS by 2008. A draft regulation to implement the GHS is in preparation.

7.4.1 Registration

All substances manufactured in or imported into the EU at ≥1 tonne per annum will be registered with the ECA, who will assign a registration number and perform a completeness check using an automated process, normally within three weeks. The registrations are forwarded to member state competent authorities and entered on to a database of registered substances. Under the compromise text, at least 5% of registration dossiers are to be checked in more detail by the ECA.

Registration will be needed before new substances are manufactured or imported. Manufacture or import of new substances can begin three weeks after the registration date, unless the ECA informs the registrant that the registration is incomplete.

All 'phase-in' substances, that is those listed in the European Inventory of Existing Chemical Substances (EINECS) or manufactured in the European Union 15 years before the regulation comes into force, must be registered in a prioritized review. In 1981 *ca* 100 000 existing chemical substances were known. The deadlines for registration of such 'phase-in' substances are based on the date the new regulation comes into force (Table 7.5). The compromise text requires earlier registration of substances classified as very toxic to aquatic organisms that may cause adverse effects in the aquatic environment (i.e. labelled with R50/53). A new manufacturer or importer of a phase-in substance can participate in the review and enter the EU market under the compromise text. It is important to note that new substances already notified under the current DSD scheme are regarded as registered under the new REACh system, but further information is required under REACh if the manufacture or import quantities are triggered. It is expected that *ca* 30 000 substances will be registered, with at least 10 000 requiring new testing.

Table 7.5 Deadlines for registration of 'phase-in' substances.

• Register new substances at ≥1 t.p.a. before manufacture or import	
• Pre-registration of all phase-in substances within a 6-month period, beginning 12 months after the regulation comes into force	
• Registration for phase-in substances (from the date the regulation is in force)	
CMRs (>1 t.p.a.)	3 years
>100 t.p.a. (R50/53)	3 years
>1000 t.p.a.	3 years
>100 t.p.a.	6 years
>1 t.p.a.	11 years
• Draft decisions for phase-in substances for further testing:	
CMRs (>1 t.p.a.)	5 years
>100 t.p.a. (R50/53)	5 years
>1000 t.p.a.	5 years
>100 t.p.a.	9 years
>1 t.p.a.	15 years

7.4.2
Evaluation

It is estimated that approximately 80% of registered substances will not proceed to the next stage of evaluation. Registration information for the *ca* 5000 substances exceeding a manufacture or import volume of 100 tonnes per annum will, however, have to be evaluated. The registration dossier for substances at 100 tonnes per annum includes a proposal for so-called 'Annex VII testing', and some substances at levels below this tonnage will also have additional proposed testing. For new substances, the ECA will evaluate the proposal and produce a draft decision within 180 days. Evaluation of testing proposals for phase-in substances must be completed as shown in Table 7.5, with priority being given to substances with CMR, PBT, vPvB or sensitizing properties. The registrant is set a deadline to submit the additional studies for examination by the ECA. At 1000 tonnes per annum an equivalent procedure is followed for 'Annex VIII' testing.

There is also a procedure for evaluating substances. The ECA select substances for evaluation on the basis of a risk-based approach, taking into account the hazardous properties, including of analogous substances, exposure, and tonnage, including aggregated tonnage from all registrants. Such substances are evaluated by Rapporteur Member States, who select substances from the EU rolling action plan. The outcome may be EU-harmonized classification, restrictions or adding to the list for authorization.

7.4.3
Authorization

Substances of very high concern will have to be authorized before being used for specific purposes demonstrated to be of negligible risk. It is estimated that *ca* 1400 substances will be subject to authorization. There will be a published list of these very-high-concern substances that are candidates for authorization. Very high concern substances are substances classified as category 1 or 2 carcinogens, mutagens or toxic for reproduction (CMRs), persistent, bioaccumulative, and toxic (PBT), or very persistent and very bioaccumulative (vPvB). The PBT and vPvB criteria given in Annex XII of the Regulation are summarized in Table 7.6. Endocrine disruptors not covered by these criteria will be added to the list of very high concern substances on an *ad hoc* basis.

The first step is to identify existing substances, or particular uses of substances, requiring authorization, and to decide on a deadline for authorization and any uses exempted from authorization. When additional very-high-concern substances are identified, largely from testing for registration and evaluation, they will be fed into the authorization system.

Particular uses of very-high-concern substances will be authorized in the second step on the basis of a risk assessment covering all stages of the lifecycle for that particular use submitted by industry. The risk assessment will focus on exposure assessment in use, and no new studies would usually be required.

Table 7.6 Criteria for identification of PBT and vPvB.

Criterion	PBT criteria	vPvB criteria
P	Half-life >60 d in marine water, or >40 d in fresh or estuarine water, or >180 d in marine sediment, or >120 d in fresh or estuarine water sediment, or >120 d in soil	Half-life >60 d in marine, fresh or estuarine water, or >180 d in marine, fresh or estuarine water sediment or >180 d in soil
B	BCF >2000 in fresh or marine aquatic species	BCF >5000
T	Chronic NOEC <0.01 mg l^{-1} for fresh or marine water organisms, Category 1 or 2 carcinogen or mutagenic or Category 1, 2 or 3 toxic for reproduction or chronically toxic (i.e. classified as T or Xn with R48)	Not applicable

- BCF is bioconcentration factor, NOEC is no-observed effect concentration, and CMR is a substance classified as carcinogenic, mutagenic, or toxic for reproduction.
- For marine environmental risk assessment, half-life data in freshwater sediment can be superseded by data obtained under marine conditions.
- Substances are classified when they fulfil the criteria for all three inherent properties for P, B and T. There is some flexibility, however, for instance when one criterion is marginally not fulfilled but the others are exceeded substantially.

There is the possibility of authorization based on an adequate control of exposure, but not for PBTs, vPvBs, and 'non-threshold' CMRs. The ECA can, however, take into account socioeconomic factors when deciding if use of the substance can nevertheless be authorized in the European Union The compromise text gives greater emphasis to the substitution principle, and applications for authorization must be accompanied by analysis of possible alternatives with their risks and the technical and economic feasibility of substitution.

The compromise text introduced an amendment to require authorizations to be subject to time-limited review, to enable further consideration of alternative substances. Authorization is also reviewed if information on possible substitutes is submitted to the ECA.

Restrictions for persistent organic pollutants (POPs) required under the Stockholm Convention will also be implemented through the restrictions provisions of the Regulation.

7.4.4
Registration Obligations

Substances manufactured or imported, either neat or in a preparation, at >1 tonne per annum have to be registered, unless exempted. There is the option for a non-EU manufacturer to appoint an EU representative to register the substance an behalf of the EU importer(s).

Some substances in articles are subject to registration, and the provisions have been clarified under the compromise text. Substances in articles must be registered if they are intended to be released from the article and are supplied at >1 tonne per annum. Instead of registration, a less onerous procedure of notification applies to substances present in articles at >0.1% unless release of the substance is excluded. Under the compromise text, articles manufactured in the EU are treated the same as imports. The ECA can, however, require a substance in an article to be registered if it poses a risk to human health or the environment.

Substances notified under the DSD are regarded as having been registered, as are active substances used only for products covered by the Plant Protection Products Directive or the Biocidal Products Directive (98/8/EC) or for coformulants of Plant Protection Products.

The regulation as a whole does not apply to radioactive substances, substances under customs supervision, substances in transit and non-isolated chemical intermediates or waste. Substances are also exempt from registration if regulated by equivalent EU legislation (human and veterinary pharmaceuticals, food additives and flavourings, animal feed and substances used in animal nutrition). Some categories of substance are, furthermore, exempt from registration (Table 7.7).

Substances needed in the interests of defence and non-isolated chemical intermediates do not have to be registered. Site-isolated intermediates at >1 tonne per annum are registered with information about the identity of the manufacturer and substance, classification and available test data. Registration also applies for isolated intermediates, which are transported between or supplied to other sites under contractual control (including for toll or contract manufacture) and for which there are strict conditions for manufacture and use to ensure only limited exposure. When transported at ≥1 tonne per annum these are registered

Table 7.7 Categories of substances exempt from registration.

- Specific substances listed in Annex II of the Regulation
- Substances covered by Annex III:
  Substances produced by environmental degradation
  Substances produced by degradation during storage
  Substances produced during use
  Products from reaction with additives
  By-products
  Hydrates, if the anhydrous forms are registered
  Non-dangerous natural substances
  Hydrogen, oxygen, nitrogen and the noble gases
  Minerals, ores, and ore concentrates, natural gas, crude oil and coal
- Monomers bound in polymers, but note that registration is required if the monomer is present at >2% (w/w) in the polymer and is at >1 tonne per annum
- Polymers
- Food and food ingredients
- Waste and some recycled materials

with the same information as site-isolated intermediates, but at >1000 tonnes per annum the basic Annex V test data are needed.

Although polymers do not have to be registered, if a polymer contains a monomer or other starting substance at ≥2% (w/w) in the chemically bound form at ≥1 tonne per annum that has not been registered by another registrant, this monomer or starting substance must be registered by the polymer manufacturer or importer.

Substances used only for process-orientated research and development (PORD) are exempt from registration for 5 years (further extendable in exceptional circumstances). The manufacturer or importer must inform the ECA of the identity of the substance, its labelling and its quantity, and must list the customers using it. Those customers only can use the PORD substance and it cannot be supplied to the public.

The registration dossier and chemical safety report is specified in Annexes IV–VIII of the regulation. The general technical, commercial and administrative information needed for all registrations for the technical dossier is specified in Table 4 of Annex IV.

The technical dossier, including robust summaries of the study reports, for registration of chemicals under REACh is to be submitted to the ECA electronically using the International Chemical Information Database (IUCLID) format, which is an established database format for communicating and storing information an chemicals.

7.5 Globally Harmonized System of Classification and Labelling (GHS)

The use of chemical products to enhance and improve life is a widespread practice worldwide. Alongside the benefits of these products, however, there is also the potential for adverse effects on people or the environment. As a result, several countries and organizations have developed laws or regulations over the years that require information to be prepared and transmitted to those using chemicals by use of labels or safety data sheets (SDS). Given the large number of chemical products available, individual regulation of all of them is simply not possible for any entity. Provision of information gives those using chemicals the identities and hazards of these chemicals, and enables the appropriate protective measures to be implemented locally

Although these laws or regulations are similar in many respects, differences are sufficiently significant to result in different labels or SDS for the same product in different countries. As a result of variations in definitions of hazards, a chemical may be regarded as flammable in one country, but not in another. Or it may be believed to cause cancer in one country, but not in another. Decisions on when or how to communicate hazards on a label or SDS thus vary around the world, and companies wishing to be involved in international trade must employ many experts who can follow the changes in these laws and regulations and prepare different labels and SDS. In addition, given the complexity of developing and maintaining a comprehensive system for classifying and labelling chemicals, many countries have no system.

Given the reality of the extensive global trade in chemicals and the need to develop national programs to ensure their safe use, transport and disposal, it was recognized that an internationally harmonized approach to classification and labelling would provide the foundation for such programs. When countries have consistent and appropriate information on the chemicals they import or produce in their own countries, the infrastructure to control chemical exposures and protect people and the environment can be established comprehensively.

In this sense, the new 'Globally Harmonized System of Classification and Labelling of Chemicals' (GHS) addresses the classification of chemicals by types of hazard and proposes harmonized hazard-communication elements, including labels and safety data sheets. Its objective is to ensure that information on physical hazards and toxicity of chemicals is available to enhance the protection of human health and the environment during the handling, transport and use of these chemicals. The GHS also provides a basis for harmonization of rules and regulations on chemicals at national, regional and worldwide levels, an important factor also for trade facilitation.

The GHS was developed as a consequence of the 'Rio Earth Summit' in 1992 and provides a basis for classifying and communicating the hazards of chemical products at different stages of their lifecycle from raw materials to recycling or disposal. The target beneficiaries of the new system include consumers, workers, and emergency responders and transport workers. The World Summit on Sustainable Development in 2002 requested all countries to implement the GHS and have it fully operational by 2008. Heads of Governments have signed up to this, and progress is being monitored by the UN.

Although governments, regional institutions and international organizations are the primary audiences for the GHS, it also contains sufficient context and guidance for those in industry who will ultimately be implementing the requirements which have been adopted.

The European Commission has outlined its timetable to implement the GHS in the European Union. This envisages that it will be implemented at the same time as REACh, which is currently foreseen for mid 2007. The GHS will replace the current EU classification legislation and will be the basis for classification under REACh.

The GHS will be implemented in the European Union by means of a regulation, which will be separate from REACh but will provide the basis of classification for REACh. The European Commission is currently finalizing the draft text of the proposed legislation.

It is expected that the GHS will be implemented in two phases, with, first of all, substances having to be reclassified according to the GHS by the time registration under REACh commences. When this is complete, reclassification of preparations/mixtures will start.

There will be benefits for industry resulting from the global adoption of the GHS, because companies that trade multinationally or globally should eventually have a common basis for the classification and labelling of their products, irrespective of the countries in which they operate or with which they trade. The GHS thus provides the potential to facilitate international trade and simplify some of the business operations of companies.

The GHS contains the following two elements:

- Harmonized criteria for classifying substances and mixtures according to their health, environmental and physical hazards.
- Harmonized hazard communication elements, including requirements for labelling and safety data sheets.

The information in the SDS will use the following16 headings:

1) Identification
2) Hazard(s) identification
3) Composition/information on ingredients
4) First-aid measures
5) Fire-fighting measures
6) Accidental release measures
7) Handling and storage
8) Exposure controls/personal protection
9) Physical and chemical properties
10) Stability and reactivity
11) Toxicological information
12) Ecological information
13) Disposal considerations
14) Transport information
15) Regulatory information
16) Other information.

7.6 Environmental Legislation 2: Classification and Labelling of Chemicals

7.6.1 Dangerous Preparations Directive (1999/45/EC)

The 'Dangerous Preparations Directive' 1999/45/EC effectively came into force in the European Community on 30 July 2002 and requires preparations to be evaluated for classification as 'Dangerous for the Environment' (DFE) and, if necessary, to carry the 'dead fish/dead tree' hazard symbol, based on the amount of DFE classified components they contain, or their intrinsic properties. DFE classifications have applied to substances since 1994, with the introduction of the 18th Amendment to Technical Progress.

Until now only single substances have had to be labelled as 'dangerous for the environment' (symbol 'N'), and no preparations. This has now changed – preparations must also be assessed according to exact regulations. Depending on the amount of hazardous substances it contains, the preparation must be declared with relevant R-phrases or – in the worse case – classified as 'dangerous to the environment'.

Thus, directive 99/45/EC affects Material Safety Data Sheets for preparations. Article 14 extends the scope for the provision of SDS, extending the right of professional users to request SDSs containing 'proportionate' information for some non-dangerous preparations. In any event, the relevant R-phrases, respectively the symbol 'N', must be mentioned. Some details of the regulations are given below.

With directive 1999/45/EC limits for the concentrations of dangerous substances in mixtures are introduced:

Labelling of substance	Concentration (%)	Labelling of mixture
N, R50-53	>25	N, R50/53
	2.5–24.9	N, R51/53
	0.25–2.4	R52/53
N, R51-53	>25	N, R51/53
	2.5–24.9	N, R52/53
R52/53	>25	R52/53

R50 very toxic to aquatic organisms

R51 toxic to aquatic organisms

R52 harmful to aquatic organisms

R53 may cause long-term adverse effects in the aquatic environment

R52, R53 and the combination R52/53 are not given a symbol (e.g. 'N').

The limits for mentioning under point 2 in the SDS are given by:

0.1% for R50, R51, R53 (in combination with R50 or R51)

1.0% for R52, R53.

This is the 'calculation respectively conventional method'. Alternatively, the health and environmental hazards of a preparation can be evaluated by testing the preparation using experimental methods, for example OECD. In general, classification derived using test data of the finished product override those given by the 'conventional method', although there are several exceptions to this. Any preparation containing more than the specified amount of a component classified as a carcinogen, mutagen or reproductive toxicant must be classified using the conventional method – testing of preparations for biodegradability or bioaccumulation is not allowed. For acute toxicity, preparations must, furthermore, be tested on all three aquatic species (e.g. OECD 201, 202, 203) with the limit of $LC/EC_{50}>100\,mg\,L^{-1}$. After successful testing, such a preparation is not awarded a symbol, even if the calculation scheme would require labelling.

Responsibility for determining whether the labelling must change as a result of the new directive lies with the supplier of a material. Individual companies are not obliged to require updated SDS from their supplier and they can legally rely on the SDS they hold in hands for raw materials purchased within the last year.

7.6.2 Globally Harmonized System of Classification and Labelling of Chemicals (GHS)

The follow-up of the *Dangerous Preparations Directive* is the *CLP Directive*, which is derived from the GHS Directive. The challenge was replacing of the European classification of hazardous substances via R phrases by the (UNO-) *Globally Harmonized System of Classification and Labelling of Chemicals* (GHS):

'Regulation (EC) No 1272/2008 of the European Parliament and the Council of 16 December 2008 on classification, labelling and packaging of substances and mixtures, amending and repealing Directives 67/548/EEC and 1999/45/EC, and amending Regulation (EC) No 1907/2006' (Table 7.8).

Some consequences in regard to lubricants are as follows:

- The German WHC classification, based on classification and labelling according to the European Chemicals Regulation, loses its validity, because the R phrases will be substituted by H statements.

Table 7.8 REACh and GHS – a comparison.

REACh	GHS
Registration, evaluation, authorization	Classification, labelling, SDS
Risk (probably hazard-based within authorization)	Hazard
Substances produced	Substances marketed
Hazardous and nonhazardous	Hazardous
> 1 t per manufacturer/importer	Any volumes
Harmonized classification (CMRs at EU level)	Self classification
European Union	Global
Supply	Supply and transport

- The H statements refer to the same end points (concerning the assessment of results) as the previous classification system; the classification limits in both systems differ mostly only in details.
- The translation table from classification under Directive 67/548/EC to classification under Regulation 1272/2008/EC can be found under Annex VII of Regulation 1272/2008/EC.
- The hazard pictograms, signal words, hazard statements and precautionary statements have changed in some regards. They can be looked for in Annexes I, III and IV of Regulation 1272/2008/EC.
- The criteria in regard to the classification as 'Dangerous for the Environment' are comparable; especially the alternative classification by calculation scheme or by testing the preparation using experimental methods is permitted again, as known from the former Dangerous Preparations Directive (1999/45/EC).

7.7
Environmental Legislation 3: Regular use

All industrialized countries have laws which are designed to protect waters, the ground, working places and the air from pollution. However, the only ban on the use of mineral oil-based lubricants exists in Austria and this is only for chain saw oil. Austria banned mineral oil-based chain saw oils effective 1 May 1992, following a resolution (No. 647) passed on 16 October 1990. Germany and a number of other countries have implemented provisions or quasi-legal procedures which are designed to promote the use of this new generation of products (e.g. the Environmental Seal in Germany). There is also political pressure to channel considerable funds into environmental protection. The following summarizes the laws and legislative initiatives in Germany which have an effect on the development and use of lubricants. Even if European legislation more and more dominates national legislation, a lot of individual laws and directives are to be considered.

The fundamental point is that 'if lubricants are released into the environment during or after use, water, the ground and air are endangered'.

The relevant ecological laws and regulations especially in Germany are as follows:

- The German Water Law
- The Drinking Water Directive
- The German Ground Law (in draft)
- The Federal Emissions Law
- The Recycling and Waste Law
- The Chemicals Law
- The Environmental Liability Law
- German Parliament initiatives
- The 'Blue Angel' environmental seal.

The following sections concentrate on German legislation.

7.7.1 Environmental Liability Law

Following the introduction of the Environmental Liability Law in Germany, the environmental risks for a company have increased significantly.

This law states that causer is liable for environmental pollution regardless of blame. This ensures that it is easier for the 'victim' to claim compensation.

The law names about 100 types of factory which pose a potential risk in terms of liability, for example warehouses, machinery, vehicles, gearboxes.

Important Articles of the law are as follows:

- *Article 6*: Causer principle.
- *Article 7*: Cause assumption (Liability on suspicion).
- *Article 8*: Obligation to inform.
- *Article 17*: Funding of clean-ups.
- *Article 19*: Preventative obligations.

The effects on lubricants are the reversal of the innocent until proven guilty principle increases the risks for operators using 'doubtful' lubricants. This risk can be reduced by the use of environment-friendly products.

7.7.2 The Chemicals Law, Hazardous Substances Law

The German Chemicals Law is orientated to safety-at-work considerations while its links with other environmental laws are still wanting. The European harmonization of the toxicological evaluation of chemicals, which is well advanced, has led to the labelling of lubricants if hazardous substances are contained and the uniform design of EU Safety Data Sheets.

According to the Hazardous Substances Law, the Definition of hazardous substances and mixtures (threshold concentrations of chemicals and additives) leads, in practice, to an increasing rejection of lubricants which contain hazardous substances even if the concentration of such is below the corresponding threshold. This means that hazardous substances are also rejected for mixtures (e.g. lead compounds in lubricants have largely been replaced by new non-toxic additives).

A recent addition to the German Chemicals Law also defines hazardous substances and mixtures according to hazard parameters. New to the list is the Parameter 'Environmentally dangerous'. Article 4 of the Hazardous Substances Law explains this as follows: 'Substances or mixtures are environmentally hazardous if they or their decomposition products can alter the nature of water, the ground, the air, animals, plants or microorganisms in such a way that they suffer immediate or delayed harm'. Additives which display these features are thus unsuitable for the development of ecologically friendly lubricants.

The effects on the use of lubricants are, for example:

- A reduction in the substances which pollute workplaces.
- The use of fewer 'risky' substances (e.g. oil changing, maintenance and production materials, the use of lubricants as production materials, contact with lubricants, during the servicing of machinery).

This limitation in the Chemicals Law and those in the Water Law severely restricts the number of additives which are suitable for environment-friendly, rapidly biodegradable lubricants.

A further Chemicals Law restriction on additives for rapidly biodegradable lubricants results from the complicated approval procedures which are necessary for new substances.

7.7.3
Transport Regulations

In Germany, these regulations focus primarily on the transport of hazardous substances by road and rail.

These hazardous substances regulations classify hazardous substances into various groups. Category 3 is of greatest importance to lubricants. This category includes ignitable fluids with a flashpoint less than 100 °C (e.g. low-viscosity mineral oils). Environment-friendly products based on rapeseed or ester oils normally have a considerably higher flashpoint than equi-viscous mineral oils.

7.7.4
Disposal (Waste and Recycling Laws)

The Law on Waste Oils had a special status in Germany before it was integrated into the Wastes Law in 1986. This was then combined with a Recycling Law in 1996. Separate regulations determine the Waste Codes for waste products.

The integration into the Wastes Law obliged lubricant manufacturers to take back used engine and gear oils from consumers. As a rule, the lubricant manufacturers subcontracted the work to waste oil collection companies. In practice, the lubricant consumer pays for the collection and disposal or recycling. Moreover, the law has defined a number of waste oil categories with chlorine content (0.2%) and PCB content (less important today) as major classification parameters. The great expense of disposing of products containing chlorine led to the development of chlorine-free products.

The most important methods and options for waste are listed in Table 7.9.

Biodegradable lubricants based on synthetic esters are included in the term waste oil while vegetable oil-based products are not, and are thus discriminated against by the law. As they are not included in the recycling and the associated waste oil laws, they have to be treated differently as their Waste Disposal Code

Table 7.9 Methods and options for waste.

Product group	Methods + options
Mineral oil-based hydraulic oils	Reconditioning (disposal)
Ester-based hydraulic oils	Reconditioning (disposal)
Mineral and ester oil-based engine and gear oils	Reconditioning (disposal)
Vegetable oils	Special waste, disposal

indicates. The Code 12 102 requires very careful monitoring and such wastes have to be stored and transported separately. In the reality of collecting used lubricants and hydraulic fluids, a certain normalization has taken place as regards vegetable oil-based products. However, the waste oil collectors still charge significantly more to dispose of vegetable oils than they charge for mineral oil-based products.

The chances are good that a revision of the Law on Wastes and other subordinate provisions and regulations will include vegetable oil-based lubricants in the term waste.

7.7.5
Disposal Options for 'Not Water Pollutant' Vegetable Oils

The waste catalogue presents the following options:

- Wastes which can be burnt or dumped together with household wastes.
- Industrial or factory wastes which cannot be disposed of in household waste disposal plants but in special waste disposal plants.
- Wastes, which due to their toxicity, can only be incinerated in special plants fitted with exhaust filters or at sea.

A used vegetable oil-based hydraulic oil cannot be allocated to any of these three groups because vegetable oil used as lubricants are not included in the term 'waste oils'. Instead, used vegetable oil-based lubricants have to be disposed of in line with the waste catalogue. This catalogue again defines three groups of wastes. The first includes wastes which can be disposed of or dumped together with household waste. However, used vegetable oil-based lubricants have been allocated Waste Code 12 102 and thus are covered by the second group of wastes. This means that they cannot be disposed of together with household waste but must be incinerated in special plants (Waste Code 13503 applies to non-contaminated animal and vegetable oils). As regards their collection and storage, they are also special. This is because as lubricating oils (e.g. hydraulic oils), vegetable oils are not subject to the law on waste oils but to the general law on wastes and cannot at present be mixed with other waste oils which are used, for example for heating purposes. In practice, this means that mineral-based oils and vegetable oils have to be separately collected and stored.

Disposal possibilities for rapeseed-based hydraulic fluids are as follows:

- Burning for heating purposes.
- Use as a concrete mould release oils.
- Flux oils for bitumens (e.g. roofing felt and sealants).

Official reconditioning methods to convert used vegetable oil-based lubricants into a type of re-refined oil do not exist at present so that the above-mentioned disposal possibilities for such re-refined oils like flux oils or concrete mould release oils are not available until suitable refining technologies have been developed for used vegetable oil-based lubricants.

7.8 Environmental Legislation 4: Emissions

7.8.1 Air Pollution

Most air pollution is caused by the fossil fuels we burn in our vehicles, homes, thermal power plants and factories.

Many chemicals have been identified in urban air pollution. A small number of these have been found to contribute to a range of air quality problems. These pollutants include nitrogen oxides (NO_x), carbon monoxide (CO), sulfur dioxide (SO_2), particulate matter (PM) and volatile organic compounds (VOC). When some of them combine, they produce smog or acid rain. Ground-level ozone, the major component of smog, is formed when NO_x) and VOC react in the presence of warm temperatures and sunlight. Another key element of smog is particulate matter.

7.8.2 Water Pollution

The quality of freshwater and marine areas is affected by three important water pollution problems: toxic substances, excess nutrients and sedimentation.

Toxic substances from industrial, agricultural and domestic use are some of the main pollutants in our water. These include trace elements, polychlorinated biphenyls (PCBs), mercury, petroleum hydrocarbons, dioxins, furans and some pesticides. Some of these substances accumulate through the food chain rather than break down in the environment.

These substances enter the water in a variety of ways, including: industrial sources such as mining, steel production, the generation of electricity and chemical production; accidents such as oil or chemical spills; municipal wastewater effluents and agricultural run-off.

Excess nutrients such as nitrogen and phosphorous compounds come mainly from municipal sewage and farm run-off containing fertilizers and animal waste. These nutrients can cause excess growth of aquatic plants, which then die and decay, depleting water of dissolved oxygen and killing fish.

Sedimentation is an increase in the amount of solid particles in water, caused primarily by human activities such as forestry, farming and construction. When sediment settles, it can smother the feeding and spawning grounds of fish and kill aquatic organisms (i.e. toxic substances in the food chain).

The impact of lubricants to the environment, that means to soil, water and air, is restricted by different types of laws. Again the German situation will give an example for other countries.

7.8.3
German Law for Soil Protection

In March 1999, a German Federal law to protect the ground came into force. This law is a framework law, similar to the Water Law and the Federal Emissions Law and implementation will be the responsibility of the Federal States. While the Water Law only has an indirect effect on the ground (in the sense of clean water does not pollute), the Ground Law should avoid the build-up of ecologically harmful substances in the ground.

The law is aimed at private and industrial landowners along with the operators of factories (avoidance and preventative, clean-up and re-cultivating measures). The law will provide greater precision regarding environmental damage and a further restriction on the use of ecologically harmful substances.

Effects on the use of lubricants are: as a significant proportion of lubricants pollute the ground, the Soil Law will have profound effects on lubricant applications. For example, the clean-up measures for mineral oil-polluted ground.

Because threshold values are not part of national laws designed to protect the ground, most European countries use the 'Holland List' which stipulates when mineral oil contamination requires cleaning-up. According to this list, the thresholds which then require clean-up measures are as follows:

1) $>500\,mg\,kg^{-1}$ in residential areas and in Protected Water Zones.
2) $>1000\,mg\,kg^{-1}$ for general cleaning-up.
3) In individual cases and when confirmed by an independent expert, up to $5000\,mg\,kg^{-1}$ can be tolerated (such as in industrial areas with no ground water relevance).

It is assumed that the necessary disposal and cleaning of $1\,m^3$ earth costs about 1000 \$EUR\$. This fact also promotes the development of environment-friendly products which may generate far lower clean-up costs.

7.8.4
German Water Law

German water law offers direct protection for waters and only indirect protection for the ground. At present, a law (titled: German Soil Law) directly aimed at protecting the ground is being draughted.

Important articles of the German Water Law are as follows:

- *Article 19 G* – Plant and equipment for handling water-polluting substances (mineral oils as well as their derivatives): 'Plants and equipment designed to store, fill, manufacture or treat water-polluting substances as well as plant and machinery using water-polluting substances in trading companies and the public sector must be designed, constructed, installed, maintained and operated in such a way that no contamination of waters or any other disadvantageous changes to the characteristics of the water can occur. The same applies to pipework within the boundaries of a company'.
- *Article 22* – Liability for changes to the characteristics of the water: 'Any party which allows substances to contact water which change the physical, chemical or biological characteristics of the water is liable for any damage caused. If more than one party is involved, then all are liable'.

These articles and German building regulations are important guidelines for lubricant manufacturers and customers alike. Applying these guidelines to the storage and use of lubricants could lead to restrictions if larger quantities of oil are involved (central warehouses, large mobile plant etc.). In the past, these regulations were handled differently in different parts of Germany. The States working party on water, which may form the basis of a national model, created a uniform definition of Hazard Categories and the accompanying preventive measures. Applying building regulations to the design and spillage protection of production plants would be enormously expensive.

According to the Environmental Agency's 'Guideline on Handling Water-Polluting Substances', the potential hazard depends on the volume of the plant and the Water Pollution Category of the substances used therein (see Section 7.3.4). The aim is to reduce this potential and thus avoid the massive cost of renewing plant and equipment by the following:

- Setting-up a plant and equipment register.
- Securing and sealing flooring.
- Setting-up containment capacity.
- Infrastructure measures.

This is only possible with biodegradable lubricants. The German Water Law details biotest procedures (e.g. bacteria–algae toxicity) which are mandatory for the monitoring of wastes routed into public sewers. In addition, threshold concentrations of heavy metals are measured in wastes routed into public sewers (these can originate from cutting fluid emulsions which are contaminated with hydraulic oils containing zinc).

7.8.5
Wastewater Charges

The charges for directly routing wastes into public sewers depends on the degree of contamination. The corresponding law defines:

- the definition of hazardous substances, for example heavy metals such as zinc and barium in lubricant additives, organic halogen compounds, and
- the charges for hazardous substances.

During use, such as in machine tools, lubricants come into contact with metalworking fluids, slideway oils and hydraulic oils. When water-miscible cutting fluids are split, parts of the hydraulic oil's additives can remain in the aqueous phase and in unfavourable circumstances, this can lead to increased waste water charges. This problem has lead to the development of zinc- and ash-free hydraulic oils. The use of ecologically friendly lubricants can also make a positive contribution to reducing pollution as well as lowering waste water charges.

7.8.6
Clean Air: German Emissions Law

The German Emissions Law focuses on keeping the air clean and has had an increasing influence on the manufacture, use and disposal of lubricants in recent years.

The Emission Law affects lubricants if total emissions from a plant reach a critical level. It should be remembered that in certain applications, over 10% of the lubricant can evaporate or form oil mist. This in turn, can exhaust into the atmosphere as emissions. Lubricants which evaporate or mist significantly less than conventional products are already views as being environment friendly. Compared to equiviscous mineral oils, rapidly biodegradable, rapeseed oil- or ester-based products reduce emissions by up to 90%. According to the Water Law, if the state-of-the-art is technically and economically feasible, all methods, installations and procedures should be employed to reduce emissions. According to the Emissions Law, all methods, installations or procedures which are practicable should be used to limit emissions. Regarding the determination of the state-of-the-art, comparable methods, equipment and processes were selected and tested for practical success.

7.8.7
Drinking Water Directive

The cleanliness of drinking water is already an objective of the Water law, particularly with regard to the restrictions applicable in Protected Water Zones. An important aspect of laws concerning the safety of machinery is the classification of lubricants as combustible fluids. These law lay-down measures for the storage of flammable fluids.

A lubricant which is not classified as being a combustible fluid and which is 'not water hazardous' is the least costly to store. This applies to a number of environment-friendly lubricants.

7.9
Standardization of Environment-Compatible Hydraulic Fluids

The most known standardization of environment-compatible lubricants concerns to hydraulic fluids because this was the application with the greatest amount of biodegradable alternatives up to now. The most important regulations for environment-compatible hydraulic fluids are shown in Table 7.10.

7.9.1
The German Regulation VDMA 24568

The well known minimum requirements of mineral oil-based hydraulic fluids of DIN 51 524 do not satisfy the specifications of rapidly biodegradable hydraulic fluids. In order to do justice to the technical performance of different fluids, the VDMA (Association of German Machinery and Plant Manufacturers) created the specifications 24 568 (Minimum Technical Requirements) and 24 569 (Change-Over Guidelines). These specifications detail the minimum technical requirements of the ecologically safe product families, HETG (environment-compatible hydraulic fluids based on triglycerides, i.e. vegetable oils), HEES (based on synthetic esters) and HEPG (based on polyglycols).

7.9.2
ISO Regulation 15380

The increasing importance of environment-friendly fluids is reflected at an international level by the ISO: ISO 15380 'Lubricants, industrial oils and related products (class L) – Family H (hydraulic systems) – specification for categories HETG, HEES, HEPG, HEPR' (Table 7.11). The European activities of CETOP (European Committee for Oil Hydraulics and Pneumatics) also reflects this activity.

The ISO standard 15380 deals with ecologically harmless fluids (ref. VDMA 24568 and 24569, enlarged to HEPR). It includes, in addition to technical requirements (Table 7.10), specific demands regarding the environmental impact of the hydraulic fluid.

ISO 15380 is the 'technical standard' for European Eco-Labelling of Lubricants and for the German 'Blue Angel'.

Numerous specifications have also been issued by leading hydraulic component manufacturers and their direct customers (Mannesmann–Rexroth, Sauer Sundstrand, Caterpillar, Komatsu, etc.).

Table 7.10 Regulations for environment-compatible hydraulic fluids – minimum requirements HEES 46

	ISO 15380	EU Margerite 2005/360/EC	Blue Angel RAL-UZ 79	White Swan Version 4.2	Swedish Standard 15 54 34 ed. 4
Technical requirements	The technical criteria of EC Directive 2005/360/EC ('EC Ecolabel for Lubricants'), RAL-UZ 79 ('Blue Angel') and Nordic Ecolabelling for Lubricants (Version 4.2, 'White Swan') for hydraulic fluids refer to ISO 15380; the German VDMA Guideline 24568 is obsolete.				
Foam	ISO 6247 (24/93/24)	ISO 6247 (24/93/24)	ISO 6247 (24/93/24)	ISO 6247 (24/93/24)	ISO 6247 (24/93/24)
Requirement	150/0, 75/0, 150/0	150/0, 75/0, 150/0	150/0, 75/0, 150/0	150/0, 75/0, 150/0	150/0, 75/0, 150/0
Air release	ISO 9120	ISO 9120	ISO 9120	ISO 9120	ISO 9120
Requirement	Max 10	Max 10	Max 10	Max 10	Max 10
Seal swell	ISO 1817	ISO 1817	ISO 1817	ISO 1817	ISO 1817
Requirement	Volume −3 to +10%; elongation −30%; tensile strength −30%; hardness ±10%	Volume −3 to +10%; elongation −30%; tensile strength −30%; hardness ±10%	Volume −3 to +10%; elongation −30%; tensile strength −30%; hardness ±10%	Volume −3 to +10%; elongation −30%; tensile strength −30%; hardness ±10%	Volume −3 to +10%; elongation −50%; tensile strength −50%; hardness ±10%
Hydrolytic stability	Not specified	Not specified	Not specified	Not specified	Not specified
Requirement	None	None	None	None	None
Cu corrosion	ISO 2160 (100 °C, 3 h)	ISO 2160 (100 °C, 3 h)	ISO 2160 (100 °C, 3 h)	ISO 2160 (100 °C, 3 h)	ISO 2160 (100 °C, 3 h)
Requirement	Rating max 2	Rating max 2	Rating max 2	Rating max 2	Rating max 1b
Rust prevention	ISO 7120, procedure A	ISO 7120, procedure A	ISO 7120, procedure A	ISO 7120, procedure A	—
Requirement	Pass	Pass	Pass	Pass	—
Demulsification	ISO 6614	ISO 6614	ISO 6614	ISO 6614	ISO 6614
Requirement	None	None	None	None	Max 30 min

(*continued*)

Table 7.10 *(Continued)*

	ISO 15380	EU Margerite 2005/360/EC	Blue Angel RAL-UZ 79	White Swan Version 4.2	Swedish Standard 15 54 34 ed. 4
Oxidation stability I	ASTM D 943 – dry	ASTM D 943 – dry	ASTM D 943 – dry	ASTM D 943 – dry	ASTM D 943 – dry
Requirement	Report	Report	Report	Report	Min 1000 h
Oxidation stability II	DIN 51554-3 (95 °C/72 h)	DIN 51554-3 (95 °C/72 h)	DIN 51554-3 (95 °C/72 h)	DIN 51554-3 (95 °C/72 h)	ASTM D 943 DIN 51554-3
Requirement	Visc. max +20%	Visc. max +20%	Visc. max +20%	Visc. max +20%	Visc. max +20%
Low temperature properties	ISO 3104 (−20 °C/7 d)	ISO 3104 (−20 °C/7 d)	ISO 3104 (−20 °C/7 d)	ISO 3104 (−20 °C/7 d)	ISO 3104 (−20 °C/3 d)
Requirement	No precipitates/no particles	No precipitates/no particles	No precipitates/no particles	No precipitates/no particles	Max. 2400 $mm^2 s^{-1}$
Filterability	Not specified	Not specified	Not specified	Not specified	ISO 13357-2
Requirement	None	None	None	None	None
FZG gear test	DIN 51354	DIN 51354	DIN 51354	DIN 51354	DIN 51354
Requirement	Min 10	Min 10	Min 10	Min 10	Min 10
Vane pump	ASTM D2882/IP 281	ASTM D2882/IP 281	ASTM D2882/IP 281	ASTM D2882/IP 281	DIN 51389 (=ASTM)
Requirement	120/30	120/30	120/30	120/30	120/30
Shear stability	Not specified	Not specified	Not specified	Not specified	DIN 51350-6/ASTM D445
Requirement	None	None	None	None	>6.5 $mm^2 s^{-1}$ @100 °C
			Occupational safety aspects		
Requirements concerning single substances		Not permitted if appearing in OSPAR list, organic halogen + nitrite compounds, metals or metallic compounds with the exception of sodium, potassium, calcium and magnesium	Not permitted if classified as • T+, T, R-45, R-46, R-48 or R-68 or,	• Base fluids should not be carcinogenic	• R-42 or R-43 are only permitted up to 1%

		• According to MAK 905, TRGS or EC cat.1,2,3 – Carcinogenic or – Mutagenic or – Toxic to reproduction		
Requirements concerning the final formulated product	Product shall not have been assigned any R-phrase indicating environmental and human health hazards according to EC Directive 99/45/EC	• Not classified as Xi or N according to EC Directive 99/45/EC • Max 50% concentration leading to classification of the formulation as Xn according to 99/45/EC	Formulated hydraulic oil should not • be classified as N • have health hazard • have explosion hazard	Formulated hydraulic oil should not • be harmful
Environmental aspects base fluids	Base fluids (>5%) each • Aquatic toxicity in OECD 201 and 202 ≥ $100\,mg\,l^{-1}$	Base fluids (>5%) each • 70% BOD/ThOD resp. CO_2 within 28 d • 80% in BODIS test • LC/EC_{50} not <$100\,mg\,l^{-1}$	Base fluids each • Readily degradable and not R-50/53, not R-51/53, not R-50, not R-52/53, not R-53	Base fluids (> 5 %) each • >60% BOD/ThOD resp. CO_2 within 28 d (if solubility <$100\,mg\,l^{-1}$) • >70% COD in 28 d (if solubility >$100\,mg\,l^{-1}$) • LC/EC_{50} >$100\,mg\,l^{-1}$

(continued)

Table 7.10 (*Continued*)

	ISO 15380	EU Margerite 2005/360/EC	Blue Angel RAL-UZ 79	White Swan Version 4.2	Swedish Standard 15 54 34 ed. 4
Additives	• No limitation on individual substances	• Cumulative mass concentration of substances has to be – ≥90% ultimately biodegradable (e.g. ≥60% in OECD 301) – ≤5% inherently biodegradable (f.e. >70% in OECD302C) – ≤5% non-biodegradable • Substances that are non-biodegradable and bioaccumulative are not permitted	• Max 5% inherently or not biodegradable substances, max 2% non-biodegradable • Not R-51/53 and bioaccumulative • Not R-52/53 and bioaccumulative • Not EC_{50} <100 mg l^{-1} in OECD 208 • Not IC_{50} <100 mg l^{-1} in OECD 209	• No limitation on individual substances, see below	• No limitation on individual substances, see below
Polymers		Substance does not bioaccumulate if its MM >800 or has a molecular diameter >1.5 nm	• Not mobile • LC/EC_{50} >100 mg l^{-1} in OECD 201, 202 or 203 • EC_{50} >100 mg l^{-1} in OECD 208		

Formulated product	• Biodegradability according to ISO 14593 or ISO 9439: >60% • Fish ISO 7346-2: LC_{50} $>100\ mg\ l^{-1}$ • Daphnia ISO 6341: $EC_{50} >100\ mg\ l^{-1}$ • Inhibition ISO 8192: EC_{50} $>100\ mg\ l^{-1}$	Aquatic toxicity in all three OECD tests 201, 202 and 203 $\geq 100\ mg\ l^{-1}$. Alternatively: special requirements for each constituent substance	• Max 2% non-degradable • No AOX, NO_2-salts, metals, except 0.1% Ca	• Max 3% R-53 or R-52/53 • Max 2% R-50 or R-50/53 • Max 1% R-51/53	• Max 1% of components with LC_{50} $<1\ mg\ l^{-1}$ • Max 5% of components with $LC_{50} = 1–100\ mg\ l^{-1}$
Renewable material	No requirements	>50% in hydraulic fluids	No requirements, but the German Market Introduction Program requires 50% renewable components	Min 65% renewable substances in the product required	No requirements, but the environmental criteria laid down in SS 155434 correspond to the B level of the 'Clean lubrication' definition of the City of Göteborg. The A level of the 'Clean lubrication' definition of the City of Göteborg requires that only material from renewable resources are used

Table 7.11 Rapidly biodegradable hydraulic fluids according to ISO 15380.

HEPG	Polyalkylene glycols soluble in water	Hydrostatic drives, for example locks, 'water hydraulics', to ≤90 °C
HETG	Triglycerides (vegetable oils), not soluble in water	Hydrostatic drives, for example mobile hydraulic systems, −20 to ≤70 °C
HEES	Synthetic esters, not soluble in water	Hydrostatic drives, mobile and industrial hydraulic systems, −30 to ≤90 °C
HEPR	Polyalphaolefins and/or related hydrocarbons, not soluble in water	Hydrostatic drives, mobile and industrial hydraulic systems, −35 to ≤80 °C

7.10 Environmental Seal

To combine the environmental behaviour and the technical properties of lubricants, a lot of countries have introduced so called 'eco-labels' or 'eco-logos'. The labels should give a sense of security to the users of environmental compatible products. Amongst many household appliances the most countries include lubricants in the system of eco-labelling (Table 7.12).

7.10.1 Global Eco-Labelling Network

Table 7.12 Market presence of important European Eco-Lables.

Standard	Number of companies	Number of products
Swedish Standard: hydraulic oils (SS 15 54 34)	34	80
Swedish Standard: lubricating greases (SS 15 54 70)	13	20
Nordic Swan	0	0
Blue Angel: hydraulic fluids (RAL-UZ 79)	24	75
Blue Angel: total loss lubricants (RAL-UZ 64)	24	49
EU Ecolabel for Lubricants (2011/381/EU)	28	97

The Global Eco-Labelling Network (GEN) is a new international initiative of the national environmental labelling organizations with the purpose of creating a forum for information exchange and promotion of eco-labelling. GEN is a non-profit association of organizations from around the world. To date, eco-labelling agencies from Spain, the United States, Canada, Sweden, Finland, Norway, Taiwan, Japan and the United Kingdom have committed to membership. India and Greece are also in the process of commitment.

7.10.2
European Eco-Label for Lubricants (EEL)

The current EU Eco-Label ('EU Margerite') award scheme has been in operation since 1993, when the first product groups were established. For all product groups, the relevant ecological issues and the corresponding criteria have been identified on the basis of comprehensive studies of environmental aspects related to the entire lifecycle of these products.

EEL Directive 2005/360/EC

With publication in the Official Journal of the European Community on 5 May 2005 the directive 2005/360/EC came into force, establishing ecological criteria and related assessment and verification requirements for the award of the Community Eco-Label to Lubricants. The 'Competent Body' for the eco-labelling process was the Netherlands' Stichting Milieukeur, the task of preparing the criteria was given to the consulting-bureau IVAM.

The eco-labelling process started 2 years ago with discussions with the stakeholders. Because lubricants vary substantially, depending on their applications, it was clear from the very beginning that this product group had to be divided in special subgroups. Final selection of the first version of the EU Eco-Label for Lubricants was driven by the environmental relevance of the following products:

- *Category 1*: Hydraulic fluids.
- *Category 2*: Chainsaw oils, concrete release agents, other total-loss lubricants.

- *Category 3*: Two-stroke engine oils.
- *Category 4*: Lubricating greases.

Automotive lubricants were not included, because the most relevant environmental issues for these type of lubricant differ from those selected – according to some Life Cycle Assessments fuel efficiency improvement by use of advanced engine and gear oils is of greater importance to the environment than biodegradability. This does not exclude the possibility of developing criteria for automotive lubricants in the future.

Revised EEL Directive 2011/381/EU

By the Commission Decision of 24 June 2011 on establishing the ecological criteria for the award of the EU Ecolabel to lubricants the first revision came into force (EU Directive 2011/381/EU). These criteria will be valid till June 2015, with the option to extend this period.

Now five categories are defined; additionally industrial gear oils are now included:

- *Category 1*: Hydraulic fluids and tractor transmission oils.
- *Category 2*: Greases and stern tube greases.
- *Category 3*: Chainsaw oils, concrete release agents, wire rope lubricants, stern tube oils and other total loss lubricants.
- *Category 4*: Two-stroke oils.
- *Category 5*: Industrial and marine gear oils.

National Competent Bodies' are responsible for implementation. The seven criteria for European eco-labelling of lubricants according to EU directive 2011/381/EU are described briefly below.

Criterion 1: Excluded or limited substances and mixtures

The product shall not have been assigned any R-phrase at the time of applying for the eco label, indicating environmental and human health hazards according to European Preparation Directive 1999/45/EC.

Criterion 2: Exclusion of specific substances

The following stated substances are not allowed in quantities exceeding 0.010% (w/w) of the final product – in other words, they shall not be intentionally added to the product:

- Substances appearing in the Union List of priority substances in the field of water policy and the OSPAR List of Chemicals for Priority Action.
- Organic halogen compounds and nitrite compounds.
- Metals or metallic compounds with the exception of sodium, potassium, magnesium and calcium. In the case of thickeners, also lithium and/or aluminium compounds may be used up to concentrations limited by the other criteria.

Criterion 3: Additional aquatic toxicity requirements

The applicant shall demonstrate compliance by meeting the requirements of either criterion 3.1 or criterion 3.2. Hence, for aquatic toxicity two approaches are allowed – one for the fully formulated preparation and one for the single components.

Criterion 3.1: Requirements for the lubricant and its main components

Acute aquatic toxicity data shall be provided for the fully formulated product and the main components:

- Main components are all substances >5%, that is base oils and thickener.
- Additives are allowed up to a limit of 5%.

Acute aquatic toxicity data for each main component shall be stated on each of the following two trophic levels: algae and daphnia. The critical concentration for the acute aquatic toxicity for each main component shall be at least 100 mg l^{-1}.

Acute aquatic toxicity data for the applied, fully formulated lubricant shall be stated on each of the following three trophic levels: algae, daphnia and fish. The critical concentration for the acute aquatic toxicity for a lubricant in Category 1 and 5 shall be at least 100 mg l^{-1} and for a lubricant in Category 2–4 at least 1000 mg l^{-1}.

Criterion 3.1: Aquatic toxicity, first approach	**Category 1**	**Category 2**	**Category 3**	**Category 4**	**Category 5**
Aquatic toxicity for the fully formulated product in all three of the acute toxicity tests – OECD 201, 202 and 203	>100 mg L^{-1}	>1000 mg L^{-1}	>1000 mg l^{-1}	>1000 mg l^{-1}	>100 mg l^{-1}
Aquatic toxicity for each individual main component in OECD 201 und 202	>100 mg l^{-1}	>100 mg l^{-1}	>100 mg l^{-1}	>100 mg l^{-1}	>100 mg l^{-1}

Criterion 3.2: Requirements for each stated substance present above 0.10% (w/w)

Chronic toxicity test results in the form of No Observed Effect Concentration (NOEC) data shall be stated on each of the following two aquatic trophic levels: daphnia and fish.

This has to be done for each constituent substance, for example according to OECD 201 and 202 – without differentiation of the main components and additives. Single component and constituent substance means any substance which has been deliberately added and which constitutes more than 0.10% of the product's content, as measured both before and after any chemical reaction has occurred between the substances mixed to produce the lubricant preparation.

Criterion 3.2 Aquatic toxicity, second approach	Maximum treat rates of substances with some aquatic toxicity: cumulative mass concentration of substances (%) in:				
	Category 1	Category 2	Category 3	Category 4	Category 5
Harmful: 10 mg l^{-1} < acute toxicity[a)] ≤ 100 mg l^{-1}	≤20	<25	≤5	≤25	≤20
Toxic: 1 mg l^{-1}< acute toxicity[a)] ≤ 10 mg l^{-1}	≤5	≤1	≤0.5	≤1	≤5
Very toxic: Acute toxicity ≤ 1 mg L^{-1}	≤0.1/M[a)]	≤0.1/M[a)]	≤0.1/M[a)]	≤0.1/M[a)]	≤1/M[a)]

a) M is the multiplication factor of 10 for substances that are very toxic to the aquatic environment as from Table 1b in Commission Directive 2006/8/EC (OJ L 19, 24.1.2006, p. 12).

Criterion 4: Biodegradability and bioaccumulative potential

Requirements with regard to biodegradability must be measured for each component. For this criterion determination for the fully formulated product is not of interest. Stipulated amounts of some non-biodegradable substances are allowed, in principle, but only if they are not bioaccumulative. *Not bioaccumulative* are higher molecular weight or minor amounts of bioactive substances (molecular weight >800 *or* molecular diameter >1.5 nm *or* log K_{ow}< 3 or >7 *or* BCF ≤ 100). In other words, substances which are both non-biodegradable and bioaccumulative are only permitted up to 0.1%.

Criterion 4: Biodegradability and bioaccumulative potential	Cumulative mass concentration of substances (%) in:				
	Category 1	Category 2	Category 3	Category 4	Category 5
Ultimately aerobically biodegradable, for example according to OECD 301, 306, 310	>90	>75	>90	>75	>90
Inherently aerobically biodegradable, for example according to OECD 302	≤5	≤25	≤5	≤20	≤5
Non biodegradable and non-bioaccumulative	≤5	—	0	≤10	≤5
Non-biodegradable and bioaccumulative	≤0.1	≤0.1	≤0.1	≤0.1	≤0.1

Criterion 5: Renewable raw materials

The demand for a minimum content of renewable raw material within the finished lubricant is motivated by the European Climate Change Program

(ECCP). The formulated product shall have a carbon content derived from renewable raw materials.

Criterion 5	Category 1	Category 2	Category 3	Category 4	Category 5
Carbon content derived from renewable raw materials (%,m/m)	≥ 50%	≥ 45%	≥ 70%	≥ 50%	≥ 50%

The bio-based carbon content has to be calculated according to the following scheme: Carbon content derived from renewable raw material means the mass percentage of component A × (number of C-atoms in component A, which are derived from (vegetable) oils or (animal) fats divided by the total number of C-atoms in component A) plus mass percentage of component B × (number of C-atoms in component B, which are derived from (vegetable) oils or (animal) fats divided by the total number of C-atoms in component B) plus the mass percentage of component C × (number of C-atoms in component C, which are derived from (vegetable) oils or (animal) fats divided by the total number of C-atoms in component C), and so on.

The direct measurement of the bio-based carbon content of the fully formulated product via a radiocarbon method (like ASTM D-6866) is not intended.

Criterion 6: Minimum technical performance

At the beginning of the eco-labelling process it was recognized that minimum technical requirements are essential for acceptance of biolubricants. Thus, for hydraulic fluids, chain-saw oils, two-stroke engine oils and industrial gear oils the technical basis is described by accepted standards; for greases and total-loss lubricants, only, a general 'fit for use' is required.

Criterion 6	Category 1	Category 2	Category 3	Category 4	Category 5
Technical requirements	ISO 15380	'Fit for purpose'	Chain saw oils: 'Blue Angel' UZ 48 All others: 'Fit for purpose'	Marine: NMMA TC-W3 Terrestrial: EGD level of ISO13738:2000	DIN 51517 – the section (I, II or III) has to be declared.

Criterion 7: Information appearing on the eco-label

This criterion is related to the special appearance of the eco-label with regard to the product group 'Lubricants'. The information appearing in the 'Box 2' of the eco-label shall contain the text:

' – Reduced harm for water and soil during use.

– Contain a large fraction of bio-based material.

LuSC-List

A relevant relief for applying the EU Eco-Label was the introduction of the so-called 'Lubricant Substance Classification List' (LuSC-List). This non-limitative list comprises those substances and brands that have been assessed on its biodegradation/bioaccumulation, aquatic toxicity, renewability and exclusion lists of substances by a competent body. The assessment is only based on a maximum treat rate allowed in a lubricant. The list is published on the EU Ecolabel Web site and the data can be used directly in the application form. Companies are not obliged to use one of these substances or brands; but if used, the listed products can be taken directly on the application form without requesting the underlying documents. The list consists of two parts. Part 1 consists of substances and part 2 consists of brands. These are commercially available brands and are therefore indicated by their commercial name.

In summary, the main target of the (voluntary) EU Eco-Label for Lubricants is to draw attention to products which have the potential to reduce negative environmental impact. However, a European approach for ending the confusion of different national eco-labels in Europe is not yet recognizable.

7.10.3
The German 'Blue Angel'

The world's first eco-labelling program, the German 'Blue Angel', was created in 1977 to promote environmentally sound products, relative to others in the same group categories. The authorities hoped it would be seen as a positive step not only for individual consumers but for the retailers and manufacturers as well. Anyone can propose product groups for the 'Blue Angel' award. It is encouraged, apparently, 'in the interests of global warming management'.

The institution responsible for assessing such proposals is the Umweltbundesamt (German Environmental Protection Agency, UBA) guided by an Ecolabel Jury, which considers these proposals. The Jury is composed of representatives from industry, environmental organizations, consumer associations, trade unions, the Churches and public authorities, in order to ensure that the interests of various groups in society are taken into consideration.

The label is composed of a blue figure with outstretched arms surrounded by a blue ring with a laurel wreath. A standard inscription, 'Umweltzeichen' (environmental label) is at the top of the logo, and a second inscription for the individual product group is found at the bottom, for lubricants it is '. . . because rapid biodegradable'.

This eco-label relies on information and voluntary cooperation, as well as on the motivation and the willingness of each individual to make a contribution towards environmental protection. The Blue Angel is addressed at all market players, enabling retailers and consumers to make deliberate choices in favour of environmentally sound alternatives. Once approved, eco-labelled products are reviewed every 2 or 3 years to reflect state-of-the-art developments in ecological technology and product design.

The Eco-Label Jury scrutinizes product groups twice yearly. The criteria for awarding the Blue Angel includes: the efficient use of fossil energy, alternative products with less of an impact on the climate, reduction of greenhouse gas emission and conservation of resources.

The German environmental award should highlight products which are more environmentally compatible than others. The following environmental awards have been issued for lubricants since 1988:

- RAL UZ 48 for chainsaw oil (1988).
- RAL UZ 64 for rapidly biodegradable lubricants and shuttering oils (1991).
- RAL UZ 79 for rapidly biodegradable hydraulic oils (1996).

For example, RAL UZ 79 for rapidly biodegradable hydraulic oils focuses on the following points:

- *Objective*: To reduce environmental pollution.
- *Application*: Hydraulic systems.
- *Requirements*: Formulation, ingredients, disposal, performance.
- *Proof*: Formulation, expert evaluation, statement, ISO 15380.
- *Issued by*: Product manufacturer, RAL, Federal Environment Agency.
- *Award use*: Contract, duration.
- *Wording*: Awarded because rapidly biodegradable.

Together with the German Ministry for the Environment, the awarding committee has decided that the conditions for issuing this award will consider results obtained by the RAL organization consisting of a commission of consumer experts, manufacturers and Federal Environment Agency. The actual issuing of the award is performed by the RAL organization.

The third issuing guideline represents a tightening-up of the first two guidelines in which it covers the use of hazardous substances (additives). The guideline combines demands originating in chemical law, the law on water, biodegradability and eco-toxicological evaluations.

The 'Blue Angel' combines the performance level of the requirements (in case of hydraulic fluids, for example according to ISO 15380) with environmental relevant criteria, which are not compatible with the requirements of the European Eco-label.

Revision of the German Ecolabels for Lubricants RAL UZ 48, RAL UZ 64, RAL UZ 79
The latest revision of the 'Blue Angel' for lubricants (Edition July 2014) has introduced a new approach. Instead of three single basic criteria for award of the

Environmental Label, the new RAL UZ 178 for 'Biodegradable Lubricants and Hydraulic Fluids' includes the former awards UZ48, UZ64 and UZ 79, and adds further applications for lubricants.

Lubricants for the following fields of application are considered:

- Products for areas in which lubricant loss occurs during their intended use. This includes
  - lubricants that primarily escape into the environment during their intended use, for example point and rail lubricants and lubricants for open bearings, guides or sealing purposes (incl. stern tube greases),
  - lubricants for the glass industry,
  - concrete release agents for use in formwork and
  - release agents for use in asphalt paving work.
- Hydraulic fluids (pressure fluids) particularly in environment-sensitive hydraulic systems and tractor transmission oils.
- Chain lubricants for motor saws.
- Gear lubricants for industry and shipping.
- Greases.

It may be remarkable that two-stroke engine oils are not considered in the range of applications, even if they are used as total loss lubricants. They are excluded due to the valuation that two-stroke engines itself cannot be environmentally acceptable.

UZ 178: Requirement

Even if the requirements of UZ 178 are orientated at the actual European regulations, in particular observing the EU Chemicals Regulation REACh (1907/2006/EC) and the CLP Regulation (1272/2008/EC), the new criteria are not compatible with the criteria of the European Ecolabel for Lubricants (2011/381/EU).

The Environmental Label may be used for the labelling of final products, provided that they comply with the defined requirements, structured as follows in UZ 178:

3.1 Substance restrictions due to intrinsic properties of the substances according to European chemical law (REACH, CLP).

3.1.1 Substance restrictions due to a harmful effect on human health.
- Carcinogenic, reprotoxic and mutagenic substances
- Substances of Very High Concern (SVHC)
- Additional requirements for substances with harmful effects on human health

3.1.2 Substance restrictions due to a harmful effect on the environment.

3.2 Substance restrictions for other relevant substance groups.

3.2.1 Substance restrictions based on other regulations.

3.2.2 Substance restrictions due to belonging to certain substance groups.

3.3 Additional requirements for aquatic toxicity.

3.3.1 Requirements for the final product.

3.3.2 Requirements for components.

3.4 Biodegradability and bioaccumulation potential of the substances.

3.4.1 Biodegradability.

3.4.2 Bioaccumulation potential of the substances.

3.5 Disposal information.

3.5.1 Hydraulic fluids (pressure fluids) particularly in environment-sensitive hydraulic systems and tractor transmission oils.

3.5.2 Gear lubricants for industry and shipping.

3.5.3 Greases.

3.6 Technical requirements and fields of application.

3.6.1 Lubricants for areas in which lubricant loss occurs during their intended use.

3.6.2 Hydraulic fluids (pressure fluids) particularly in environment-sensitive hydraulic systems and tractor transmission oils.

3.6.3 Chain lubricants for motor saws.

3.6.4 Gear lubricants for industry and shipping.

3.6.5 Greases.

3.7 Advertising messages.

- The type of lubricant in accordance with Paragraph 2 is to be named in combination with the product description on the container and in the technical data sheet.
- Advertising messages may not display any information that downplays the dangers in the sense of Article 48 of Regulation (EC) 1272/2008 (e.g. 'not toxic', 'not damaging to health', 'free of . . .').
- Advertising messages are not permitted to contain any vague and unspecific environmental statements. Product labels that contain statements as part of the name or description such as 'environmentally safe', 'environmentally friendly', and so on are not permitted.
- The term 'bio' can be used in accordance with the requirements in DIN CEN/TR 16227:2011-10. This requires the determination of the bio-based carbon content in accordance with ASTM D-6866 or DIN CEN/TS 16137 (DIN SPEC 91236) and its calculation as bio-based carbon content in proportion to the total carbon content of the lubricant as well as its specification in steps of 5%.
- The use of the advertising prefix 'bio' on mineral oil-based lubricants and on lubricants having a biomass content of less than 25 mass percent in the final product shall not be permitted. In this case, there will be no need to determine the bio-based carbon content.

Even if both EU Ecolabel and CEN/TR 16227 include a criterion with regard to the bio-based character of environment-compatible lubricants, the UZ 178 does not consider this approach.

Only in order to establish the bio-based carbon content, the applicant shall submit a final product test report in accordance with ASTM D 6866 or DIN

CEN/TS 16137 (SPEC 91236):2011-07, without fixing of a minimum value for the bio-based content.

This sight is stated as long as safely excluding the possible negative environmental effects of the cultivation and processing of renewable raw materials using a targeted set of criteria and verification obligations (e.g. via a suitable certificate) are not comprehensively realized.

However, the German 'Initiative for the Sustainable Provision of Raw Materials for the Material Use of Biomass' (INRO) already has prepared sustainability criteria (including ecological, social and economic criteria), agreed upon by the INRO members, comprising industrial federations and associations, ministries and authorities, science, environmental and development organizations (NGO), industrial companies (2013).

Further harmonization with other national environmental labelling programmes is announced for possible future revisions of UZ 178.

7.10.4 Nordic Ecolabel (Norway, Sweden, Finland, Iceland): 'White Swan'

The Nordic Ecolabel criteria were first established in 1998. A revised set of criteria were published lastly in 2011. Separate requirements are imposed to chain oils, concrete release oils, hydraulic fluids, lubricating greases, two-stroke oils, metal cutting fluids and gear- and transmission fluids. An important feature of the scheme is a relatively high requirement for renewability. It is not considered to be a successful scheme since actually no lubricant product has been approved for the Nordic Swan criteria. The reasons for this lack of success are not clear. Some claim that the problem lies in the high renewability requirements but others point out to the high costs related to the procedure of gaining the eco-label.

The Nordic Swan Label claims to be neutral, independent and the world's first multinational ecolabeling scheme. Only products that satisfy strict environmental requirements on the basis of objective assessments are allowed to display the environmental product label.

The criteria for the eco labelling of lubricating oils (Nordic Ecolabeling 05.09.1997) encompass lubricating oils with a lubricating and pressure-transmitting effect. The product group encompasses engine oil, transmission oil, hydraulic oil, lubricating grease, two-stroke oil, metal cutting fluid, saw chain oil and mould oil.

Eco labelled lubricating oils must not be subject to classification under regulations concerning hazards to health, environmental hazards, fire hazards or explosion hazards, except products classified in accordance with the 'Mättligt hälsoskadlig' system in Sweden.

The packaging must not contain chlorinated plastics and must be designed in such a way that as little oil as possible is able to remain in the packaging. The type of plastic used in plastic parts must be marked on the parts concerned.

In the case of the products engine oil, transmission oil, hydraulic oil, lubricating grease and two-stroke oil strict requirements are imposed with regard to the permitted level of non-renewable raw materials, re-refined oil, environmentally harmful components and potentially non-degradable components. These parameters are assessed on an overall basis so that a product with a high score in one area may score somewhat lower in another area.

Separate requirements are imposed with regard to saw chain oil, mould oil and metal cutting fluids. Products of this type can not be based on mineral oil.

Documentation must be provided on lubricating oils demonstrating that the lubricating oil in question is of the same quality as an average of competing products available on the market.

Environmentally adapted lubricating oils are produced on the basis of either vegetable or synthetic base oils. Traditional mineral base oils will not satisfy the requirements in the criteria since they have a higher toxic effect on waterborne organisms and poorer biological degradability compared with vegetable and synthetic base oil.

Vegetable base oils and certain synthetic base oils which are extracted from renewable resources (plants) are rewarded in the criteria.

The policy in Norway is that used lubricating oils should be collected and to this end the Government has established a tax refund scheme on returned waste oil. According to this scheme, people or organizations in possession of waste oil receive oil delivered free of charge or are paid for each litre that is returned. As a result of this system considerably lower quantities of oil go astray.

A similar return waste oil scheme is being established on Iceland, according to which waste oil is returned to the oil companies for incineration under controlled conditions (cement production plant). Finland, Sweden and Denmark have no corresponding return oil scheme but do have systems for collecting waste oil.

The criteria also take account for the fact that certain lubricating oils require a higher percentage of components that are considered environmentally hazardous. Permitting a higher percentage of environmentally harmful components in certain products can be justified on the grounds that these components may improve the performance of the products.

Weights have been ascribed to the parameters on the basis of their environmental significance. For a product to qualify for eco-labelling it must attain a minimum score.

7.10.4.1 Requirements Concerning Renewable Resources

The environmental matrix regulates the use of renewable and non-renewable sources by means of a maximum permitted quantity (% w/w of the product).

The purpose of this requirement is to permit some synthetic base oils.

7.10.4.2 Requirements Concerning Re-Refined Oil

The environmental matrix permits the use of a limited proportion re-refined waste oil in motor oil and transmission oil.

7.10.4.3 Requirements Concerning Environmentally Harmful Components

Components which are classified as environmentally harmful according to current regulations in Denmark, Finland, Iceland, Sweden or Norway or EU Directive 67/548/EEC 18th adaptation with risk allocations R50, R53, R50 + R53, R51 + R53 or R52 + R53 are subject to limits in the environmental matrix with a maximum threshold value for each individual product.

The same requirements concerning environmental harmfulness apply to substances that have been shown to form persistent environmentally harmful degradation products in relevant conditions.

7.10.4.4 Requirements for Hydraulic Fluids, Mould Oil, Metalworking Fluids

Requirements A

Data sheet/confirmation to show which components constitute renewable resources.

Requirements C

Test results for all components in the product (base oil and additives) performed in accordance with the following OECD test methods:

- Rapid biodegradability: OECD 301 A–F
- Bioaccumulation: OECD 107, 117 or if applicable 305
- Ecotoxicity: OECD 201, 202 and 203.

If corresponding test methods are used, the similarities/differences of the method must be verified by an independent body. This is necessary to ensure credible test results.

Requirements concerning potential degradability (hydraulic fluids, mould oil): the test result for all components in the product (base oil and additives) performed in accordance with OECD test method 302 A–C.

7.10.5
Sweden Standard

The Swedish City of Gothenburg had started in 1992 a technology programme for lubricant products with the aim to encourage the manufacturing industry to switch to biolubricants. This so-called 'Ren Smörja – Clean Lubricants' project was a co-operation between municipal authorities, consultants, industries and the Swedish National Chemicals Inspectorate, resulting in 1998 in environmental criteria for lubricating greases and hydraulic fluids. These criteria have been elaborated and are now parts of the Swedish Standard. In addition, in the Scandinavian countries a tax exemption on biolubricants is in place.

Lubricants examined by SP Technical Research Institute of Sweden on behalf of the producer and fulfil the environmental requirements, are listed publically. Each examination presumes access to the formulation, including the chemical composition of the base fluids and the additives. All information about the product and the test results are given under a written personal confidentiality agreement. The technical performance level is set by the standard and is fulfilled by self-claim by the manufacturers.

Hydraulic Fluids – Requirements and Test Methods – SS 15 54 34

The version 4 of SS 155434 came into force in July 2000 and introduced more stringent ecological requirements. From an international point of view these requirements are rigorous and require that a hydraulic fluid should be biodegradable, with minimal aquatic toxicity. Further an examination of chemical compounds with sensitizing properties is required.

Whereas the Blue Angel is optional, the Swedish standard 15 54 34 is a legal requirement. Hydraulic fluids not fulfilling these criteria are not permitted on the Swedish market.

Lubricating Grease – Requirements and Test Methods – SS 15 54 70

Environmental criteria for lubricating grease were compiled within the Gothenburg project 'Ren Smörja' in 1998. These criteria have been elaborated and are now a part of the Swedish Standard SS 15 54 70. The standard was issued in 2002.

The environmental criteria, as given in Section 4.2 in the standard, are rigorous and require that the base fluid in the lubricating grease should be biodegradable, with minimal aquatic toxicity. Furtheron an examination of chemical compounds with sensitizing properties is required. Lubricating grease conforming to SS 15 54 70 is classified in one of the environment-adapted classes A, B or C differentiated by the maximum level of substances allowed. Throughout the whole standard class A implies the strictest requirements and with respect on renewable resources class A stipulates a mass content of more than 65%. The content of renewable resources in class B should be more than 45%. In contrast, class C, do not include any demands on a given content on renewable materials.

7.10.6
The Canadian Environmental Choice' (Maple Leaf)

In 1994, total sales of lubricant products in Canada equalled almost 1 billion liters, of which 200 000 l were vegetable oil-based lubricants.

In all the usage of vegetable oil-based lubricants is increasing in Canada because many companies prefer environment-friendly products. By 2000, sales of vegetable oil lubricants are expected to 1 million l, annually.

Vegetable oil lubricants are priced twice as high as conventional petroleum lubricants. The willingness of Canadian companies to pay very high prices for vegetable oil lubricants indicates a preference for these products in environment-sensitive areas.

The 'Environmental Choice Program' (ECP), Environment Canada's ecolabeling program, provides a market incentive to manufacturers and suppliers of environmentally preferable products and services, and thereby helps consumers identify products and services that are less harmful to the environment. Established in 1988, the ECP was the second national ecolabeling initiative undertaken.

Canada's 'Environmental Choice' Eco-Logo symbol of certification features three stylized doves intertwined to form a maple leaf, representing consumers, industry and government working together to improve Canada's environment.

A product or service may be certified because it is made or offered in a way that improves energy efficiency, reduces hazardous by-products, uses recycled materials or because the product itself can be reused. Product manufacturers, importers or purveyors of services may apply for a license to use the Eco-Logo once a guideline containing criteria relevant to the product or service type has been approved. Environmental Choice guidelines are based on the best information available at the time and are upgraded as new information and technology make higher standards possible. Guidelines are developed in consultation with industry, environmental groups, universities and independent technical and scientific advisors.

Environmental Choice issues kits to potential licensees, and Environment Canada's independent Technical Agency assists companies through the application process. Agreements granting use of the Eco-Logo are renewed annually and continued compliance with the guideline is monitored. Certified products and services must continue to meet all applicable safety and performance standards; specifically, they must be as good in every other respect as is generally expected of that type.

Currently, Environmental Choice has more than 1400 approved products, with 119 licensees and 29 guidelines under which companies may be licensed and their products certified. In the area of lubricants there are listed anticorrosion products, automotive engine oil, synthetic industrial lubricants, vegetable-based industrial lubricants.

'Vegetable-based industrial lubricants' must fulfil the following criteria:

- Not toxic to fish by demonstrating an LC_{50} not lower than 40 000 mg l^{-1} when tested according to the Acute Lethality Test Using Rainbow Trout, Report EPS 1/RM/9, July 1990, Environment Canada, biological test method.
- Be biodegradable, according to CEC L-33-A-93 or one of OECD 301 A–F.
- Donot contain more than 5% (w/w) additives.
- Donot contain more than 3% (w/w) of an additive that is not proven to be biodegradable.
- Donot contain petroleum oil or additives containing petroleum oil as confirmed by EPA TPH 418.1 with a measured reading no greater than 10.6 g kg^{-1}.
- Donot contain organic chlorine or nitrite compounds or lead, zinc, chromium, magnesium or vanadium.
- Donot have to be labelled according to class D, Poisonous and Infectious Material, as set out in the Controlled Products Regulations of the Hazardous Products Act.
- Yield pass results when tested against ASTM D 665 (Standard Test Method for Rust Preventing).
- Characteristics of inhibited mineral oil in the presence of water.
- Donot have a flash point lower than 200 °C, if ISO grade VG 32 and higher, and not lower than 190 °C, if ISO grade VG 15–22.
- Produce a minimum fire point of 311 °C as per ASTM D 92.
- Demonstrate a viscosity index of at least 200 as per ASTM D 2270.
- Demonstrate a capacity to produce a peroxide value no greater than 15 milliequivalents after 1000 h.

The Canadian Environmental Protection Act (CEPA) gives support to the use of labelled lubricants. The CEPA was proclaimed into law in 1988 and is designed to protect Canadians from pollution caused by toxic substances. It provides the power to regulate the entire life cycle of toxic substances.

7.10.7 Other Eco-Labels

7.10.7.1 Austria

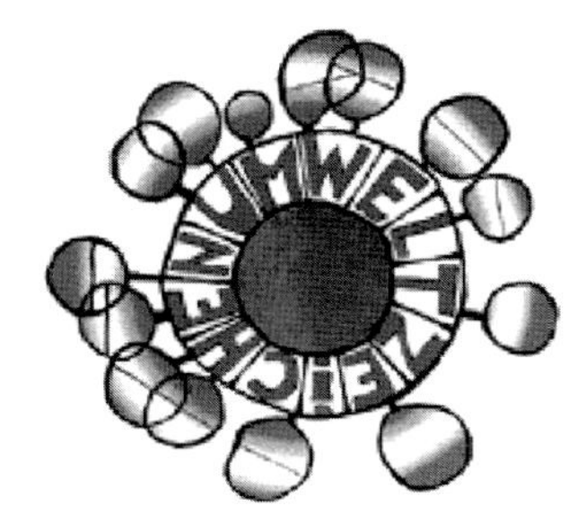

The Austrian eco-label was created by Friedensreich Hundertwasser. It is the sign for products and services with low impact to the environment for product lifetime.

The lubricant guideline is published under no UZ 14. This guideline is harmonized with the European Ecolabel for Lubricants (2011/381/EG), with a supplement in regard to the Austrian legislation about the permit of certain additives for lubricants and the special (obligatory) use of chain saw oils.

7.10.7.2 France

The 'NF Environnement mark' is the Eco-Logo for France. Created in 1992, it features a single leaf covering a globe.

The mark means that the product has less impact on the environment while achieving the same level of service as other products on the market. Industries who wish to highlight their environmental efforts can voluntarily apply to use the Eco-Logo on their products.

Certification is based on a multi-criteria approach, combining technical and environmental requirements. Since 2003, for chain saw oils a French Eco-label is available, as first and only luibricant group. The identification number is NF375: 'Lubrifiants pour Chaines de Tronconneuse'.

7.10.7.3 Japan

Since 1989, the Japanese Environmental Association (JEA) has administered the Eco Mark Program with the goal of disseminating environmental information on products and encouraging consumers to choose environmentally sound products. The symbol itself represents the desire to protect the earth with our own hands, using the phrase 'Friendly to the Earth' at the top of the symbol and the product category below it.

This goal will be accomplished by authorizing the Eco Mark to be displayed on products that reduce the environmental load caused by everyday activities, thereby contributing to the preservation of the environment.

In principle, products must meet the following criteria: impose less environmental load than similar products in their manufacture, use and disposal; and reduce the environmental load in other ways, thus contributing significantly to environmental conservation. This scheme also is applicable to lubricants.

The Eco Mark Program is intended as a means of offering a choice of products with a lower environmental impact.

7.10.7.4 **USA**

Green Seal is an American non-profit environmental labelling organization that awards a 'Green Seal of Approval' to products found to cause significantly less harm to the environment than other similar products. The Green Seal certification mark identifies those products which are environmentally preferable, empowering consumers to choose products based on their environmental impacts.

Green Seal develops environmental standards for consumer products through a public review process involving manufacturers, environmental organizations, consumer groups and government agencies. Products are certified only after rigorous testing and evaluation. Underwriters Laboratories Inc. (UL) is the primary testing contractor for Green Seal.

To date, Green Seal has awarded its seal of approval to 234 products and certifies products in over 50 categories, including, in addition to major household appliances, re-refined engine oil.

7.10.7.5 **The Netherlands**

The Netherlands have a special system to give support to environmental compatible products – the accelerated depreciation of environmental investments, the so called VAMIL regulation.

The VAMIL measure has been in effect since 1 September 1991. It is a tax facility offering companies the opportunity to apply accelerated depreciation on environment-friendly operating assets. If the asset is operational and fully paid for, it even allows depreciation of the full purchase price in the year an asset is acquired. This provides an attractive liquidity and interest gain for these companies. Eligible operating assets appear on a special 'VAMIL list'. The measure is not aimed at a specific environmental problem or region, but has a very wide operating ambit. The 1993 list, for example contained elements aimed at reducing water, soil and air pollution, noise emissions, waste production and energy use.

To be eligible for the VAMIL list, operating assets should:

- be clearly defined for fiscal purposes;
- have relatively good environmental impacts;
- not yet be widely accepted in the Netherlands;
- have no negative side effects, such as excessive energy use; and
- have a substantial potential market.

Periodically – in principle, once in every year the VAMIL list is replaced by a new one. Adaptations include the removal of operating assets that have become widely accepted and the addition of new environment-friendly technologies. The list is prepared by Bureau Vervroegde Afschrijving en Milieu, Ministry of Finance and Department of Environmental Investments (VROM–Department of Environmental Investments). It is hoped that eventually the VAMIL list will correspond to approximately 30% of all investments in environment-friendly operating assets.

All companies and persons liable to pay income or corporate taxes in the Netherlands can make use of the measure. However, the measure aims mainly at small- and medium-sized companies. The government determines a budget for the VAMIL measure once every year, setting an upper limit for tax allowances. The budget does not reflect government expenditures, as the reduced tax revenue in a given year is followed by increased tax revenue in later years. Therefore, the cost to the government consists only of lost interest. Additional costs necessary to make the asset operative are also eligible for accelerated depreciation. If an asset is developed and produced within a company, own-production costs can be depreciated in an accelerated manner.

7.10.7.6 US Regulation VGP and Definition of Environmentally Acceptable Lubricants (EAL)

The US Environmental Agency (EPA) has introduced a 'Vessel General Permit for discharges incidental to the normal operation of vessels' (VGP). The first VGP was issued in 2008 and effective until 19 December 2013. On 28 March 2013, EPA re issued the VGP for another 5 years.

The VGP regulates the use of substances in commercial vessels greater than 79 ft in length and operating as a means of transportation. Hence, the use of lubricants is restricted in those applications.

For these reasons, EPA has defined the so-called Environmentally Acceptable Lubricants (EAL). The main requirements refer to biodegradability, toxicity and bioaccumulation.

Biodegradability: Regarding environmentally acceptable lubricants and greases, biodegradable means lubricant formulations that contain at least 90% (w/w (weight in weight concentration)) or grease formulations that contain at least 75% (w/w) of a constituent substance or constituent substances (only stated substances present above 0.10% shall be assessed) that each demonstrate either the removal of at least 70% of dissolved organic carbon, production of at least 60% of the theoretical carbon dioxide, or consumption of at least 60% of the theoretical oxygen demand within 28 days. Acceptable test methods include: Organization for Economic Co-operation and Development Test Guidelines 301 A-F, 306 and 310, ASTM 5864, ASTM D-7373, OCSPP Harmonized Guideline 835.3110 and International Organization for Standardization 14593:1999. For lubricant formulations, the 10% (w/w) of the formulation that need not meet the above biodegradability requirements, up to 5% (w/w) may be non-biodegradable (but not bioaccumulative) while the remainder must be inherently biodegradable. For grease formulations, the 25% (w/w) of the formulation that need not meet the above biodegradability requirement, the constituent substances may be either inherently biodegradable or non-biodegradable, but may not be bioaccumulative. Acceptable test methods to demonstrate inherent biodegradability include: OECD Test Guidelines 302C (>70% biodegradation after 28 days) or OECD Test Guidelines 301 A–F (>20% but <60% biodegradation after 28 days).

Toxicity: 'Minimally toxic' means a substance must pass either OECD 201, 202 and 203 for acute toxicity testing, or OECD 210 and 211 for chronic toxicity testing. For purposes of the VGP, equivalent toxicity data for marine species, including methods ISO/DIS 10253for algae, ISO TC147/SC5/W62 for crustacean and OSPAR 2005 for fish, may be substituted for OECD 201, 202 and 203. If a substance is evaluated for the formulation and main constituents, the LC50 of fluids must be at least 100 mg l^{-1} and the LC50 of greases, two-stroke oils, and all other total loss lubricants must be at least 1000 mg l^{-1}. If a substance is evaluated for each constituent substance, rather than the complete formulation and main compounds, then constituents comprising less than 20% of fluids can have an LC50 between 10 and 100 mg l^{-1} or a no observed effect concentration (NOEC) between 1 and 10 mg l^{-1}, constituents comprising less than 5% of fluids can have an LC50 between 1 and 10 mg l^{-1} or a NOEC between 0.1 and 1 mg l^{-1} and constituents comprising less than 1% of fluids can have an LC50 less than 1 mg l^{-1} or a NOEC between 0 and 0.1 mg l^{-1}.

Bioaccumulation: 'Not bioaccumulative' means:

- the partition coefficient in the marine environment is log KOW <3 or >7 using test methods OECD 117 and 107,
- molecular mass >800 Da,
- molecular diameter >1.5 nm,
- BCF or BAF is <100 l kg^{-1}, using OECD 305, OCSPP 850.1710 or OCSPP 850.1730 or a field-measured BAF or
- polymer with MW fraction below 1000 g mol^{-1} is <1%.

In contrast to actual European understanding of the term 'biolubricant' (CEN/TR 16227, EU Ecolabel), the EAL definition does not include a criterion in regard to renewable raw materials.

7.11
Base Fluids

7.11.1
Biodegradable Base Oils for Lubricants

The main 'chemistry' concerning biodegradable lubricants is different types of ester oil, for example:

- vegetable oil from harvestable raw materials, for example rapeseed or sunflower oil
- semi-saturated, transesterified ester oils with natural fatty acids, for example trimethyclepropanetrioleate
- fully saturated, synthetic esters based on chemical modified vegetable oils or mineral oil for example diisotridecyladipate.

Other, in principle, biodegradable base oils mostly are not present on the market of 'environment-compatible lubricants' due to the following technical or environmental reasons:

- Polyalkyleneglycols (PAG)
- Low viscosity polyalphaolefins (PAO2)
- Some special types of synthetic hydrocarbon up to a viscosity of 6 $mm^2 s^{-1}$ at 100 °C.

The lubricants industry has invested significant sums of money in developing and marketing biodegradable lubricants.

Natural fatty oils such as castor oil, palm oil, rapeseed oil, soybean oil, sunflower oil, lard, tallow and sperm oil have been used in lubricants for years. They are so-called triglycerides of more or less unsaturated fatty esters. This type of base is biodegradable and, compared to mineral oils, will show excellent tribological qualities (low friction coefficient, good wear protection). Their range of use is limited by lower stability against thermal oxidative and hydrolytic stress and partly inferior cold flow properties. These limits can be improved gradually either with additives, or with the selection, cultivation or genetically modification of new types of plants. With new types of 'high oleic sunflower oils' (HOSO) with an amount of oleic acid of more than 90% it seems possible to formulate oils for higher performance levels.

7.11.2
Synthetic Esters

The wider use of natural base oils for additional, large volume lubricant technologies highlighted the dilemma that unaltered natural oils cannot satisfy a number of technical requirements while defined and highly specialized modification in a number of manufacturing stages prices them out of competition.

The collective name 'synthetic esters' covers a broad range of chemicals with different qualities and prices. For the development of environmentally acceptable lubricants, esters have to be selected which fulfil the ecological requirements and have more favourable properties than natural fatty oils.

These properties are mainly thermal-oxidative resistance, better low temperature behaviour and better resistance to hydrolysis. Chemistry offers a wide range of possibilities in the area of synthetic esters.

At present, polyolesters such as trimethylolpropane esters (TMP-esters) dominate. The basis of these are mainly alcohols from petrochemical and oleochemical industries and fatty acids derived from natural oils.

In regard to hydrolytic stability, 'normal' polyolesters differ only slightly from rapeseed oil; the difference in oxidation resistance is much greater. Both characteristics are significantly improved with complex esters. Normally, an improvement in hydrolytic stability worsens the base fluid's biodegradability.

The most important chemical reactions to improve the properties of esters are transesterification, (selective) hydrogenation, ozonolysis, dimerization.

However, there are complex esters (medium chain, saturated fatty acids on trimethylolpropane or other polyols) which combine excellent thermal-oxidative characteristics with good hydrolysis resistance and good biodegradability.

7.11.3 Polyglycols

Lubricants based on PAG can have very good technical properties and are well-known in long-term practical use.

Polyethylene glycols (PEG) are mostly biodegradable. They are not miscible with mineral oils or esters, but are water-soluble which, at the present time, is considered to be a disadvantage. If PEG is released through a leak or an accident, it will migrate quickly in the ground or in the water.

Polypropylene glycols (PPG) are (partially) miscible with mineral oils or esters, but in general not easily biodegradable.

The chemical industry is attempting to develop biodegradable, non-water soluble PAGs. These could be alternatives to esters.

7.11.4 Polyalphaolefins

Low viscous polyalphaolefins (PAO 2) are biodegradable. These base stocks have only limited application in the formulation of lubricants. The new types of biodegradable, higher viscosity synthetic hydrocarbons may have a greater influence in the future.

7.11.5 Relevant Properties of Ester Oils

7.11.5.1 Evaporation Loss

The evaporation loss of esters and vegetable oils is excellent as can be seen in Figure 7.3. The evaporation loss of various base fluids have been determined

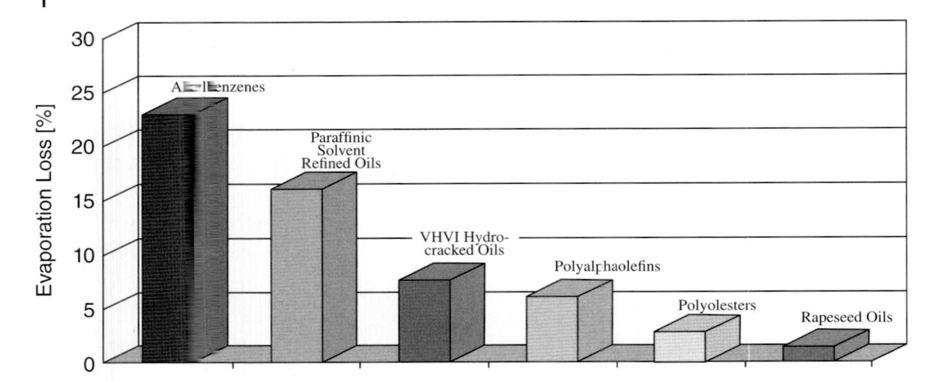

Figure 7.3 Dependence of evaporation loss on base-oil chemistry.

with the Noack test method according to DIN 51 581. It is apparent that vegetable oils such as rapeseed oil give the best results.

The lower evaporation loss of ester oils has great advantages, for example in regard to emissions from machine tools in metalworking and for the emissions, especially the particle emissions, of internal combustion engines.

7.11.5.2 Viscosity–Temperature Behaviour

Using an ester instead of a mineral oil improves the viscosity–temperature behaviour of a lubricant. The higher viscosity index (VI) of an ester results in a wider temperature range in application but with the recommended working viscosity. Also the higher VI can lead to polymer-free multigrade lubricants with improved shear stability.

7.11.5.3 Boundary Lubrication

Most vegetable oils, synthetic esters and glycols display excellent lubricity in boundary lubrication conditions. The high degree of polarity of these lubricants results in superiority against mineral oil-based lubricants. This has been proven by a series of tests. Experimental investigations on twin disc test rigs showed that the friction coefficient of vegetable oils, synthetic esters and glycols is half that of mineral oils.

Figure 7.4 compares a mineral oil and a TMP-ester of the same viscosity grade. The experimentally determined friction coefficients are represented depending upon the slip of the two-disk test rig. It can be stated that, for these measurements, the friction coefficient for this ester is half that of a mineral oil.

7.12 Additives

This section contains a brief summary only. Further information is given in Chapter 6.

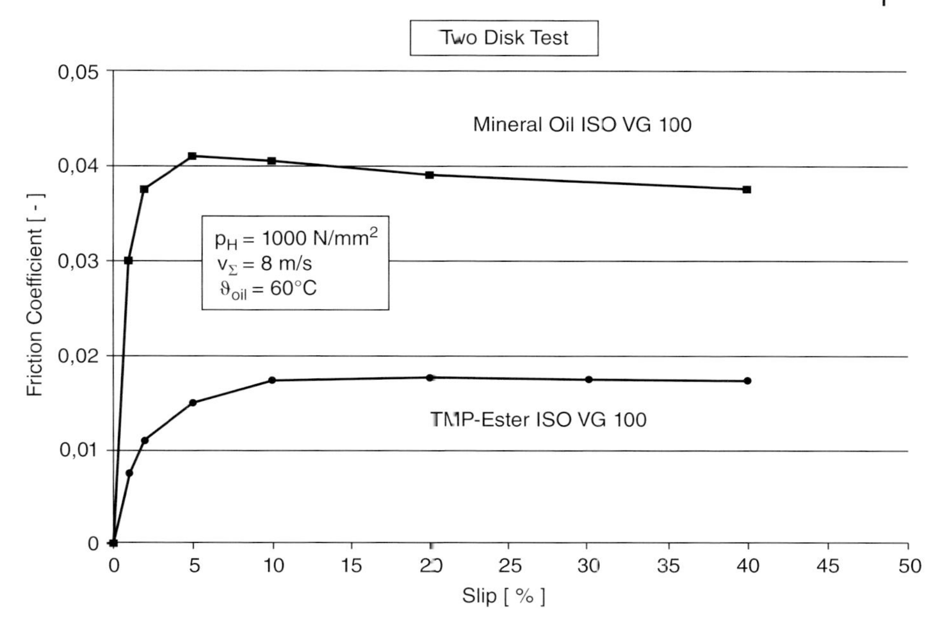

Figure 7.4 Friction coefficients of two base fluids as function of slip.

7.12.1
Extreme Pressure/Antiwear Additives

Most vegetable oils and synthetic esters used in lubricants display have a high degree of polarity. This characteristic results in better lubricity than mineral oils in boundary conditions. Sulfurized fatty materials are environment-friendly EP additives. Sulfur as an additive in esters provides an AW effect. Sulfur carriers with 15% total sulfur and 5% active sulfur have proven effective in rapidly biodegradable esters and vegetable oils.

7.12.2
Corrosion Protection

Vegetable oils and synthetic esters show a high polarity. This also applies to corrosion inhibitors and can result in a competitive reaction on the metal surface.

Special calcium sulfonates, succinic acid derivatives or ashless sulfonates can be used to provide corrosion prevention.

7.12.3
Antioxidants

Special phenolic and aminic materials are suitable antioxidants for the formulation of biodegradable lubricants.

7.13
Products (Examples)

With some examples of environment-compatible lubricants the special advantages of ester-based products will be illustrated.

7.13.1
Hydraulic Fluids

Viewed against the overall lubricants market (greases, engine oils, etc.), hydraulic oils have a very large market volume. Of about 150 000 tonnes of hydraulic oils (Germany 2005), about 40% were used in mobile and 60% in stationary systems. In 2005, about 6% of all hydraulic fluids were rapidly biodegradable, 88% were mineral oil-based (HLP, HLPD, etc.) and about 6% were fire-resistant products (HFC, HFD, etc.).

Especially hydraulic fluids are of great interest for biodegradable substitutes because the leakages and uncontrolled losses are well known since a long time. From a technical point of view today, it should be possible to substitute 90% of all mineral-based hydraulic oils by ester-based products. But there is a problem with the wide range of cost/performance ratio: Rapeseed-based, low-cost hydraulic fluids only are suitable up to 70 °C tank temperature. The high-performance ester oils, with four–five times higher prices, sometimes more than fulfil the requirements–the best oil for the application is to be chosen individually for each unit.

7.13.2
Metal Working Oil

Metal working/cutting oils are used to reduce heat when cutting or forming metal. As well, they lubricate the cutting area, remove contaminants and prevent corrosion. There are two broad categories of conventional metal working oils: neat- and water-based. Neat oils can be derived from animal, petroleum, vegetable or synthetic sources; however, most are petroleum-based. Water-based or aqueous metal working oils contain water or must be mixed with water after purchase. There are three types of aqueous oils: soluble, semi-synthetic and synthetic.

Potential consumers of metal working oils include companies involved in the manufacture of automotive parts, aerospace equipment, heavy equipment and electronics.

Having achieved success in Europe, vegetable oil-based metal working oils are now being introduced in other countries like Australia, Brazil and Canada. Because vegetable oil cutting fluids are more environment friendly, less of a health hazard, and superior in performance, they could become the oil of choice.

Metal working oils are sold to the end users directly from the manufacturer or through machine tool shops. Machine tool shops also sell the actual metal cutting tools. Ester-based metal working oils typically sell for twice that of conventional oils.

Because much technical testing is required to determine which applications are feasible for ester-based cutting oil, researching this product idea may be very time consuming and complex. In order to achieve success, the ester product would have to perform better, have a universal use, or be less expensive in an overall analysis of manufacturing cost.

7.13.3 Oil-Refreshing System

Because of their physical properties, refined, chemically unchanged vegetable oils can be used as alternatives to petroleum products in a large number of lubricants. Compared to mineral oils, their performance in most applications is limited by their relatively poor ageing resistance. The reasons for this are the unsaturated hydrocarbon chains in the natural fatty acids and the easy hydrolysis of the ester compounds.

Vegetable oils, at present, are primarily used for total-loss applications and products with relatively low technical specifications. At face value, using rapeseed oil as a basis for engine oils does not appear very promising – the present operating conditions of engine oils cannot be satisfied by chemically unmodified vegetable oils.

Over many years the development in the lubricants market had the main aim to formulate products for 'long oil drain intervals' or for lifetime. This was not possible with vegetable oil. But today we can recognize the opposite development in the following applications – the substitution of circulating oils by total-loss lubricants:

- *Example 1*: Minimum quantity lubrication in metalworking
- *Example 2*: Die-casting and forming instead of cutting and grinding
- *Example 3*: Oil-refreshing systems for lubrication of four-stroke engines.

Those ideas primarily are not motivated to use chemically unchanged vegetable oils in new applications. The main idea behind all examples is the lower overall impact of lubricants and its degradation products to environment – with total-loss lubricants.

For four-stroke engines, a novel lubrication concept for diesel engines was developed which counteracts the relatively poor ageing resistance of vegetable oils by continuous refreshing (Plantotronic system). This has permitted the first ever use of rapeseed and sunflower oils for engine lubrication. In this lubrication system, used oil is gradually burnt along with the fuel itself without negatively influencing emission values.

The concept can be seen as a way of continuously refreshing a vegetable-based engine oil or as a way of performing a continuous oil change. Combustion in the engine was only made possible by the use of low-additive vegetable oils – burning conventional oils seems highly contentious considering the additives commonly used in such oils. The use of vegetable oils for engine lubrication depends on the progressive burning of used engine oil with the fuel. The advantage of the oil refreshing method is that chemically unmodified, low additivated vegetable oils can be used and that no waste oil has to be disposed of. Another advantage is the automated oil-change, which helps to reduce the maintenance costs, especially for stationary engines (power plants).

7.14
Safety Aspects of Handling Lubricants (Working Materials)

Lubricants are working materials when personnel contact with these substances via contact with the skin or clothes or by inhalation or swallowing. Toxicological, medical and industrial hygiene aspects thus have to be considered.

With regard to the protection of personnel and material goods, possible fire and explosion damage should be evaluated.

Particularly in the field of metalworking, the properties of lubricants which effect the workplace are of relevance. Metalworking lubricants have to be allocated to the group of environmental chemicals which may be released from their direct application area and spread through the environment, that is water, the ground or air, during or after their application. This applies in particular to the disposal of used fluids.

The Law on Working Substances significantly effects health and safety-at-work demands on the commissioning, disposal, labelling and monitoring of hazardous substances. Products which are explosive, flammable, easily ignitable, highly poisonous, slightly poisonous, corrosive or irritating are all listed.

The focal point of the Law on Working Substances is substances themselves and not mixtures such as lubricants. Mixtures are only included if they are poisonous or if they contain potentially harmful solvents in certain concentrations. Furthermore, a part of the law specifically deals with hazardous mixtures used as paints or coatings. Dangerous substances and their concentration in a hazardous working substance are defined. Certain analogies for metalworking lubricants could be drawn from this. According to this definition, mixtures with >0.5% pentachlorophenol or >5% formaldehyde (both biocides in cutting fluids) would have to be classified.

All in all, the list of hazardous substances contained in the Law on Working Substances also offers a guide for mixtures in which these substances are used.

When lubricant manufacturers develop and formulate metalworking fluids, the above-mentioned safety aspects are a high priority. However, a number of technical and economical aspects require the use of some hazardous substances. In these cases, all available measures to protect personnel must be used and every effort made to avoid personnel contact with these lubricants.

7.14.1
Toxicological Terminology and Hazard Indicators

7.14.1.1 Acute Toxicity

This is the toxicity of a substance after a single exposure. In animal tests, the LD50 (lethal dose) figure is given to characterize the substance. The acute LD50 is the value in g or mg kg^{-1} body weight of the animal after which 50% of the animals die after 1 dose. In animal tests, the substance can be given orally (in the mouth), dermally (via the skin) or by inhalation. For the latter, the lethal concentration in air, LC50 is given in mg l^{-1} air over a defined period of time.

7.14.1.2 Subchronic and Chronic Toxicity

Health hazards are often only noticed after repeated exposure. Animal tests to determine these long-term effects usually last 28 days (subacute), 90 days (subchronic) and over a half-year (chronic).

The complexity of the individual tests varies greatly. Determining subacute toxicity costs about 10 times more, and chronic toxicity about 100 times more than the cost of an acute toxicity test.

7.14.1.3 Poison Categories

The German Law on Chemicals lists three poison categories for substances and mixtures; the LD_{50} values for rats are listed in Table 7.13.

Some countries use a larger number of poison categories. At one extreme, oral LD50 can be maximum from 5 mg kg^{-1} to 15 g kg^{-1}.

The vast majority of lubricants used in metalworking do not fall into these poison categories. Most products have an acute oral LD_{50} of >10 g kg^{-1} and an acute dermal LD_{50} of >2 g kg^{-1}. Most additives such as EP agents, polar additives and emulsifiers are also in this range. In a few cases, additive values below 2 g kg^{-1} are found (some biocides). When these substances are used, their concentration in the finished product must be considered. The large dilution factor in water-miscible products is a case in point. Water-miscible concentrates must be evaluated differently to water-miscible finished products.

7.14.1.4 Corrosive and Caustic

Substances are corrosive if 0.5 g or 0.5 ml of the substance in contact with the skin of a rabbit in defined test conditions leads to its destruction after 30 min.

Substances are caustic if they cause inflammation under the above-mentioned conditions.

7.14.1.5 Explosion and Flammability

This applies in particular to the flammable substances or fluids with flashpoints between 21 and 55 °C. This applies to lubricants if, as an application aid, they

Table 7.13 German Law on Chemicals LD50 values for rats.

	Highly poisonous	Poisonous	Slightly poisonous
Oral LD50 (mg kg^{-1})	≤25	25–200	200–2000
Dermal LD50 (mg kg^{-1})	≤50	50–400	400–2000
LC50, inhaled (mg l^{-1} in 4 h)	≤0.5	0.5–2	2–20

contain a Hazard Category AII white spirit. Substances which the German Law on Chemicals defines as being explosive, flammable or easily ignitable are generally not used in metalworking fluids.

7.14.1.6 Carcinogenic

A recurring topic in the field of metalworking lubricants is carcinogenic effect. In recent years, this was the case with some corrosion inhibitors, especially $NaNO_2$ together with amines (nitrosamines) and with polycyclic aromatic hydrocarbons.

7.14.1.7 Teratogens and Mutagens

Teratogens are chemicals which produce malfunctions to human life. When these alter genetic structures, these are called mutagens.

7.14.2 MAK (Maximum Workplace Concentration) Values

This is the maximum permissible workplace concentration for working substances in the form of gas, vapour or suspensions which are found in the air. These maximum permissible values are contained in a list of hazardous substances and this list is constantly being expanded. However, it is accepted that long-term exposure to the maximum concentrations does not pose any health risks.

As regards metalworking lubricants, it should be noted that some contain solvents. Table 7.14 shows the MAK values for the most commonly used solvents in metalworking fluids. Evaporation tendency which is characterized as vapour pressure (in mbar) or as an evaporation figure (see DIN 53170) is also of interest to the user. Solvents with low MAK values and high evaporation tendencies are particularly critical.

Apart from the MAK values, substances for which no toxicological or health and safety criteria exist to determine a MAK value, are given a TRK (technical guideline concentration) value.

Table 7.14 Solvent data.

MAK value (ppm)	Vapour pressure at 20 °C (mbar)	mg m^{-3}	Boiling point or range (°C)	Evaporation figure DIN 53170
(500)	6	(2000)	155–185	60
(500)	1	(2000)	180–210	165
50	77	260	87	3.8
200	133	1080	74	3.0
100	19	670	121	9.5
200	453	720	40	1.8
1000	389	5600	24	1.0
1000	380	7600	47	1.3

A number of carcinogenic substances are included in the TRK list. A list has also been published for the components and concentrations thereof in cutting fluids.

The subject of permissible oil mist concentration is constantly being discussed, in particular in connection with neat cutting oils (see Chapter 14). Oil mist as a dispersed phase must be treated differently to gas-like oil part. Based on the US threshold limit values (TLV) for oil mists, a MAK value of $5\,mg\,m^{-3}$ and total mist and vapour of $20\,mg\,m^{-3}$ is often proposed. The TLV value of $5\,mg\,m^{-3}$ is based on a toxicologically barely relevant white oil with no additives.

7.14.3
Polycyclic Aromatic Hydrocarbons (PAK, PAH, PCA)

The largest proportion of raw materials in metalworking lubricants are high-boiling point petroleum cuts which contain very few polycyclic aromatics (PAH (polycyclic aromatic hydrocarbons), toxicologically relevant aromatics). However, it is already known that in the area of aromatic hydrocarbons with four–six aromatic rings, some high carcinogenic substances exist. Benzo(*a*)pyrene (BaP, 1,2-benzopyrene, formerly 3,4-benzopyrene) was identified as such. This is often the key substance used to characterize the risk of environmental carcinogenicity of this group of substances.

While mineral oil hydrocarbons were evaluated for PAH content in the past by total aromatic content or UV adsorption in some areas, these days considerably analytical effort is used to quantitatively identify groups (such as isomers) or individual chemical components with known carcinogenic characteristics. In this field, special enriching processes with gas chromatographic identification have gained acceptance (e.g. Grimmer aromatics).

Neat metalworking oils are in the forefront of the PAH discussion because these can directly contact a comparatively large number of people via the inhalation of oil mists. Efforts were already made in the past to evaluate the cancer risks posed by neat cutting oils by determining their concentration of polycyclic aromatic hydrocarbons.

In recent times, precise analytical tests similar to the Grimmer method were performed on fresh and used neat cutting oils. These also determined PAH contents in air based on 5 mg oil mist per cubic metre air. Values were recorded for benzo(*a*)pyrene (BaP) which were below or at the average value measured in various locations (e.g. $0.02\,mg\,m^{-3}$). The increase in PAH content with cutting or grinding oil use can be dramatic but the absolute concentration still remains in a comparatively insignificant area. The conclusion drawn from this is that the cancer risks posed by polycyclic aromatic hydrocarbons in neat cutting oils is generally overestimated. And finally, new refining methods, better workplace hygiene and a reduction in oil misting will further lower potential risks.

7.14.[illegible]
Nitrosamines in Cutting Fluids

Discussions on this subject were triggered by the use of alkali metal nitrates as corrosion inhibitors. The most prominent was and is sodium nitrite, a widely available and cheap substance with good inhibitor properties. In certain circumstances, the effect of amines on nitrites can create nitrosamines. It must be noted that 80% of nitrosamines are carcinogenic. Sodium nitrite ($NaNO_2$) can be found, above all, in concentrations of low-mineral oil (semi-synthetic) or hydrocarbon-free (fully synthetic) solutions. These concentrations are so high that water-miscible application concentrations of 0.05–0.2% can be found. If other, mostly organic inhibitors are used, significantly lower concentrations can be used. Apart from the use of sodium nitrite in water-miscible cutting fluid concentrates, it can also be directly added to water-miscible cutting fluids by consumers.

A particular problem is the bonding of amines with nitrites in water-miscible cutting fluids and especially the combination of sodium nitrite and alkanolamines. Nitrosamines are the general term for *N*-nitroso compounds with the typical N–NO structures. They are created primarily by the reaction of nitric acid and its salts (nitrites) on secondary amines (e.g. diethanolamines). *N*-Nitroso compounds are created in the acidic area with primary, secondary and tertiary amines but only secondary amines are stabilized to form nitrosamines. An *N* nitroso agent is the anhydride of nitric acid. The reaction runs as follows.

While the carcinogenic effect of some nitrosamines has been known since the mid-1950s, it was assumed that the reaction between nitrite and amines could not take place in cutting fluids because these were alkaline and the reaction requires an acidic environment. If cutting fluid mists are swallowed, nitrosamines can be formed in the acid areas of the stomach. This was proven with a grinding fluid along with sodium nitrite and diethanolamine. The reaction was also performed in a human stomach with a non-carcinogenic nitrosamine. Interestingly, the formation of nitrosamines was also proven in the alkaline area but the catalytic effects of contaminants and the biochemical reactions of bacteria may have been the cause.

High-performance liquid chromatography (HPLC) and other analytical methods were used to prove the presence between 0.02 and 2.99% of diethanol nitrosamine in cutting fluid concentrates of fully-synthetic products. The pH of the products was between 9 and 11.

Although nitrosodiethanolamine is the most commonly found nitrosamine in cutting fluids the other types should not be ignored. It should be noted, for example that the carcinogenic effect of diethanolnitrosamine is 200 times less than that of diethylnitrosamine.

The nitrosamine discussions have led a number of cutting fluid manufacturers to develop nitrite-free cutting fluids. However, it must be remembered that in the past, a much larger proportion of water-miscible cutting fluids were free of nitrite.

7.14.5
Law on Flammable Fluids

The German law on the installation and operation of plants to store, fill and transport flammable fluids (Law on Flammable Fluids) does not directly effect the application of lubricants containing flammable solvents. However, the corresponding technical guidelines offer a series of practical tips concerning the handling of flammable fluids. The Law on Flammable Fluids allocated flammable fluids into two Groups, Group A (neat) and Group B (water-miscible). For the flammable solvents in metalworking lubricants, these are mainly Hazard Category A II (Flashpoint 21–55 °C) and A III (Flashpoint 55–100 °C) white spirits. Exceptions apply to lubricants with high consistencies and with low amounts of solvents.

The above mentioned technical guidelines deal with the demands on containers and their labelling (from Hazard Category A II). These technical guidelines also define hazardous conditions.

As to the application of lubricants containing solvents, advice offered by professional associations should be consulted. Similar advice has been published for the application of solvent-based paints.

Regarding electrical standards, special guidelines apply when lubricants containing flammable solvents are applied. In Germany, these are VDE 0165, 0170, 0171 and 0100. Apart from the flashpoint itself, any possible heating during application must also be considered.

7.15
Skin Problems Caused by Lubricants

Skin problems are second only to noise-related hearing problems according to a 1978 report by the German Association of Professional Associations. In the metalworking industry, a large proportion of the problems are caused by contact with lubricants. Chip-forming machining operations with water-miscible and neat cutting oils are in the forefront because of the large number of contact possibilities.

7.15.1
Structure and Function of the Skin

With an area of about 2 m^2, the skin is the largest human organ. It serves as the final barrier between the body and its environment. The outer skin should protect against external effects such as radiation, heat, cold and dryness at the same time as keeping vital substances within the body. In spite of these important protective functions, it is very thin with a thickness of less than 1 mm. The outer skin is constantly being regenerated by the production of new cells. Dead skin cells, sebum and perspiration form an additional barrier, the protection layer. Its acid nature protects the skin from diseases.

The skin also offers the body protection against thermal, mechanical and chemical attack. The skin thus also plays an important role in the our working lives. The most important and most common skin damage is to the outer layers of the skin, the epidermis. Under this lies the dermis (corium) and the subcutis. The epidermis consists of three parts, the corneal layer, the basal or germinative layer and the prickle cell layer.

As long as the corneal layer is intact, that is elastic and smooth with no cracks, all deeper layers of the skin are protected from harmful influences. The skin's fats play an important role, at the outer surface it has a direct protection role and deeper down it keeps the corneal layer supple and stops the skin drying out by regulating the skin's moisture. Degreasing substances such as low-molecular weight petroleum cuts and also solvents can destroy the surface oil and the skin's oil (lipids) in the walls of the cells. The result is a drying-out of the skin with cracks in the skin and the loss of the upper corneal layer cells. In this state, the ingress of harmful substances is relatively easy. Degreasing substances can also be absorbed through the tallow glands (follicles).

A further skin defence mechanism is the ability of the barrier layer to partly neutralize alkaline and acidic substances with certain amphoteric amino compounds. The skin's acidic barrier can also be seriously damaged by the ingress of alkaline substances.

7.15.2
Skin Damage

For ordinary people, the nomenclature of skin damage is difficult to access because genesis terminology mixes with appearance terminology. Particularly regarding water-miscible and neat cutting fluids, apart from the influence of the fluid itself, a number of possible harmful effects can occur which interact and multiply. An example of this is the mechanical damage caused by the workpiece or the chips which are transported by the cutting fluid. The appearance of skin eczema often makes an allocation to genesis difficult and an allergic eczema cannot be differentiated from a degenerative eczema.

7.15.2.1 Oil Acne (Particle Acne)

This is one of the most common skin problems caused by contact with neat cutting and grinding oils. Usually it is not the oil itself but small particles such as metal fragments which are the cause. This is why this problem is often referred to as particle acne. Oil acne can appear wherever oil directly contacts the skin and this can include oil-stained clothing. Body areas particularly at risk are the lower and upper arms, the backs of hands, the face, thighs and waist.

The drying out of the skin is normally accompanied by the blocking of tallow glands (blackheads, spots). If these become infected, they form pussy blisters. If a number of these pussy infections join together, a large pyodermous is formed. Particularly at risk are people with Seborrhoea with greasy skin. Such people should be kept well away from wet metalworking operations where exposure to oil cannot be avoided.

Improvements in machine tools such as machine encapsulation, fume extraction, the avoidance of oil mists and general automation means that personnel have less contact with oil and the cases of oil acne have fallen significantly in recent years. The frequency of oil acne cases in factories is a measure of the hygiene standards in force and the personal cleanliness of machine room personnel.

The persistent occurrence of oil acne may also be the result of certain substances in the oil. In many cases, these are EP additive substances. A specific form of this problem is chlorine acne which can be triggered by chlorine compounds in the oil. It has been shown that some chlorinated aromatics, but not chloroparaffin, can also cause this type of acne.

7.15.2.2 **Oil Eczema**

The term 'oil eczema' encompasses several skin problems. These look like scaly or wet reddened areas, sometimes with a cracked surface.

Acute toxic eczema is caused by the direct effect of the substance on the skin and causes a particular appearance.

Degenerative eczema is the most important skin problem caused by water-miscible cutting fluids. Long-term contact between the cutting fluid and the skin produces signs of degeneration and a breakdown of the skin's defenses. Key factors are the alkalinity of the fluid, the long-term wetness and specific product ingredients and particularly boundary-active substances from emulsifiers. On the subject of alkalinity, there are a number of evaluations of acute alkalinity (pH value) or potential alkalinity (reserve alkalinity). However, there is a consensus that pH values of over 9.0 (9.5) accelerate degenerative eczemas following alkaline damage to the skin. It must be noted that it is disproportionately difficult to develop and manufacture stable emulsions with a pH of 8.5 instead of 10.0.

In microbiological eczemas, pathogenic and apathogenic germs play a part. There are considerable differences in opinion concerning the significance of microbiological eczemas in metalworking. One question is whether the large number of germs in water-miscible cutting fluids promote this type of eczema. Some experience indicates that no hygienic risk is posed by germ numbers of 10^5 or $10^6\,ml^{-1}$ if no particular pathogenic germs are present.

Degenerative eczemas and degenerative dermatitis are faced by facultative, allergic eczemas. In these cases, the person is oversensitive to one or more substances. Such typical allergies must be kept away from allergens. Apart from this highly individual form of allergic eczema which effects some people, long contact with a larger group of people can also cause oversensitivity and ultimately to an allergic eczema. While degenerative dermatitis can be combated with skin protection and other measures, oversensitivity can occur wherever the allergen contacts the skin (contact dermatitis). Oversensitivity can last a number of years or even for life so that the allergen must either be removed from the process or the person in question has to be transferred to other work.

Long-term degeneration, for example caused by exposure to alkalis in water-miscible cutting fluids can ease the ingress of allergens into the skin and thus start the oversensitivity process.

Allergic skin diseases are relatively seldom in metalworking companies. However, the increasing number of chemically active substances in cutting fluids may increase the oversensitivity of skin. Chemically active means suspiciously allergic to dermatologists. Such suspiciously allergic substances are often found in biocides, EP additives and corrosion inhibitors. Among the biocides, formaldehyde and formaldehyde-splitting substances are viewed as potential allergens. However, free formaldehyde is often overestimated as an allergen.

Apart from the contents of the cutting fluid, contaminants and particles can often act as allergens. An example is chrome from chrome workpieces and tools. Chrome suspended in the cutting fluid is less dangerous than chrome solutions with their oil-soluble, lipolytic compounds.

In general, it can be assumed that cutting fluid manufacturers together with skin specialists select product formulations which keep the risk of acute toxic or allergic eczemas to an absolute minimum. Furthermore, the increasing automation of metalworking processes means that operating personnel have less and less contact with such fluids.

7.15.3
Testing Skin Compatibility

Preliminary tests on additives or finished metalworking fluids can be performed on animals. A group of tests involve the substance being applied to the shaved skin of animals at defined intervals. For example, one test involves the substance being applied 24 times to 10 mice for a period of 8 weeks. The substance can also be tested for compatibility if 0.05 ml is injected into the underbody skin of white mice (intracutan method).

Most commonly, cutting fluids or their components are tested for skin compatibility on persons because the direct transfer of animal results to humans is not possible. In such cases, preliminary animal tests as described above are still performed.

Several skin tests are used to determine oversensitivity to potentially allergic substances. These tests include the rubbing test, prick test, scratch test, intercutan test and the epicutan test. These tests are practically painless. A positive test reaction in the case of the rubbing, prick, scratch or intracutan tests (in the case of an immediate allergic reaction) results in an itching rash similar to that caused by nettles. Such a rash takes a few minutes and up to half an hour to develop. The epicutan test serves to determine eczemic reactions. If the test is positive, the test area reddens and blisters or spots appear. In these tests, the skin reacts very slowly so that results can only be drawn after 2 or 3 days.

Rubbing test: An allergen is applied to the skin but repeatedly rubbing it on the forearm skin. This test serves as a sort of preliminary test of high sensitivity.

Prick test: Standard allergen solutions are applied to the upper layers of the forearm skin by lightly pricking the skin with a pricking needle. Anti-allergic medication should not be taken less than 3–5 days before the test because this can influence the result. This is the most common method of testing inhaled or foodstuff allergens.

Scratch test: The skin is lightly scratched and the allergen is applied to the scratched area. This test is more sensitive than the previous prick test. This test is suitable for the evaluation of medicines.

Intracutan test: This test is also considerably more sensitive than the prick test. The test substance is injected with a 1 ml syringe into the upper skin. This test is ideal for testing medicine allergies.

Epicutan test: The allergens, mixed with Vaseline and in small receptacles, are placed on the skin and held in place with a plaster. During the test which normally lasts 4 days, the patient must not shower or bathe, abstain from sport and stay out of the sun. The test is ideal for determining allergic contact eczemas. When modified with light, the test serves to examine photoallergic reactions.

In the epicutan test, cloths soaked with the test fluid are placed on attacked areas of the skin and covered with non-waterproof materials for 24 h. The concentration of the test fluid is higher than normally used to artificially induced skin irritation. The test area is then examined for skin irritations. The degree of skin irritation is evaluated by various criteria and a scale of points which are based on a comparison of test fluids and a standard reference substance.

The results are again read after 48 h, 72 h and 96 h. In the case of water-miscible products, the normal application concentration is used but it can also be tested at a uniform 1%. When biocides are tested, the substance is often tested at the application concentration together with water and compared to the water-miscible cutting fluid. Epicutan tests are normally performed on 40–80 persons. People with skin problems are more sensitive than people with healthy skin. This test also identifies individual allergic reactions but the method provides no information on the sensitizing effect of the test fluid.

For some critical groups of substances such as biocides, mucus tissue compatibility tests are performed in addition to the above-mentioned skin compatibility tests. This is performed by injecting the substance into the conjunctiva bag of a rabbit's eye. After 30 s the substance is flushed out and the reaction is evaluated.

Blood tests: When allergens are in the body, antibodies (specific body defenses) are formed in the blood. Many other substances such as proteins are formed when atypical dermatitis or a bronchial asthma occurs. Other methods are available to isolate immune cells in the blood and to measure their sensitivity to allergens.

Total IgE: All immune globulin E antibodies in the blood are determined.

Specific IgE: A number of immunological tests can be used to determine the concentration of specific IgE antibodies that attack allergenic substances.

Although specific IgE is easier to determine than total IgE, it does not always cause allergic symptoms in patients. The interpretation of the results requires a great deal of medical experience.

Leucotriene stimulation test: The rate at which these are released is determined. After the white blood corpuscles have been isolated, the reaction to allergens can be measured.

Histamine release test: Similarly to the Leucotriene test, white blood corpuscles are isolated and an allergen is added. In this case, the histamine released is measured.

Basophohile degranulation test: All the blood is exposed to an allergen and certain blood cells, the basophiles, are examined with a microscope.

7.15.4
Skin Function Tests

Skin damage can be evaluated with physicochemical apparatus or chemicals alone. Skin function tests can be used to diagnose an illness as well as testing the effectiveness and compatibility of dermatological products and care products.

Transepidermal water loss (*TEWL*): If the upper reaches of the skin are damaged, the barrier function of the skin no longer works and the body loses water. Damage to the skin's barrier function can be measured in a simulated transepidermal water loss (TEWL) test. A sensor directly above the skin measures the amount of water which is released through the skin and evaporated (evaporemetrics).

After removal of the blister roof, there is a strong increase in water loss as a result of the damaged barrier. It is generally accepted that the TEWL is well correlated with the degree of epidermal damage. The highest TEWL values are measured in fresh wounds with a continual decline until values for intact skin are reached at the end of healing. Therefore, the measurement of TEWL is a suitable parameter to determine the degree of re-epithelization. The critical phases of epithelial regeneration already occur during the first days following experimental wound induction.

Normally, the Duhring–Kammer test is used. The randomized, double-blind, intra-individual comparison of treatment fields (10–40 healthy volunteers) should include a placebo and a field treated with NaCl or Ringer's solution. The Duhring–Kammer test measurements are made over a 5-day treatment period following the settling of the blisters.

The physical basis of the measurement of TEWL is the diffusion law discovered by Adolf Fick in 1855. The diffusion flow dm/dt (with m = water transported, t = time) indicates the mass per cm^2 being transported in a period of time. It is proportional to the area (surface) and the change of concentration per distance dc/dx (with x = distance from skin surface to point of measurement). The constant factor D is the diffusion coefficient of water vapour in air.

This law is only valid within a homogeneous diffusion zone which is approximately formed by a hollow cylinder. The resulting density gradient is measured indirectly by two pairs of sensors (temperature and relative humidity).

Skin elasticity: The ability of the skin to regain its shape is evaluated with visual methods.

Blood circulation in the skin: The flow of blood through the skin can be measured with a laser and optical detectors.

Skin thickness: The thickness of the skin can be measured ultrasonically and the special structure of the skin can be illustrated in the same way.

Skin colour: The colour spectrum of the skin can be measured with colorimetric analysis.

Skin moisture: Moisture can be determined by measuring the electrical resistance of the skin.

Skin oiliness: The oiliness of the skin can be visually evaluated.

pH of the skin: Upper skin pH can be measured with surface electrodes.

Alkali resistance test: Caustic soda can be used as an irritant and various methods are available to determine damaged corneal layer.

In vitro method according to the BUS model: The *in vitro* isolated perfused bovine udder skin (BUS) model was developed for use in pharmaceutical research and is used in the cosmetics and chemical industries for testing efficacy and safety. The cow udders are obtained from slaughterhouses. After pretreatment they are cleaned carefully and shaved in the laboratory. The test substances are applied, after approximately 1 h of aerobic adaptation of the metabolism, by perfusion with oxygenated, warmed Tyrode's solution. The continuous perfusion keeps the udder, including the skin, viable for more than 8 h, during which the horny layer barrier and the skin metabolism remain active. Other advantages include the large application surface (up to 400 cm^2 per udder side), on which numerous product tests can be conducted comparatively (e.g. fresh and used process chemicals) and cost-efficiently. The hirsute skin on the side of the udder is histologically and functionally similar to human skin.

7.15.5 Skin Care and Skin Protection

Cleaning the skin is an important aspect of hygiene practices in the metalworking industry. The cleaners used must be matched to the type of dirt involved. Under no circumstances should cutting fluids be used to clean the skin. Although water-miscible fluids or low-viscosity honing or grinding oils may clean skin effectively, the metal chips, abrasive grinding debris or other contaminants suspended in the fluid can cause considerable damage to the skin. The best cleaners are emulsions which are either slightly alkaline, neutral or slightly acidic.

Dirt which is embedded in the corneal layer can only be removed physically. These should not be so aggressive that they damage the skin (e.g. pumice, sand). Ideal are certain types of sawdust.

Cleaning the skin always means that some of the skin's oils are lost. This is the function of skin care which should allow the skin to regenerate during work breaks and above all, replace the dissolved lipophilic components. All skin care products should be capable of penetrating the corneal layer.

Skin protection products should be used to shield the skin from harmful substances. They must anchor well to the corneal layer and should not be easily dissolved by cutting fluids (they should resist washing-off for at least half a working shift). Skin protection products should not hinder the work process (by making things slippery) or by interfering with subsequent process (e.g. silicones). They should be easy to apply and easy to wash-off. As regards, the principal function of skin protection emulsions, water-in-oil emulsions can best protect against water-soluble substances in water-miscible cutting fluids. On the other hand, oil-in-water emulsions offer the best protection against harmful oil-soluble substances.

Further Reading

Administrative Regulation on the Classification of Substances Hazardous to Waters (1999) (in German "Verwaltungsvorschrift wassergefährdende Stoffe" – VwVwS), Germany, 29 May 1999.

Anlag enverordnung, e.g. in Schiegl, Betrieblicher Umweltschutz, Juni 1992.

ASTM D6866-12 (2012) Standard Test Methods for Determining the Biobased Content of Solid, Liquid, and Gaseous Samples Using Radiocarbon Analysis. doi. 10.1520/D6866-12.

Backé, W. (1992) Bestandsaufnahme und Trends der Fluidtechnik. *Ölhydraulik Pneumatik*, **36**, 16–19.

Bericht über den Einsatz biologisch schnell abbaubarer Schmierstoffe und Hydraulikflüssigkeiten und Maßnahmen der Bundesregierung.

Coordinating European Council (1994) Biodegradability of two-stroke cycle outboard engine oils in water, CEC-L-33-A-93, London.

Coordinating European Council (2012) Biological degradability of lubricants in natural environment, CEC-L-103-12.

Blauer Engel UZ-173 (2014) *Vergabegrundlage für Umweltzeichen – Biologisch abbaubare Schmierstoffe und Hydraulikflüssigkeiten*, RAL gGmbH, https://www.blauer-engel.de/produktwelt/gewerbe/biologisch-abbaubare-schmierstoffe-und-hydraulikfl-ssigkeiten.

BMFT-Project Emissionsarme Schmierstoffe 01 HZ 8821/5.

Bundesverordnung über Anlagen zum Umgang mit wassergefährdenden Stoffen (AwSV) in Bezug auf die Einstufung wassergefährdender Stoffe (Draft) http://www.bmub.bund.de/fileadmin/Daten_BMU/Download_PDF/Binnengewaesser/awsv_verordnung_bf.pdf.

CEN/TR 16227:2011:E (2011) Liquid Petroleum Products – Bio-Lubricants – Recommendation for Terminology and Characterisation of Bio-Lubricants and Bio-based Lubricants, Beuth-Berlin.

Cheng, V.M., Wessol, A.A., Baudouin, P., BenKinney, M.T. and Novick, N.J. (1991) Biodegradable and nontoxic hydraulic oils, SAE Technical Paper Series. 42nd Earthmoving Industry Conference, 9–10 April 1991, Peoria, IL.

Collection and disposal of used lubricating oil, CONCAWE-Report no. 5/96.

Directive 2011/381/EU (2011) EU Commission decision on establishing the ecological criteria for the award of the EU Ecolabel to lubricants, Official Journal of the European Union Law, 169/28, http://eur-lex.europa.eu/legal-content/EN/TXT/PDF/?uri=CELEX:32011D0381&from=EN.

Directive 2011/381/EU Lubricant substance classification list (LuSC-list), http://ec.europa.eu/environment/ecolabel/documents/lusclist_version_01102014.doc.pdf.

Dresel, W.H. (March 1990) Schmierfette auf pflanzlicher Basis, in *Biologisch abbaubare Schmierstoffe und Arbeitsflüssigkeiten*, Technische Akademie Esslingen.

Eichenberger, H.F. (1991) Biodegradable hydraulic lubricants – an overview of current developments in Central Europe, SAE Technical Paper Series. 42nd Earthmoving Industry Conference, 9–10 April 1991, Peoria, Illinois.

Environmental Protection Agency (EPA) (2011) Environmentally acceptable lubricants, EPA 800-R-11-002, http://nepis.epa.gov/Exe/ZyPDF.cgi/P100DCJI.PDF?Dockey=P100DCJI.PDF.

Gesetzesblatt für Baden-Württemberg (1994) Verordnung des Umweltministeriums über Anlagen zum Umgang mit wassergefährdenden Stoffen und über Fachbetriebe (Anlagenverordnung-VAwS) vom 11 February 1994.

Deutsches Institut für Gütesicherung und Kennzeichnung e.V. (RAL) (1995) Grundlage für Umweltzeichenvergabe: Biologisch schnell abbaubare Hydraulikflüssigkeiten. RAL UZ 79, Bonn, Dezember 1995.

Deutsches Institut für Gütesicherung und Kennzeichnung e.V. (1988) Grundlage für Umweltzeichenvergabe: Biologisch schnell abbaubare Kettenschmierstoffe für Motorsägen. RAL UZ 48, Bonn, März 1988.

Deutsches Institut für Gütesicherung und Kennzeichnung e.V. (RAL) (1991) Grundlage für Umweltzeichenvergabe: Biologisch schnell abbaubare Schmierstoffe und Schalöle. RAL UZ 64, Bonn, Juni 1991.

Hübner, J. (1994) Das Problem der N-Nitrosamin-Bildung in wassergemischten Kühlschmierstoffen und die TRGS 611, Praxis-Forum.

ICOMIA (International Council of Marine Industry Association) Standard 33-88 (1988) Lubricating Oil for Two-Stroke Cycle Outboard Motors – Ecologically Friendly.

Ihrig, H. and Spilker, M. (1993) Umweltschonende Schmierstoffe in der Antriebs-Technik–Entwicklung und Ausblick, report at the Antriebstechnische Kolloquium ATK'93, 05.05.1993, published in symposium's report.

Ihrig, H. (1990) Erfahrungen mit Schmierstoffen und Hydraulikflüssigkeiten auf Pflanzenölbasis in Verlustschmierung und Langzeiteinsatz. *Mineralöltech.*, **8**, 180–185.

Ihrig, H. (1994) *Schmierstoffe und Umwelt*, TAE, Esslingen.

Ihrig, H. (1992) Umweltverträgliche Schmierstoffe in den 90er Jahren, *Tribol. Schmierungstech.* **3**, 121–125.

Luther, R. (1999) Einsatz von Pflanzenölen in Motorenölen, in: Gülzower Fachgespräche – Statusseminar "Biologisch schnell abbaubare Schmier- und Verfahrensstoffe" pp. 125–141.

Luther, R. (1998) Motorenöle auf der Basis von Pflanzenölen – ein neuer Ansatz. *Fett/Lipid*, **100**, 7–12.

Luther, R. (2009) Lubricants based on renewable raw materials, Biorefinica Osnabrück www.biorefinica.de/Luther_Biorefinica2009.pdf.

Luther, R. (2014) Biodegradable & biobased – standardisation activities regarding bio-lubricants. Workshop on Biolubricants Marketing Uses and Chemistry Milan.

Luther, R. (2014) Normierung von Bio-Schmierstoffen im Spannungsfeld der eingesetzten Rohstoffe, FNR Bioschmierstoff-Kongress, http://veranstaltungen.fnr.de/fileadmin/veranstaltungen/2014/Bioschmierstoffe2014/04_Luther.pdf.

Mang, T., Turek, R. and Ihrig, H. (1989) Rapsöl als Basis von Schmierstoffen. *RAPS – Fachzeitschrift Öl- Eiweißpflanzen*, **7**, 192–194.

Mang, T. (1983) Die Schmierung in der Metallbearbeitung, Vogel, Buchverlag Würzburg.

Mang, T. (1990) Legislative influences on the development, manufacture, sale and application of lubricants in the Federal Republic of Germany. Report on the CEC Symposium, 19–21 April 1990, Paris.

Mang, T. (1989) Lubricants and legislation in the Federal Republic of Germany. *Erdöl Kohle*, **42**, 406–407.

Mang, T. (1993) *Rechtsrahmen und wirtschaftliche Verwendung umweltfreundlicher Schmierstoffe*, TAE, Esslingen.

Mang, T. (1990) Schmierstoffe und Funktionsflüssigkeiten auf Pflanzenölbasis, Erfahrungen eines Herstellers. Schriftenreihe des Bundesministers für Ernährung Landwirtschaft und Forsten, Reihe A, Issue 391.

Mang, T. (1992) Schmierstoffe und Funktionsflüssigkeiten aus Pflanzenölen und deren Derivaten. DECHEMA Kolloquium 6 March 1992, Frankfurt.

Mang, T. (1991) Umweltschonende und arbeitsfreundliche Schmierstoffe. *Tribol. Schmierungstech.*, **38**, 231–236.

Möller, U.J. (1990) Entsorgung umweltschonender Flüssigkeiten. *Ölhydraul. Pneumatik*, **34**, 6.

Nordic Ecolabelling, Ecolabelling of Lubricants (2011) http://www.ecolabel.dk/kriteriedokumenter/002e_4_4.pdf.

Odi-Owei, S. (1988) Tribological properties of some vegetable oils and fats. *Lubr. Eng.*, **45**, 685–690.

OECD (1992) OECD Guidelines for the Testing of Chemicals, Section 2: Effects on Biotic Systems doi. 10.1787/20745761. 1992–2011 Section 3 Degradation and Accumulation, doi: 10.1787/2074577x, 1992–2014 http://www.oecd.org/chemicalsafety/testing/oecdguidelinesforthetestingofchemicals.htm.

OSPAR Commission (2007) Convention for the protection of the marine environment of the North-East Atlantic, http://www.ospar.org/html_documents/ospar/html/ospar_convention_e_updated_text_2007.pdf.

OSPAR Commission (2004) OSPAR list of chemicals for priority action, Reference number 2004-12.

Pelzer, E. (1992) *Normung und Vergaberichtlinien des Umweltzeichens von umweltschonenden Hydraulikflüssigkeiten*, Technische Akademie Esslingen.

German Society for Tribology (1993) Pflanzenöle als Schmierstoffe (Vegetable oils as lubricants–working sheet).

Pittermann, W. (1999) Tierversuchsfrei forschen mit dem Rinder-Eutermodell. In-vitro-Haut- und Schleimhauttests im Fokus kosmetischer Forschung. *Parfümerie Kosmetik*, **80**, 38–41.

Pittermann, W., Holtmann, W. and Kietzmann, M. (2003) Systematic in vitro studies of the skin compatibility of cutting fluids. *Dermatologie Beruf Umwelt*, **51** (2), 56–66.

Regulation (EC) No 1272/2008 (2008) On classification, labelling and packaging of substances and mixtures (CLP), Official Journal of the European Union Law 353/1, http://eur-lex.europa.eu/LexUriServ/LexUriServ.do?uri=OJ:L:2008:353:0001:1355:en:PDF.

Regulation (EC) No 1907/2006 (2006) Concerning the Registration, Evaluation, Authorisation and Restriction of Chemicals (REACh), Official Journal of the European Union Law 396/1, http://eur-lex.europa.eu/LexUriServ/LexUriServ.do?uri=OJ:L:2006:396:0001:0849:EN:PDF.

Scott, S.D. (1991) Biodegradable fluids for axial piston pumps & motors–application considerations, SAE Technical Paper Series. 42nd Earthmoving Industry Conference, 9–10 April 1991, Peoria, Illinois.

Stempfel, E.M. and Schmid, L.A. (1990) Biodegradable lubricating greases. 57th Annual Meeting of NLGI, October 1990, Denver, Colorado.

SP Technical Research Institute of Sweden, Swedish Standard 15 54 34 (2014) Hydraulic Fluids – Requirements and Test Methods

SP Technical Research Institute of Sweden, Swedish Standard 15 54 70 (2014) Greases – Requirements and Test Methods

VDMA-Einheitsblätter 24568 und 24569 Biologisch schnell abbaubare Druckflüssigkeiten.

Vessel General Permit for discharges incidental to the normal operation of vessels (VGP), EPA United States Environmental Protection Agency: http://water.epa.gov/polwaste/npdes/vessels/upload/vgp_permit2013.pdf.

Wagner, H., Luther, R. and Mang, Th. (2001) Lubricant base fluids based on renewable raw materials: There catalytic manufacture and modification. *Appl. Catal. A*, **221**, 429–442.

8
Disposal of Used Lubricating Oils

Theo Mang

Figure 8.1 shows what happens to the lubricants sold every year in Western Europe. Only 49% are collectable and only 28% are actually collected [1]. The chart also includes process oils which are not lubricants. It is possible to identify the following objectives from this figure: the intensive gathering of collectable oils and an improvement in the environmental compatibility of the lubricants, of which more than 50% pollute the environment by way of total-loss applications, leaks, evaporation and other routes.

8.1
Possible Uses of Waste Oil

Used lubricants represent a problem for the environment. Their ecologically compatible use is therefore an important environmental protection measure.

Used lubricants are created when all mechanical possibilities in a machine or at the user's premises no longer suffice to maintain the performance of the lubricant and especially when chemical additives have been used up and ageing by-products are present in the oil. The demand that new lubricants should be made of used products is based on the erroneous notion that re-refining can restore the original condition of a lubricant. In fact, lubricants lose value during use and re-refining, at best, can only restore the value of base oil. In the case of conventional mineral oils, this value is only slightly higher than fuels or heating oils. This is also the reason why re-refining is hardly economical without legislative provisions or subsidies. From a global competition point of view, other disposal options include the direct incineration of untreated waste oils, the simple pre-treatment (cleaning) and alternative uses such as flux oils for bitumen or for the manufacture of secondary feeds in sec-feed plants for catalytic crackers and as blending stock for high-sulfur fuels.

The different possibilities of regeneration, recycling and waste-to-energy application of used lubricants have been illustrated by Krishna (2006) for an integrated steel plant as an important consumer of lubricants [3].

Lubricants and Lubrication, Third Edition. Edited by Theo Mang and Wilfried Dresel.

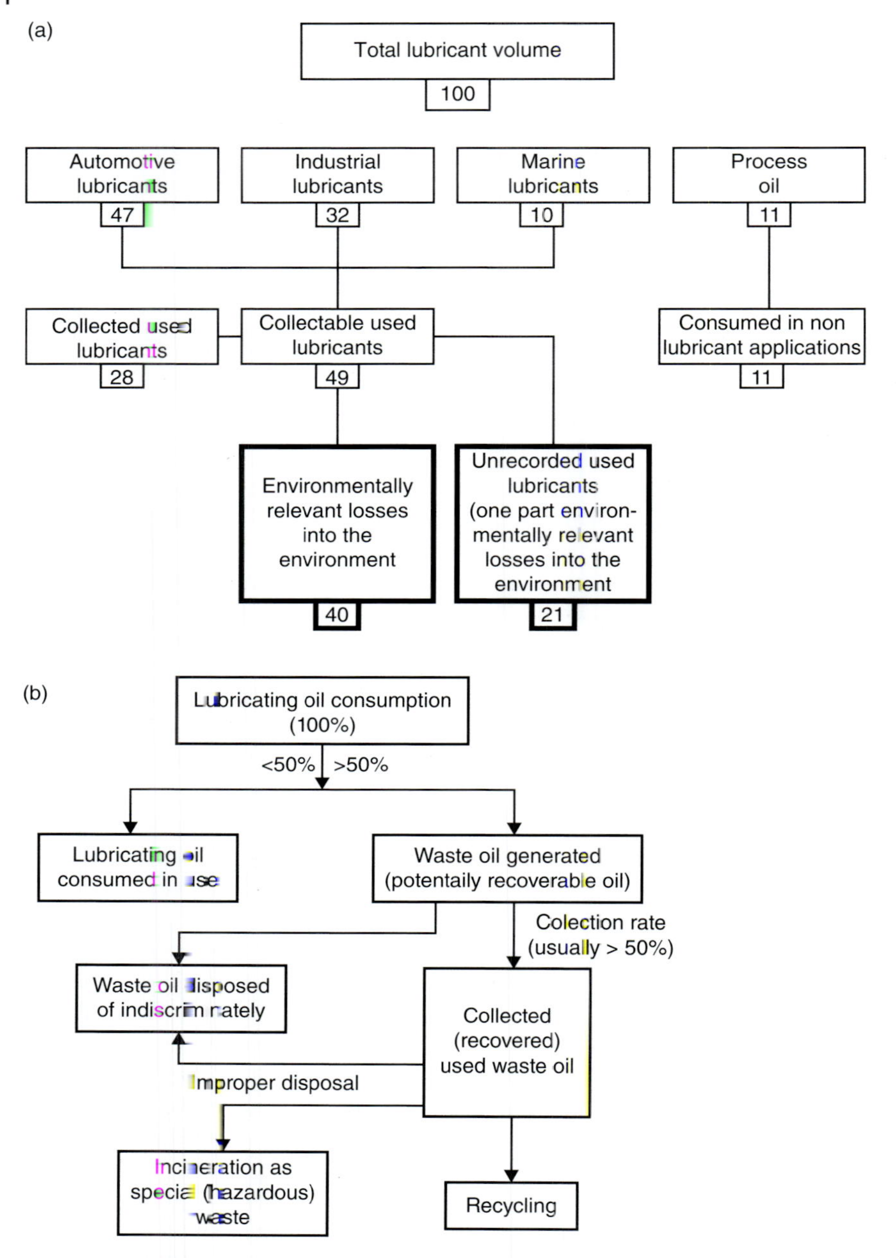

Figure 8.1 (a) Lubricating oil supply, use and disposal in Western Europe (CONCAWE 1996) [1]. (b) Generation of waste oil according to C. Kajdas [2]. (c) Waste oil recycling (taken from Ref. [2]).

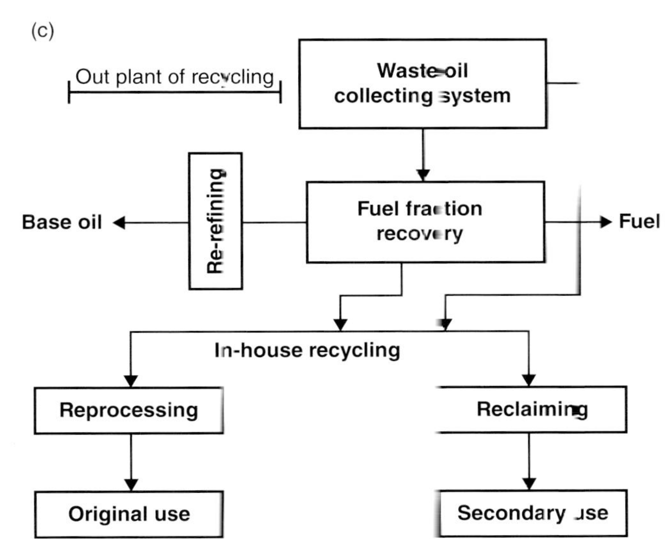

Figure 8.1 (*Continued*)

8.2 Legislative Influences on Waste Oil Collection and Reconditioning

The EC Directive 87/101 contains a recommendation to all member countries concerning the regeneration of used oils insofar as economic, technical and organizational conditions allow. Emission thresholds for incineration plants (<3 MW) make the burning of untreated used oils difficult. However, in some European countries, incineration in smaller incineration plants is still possible. Legislation permits incineration in high-temperature furnaces and by the cement manufacturing industry. In some countries, fuels and heating oils reclaimed from used lubricants are not taxed and are thus subsidized.

The PCB problem which surfaced in Europe in 1983 significantly influenced European legislation on wastes. Polychlorinated biphenyls (PCB), which enter the re-refined oil chain as fire-resistant hydraulic oils or condenser oils, have changed German waste oil legislation. While the PCB problem has practically disappeared, the division of waste oils into two groups has have considerable consequences. Used oils which contain more than 0.2% chlorine cannot be re-refined and are subject to expensive disposal procedures. This in turn has promoted the development of chlorine-free lubricants.

Re-refined used oils are subsidized in Italy. In Germany, the manufacturers of lubricants (including distributors) have transferred their legal requirement to properly dispose of waste oils to collection organizations. In 1999, these received about US$ 90 per tonne from the lubricant consumers.

In the United States, state law on this subject differs. Since 1986, used oils have been classified as hazardous wastes in California and other states have since followed. In some states, the collector is paid up to 20 cents per gallon by the oil user, in other states the collector has to pay.

Viewed globally, some extremely differing situations exist. While some countries do not regulate the collection and disposal of used oil and used oil is generally not collected, other countries can point to high collection and disposal rates (in 1996, 99% of used oils were collected or properly incinerated in Germany; but only 60% in the United States) [4].

C. Kajdas [5,6] gives the following definitions for this field:

Re-refining is the used/waste oil processing aiming at recovering/reusing a valuable resource of mineral base oil, being as good as or better than the virgin base stock and from which any petroleum-based lubricant can be produced.

Recycling is the processing of used oil in order to regain useful material for reuse; it may also concern a re-refining process to produce base oil for lubricants.

Single-use recycling is special used oil treatment generating recycled product for a single finite reuse.

Down-cycling is the processing producing recycled product of lower quality than the quality of the original material.

–

Used waste oil is any liquid lubricating material of which physical and chemical properties have changed in use in such a way that it is not fit for its original purpose.

Unused waste oil is any liquid lubricating material that has become contaminated when mixed with other waste/hazardous material, or that has failed to meet performance specifications.

Primary used oil biodegradation is the loss of one or more active groups of the oil molecules, making them inactive with respect to a particular function.

Ultimate used oil degradation is the complete breakdown of the material to CO_2, H_2O and mineral salts.

Reclaiming or reconditioning is the processing that mostly removes solids and water by simple physical methods (settling, heating, filtration, centrifuging, dehydration) but does not remove unwanted oil-soluble contaminants.

Reprocessing removes solids, water and some soluble contaminants using chemicals. Use of adsorbents may also be a part of reprocessing. Reprocessing also relates to fuel oil production from waste oil with mild cleaning.

8.3 Re-Refining

The re-refining of used oils to lube base oils started in 1935 [5]. The principal reasons why re-refining was unable to find acceptance were high process costs and therefore high selling prices compared to relatively low virgin oil prices, in

Table 8.1 General process stages for re-refining of used oils.

1	Separation of larger solid impurities along with most of the water. This is normally achieved by sedimentation.
2	Separation of the volatile parts (fuel residues in engine oils, solvents and low boiling-point lubricant components). This normally happens by atmospheric distillation. The separated light hydrocarbons can usually be used in-house for energy creation.
3	Separation of the additives and ageing by-products. This can occur by acid refining, solvent (propane) extraction, vacuum distillation or partly also by hydrogenation.
4	Finishing process to separate any remaining additives, ageing by-products and refining reaction products. This normally happens by hydrofinishing, with absorbents such as bleaching clay or mild, selective solvent extraction (e.g. furfural).

inadequate removal of carcinogenic polycyclic aromatics, the negative image of such oils in most markets and the increasing complexity of base oil blends in engine and other lubricants. In Western Europe, only 7% of base oil demand was satisfied by re-refined products in 1993.

Numerous re-refining technologies have been developed over the last 20 years. Many were patented but only few were suitable for large-scale application [7–12].

In general, the process stages shown in Table 8.1 are common to all the different methods.

8.3.1
Sulfuric Acid Refining (Meinken)

The sulfuric acid refining process was mostly developed by Meinken. Compared to older acid-based methods, various process stages reduce the amount of acidic sludge and used bleaching clay generated as well as increasing the lube oil yield. Besides the sulfuric acid treatment, bleaching clay is an essential part of the process. The treatment with bleaching clay as shown in Figure 8.2a can be seen as a processor of the process.

Due to the acidic sludge problem, acid refining has largely been replaced by other methods. However, numerous such plants were still in operation in 1999. Figure 8.2b shows a Meinken flow plan.

8.3.2
Propane Extraction Process (IFP, Snamprogetti)

Of the principal extractive refining processes, the IFP (Institut Fran¢ais de Pétrole) technology is worth mentioning. This technology initially used propane extraction together with acid refining and later with hydrofinishing. Propane extraction is also used by Snamprogetti (Italy) as the main refining step before and after vacuum distillation. Figure 8.3 shows the process with propane extraction [13,14].

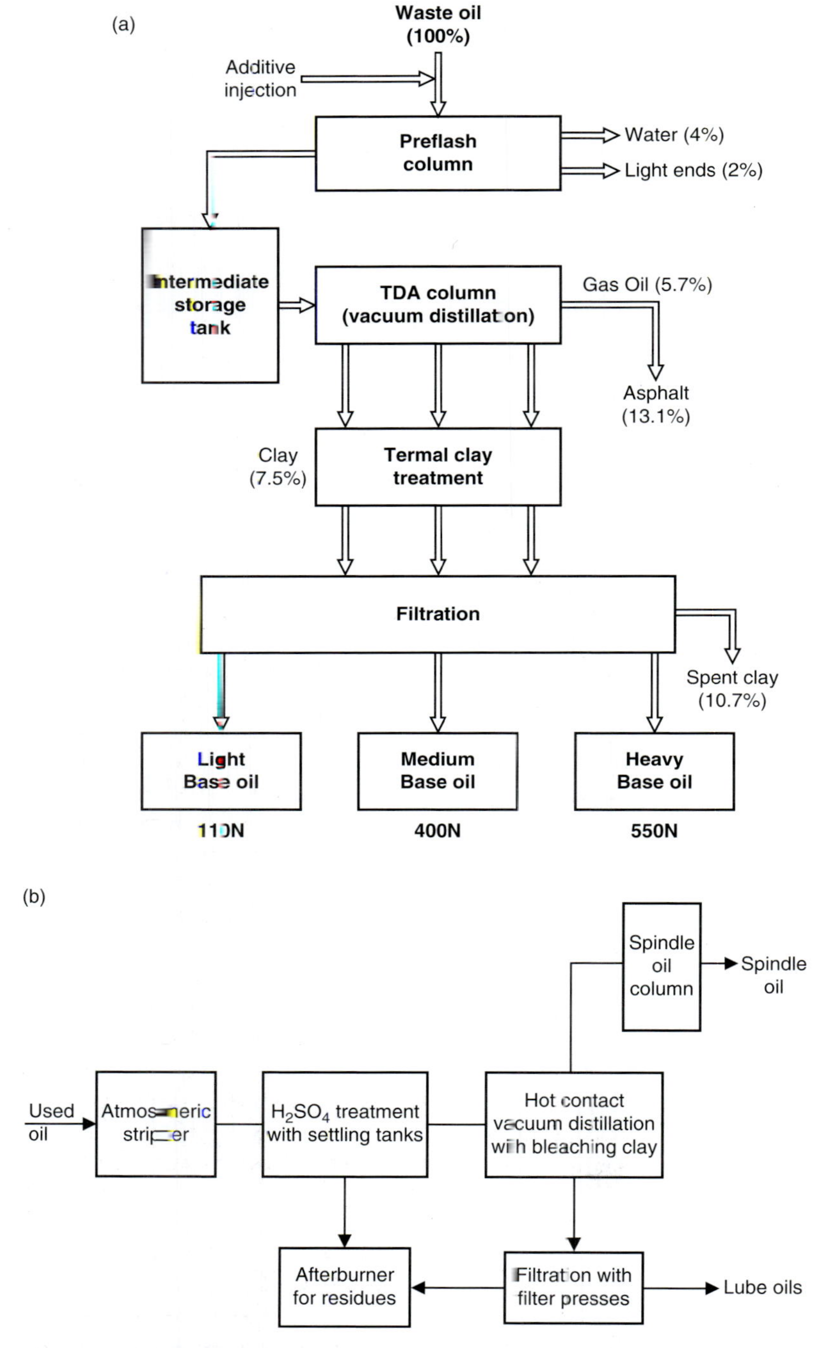

Figure 8.2 (a) Re-refining Method with thermal clay treatment (taken from Ref. [2]). (b) Sulfuric acid re-refining flow chart of the Meinken process) [13].

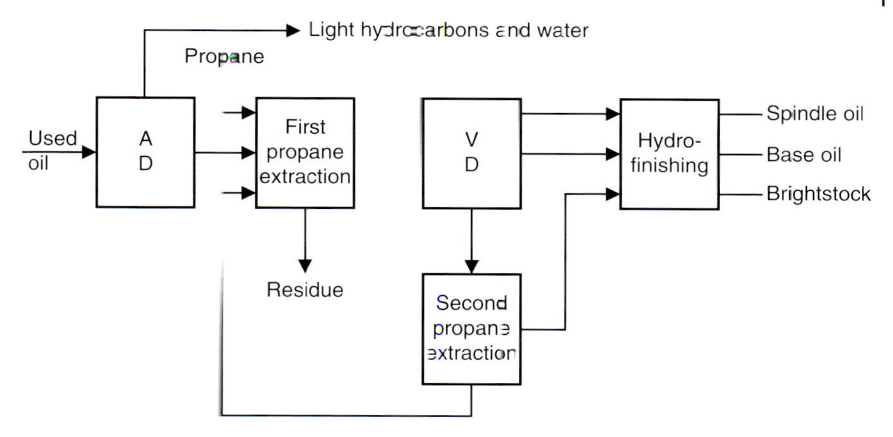

Figure 8.3 Re-refining by propane extraction (IFP, Snamprogetti).

8.3.3
Mohawk Technology (CEP–Mohawk)

The Mohawk process (subsequently CEP–Mohawk) using high-pressure hydrogenating was introduced in the United States at the end of the 1980s. The process begins with thin-film vacuum distillation (after flashing the light hydrocarbons and water). This is followed by hydrogenation of the distillate at 1000 psi over a standard catalyst. Special steps realized catalyst life of 8–12 months, which was essential for the economy of the process.

A marked reduction in the amount of water which must be treated as effluent as well as the cheaper materials for construction (absence of corrosion) is further advantages. The Mohawk process which is based on the KTI process has been licensed for Evergreen Oil (the United States and Canada).

8.3.4
KTI Process

The KTI (Kinetics Technology International) process combines vacuum distillation and hydrofinishing to remove most of the contamination and additives. The key to the process is the thin-film vacuum distillation to minimize thermal stress through mild temperatures not exceeding 250 °C.

The hydrofinisher removes sulfur, nitrogen and oxygen. The yield of finished base oils is high (82% on a dry waste oil basis). Figure 8.4 shows the flow chart of this process.

8.3.5
PROP Process

PROP technology was developed by Phillips Petroleum Company. The key elements of the process are the chemical demetalization (mixing an aqueous

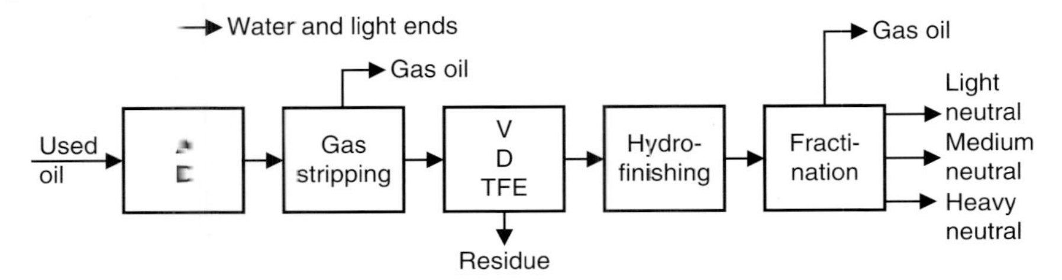

Figure 8.4 Flow chart of the KTI Process: thin-film evaporator (TFE) with hydrotreatment.

solution of diammonium phosphate with heated base oils) and a hydrogenation process. A bed of clay is used to adsorb the remaining traces of contaminants to avoid poisoning of the Ni/Mo catalyst.

Figure 8.5 shows the PROP process.

8.3.6
Safety Kleen Process

This process uses atmospheric flash for removing water and solvents, a vacuum fuel stripper, vacuum distillation with two thin-film evaporators, hydrotreater with fixed bed Ni/Mo catalysts. When using high severity, the hydrotreater is in the position to reduce polynuclear aromatics; it also removes higher boiling chlorinated paraffins.

Figure 8.6 shows a simplified block diagram. In 1998, the Safety Kleen process was used in the largest waste oil re-refinery in the world (East Chicago, Indiana, USA, plant capacity 250 000 tonne year^{-1}).

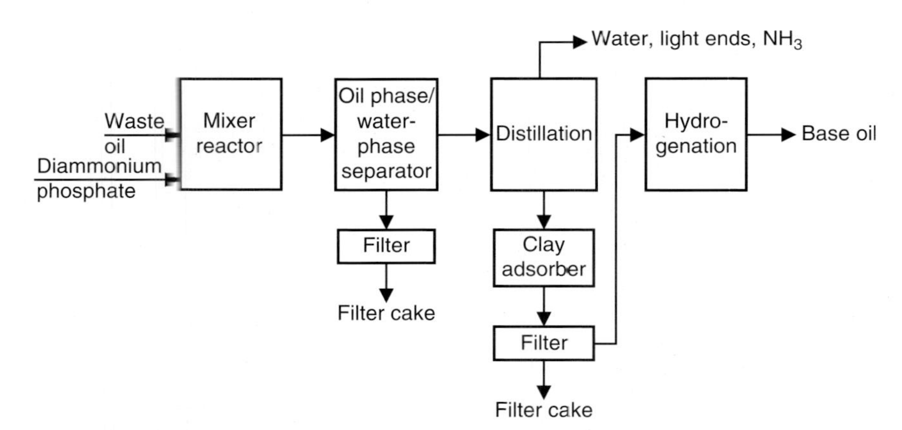

Figure 8.5 Flow chart of the PROP process.

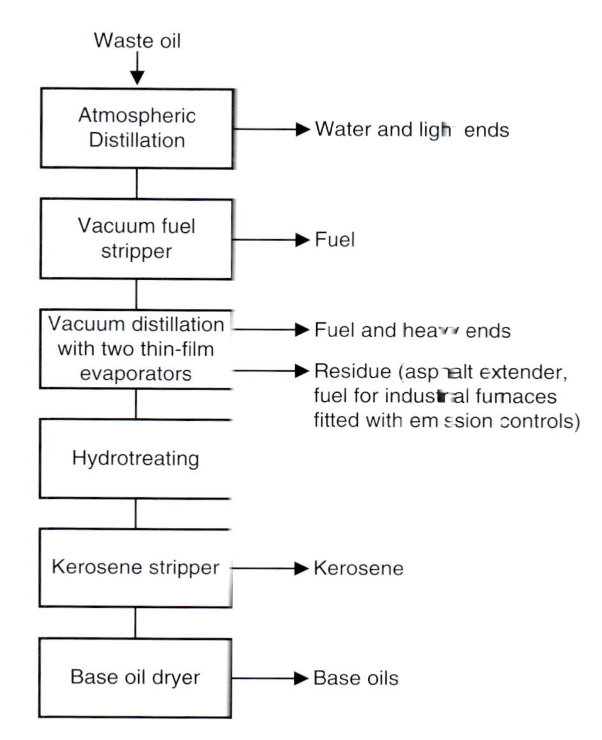

Figure 8.6 The Safety-Kleen process.

8.3.7
DEA Technology

The best results with regard to the technical and environmental quality of the re-refined oil and the elimination of PAH are provided by a combination of thin-film distillation followed by selective solvent extraction. In this process, the distillate from vacuum thin-film distillation towers equipment at the re-refinery (Dollbergen/Germany) is finally treated in a lube refinery solvent extraction plant followed by hydrofinishing (DEA, Hamburg/Germany). After this extraction process, the PAH content is lower than that of virgin solvent neutrals. Figure 8.7 shows the corresponding flow chart [14].

8.3.8
Other Re-Refining Technologies

Vaxon (Enpotec fabrication facilities in Denmark) uses three or four vacuum cyclone evaporators and finishing treatment with chemicals.

The key step in the ENTRA Technology is the special vacuum evaporation in a vacuum linear tubular reactor (single tube). After continuous evaporation by

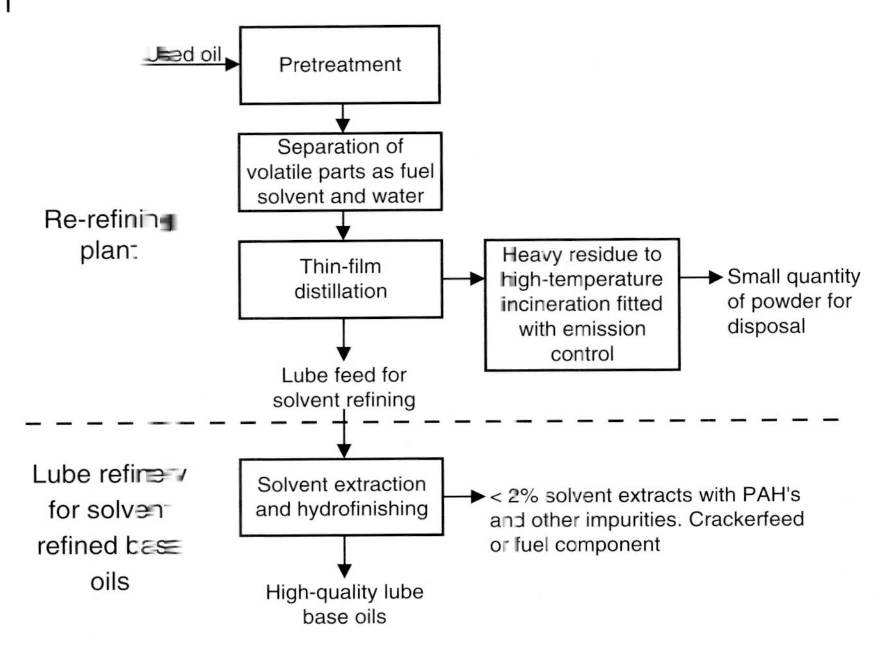

Figure 8.7 Introduction of selective solvent extraction in the re-refining process (DEA/Mineralöl-Raffinierie Dollbergen, Germany).

means of rapidly increasing temperature, vapour condensation is performed by fractional condensation. Complete dechlorination can be achieved with metallic sodium. Clay polishing is used as a finishing process.

The TDA (thermal deasphalting) process has been developed by Agip Petroli/Viscolube on the technology of PIQSA Ulibarri in Spain. The process is based on chemical treatment to facilitate subsequent deasphalting.

The de-asphalting process is combined with high fractionating efficiency (TDA unit). Finishing can be performed by clay treatment or hydrofinishing.

References

1 Collection and Disposal of Used Lubricating Oil (1996) CONCAWE Report 5 (1996), Brussels.

2 Kajdas, C. (2000) Major pathways for used oils disposal and recycling: part 1. *Tribotest*, 7 (1), 61–74.

3 Krishna, E. (2006) Recovery, re-refining and disposal of used oil of steel plants under Steel Authority of India Limited, Proceedings of the 15th International Colloquium Tribology, Automotive and Industrial Lubrication, Technische Akademie Esslingen (TAE), January 2006.

4 Betton, C.I. (2010) Lubricants and their environmental impact, in *Chemistry and Technology of Lubricants*, 3rd edn (eds R.M. Mortier, M.F. Fox and S.T. Orszulik), Springer, Dordrech.

5 Kajdas, C. (1996) *Used Oil Re-refining: Overview of Current Technologies Used,*

3rd European Congress on Re-refining, Lyon.

6 Kajdas, C. (2000) Major pathways for used oils disposal and recycling: part 2. *Tribotest J.*, **7-2**, 137–153.

7 McKeagan, D.J. (1992) *Economics of Re-Refining Used Lubricants*, Lubrication Engineering, pp. 413–423.

8 Oosterkamp, P.F.V.C. (1992) KTI Re-Refining Technology, UNIDO Workshop, Karachi.

9 Brinkman, D.W. (1991) Large grassroots lube re-refinery in operation. *Oil Gas J.*, **89**, 60–63.

10 Peel, D. (1996) Greening of America, World Base Oils '96 Conference, Elgin.

11 Schoen, C. (1996) *Einrohrreaktorverfahren zurAufarbeitung von Altoel und fluessigen Abfallstoffen, Description of the ENTRA Technology*, ENTRA Ingenieur- und Handels GmbH, Achern.

12 Giovanna, F.D., Tromeur, P., and Cohen, G. (1998) Successful re-refining in practice, International Used Oil Conference, Orlando.

13 (1980) Waste oil re-refining. *Hydrocarb. Process.*, 143.

14 Klamann, D. (1984) *Lubricants and Related Products*, Wiley VCH, Weinheim.

9
Lubricants for Internal Combustion Engines

Manfred Harperscheid

9.1
Four-Stroke Engine Oils

Technically and commercially, engine oils are the most commonly used among lubricants and functional fluids; on the global lubricants market, they account for more than 60%.

While demand in Europe is more or less stagnating and will decline slightly in the foreseeable future in spite of increasing vehicle registrations, there is still considerable growth potential in Southeast Asia and third world threshold countries (China, India, Korea, etc.) [1].

9.1.1
General Overview

Historically, the development of engine oils over the last 50 years has been focussed on the specifications issued by the international automobile industry. Starting with the first specifications published by the US Army (MIL specs), today there are three internationally recognized sets of minimum requirements. In the Europe, these are ACEA (Association des Constructeurs Europeen d'Automobiles), in the USA and Asia API (American Petroleum Institute) and ILSAC, respectively. Details of these specifications can be found in Section 9.1.3.

In principle, all specifications reflect the successive adaptation of oil qualities to developments in engine design. Back in the 1950s, monogrades dominated the engine oil market. As their name indicates, the viscosity of these oils was matched to the prevailing ambient temperature and therefore had to be changed between summer and winter. The 1960s saw the development of mineral oil-based multigrades, that is, combined summer and winter oils, initially high-viscosity types (SAE 20W-50) and later 15W-40. As time passes, base oil distillation and refining processes caught-up with the new viscosity and performance requirements (see Chapter 4). This led to new qualities with comparatively low evaporation losses and optimized cold flowing properties. Semi-synthetic, and in recent years, fully synthetic oils have dominated the premium-quality market, especially in the passenger car sector. In the recent past, a niche for environmentally friendly, biodegradable products has been created in the German-speaking market [2].

9.1.1.1 Fundamental Principles

Engine oils have to fulfil a wide range of functions in engines. The purely tribological task consists of guaranteeing the functional reliability of all friction points in all operating conditions (Figure 9.1). Apart from this classical tribological task, engine oils have to perform a number of additional functions. This begins with the sealing of the cylinder and ends with the transport of sludge, soot and abraded particles to the oil filter.

Starting with the tribological functions, the three classic sections of the Stribeck graph are satisfied, from hydrodynamic full lubrication to the elastohydrodynamic (EHL) area in bearings to the boundary friction conditions at TDC and BDC [3]. All friction pairings and a whole series of parameters are covered. Sliding friction speeds from simple linear up-and-down movement of the piston in the cylinder through to extreme rotational movements in floating needle roller bearings found in advanced turbochargers rotating at speeds of up to 200 000 rpm with micron tolerances. The temperatures encountered range from ambient in the Arctic (−40 °C) to sump temperatures of 100 °C to peak values of over 300 °C under the piston crown.

During the combustion process, the engine oil helps to seal the piston and cylinder. At the same time, it should burn off the cylinder wall without leaving

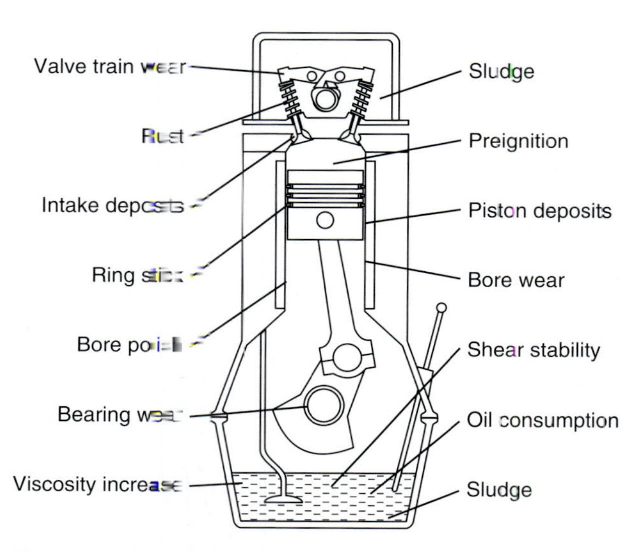

Figure 9.1 Scheme of lubrication points in an engine.

any residue. As for the piston itself, the engine oil dissipates heat from the piston and thus cools it. The blow-by gases formed when fuel is burnt and their reaction by-products have to be neutralized and held in suspension. The same applies to the soot and sludge particles caused by incomplete combustion. In particular, higher levels of soot in the engine oil can lead to a remarkable increase of wear at cams, tappets, bearings, piston rings and other highly stressed parts of the engine [4]. The oil also transports dirt and any abraded particles to the oil filter and ensures its filterability [5]. In addition, any water formed during the combustion process should be emulsified and even when higher concentrations are present and when the phases separate as temperature falls, the oil should protect against corrosion.

Acidification of the oil and corrosive wear can be in particular increased under short distance operation of the engine [6,7]

Engine oils should reduce friction and wear during extreme, low-temperature start-ups, as well as when the lubricating film is subject to high temperatures and pressures in bearings and around the piston rings. While the oil should still flow well and be pumpable without aeration at low temperatures (down to −40 °C) to avoid metal-to-metal contact during cold start-ups, the lubricating film must perform satisfactorily in bearings and hydraulic tappets [8]. At low temperatures, additives must not precipitate and the oil must not gel. At the upper end of the temperature scale, the oil must offer far-reaching resistance to thermal and mechanical ageing. And finally, the stability of the lubricating film should not be diminished by fuel dilution of up to 10%. CO_2 saving efforts lead over the last years to increased contents of ethanol in gasoline, which require even higher performance of the engine oils with regard to water compatibility and corrosion effects [9]. Diesel engine oils have to resist higher contents of biodiesel fuel, which consists of fatty acid methyl esters (FAME), that need higher thermooxidative control and corrosion protection within the engine oils [10].

9.1.1.2 **Viscosity Grades**

The viscosity of an engine oil is an indicator of how readily a load-carrying film can be formed at all lubrication points in an engine. As viscosity is a function of temperature, this applies to all potential ambient and operating temperatures. The adequate and rapid circulation of the oil at low temperatures, which is given by low viscosity [11] must be achieved at cold start cranking speeds. On the other hand, viscosity must not fall too much at high temperatures so that adequate lubricating film stability is given at high thermal loads.

As these requirements cannot be satisfactorily described with just one viscosity test method (see Section 9.1.2), corresponding threshold values were determined by the Society of Automotive Engineers (SAE) and the American Society for Testing and Materials (ASTM) as shown in Table 9.1. This table was last revised in April 2013 and a new viscosity class (SAE 16) was added with respect to the ongoing development of low viscosity engine oils for highest fuel economy demands.

Table 9.1 Engine oil viscosity classification SAE J 300 April 2013.

SAE viscosity grade	Low-temperature cranking viscosity (mPa s) at temperature in °C (CCS)	Low-temperature pumping viscosity (mPa s) at temperature in °C (MRV)	Low shear-rate kinematic viscosity at 100 °C ($mm^2 s^{-1}$)		High shear-rate viscosity at 150 °C and $10^6 s^{-1}$ (mPa s)
	Maximum	Maximum	Minimum	Maximum	Minimum
0 W	6200 at −35	60 000 at −40	3.8	—	—
5 W	6600 at −30	60 000 at −35	3.8	—	—
10 W	7000 at −25	60 000 at −30	4.1	—	—
15 W	7000 at −20	60 000 at −25	5.6	—	—
20 W	9500 at −15	60 000 at −20	5.6	—	—
25 W	13 000 at −10	60 000 at −15	9.3	—	—
16	—	—	6.1	<8.2	2.3
20	—	—	6.9	<9.3	2.6
30	—	—	9.3	<12.5	2.9
40	—	—	12.5	<16.3	3.5[a)]
40	—	—	12.5	<16.3	3.7[b)]
50	—	—	16.3	<21.9	3.7
60	—	—	21.9	<26.1	3.7

a) For 0 W, 5 W, 10 W.
b) For 15 W, 20 W, 25 W and monogrades.

According to this list, all viscosity grades can be described by their minimum kinematic viscosity at 100 °C. Additional dynamic viscosity thresholds apply to winter grades, which display the letter W. These values are determined in cold cranking simulators (CCS) or in mini-rotary viscometers (MRV). The dynamic viscosity value given by the CCS is a measure of flow properties [12] at low temperatures whereby the high shearing rate can masque paraffin crystallization. In the MRV, a so-called threshold viscosity of max. 60 000 cP has been determined for 10 °C lower temperatures to ensure that the oil pump does not draw air.

The high-temperature high-shear viscosity is an additional criterion for evaluating lube film stability at high shear rates and high temperatures in summer grades. In principle, mono- or multigrade oils could be used, depending on the climate. As already stated, modern engine oils are multigrade oils whose low temperature characteristics are indicated by the W and the high temperature viscosity by the number following the W. In central Europe, 90% of the engine oil market is accounted-for by multigrade oils. A graphical description for the VT behaviour of multigrade oils is schematically sketched in Figure 9.2. Manufacturers ensure that the correct viscosity grade for the climate considered is used in their engines by general descriptions (see Section 9.1.3) and/or by specific product approvals. In Figure 9.3 one OEM recommendation is shown as an example.

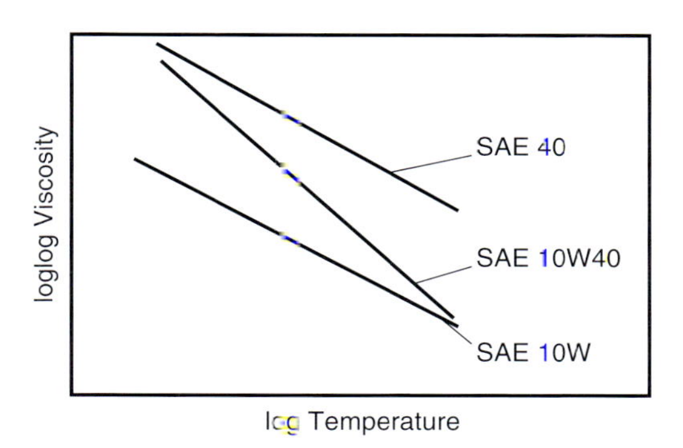

Figure 9.2 Comparison of monogrades and multigrades.

9.1.1.3 Performance Specifications

As a result of the continuing increase in specified oil performance, leading automobile manufacturers (OEMs) have discovered lubricants as constructional elements and have adopted their quality philosophy. The result is a qualitative shift in the engine oil market away from conventional products to semi-synthetic and synthetic formulations. Along with their higher price, the economic and

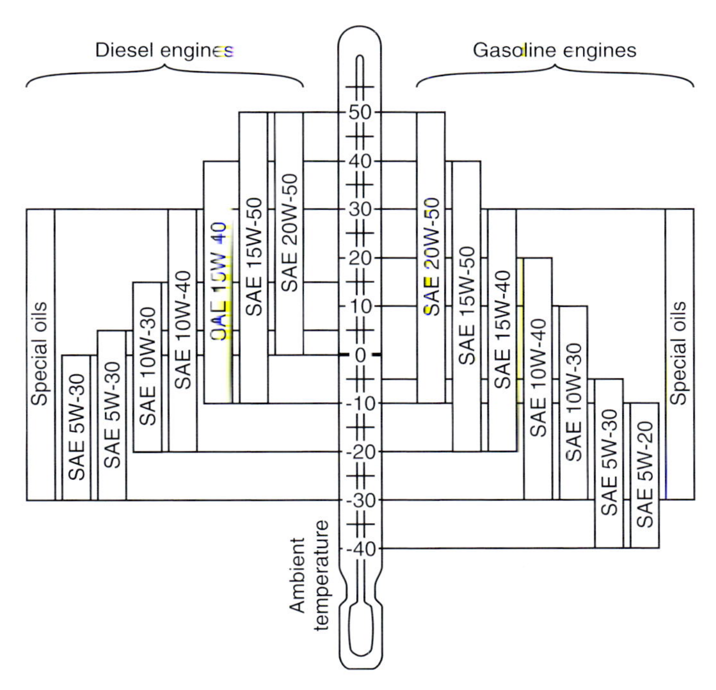

Figure 9.3 BMW engine oil recommendations.

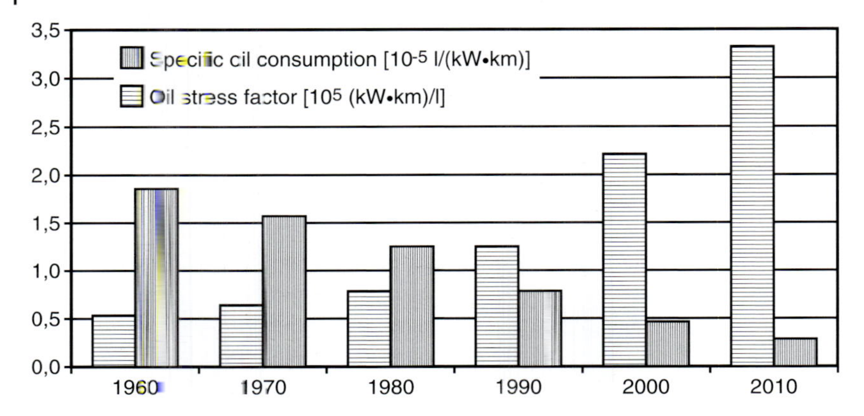

Figure 9.4 The history of oil stress.

ecological demands on these oils have also risen. Detailed demands, which are covered in Chapter 4, include

- longer life in spite of higher thermal and mechanical loads [13],
- improved emission characteristics by a cut in fuel consumption [14],
- lower oil-related particulate emissions [15],
- improved compatibility with exhaust aftertreatment devices and
- improved wear protection even in severe conditions.

Put simply, engine oils will have to offer significant potential regarding reducing fuel consumption and extending oil change intervals. At present, the oil change intervals for cars are between 10 000 and 50 000 km and 30 000 and 150 000 km for trucks. In the future, these figures could increase even further. A retrospective view of specific oil performance shows that this development has been in progress for the last 50 years [16]. Figure 9.4 shows that oil consumption per energy unit has fallen eight times during this period.

As the established CEC (Co-ordinating European Council for the Development of Performance Tests for Lubricants and Engine Fuels) engine tests do not allow comprehensive testing of all required oil properties, additionally in-house methods are now being used. A series of OEM-specific, long-term trials, which have now been taken-over for developing oils, are part of these engine tests. In total, these tests represent an enormous technical advance and financial expense because the engine tests themselves have been supplemented by radionuclide techniques (RNT) [17]. The advantages of this technology lie in the online monitoring of wear in defined running conditions, from running-in to full-throttle operation as well as the selective examination of critical engine components.

9.1.1.4 Formulation of Engine Oils

Engine oils are complex mixtures that are best described as formulations containing base oils and additives. Compared to other groups of lubricants, the base oils [18] play an important role (see Chapters 4 and 5). Without going into the characteristics and manufacturing in detail (see Chapters 4 and 19), mixtures of base oils are selected, which have the necessary viscosity and performance to correspond to a rough classification. The final products are then marketed as conventional mineral oil-based, unconventional semi-synthetic (hydrocracked) and synthetic engine oils. Precise international nomenclature divides base oils into the following five groups:

- Group 1: SN mineral oils with saturates <90%, 80 < VI < 120, S>0.03%.
- Group 2: HC oils with saturates >90%, 80 < VI < 120, S < 0.03%.
- Group 3: HC oils with saturates >90%, VI > 120, S < 0.03%.
- Group 4: Polyalphaolefins.
- Group 5: Esters and others.

9.1.1.5 Additives

Depending on the base oil used and the required engine performance, engine oils can contain up to 30 different additives whose percentage content can range from 5 to 25% in total. In the oil industry, a differentiation is made between performance additives [19], viscosity improvers and flow improvers. As a rule, the performance additives make up the largest group.

9.1.1.6 Performance Additives

The following species of chemical components are summarized under the general term of performance additives (Table 9.2).

Chemical formulation, tribological effect and theoretical background to these additive groups are examined in detail in Chapter 6. Particularly in the case of engine oils, the substance categories listed generally perform more than one function. ZnDDPs for example are primary anti-wear additives but also have a secondary antioxidant character resulting from a specific decomposition mechanism [20]. Furthermore, complex formulations of a number of individual components typically display synergistic as well as antagonistic interactions that have

Table 9.2 Performance additives.

Antioxidants	Phenols, amines, phosphites, sulfurized substances
Anti-wear agents	Metal dithiophosphate, carbamate
Detergents	Ca and Mg sulfonates, phenolates, salicylates
Dispersants	Polyisobutylene and ethylene–propylene oligomers with nitrogen and or oxygen as a functional group
Friction modifiers	MoS compounds, alcohols, esters, fatty acid amides, and so on
Anti-misting agents	Silicone and acrylate

to be matched to the application considered. The composition of the base oil components has an additional effect on these specific interactions. Considerable experience and know how is thus necessary to create an optimum formulation.

9.1.1.7 Viscosity Improvers

Viscosity improvers can be divided into two groups, the non-polar, non-dispersing and the polar, dispersing group. The first group are only really needed to set the viscosity of multigrade oils. Viscosity improvers increase viscosity and the viscosity index by altering their solubility at various temperatures (see Section 6.2). In an absolute concentration of 0.2–1.0%, they can generate a viscosity increase of between 50 and 200% depending on chemical structure and base oil solubility. Due to special modification, dispersing viscosity improvers are often ashless dispersants with additional thickening effects. Furthermore, viscosity improvers and pour point depressants have an effect on low temperature behaviour of a formulation (PP, CCS, MRV) and are essential parts of the HTHS viscosity. In the USA, additional demands are made on low-temperature stability (gelation index) that cannot be achieved without the viscosity improvers and pour point depressants being matched to the base oil. Due to their big impact on the rheological properties of engine oils, viscosity improvers can contribute to the energy saving effect of multigrade engine oils [21].

9.1.2 Characterization and Testing

To explain performance specifications and viscosity grading better, this section discusses engine oil tests in detail.

9.1.2.1 Physical and Chemical Testing

The physicochemical properties of an engine oil can be determined in the laboratory with standard test methods (Chapter 19). This characterization mainly focuses on rheological test values and the previously shown SAE classification system.

Various viscosity tests are used to determine exact low- and high-temperature viscosities [22]. The viscosity thus determined is a characteristic of the engine oil at a defined engine state. At low temperatures (−10 to −40 °C), a MRV (mini rotary viscometer) with a low shear gradient is used to determine the apparent viscosity and thus the oil's flowability in the area of the oil pump. In addition, maximum viscosity as the threshold of viscosity is determined in five graduated steps. The dynamic CCS (cold cranking simulator) viscosity, which is determined at −10 to −40 °C with a high shear gradient, is also an apparent viscosity which represents the tribological conditions at the crankshaft during cold starts. The maximum values laid-down in SAE J 300 guarantee reliable oil circulation during the start-up phase. The rheological characteristics at higher thermal loads, which occur during full-throttle operation are described by the dynamic viscosity at 150 °C and a shear rate of 10^6 s^{-1} or HTHS (high-temperature high-shear).

The corresponding threshold values also guarantee an adequate lube film even in these conditions.

Apart from the rheological characteristics, the Noack evaporation test described in Chapter 19 to test the volatility of base oils and additives as well as foaming tendency and air release can be characterized with simple methods. Furthermore, the compatibility of high-additive oils with seal materials is tested on standard reference elastomers in static swelling and subsequent elongation tests [23]. The viscosity loss resulting from mechanical load is described in Section 6.2.

9.1.2.2 Engine Testing

Because realistic engine oil tests cannot be performed only over long lasting field trials, a number of international committees have created methods of testing engine oils in defined test engines operated in reproducible and practically relevant conditions.

In Europe, the CEC is responsible for testing, approval and standardization [24]. Performance requirements are set-up in the form of ACEA oil sequences which are decided together with the additive and lubricant industries. In the USA, this task is performed by the automobile industry and the API. This institution lays down test procedures and limits. The Asian ILSAC has largely adopted the American specifications for automobiles.

In principle, all engine test procedures focus on the following general performance criteria:

- Oxidation [25] and thermal stability
- Dispersion of soot and sludge particles
- Protection against wear [26] and corrosion
- Foaming and shear stability [27].

In detail, the specification of the tests differentiate between gasoline- and diesel-powered car engines and truck engines whereby every test engine is characterized by one or a group of criteria. Tables 9.3 and 9.4 show a selection of engine tests and the relevant criteria for gasoline and diesel engines.

9.1.2.3 Passenger Car Engine Oils

Car engines include all gasoline and light diesel engines with direct or indirect injection. To ensure that the minimum requirements are met, the performance of the oils must be proven in the listed test engines irrespective of viscosity grade or the base oil used.

For gasoline engines, oxidation stability is tested in Seq. III G and in a Peugeot TU5JP engine (both at T max = 150 °C). Apart from the oxidation-related increase in viscosity, the effect of ageing-induced deposits on the piston and ring groove cleanliness is evaluated. Other standardized tests focus on sludge evaluation. This is the ability of an oil to efficiently disperse oil-insoluble ageing residues which result from the combustion process. Insoluble and inadequately dispersed particles lead to a sticky, pasty oil sludge that can block oil passages

Table 9.3 Passenger car engine tests.

Engine test	Engine manufacturer	Test procedure	Test criteria
TU3 MS	PSA	CEC L-038-94	Valvetrain wear
M111 SL	Daimler	CEC L-053-95	Black sludge
M111 FE	Daimler	CEC L-054-96	Fuel efficiency
TDI	Volkswagen	CEC-L-078-99	Piston cleanliness Ring sticking
TU5 JP-L4	PSA	CEC L-038-02	Piston cleanliness Oxidation Ring sticking
DV4 TD	PSA	CEC L-093-04	Soot handling (visc. increase) Piston cleanliness
OM646 LA	Daimler	CEC L-099-08	Valvetrain wear cylinder liner wear Ring wear Bearing wear Chain wear Piston cleanliness Oxidation Oil consumption
Sequence VG	Ford	ASTM D6593	Sludge Piston cleanliness Ring sticking
Sequence IVA	Nissan	ASTM D6891	Cam wear
Sequence IIIG	GM	ASTM D7320	Oil oxidation Wear Cleanliness
Sequence IIIGA	GM	ASTM D7320	Aged oil low temperature viscosity
Sequence VID	GM	ASTM D7589	Fuel efficiency
Sequence VIII	CLR (single cylinder)	ASTM D6709	Bearing Wear
N52 B30	BMW	BMW in-house norm	Cylinder liner wear Piston cleanliness Sludge Oil oxidation TBN depletion Oil consumption
N52 B30 aeration	BMW	BMW in-house norm	Air entrainment
N42 B18 RNT	BMW	BMW in-house norm	Valve train wear
M271 KE E[illegible] sludge	Daimler	Daimler in-house norm	Black Sludge

M271 KE E18 wear	Daimler	Daimler in house norm	Cam wear Ring wear Bearing wear Chain wear Cleanliness Ringsticking
X20XEV OP1	Opel (GM)	Opel in house norm	Oil oxidation and ageing Oil consumption Oil pressure
X20XEV aeration	Opel (GM)	Opel in house norm	Air entrainment
X20XEV RNT	Opel (GM)	Opel in house norm	Valvetrain wear
Y22DTR (DI diesel)	Opel (GM)	Opel in house norm	Cam wear Bearing wear Ring wear Chain wear Ring sticking Piston cleanliness Oil consumption
T4	Volkswagen	VW in house norm	Oil oxidation TBN depletion Piston cleanliness
FSI	Volkswagen	VW in house norm	Inlet valve deposits
TDI (RNT)	Volkswagen	VW in house norm	Valvetrain wear

and filters and thus lead to lubrication breakdowns. According to M 271 SL and M 111 SL, such sludge should be visually examined in the sump, in the crankcase and oil passages as well as by measuring the pressure increase created in filters. While the European M 271 SL and M 111 SL tests are performed ‘hot’, that is at high loads and speeds with a fuel which is sensitive to nitroxidation, sequence VG focuses on the generally lower operating temperatures in North America which lead to the formation of a so-called ‘cold’ black sludge. The Peugeot TU3MS engine test is used to check critical valve-train wear which can affect the timing of the engine. After a variable-load test program, cam scuffing and tappet pitting is evaluated.

The light diesel test engines, which are gaining popularity in passenger cars in Europe, are exclusively European engines. Again, oxidation stability and diesel-specific soot dispersion are in the forefront. Higher fuel efficiency and lower emissions have led to an increase in soot formation and thus up to 500% oil thickening. Due to higher combustion pressures combustion temperatures have also increased. These criteria as well as their influence on exhaust gases are tested in VW 1.9 l intercooler (piston cleanliness and deposits) and a Peugeot DV4TD (viscosity increase under high soot level). Also to be avoided are secondary effects on cylinder and cam wear and bore polishing which indicates that the original honing patterns have been worn away.

In 2008, the OM646 LA got an important additional multipurpose test in diesel engine oil development. This test has to be run with today’s low-sulfur diesel

Table 9.4 Heavy-duty engine tests.

Engine test	Engine manufacturer	Test procedure	Test criteria
OM501 LA	Daimler	CEC L-101-08	Piston cleanliness Ring sticking Sludge Engine deposits Cylinder liner wear Turbocharger deposits Oil consumption
T-11	Mack (Volvo)	ASTM D4485	Soot handling (visc. increase)
T-8E	Mack (Volvo)	ASTM D5967	Soot handling (visc. increase)
1 N	Caterpillar	ASTM D6750	Engine cleanliness Piston, ring and liner wear Ring sticking Oil consumption
T-12	Mack (Volvo)	ASTM D7422	Cylinder liner wear Ring wear Bearing wear Oil consumption
ISM	Cummins	ASTM D7468	Valvetrain wear Crosshead wear
ISB	Cummins	ASTM D7484	Valvetrain wear Crosshead wear
C-13	Caterpillar	ASTM D7549	Piston cleanliness Ring sticking Oil consumption
D2876 LF04	MAN	MAN in-house norm	Piston cleanliness Sludge Engine deposits Ring sticking Valvetrain wear Oil oxidation Oil consumption
D12D	Volvo	Volvo in-house norm	Piston cleanliness Ring sticking Cylinder liner wear Oil consumption

fuel and shows soot concepts up to 8% after its 300 h runtime. Such conditions need engine oils with extremely good soot handling properties to avoid large viscosity increases and wear.

Further reaching OEM-specific tests include the severe criteria of extended oil drain intervals and fuel saving. These apparently contradictory aspects of lower viscosity and less consumption on one hand and lower viscosity and greater reliability on the other represents a great challenge to oil manufacturers.

9.1.2.4 Engine Oil for Commercial Vehicles

Commercial vehicles include trucks, buses, tractors, harvesters, construction machines and stationary machinery powered by diesel engines. Apart from the pre-chamber diesel engines, which have largely been superseded in Europe, the engines are usually highly turbocharged direct injection motors. Economic and ecological aspects along with high injection pressures have improved combustion and thus reduce emissions. As an initiative of ACEA, oil change intervals have been extended up to 150 000 km for long haulage. The following highlights the fundamental differences between diesel and gasoline engines.

Long-life and reliability are the criteria for the commercial vehicle sector. The HD (heavy duty) oils have to match these requirements. The predominant requirements are the dispersion of large concentrations of soot particles as well as the neutralization of sulfuric acid combustion by-products. Performance is also judged by piston cleanliness, wear and bore polishing. Oxidation and soot-related deposits, mainly in the top ring groove lead to poor piston evaluations and an increase in wear. This, in turn, leads to the abrasion of the honing patterns in the cylinders, a problem better known as bore polishing. The result is increased oil consumption and poorer piston lubrication, because the oil cannot be trapped by the honing rings. Inadequate soot and sludge dispersion as well as chemical corrosion can lead to premature bearing wear. And finally, advanced turbocharged diesel engines have also been evaluated. Blow-by gases always carry some oil mist into the exhaust and turbochargers are very sensitive to unstable oil components.

In total, all characteristics can be found in HD oils whereby these are allocated to the following categories with increasing performance:

- Heavy duty (HD
- Severe heavy duty (SHPD)
- Extreme heavy duty (XHPD).

Despite numerous efforts to use screening tests to find the information, four–six cylinder engines are used to test the main performance criteria in runs of over 400 h and have displaced the original single-cylinder test engines (MWMB; Petter AWB).

Apart from the above-mentioned CM646 LA multipurpose test engines, European specifications demand a OM501 LA Daimler engine. Both test procedures are only used with XHPD oils (oil change intervals up to 150 000 km), piston cleanliness, cylinder wear and bore polishing are determined and evaluated. Particularly in the OM501 LA, deposits on the turbocharger as well as a pressure decrease have been recorded. The criterion soot-induced oil thickening is tested by the ASTM tests (Mack T-8E, Mack T-11).

Independent of the viscosity grade and the base oils used, classic HD oils have a high reserve alkalinity and thus a higher content of alkaline earth salts and organic acids [28]. Also regarding ashless dispersants, the oils are designed for soot dispersing. Special viscosity improvers are used generally to avoid additional deposits.

Oils for vehicle fleets pose a particular challenge. As opposed to special products, these should simultaneously satisfy as many car and truck demands as possible. Possible piston cleanliness provided by high concentrations of over-based soaps is sacrificed because gasoline engines are prone to self-ignition if high proportions of metal detergents are present. As a result, other components are selected, such as the skilful use of unconventional base oils along with detergents, dispersants, VI-improvers and antioxidants.

9.1.3
Classification by Specification

As already mentioned, physical and chemical properties are not enough to select the best lubricant for an engine. Complex and expensive practical and bench engine tests are performed to test and understand the performance of a lubricant. These requirements reappear in delivery conditions, in-house standards and general specifications.

9.1.3.1 MIL Specifications

These specifications originate from the US Forces that set the minimum requirements for their engine oils. These are based on certain physical and chemical data along with some standardized engine tests. In the past, these classifications were also used in the civilian sector to define engine oil quality. In recent years, this specification has become almost irrelevant for the German market.

MIL-L-46152 A to MIL-L-46152 E. These military specifications have now been discarded. Engine oils, which meet these specifications are suitable for use in the US gasoline and diesel engines. MIL-L-46152 E (discarded in 1991) corresponds to API SG/CC.

MIL-L-2104 C. Classifies high-additive engine oils for the US gasoline and normally aspirated and turbocharged diesel engines.

MIL-L-2104 D. Covers MIL-L-2104 C and requires an additional engine test in a highly charged Detroit 2-stroke diesel engine. In addition, Caterpillar TO-2 and Allison C-3 specifications are fulfilled.

MIL-L-2104 E. Similar in content to MIL-L-2104 C. The gasoline engine tests have been up-dated and include more stringent test procedures (Seq. III E/Seq. V E).

MIL-L-2104 F. Upgrade in heavy-duty diesel performance. Tests Caterpillar 1 K and Detroit Diesel 6V-92TA have been included.

9.1.3.2 API and ILSAC Classification

The API together with the ASTM and the SAE (Society of Automotive Engineers Inc., New York) have created a classification in which engine oils are classified according to the demands made on them, bearing in mind the varying conditions in which they are operated and the different engine designs in use (Table 9.5). The tests are standard engine tests. The API has defined a class for gasoline engines (S = service oils) and for diesel engines (C = commercial). Diesel

Table 9.5 Engine oil classification according to API SAE J 183.

Gasoline engines	Service classes
API-SA (Obsolete)	Regular engine oils possibly containing pour point improvers and/or foam inhibitors.
API-SB (Obsolete)	Low-additive engine oils low-power gasoline engines. Include additives to combat ageing, corrosion and wear. Issued in 1930.
API-SC (Obsolete)	Engine oils for average operating conditions. Contain additives against coking, black sludge, ageing, corrosion and wear. Fulfil the specifications issued by US automobile manufacturers for vehicles built between 1964 and 1967.
API-SD (Obsolete)	Gasoline engine oils for more difficult operating conditions than API-SC. Fulfil the specifications issued by US automobile manufacturers for vehicles built between 1968 and 1971.
API-SE (Obsolete)	Gasoline engine oils for very severe demands and highly stressed operating conditions (stop and go traffic). Fulfil the specifications issued by US automobile manufacturers for vehicles built between 1971 and 1979. Covers API-SD; corresponds approximately to Ford M2C-9001-AA, GM 6136 M and MIL-L 46 152 A.
API-SF (Obsolete)	Gasoline engine oils for very severe demands and highly stressed operating conditions (stop and go traffic) and some trucks. Fulfil the specifications issued by US automobile manufacturers for vehicles built between 1980 and 1987. Surpasses API-SE with regard to oxidation stability, wear protection and sludge transportation. Corresponds to Ford SSM-2C-9011 A (M2C-153-B), GM 6048-M and MIL-L 46 152 B
API-SG (Obsolete)	Engine oils for the severest of conditions. Include special oxidation stability and sludge formation tests. Fulfil the specifications issued by US automobile manufacturers for vehicles built between 1987 and 1993. Specifications similar to MIL-L 46 152 D
API-SH (Obsolete)	Specification for engines oils built after 1993. API-SH must be tested according to the CMAs Code of Practice. API-SH largely corresponds to API-SG with additional demands regarding HTHS, evaporation losses (ASTM and Noack tests), filterability, foaming and flashpoint. Furthermore, API-SH corresponds to ILSAC GF-1 without the fuel economy test but with the difference that 15 W-X multigrade oils are also permissible.
API-SJ	Supersedes API-SH. Greater demands regarding evaporation losses. Valid since October 1996.
API-SL	For 2004 and older automotive engines. Designed to provide better high-temperature deposit control and lower oil consumption. May also meet the ILSAC GF-3 specification and qualify as energy conserving. Introduced in July 2001.
API-SM	For all automotive engines currently in use. Designed to provide improved oxidation resistance, improved deposit protection, better wear protection, and better low temperature performance. May also meet the ILSAC GF-4 specification and qualify as energy conserving. Introduced in November 2004.
API-SN	This category was designed to provide improved high temperature stability, higher deposit control on pistons, lower sludge formation and better seal compatibility. Introduced in 2010 with application for all engines currently in use.

(continued)

Table 9.5 (Continued)

Gasoline engines	Service classes
Diesel engines	Commercial classes
API-CA (Obsolete)	Engine oils for low-power gasoline and normally aspirated diesel engines run on low-sulfur fuels. Corresponds to MIL-L 2104 A. Suitable for engines built into the 1950s.
API-CB (Obsolete)	Engine oils for low-to-medium power gasoline and normally aspirated diesel engines run on low-sulfur fuels. Corresponds to DEF 2101 D and MIL-L 2104 A Suppl. 1 (S1). Suitable for engines built from 1949 onwards. Offer protection against high-temperature deposits and bearing corrosion.
API-CC (Obsolete)	Gasoline and diesel engine oils for average to difficult operating conditions. Corresponds to MIL-L 2104 C. Offer protection against black sludge, corrosion and high-temperature deposits. For engines built after 1961.
API-CD (Obsolete)	Engine oils for heavy-duty, normally aspirated and turbocharged diesel engines. Covers MIL-L 45 199 B (S3) and corresponds to MIL-L 2104 C. Satisfies the requirements of Caterpillar Series 3.
API-CD II (Obsolete)	Corresponds to API-CD. Additionally, fulfils the requirements of US two-stroke diesel engines. Increased protection against wear and deposits.
API-CE (Obsolete)	Engine oils for heavy-duty and high-speed diesel engines with or without turbocharging subject to fluctuating loads. Greater protection against oil thickening and wear. Improved piston cleanliness. In addition to API-CD, Cummins NTC 400 and Mack EO-K/2 specifications must be fulfilled. For US engines built after 1983.
API-CF (Obsolete)	Replaced API-CD for highly turbocharged diesel engines in 1994. High ash. Suitable for sulfur contents >0.5%
API-CF-2 (Obsolete)	Only for two-stroke diesel engines. Replaced API-CD II in 1994.
API-CF-4 (Obsolete)	Engine oil specification for high-speed, four-stroke diesel engines since 1990. Meets the requirements of API-CE plus additional demands regarding oil consumption and piston cleanliness. Lower ash content.
API-CG-4 (Obsolete)	For heavy-duty truck engines. Complies with EPAs emission thresholds introduced in 1994. Replaced API-CF-4 in June 1994.
API-CH-4	Replaces API-CG-4. Suitable for sulfur contents >0.5%.
API-CI4	For high-speed, four-stroke engines designed to meet 2004 exhaust emission standards. Formulated to sustain engine durability where exhaust gas recirculation (EGR) is used and are intended for use with diesel fuels ranging in sulfur content up to 0.5% weight. Replaces oils with API CD, CE, CF-4, CG-4 and CH-4.
API-CJ-4 (Current)	Since 2006 for high-speed, four-stroke engines to protect year 2007 exhaust emission equipment. Limited sulfated ash and phosphorous content. In use for applications with low sulfur fuels for maximum service intervals. Exceeds the performance criteria of API CI-4.

engines in passenger cars are still outnumbered but have increased in recent years and are finding more acceptance in the USA.

9.1.3.3 CCMC Specifications

As API and MIL specifications were only tested on large-capacity, slow-running US V8 engines, and the demands made by European engines (small capacity, high-speed) were only inadequately satisfied, the CEC together with the CCMC (Committee of Common Market Automobile Constructors) developed a series of tests in which European engines were used to test engine oils (Table 9.6).

Table 9.6 Engine oil classification according to CCMC.

Gasoline engines	Gasoline engines
CCMC G1	Corresponds approximately to API-SE with three additional tests in European engines. Withdrawn on December 31, 1989.
CCMC G2	Corresponds approximately to API-SF with three additional tests in European engines. Applies to conventional engine oils. Replaced by CCMC G4 on January 1, 1990.
CCMC G3	Corresponds approximately to API-SF with three additional tests in European engines. Makes high demands on oxidation stability and evaporation losses. Applies to low viscosity oils. Replaced by CCMC G4 on January 1, 1990.
CCMC G4	Conventional multigrade oils in-line with API-SG, with additional black sludge and wear tests.
CCMC G5	Low-viscosity engine oils complying to API-SG with additional black sludge and wear tests. Greater demands than CCMC G4.
Diesel engines	Diesel Engines
CCMC D1	Corresponds approximately to API-CC with two additional tests in European engines. For light trucks with normally aspirated diesel engines. Withdrawn on December 31, 1989.
CCMC D2	Corresponds approximately to API-CD with two additional tests in European engines. For trucks with normally aspirated and turbocharged diesel engines. Replaced on January 1, 1990 by CCMC D4.
CCMC D3	Corresponds approximately to API-CD/CE with two additional tests in European engines. For trucks with turbocharged diesel engines and extended oil change intervals (SHPD oils). Replaced on January 1, 1990 by CCMC D5.
CCMC D4	Surpasses API-CD/CE. Corresponds to Mercedes–Benz Sheet 227.0/1. For trucks with normally aspirated and turbocharged diesel engines. Better protection against wear and oil thickening than CCMC D2.
CCMC D5	Corresponds to Mercedes–Benz Sheet 228.2/3. For heavy-duty trucks with normally aspirated and turbocharged diesel engines and extended oil change intervals (SHPD oils). Better protection against wear and oil thickening than CCMC D3.
CCMC PD 1	Corresponds to API-CD/CE. For normally aspirated and turbocharged diesel engines in cars. Replaced by CCMC PD 2 on January 1, 1990.
CCMC PD 2	Defines the requirements of high-performance multigrade oils for the present generation of diesel engines in cars.

These and the API tests formed the basis for the development of new engine oils. In 1995, CCMC was replaced by ACEA and ceased to be valid.

9.1.3.4 ACEA Specifications

As a result of persistent internal differences, the CCMC was disbanded and succeeded by the ACEA. CCMC specifications remained valid in the interim period. The first ACEA classifications came into force on January 1, 1996.

The ACEA specifications were revised in 1996 and replaced by 1998 versions. The 1998 specifications became valid on March 1, 1998.

Additional foaming tests were introduced for all categories and the elastomer tests were modified.

A-categories referred to gasoline, B-categories to passenger car diesel and E-categories to heavy-duty diesel engines.

The 1998 specifications were then replaced by the 1999 version, on September 1, 1999, and remained valid until February 1, 2004. Categories E2, E3 and E4 for heavy-duty diesel oils were updated, and a new category E5, was introduced; these were specifically aimed at the new demands for Euro 3 engines and the often higher soot content of such oils. A and B categories remained identical with the 1998 version.

On February 1, 2002 the ACEA 2002 oil sequences were issued to replace the 1999 sequences; these will be valid until November 1, 2006. Updates in cleanliness and sludge for gasoline engines (categories A1, A2 and A3) and a new category A5 with the engine performance of A3 but higher fuel economy were introduced. Tests for cleanliness, wear and soot handling were updated for diesel passenger cars and a new category B5 with outstanding cleanliness and increased fuel economy was introduced. For category E5 oils wear performance in respect of ring, liner, and bearings was tightened.

Since November 1, 2004 the ACEA 2004 oil sequences were in use and could be claimed by oil marketers. Oils in these categories were backward compatible with all other issues. Categories A and B were combined and could now only be claimed together in conjunction. Categories C1, C2 and C3 were introduced to refer to engine oils for use in cars with exhaust after treatment systems such as diesel particulate filters (DPF). Such oils are characterized by especially low content of ash forming components and reduced sulfur and phosphorus levels to minimize the impact on filter systems and catalysts.

On February 28, 2007 new ACEA 2007 oil sequences replaced the 2004 sequences. For light-duty diesel engines an additional category C4 was introduced with a sulfated ash limit of maximum 0.5% at higher viscosity level of HTHSV min. 3.51 mPa s.

With the 2008-revision, the sequences were again updated and due to increased use of biodiesel minimum TBN-levels for the fresh heavy-duty diesel oils were added. Concurrently, some low and heavy-duty engine tests were replaced by more recent ones with turbochargers (OM 646LA, OM 501LA).

On December 22, 2010 the ACEA 2010 oil sequences could be claimed first time. The Ax/Bx-categories now got also minimum TBN-requirements for the fresh oils. In addition also limits for minimum values of sulfated ash were added to achieve a more stringent separation from the Cx-categories. Oils tested according the 2010 sequences can still be claimed until end of 2014.

The current ACEA 2012 oil sequences became effective on December 14, 2012. Laboratory tests to ensure high temperature stability and low temperature pumpability under use of biodiesel as well as a new soot dispersion test Peugeot DV6C as future successor of the Peugeot DV4 were introduced (Table 9.7).

Table 9.7 Engine oil classification according to ACEA 2012

Passenger car and light-duty diesel engines category	Application area
ACEA 2012:	
A1/B1	Extreme fuel economy engine oil for use in dedicated gasoline and light-duty diesel engines. Include 0 W 20 grades with high temperature/high shear viscosity (HTHSV) down to 2.6 mPa s. Sulfated ash limited of maximum 1.3%.
A3/B3	Engine oils for use in gasoline and light-duty diesel engines. High temperature/high shear viscosity limited to minimum 3.5 mPa s. Sulfated ash limited from 0.9 to1.5%.
A3/B4	Engine oils for use in gasoline and light-duty diesel engines with direct injection. High temperature/high shear viscosity limited to minimum 3.5 mPa s. Sulfated ash limited from 1.0 to 1.6%. Additionally, TBN limited to min. 10 mg KOH/g. Compared to A3/B3 higher demands on piston cleanliness and wear in diesel engines.
A5/B5	Fuel economy engine oils for use in dedicated gasoline and light duty diesel engines of high performance including DI engines. High temperature/high shear viscosity limited from 2.9 to 3.5 mPa s. Sulfated ash limit of max. 1.6%.
C1	Extreme fuel economy engine oils widely based on A5/B5 type, but due to more severe restrictions on phosphorous, sulfur and ash formation optimized for cars with particulate filters and more sophisticated exhaust aftertreatment systems. Lowest limits for phosphorous (0.05 m-%) and sulfated ash (0.5 m-%) within all categories. Base number not limited. Fuel Economy minimum of 3.0% in M111 FE test
C2	Fuel economy engine oils with basic performance requirements of C1, but higher sulfur (0.03 m-%), phosphorous (0.09 m-%) and sulfated ash (0.8 m-%) levels allowed. Base number not limited. Fuel economy minimum of 2.5% in M111 FE test.
C3	Engine oils with basic performance requirements of C1, but due to HTHS limit of minimum 3.5 mPa s lower fuel economy requirements than for C1 and C2. Base number limited on minimum of 6. Fuel Economy minimum of 1% in M111 FE test. P limited between 0.07–0.09 m-%. Sulfur and sulfated ash limits like C2.

(continued)

Table 9.7 (Continued)

Passenger car and light-duty diesel engines category	Application area
C4	Engine oils with basic performance requirements of C1, but due to HTHS limit of minimum 3.5 mPa s lower fuel economy requirements than for C1 and C2. Base number limited on minimum of six. Fuel economy minimum of 1% in M111 FE test. Very low sulfated ash limit of 0.5% like C1. Phosphourous limit of maximum 0.09 m-% like C2. Lower limit for evaporation loss (max. 11% compared to max. 13% for C1-C3)
Heavy-duty engines category	Application area
ACEA 2012	
E4	Engine oils with high performance under severe conditions and significantly extended oil drain intervalls. Sulfated ash limited to 2.0 m-%. No phosphorous and sulfur limits. Base number limited to minimum 12. Preferred use in engines without particulate filter.
E6	Engine oils with improved performance to E4 and significantly reduced ash formation (sulfated ash max. 1.0 m-%) and lower phosphorous (max. 0.08 m-%) and sulfur (max. 0.3 m-%) limits. Mostly used for application in engines with particulate filters and DeNOx systems like in Euro5 equipment.
E7	Engine oils with improved performance to E6, but sulfated ash limit of maximum 2.0% and no phosphorous and sulphur limits. Preferred use in engines without particulate filter.
E9	Engine oils with basic performance of E7 and extra high soot handling performance of API CJ-4 level. Limited sulfated ash (max. 1.0 m-%), phosphorous (max. 0.12 m-%) and sulphur (max. 0.4 m-%) content. Widest combination of European and American demands.

9.1.3.5 Manufacturers' Approval of Service Engine Oils

Apart from the specifications already listed, some manufacturers have their own specifications and usually demand tests on their own engines (Table 9.8).

The European ACEA, North American EMA (Engine Manufacturers Association) and Japanese JAMA (Japanese Automobile Manufacturers Association) are working together on specifications for a worldwide classification system with consistent oil performance. The first specification of this kind – DHD-1 (diesel heavy duty) – was issued in early 2001. The testing includes a combination of engine and bench tests from the API CH-4, ACEA E3/E5 and Japanese DX-1 categories. In 2002 categories for light-duty diesel engines (DLD) also were set up (Table 9.9).

Table 9.8 Manufacturers' approvals.

OEM approval	Application area
BMW	
Longlife 01	For nearly all BMW cars from 2001 onwards. With introduction of a new test engine the oil performance increased significantly. The average service interval, given by the flexible service system increased. This category is backwards compatible also.
Longlife 01 FE	BMW introduced a new generation of gasoline engines with the capability of using engine oils with reduced high temperature high shear viscosity. Therefore, the longlife 01 FE category was introduced which provided a fuel economy benefit of minimum 1% compared with an SAE 5 W-30 longlife 01 engine oil.
Longlife 04	This category was designed for the specific requirements of exhaust gas after treatment with, for example particulate filters. Longlife 04 oils therefore have components resulting in especially low phosphorus, sulfur and ash content. They are backwards compatible for vehicles operating in central Europe.
Deutz	
DQC I-02	Specifies oils meeting ACEA E2, API CF/CF-4 or CG-4 for natural aspirated diesel engines under light and medium operating conditions.
DQC II-10	Specifies oils meeting ACEA E3 or E4 or E5 or E7 or alternatively API CG-4 or CI-4 or CI-4 plus or alternatively DHD-1 or alternatively JASO DH-1. For use in natural aspirated and turbo charged engines under medium and severe operating conditions.
DQC II-10 LA	Specifies oils meeting ACEA E6 or E9 or alternatively API CJ-4 or alternatively JASO DH-2. Oils with reduced ash formation in particular for use with particulate filters under medium and severe operating conditions.
DQC III-10	Specifies oils exceeding ACEA E7 requirements for modern engines under more severe operating conditions, for example power plant application, or closed-crankcase ventilation.
DQC III-10 LA	Specifies DQC III-10 level oils with reduced ash formation in particular for use with particle filter systems.
DQC IV-10	Specifies synthetic engine oils exceeding ACEA E4 requirements for use in high-power engines under severest conditions with closed-crankcase ventilation systems.
DQC IV-10 LA	Specifies DQC IV-10 level oils with reduced ash formation in particular for use with particle filter systems.
MAN	
MAN 270	Monograde oils for turbocharged and non-turbocharged diesel engines. Oil-drain intervals of 30 000–45 000 km. Performance level of former ACEA E2.
MAN 271	Multigrade oils for turbocharged and non-turbocharged diesel engines. Oil-drain intervals of 30 000–45 000 km. Performance level of former ACEA E2.
MAN M 3275	SHPD oils for all diesel engines with oil-change intervals of 45 000–60 000 km.

(continued)

Table 9.8 (*continued*)

OEM approval	Application area
MAN M 3277	UHPDO oils for all diesel engines with oil-drain intervals up to 120 000 km.
MAN M 3477	UHPDO oils for all diesel engines with oil-drain intervals up to 120 000 km. Reduced ash, sulfur and phosphorus content for use in trucks with advanced exhaust-treatment systems (e.g. EU 5 aftertreatment equipment).
MAN M 3575	Ash reduced SHPD oils with oil drain intervals up to 55 000 km and performance level of ACEA E9 and API CJ-4. In particular, for use in engines with high exhaust gas recirculation.
M 3271	Oils for natural gas fired engines
Mercedes Benz for heavy-duty diesel engines	
MB 228.0	Monograde oils for turbocharged and non-turbocharged diesel engines with lower severity.
MB 228.1	High-performance multigrade oils for turbocharged and non-turbo-charged diesel engines. Extended oil-drain intervals up to 30 000 km in medium/heavy duty.
MB 228.3	Super-high-performance diesel oil (SHPDO) for highly turbocharged diesel engines. Extended oil-drain intervals up to 45 000 km in medium/heavy duty.
MB 228.5	Ultra-high-performance diesel oils (UHPDO) for highly turbocharged diesel engines. Extended oil-drain intervals up to 100 000 km in heavy duty (e.g. MB Actross).
MB 228.51	UHPDO with lower ash, sulfur and phosphorus for use in trucks with advanced exhaust gas after-treatment systems. Extended oil-drain intervals up to 100 000 km in heavy duty.
Mercedes Benz for passenger car engines	
MB 229.1	Multigrade oils for gasoline and diesel engines.
MB 229.3	Multigrade fuel-economy oils for passenger car gasoline and diesel engines. Extended oil-drain intervals.
MB 229.31	Multigrade fuel-economy oils for passenger car gasoline and diesel engines. Extended oil-drain intervals. Lower ash, sulfur and phosphorus for use in cars with advanced exhaust gas-treatment systems.
MB 229.5	Multigrade fuel-economy oils for passenger car gasoline and diesel engines. Extended oil-drain intervals. Fuel economy and engine performance above MB 229.3.
MB 229.51	Multigrade fuel-economy oils for passenger car gasoline and diesel engines. Extended oil-drain intervals. Fuel economy and engine perform-ance above MB 229.31. Lower ash, sulfur and phosphorus for use in cars with advanced exhaust gas-treatment systems.
MB 229.52	Multigrade fuel-economy oils for passenger car gasoline and diesel engines. Extended oil-drain intervals. Fuel economy and engine perform-ance above MB 229.51. Lower ash, sulfur and phosphorus for use in cars with advanced exhaust gas-treatment systems of latest generation.

MTU	
Oil type 1	Specifies oil qualities generally corresponding to API-CF, CG-4 or ACEA, E2) for light- and medium-operating conditions and short oil-drain intervals.
Oil type 2	Specifies oils of higher quality levels corresponding to SHPDO like ACEA E3 for medium- and severe-operating conditions and medium length oil-change intervals.
Oil type 3	Specifies oils with highest quality levels corresponding to UHPDO types like ACEA E4-99 for medium- and severe-operating conditions of all engines. Such oils achieve the longest oil drain intervals in MTU engines and provide the highest cleanliness of air intake systems of super charged diesel engines.
Opel/GM	
dexos1™	High performance fuel economy oils for gasoline engines. Use mainly outside Europe in markets with low diesel passenger car share (e.g. North America).
dexos2™	High performance fuel economy oils for gasoline and diesel engines. Mainly used for European markets with high-diesel passenger car share.
PSA	
B71 2290	5W-30 oils with ACEA C2 basic performance, ash reduced in particular for use with diesel particle filters
B71 2294	xW-40 or xW-50 mineral-based oils with ACEA A3/B3 basic performance.
B71 2296	xW-30 or xW-40 oils with ACEA A3/B4 or A5/B5 basic performance. xW-30 oils due to ACEA A5/B5 also allowed with lower HTHS viscosity and improved fuel economy.
B71 2297	Ash reduced xW-30 or xW-40 oils with ACEA C3 basic performance in particular for use with diesel particle filters.
B71 2300	xW-40 or xW-50 oils with ACEA A3/B4 basic performance.
Porsche	
A40	0 W-40, 5 W-40 or 5 W-50 oils of ACEA A3/B4, C3 or C4 basic performance, additionally tested in VW T4 and Porsche 911 oil tests, for use in all Porsche gasoline engines. 0 W-40 and 5 W-40 also for use VW gasoline engines without longlife-service.
C30	5 W-30 oils with performance of VW 504 00/507 00 for use in all VW gasoline and diesel engines with and without longlife service.
Scania	
LDF	ACEA E5 or DHD-1 oils with special long drain field test approval. Oil-drain intervals up to 60 000 km.
LDF-2	ACEA E4, E6, E7, E9 or API CJ-4 performance level is required for this category. A field trial in a Scania engine of the Euro 3 or Euro 4 generation is required to demonstrate specific performance. These oils are required for use in Euro 4 engines with extended oil-drain interval up to 120 000 km and Scania maintenance system. Backwards compatible with LDF.
LDF-3	Specifies oil with basic performance of LDF-2 and improved piston cleanliness. Limited phosphorous and sulfur content for use in engines with advanced emission behaviour used for Euro 6 emission limits. Backwards compatible with LDF-2 and LDF.

(continued)

Table 9.8 (*Continued*)

OEM approval	Application area
LA	Specifies oil with basic performance of LDF, but reduced phosphorous and sulfur content and ash formation for use in engines with advanced exhaust aftertreatment systems and particle filters used for Euro 6 emission limits.
Renault for passenger car engines	
RN 0700	Oils with basic performance of ACEA A3/B4 or A5/B5 for use in gasoline engines.
RN 0710	5 W-40 oils with basic performance of ACEA A3/B4.
RN 0720	Oils with ash limit of 0.5% (basis ACEA C4) for use in diesel vehicles with particulate filters.
Volkswagen	
VW 505 00	Multigrade oils for turbocharged and non-turbocharged diesel engines (indirect injected and normally aspirated). Standard oil-drain intervals.
VW 501 01	Multigrade oils for gasoline and normally aspirated diesel engines. Standard oil-drain intervals.
VW 502 00	Multigrade oils for gasoline engines, higher ageing stability than VW 501 01.
VW 505 01	Multigrade oils for gasoline and diesel engines, including 'Pumpe-Düse' DI diesel engines. Standard oil-drain intervals.
VW 506 00	Multigrade, low-viscosity oils fuel-economy oils for DI diesel, except 'Pumpe-Düse' engines. Extended oil-drain intervals ('long life').
VW 504 00	Multigrade fuel-economy oils with reduced ash content for all gasoline engines. Extended oil-drain intervals ('long life').
VW 507 00	Multigrade fuel-economy oils with reduced ash content for all diesel engines. Extended oil-drain intervals ('long life').
Volvo	
VDS-2	Oils for heavy-duty diesel engines with oil-drain intervals up to 60 000 km. (Euro 2 engines).
VDS-3	Oils for heavy-duty diesel engines with oil-drain intervals up to 100 000 km. Requires basic performance of ACEA E7 or DHD-1 plus additional Volvo D12D engine test.
VDS-4	Oils for heavy-duty diesel engines with oil-drain intervals up to 100 000 km. Limited phosphorous, sulfur and ash formation. Requires ACEA E9/API CJ-4 performance plus additional Volvo D12D engine test. In use for engines with advanced exhaust aftertreatment systems and low emission levels.

9.1.3.6 Future Trends

New generations of engines using optimized technologies advance the concept of tailor-made special oils.

The continuing optimization of the combustion process to increase the efficiency of gasoline engines has led to the development of direct-injection gasoline

Table 9.9 Global performance classification for engine oils.

Category	Application area
DHD is a performance specification for engine oils to be used in high-speed, four-stroke cycle heavy-duty diesel engines designed to meet 1998 and newer exhaust emission standards world-wide. Oils meeting this specification are also compatible with some older engines. Application of these oils is subject to the recommendation of individual engine manufacturers.	
DHD-1	Multigrade oils for engines meeting emission requirements from 1998 and later. Mack T8, Mack T9, Cummins M11, MB OM 441 LA, Caterpillar 1R, Sequence III F, International 7.3 l and Mitsubishi 4D34 T4 tests are necessary to qualify such oils to a level, which can be seen as comparable with an MB228.3/ACEA E5 level of the European market.
Engine oils meeting the minimum performance requirements of Global DLD-1, DLD-2 and DLD-3 are intended to provide consistent oil performance car engines worldwide and may therefore be recommended as appropriate by individual engine manufacturers to maintain engine durability wherever their light-duty diesel engine is being used.	
DLD-1	Standard multigrade oils for light-duty diesel engines. The scope of testing includes several passenger car engine tests out of ACEA categories (VW IDI-Intercooler, Peugeot XUD11BTE, Peugeot TU5 JP, MB OM604A) plus the Japanese Mitsubishi 4D34 T4. The quality level of such oils can therefore be seen as comparable with B2-98 issue 2.
DLD-2	Standard low-viscosity multigrade oils for light-duty diesel engines with extra high fuel economy with basic engine performance like DLD-1.
DLD-3	Multigrade oils for light-duty diesel engines tested also in DI turbocharged diesel (VW TDI) with quality level comparable to ACEA A3/B4.

engines (GDI engines) which may offer considerable fuel savings On the diesel side, direct injection with common rail technology using pressure of up to 3000 bar have become the norm. As for diesel engines widely used, the application of turbochargers (TC) leads to improvements in efficiency also for gasoline engines. The latest gasoline engine generations are therefore based on combinations of GDI and TC.

Besides all hardware improvement also engine has a remarkable influence on the fuel consumption by reducing friction of the engine.

Lowering the friction losses by specific additive chemistry is already playing a much more important role, but also lowering the viscosity of the oils will be a major demand for the future. Development and use of SAE 0W-20 or 5W-20 oils for passenger cars and 5W-30 or 10W-30 oils for heavy-duty application are ongoing and their share in the engine oils market will increase.

For trucks, and for passenger cars, reducing exhaust emissions has the highest priority. The thresholds of Euro 2 and Euro 3 (from 2001) could easily be surpassed by use of special exhaust recycling and catalytic converter systems. Euro 4 for passenger car vehicles (since 2005) required many light-duty diesel engines to implement DPF, to meet the rigorously tightened threshold for particle emission (max. 0.025 g km^2). The introduction of Euro 4 for trucks and buses in October 2005 and Euro 5 for passenger cars and low-duty vehicles in 2009 required further reductions of NO_x and particulate matter which led to further optimized burning processes and/or more advanced exhaust gas-purification equipment, for example selective catalytic reduction (SCR) of nitrogen oxides or use of additional soot filter systems in most new heavy-duty vehicles. The demands on emission reduction have been increased once more into the same direction with the introduction of the Euro 5 and Euro 6 limits for trucks in 2008 and 2013. Because these new engines and exhaust gas-purification systems require low sulfur and, most suitable, sulfur-free diesel fuels (below 10 ppm S), also the engine oils need to be modified to much lower sulfur contents.

Not only the durability of the engines, but also their exhaust aftertreatment systems is getting more significant [29,30]. This led to limitations, for example of phosphorous containing additives, which can have poisoning effects on catalytic converter systems. As phosphorous is a good antiwear additive, the introduction of new and more efficient antiwear chemistry will be of growing importance to comply with the demands on engine durability.

Furthermore, the materials and construction of pistons and cylinders have improved to such an extent that oil consumption is steadily falling. The sum of all measures have permanently increased oil temperatures and specific oil loading. In addition, oil change intervals have been constantly increased. All in all, three factors will characterize engine oils of the future – fuel efficiency, increased stress to the oil, long oil drain intervals and low emissions.

9.1.3.7 Fuel Efficiency

As a result of strict limits to fuel consumption in the USA (CAFE (Californian Act for Fuel Emissions)) and the proven fuel economy effect of low-viscosity engine oils [31], this topic is attracting attention in the Europe and Asia. As a rule, engine-based savings can reach a theoretical 8–10% (see Figure 9.5) [32].

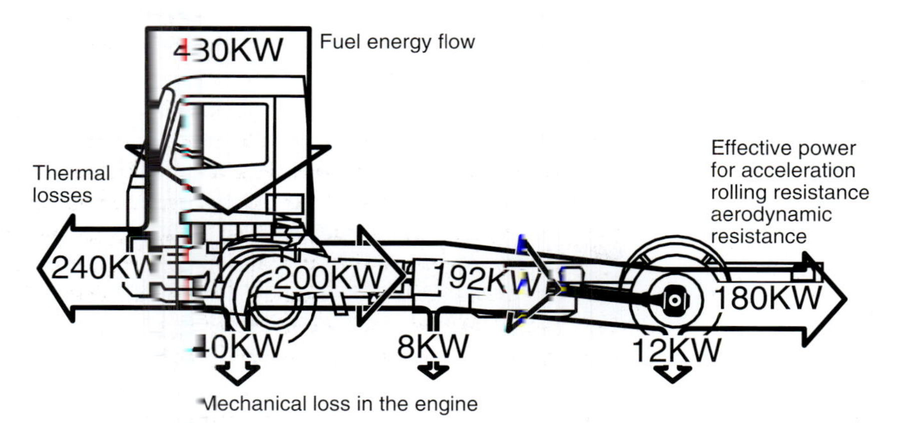

Figure 9.5 Total losses in commercial vehicles.

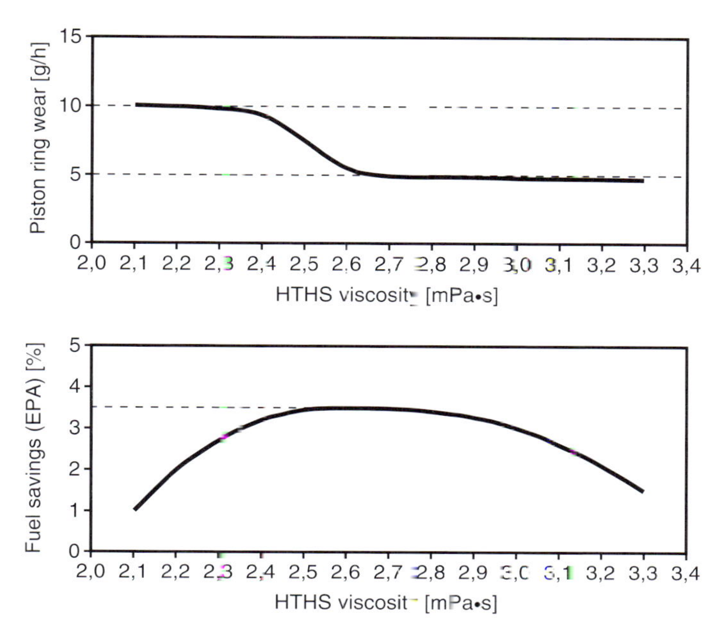

Figure 9.6 Fuel efficiency and wear versus HTHS [33].

As engine oils cannot totally eliminate frictional losses, saving potentials of 4 and 5% present enormous challenges, but they are possible today using optimized low viscosity oils.

According to the opinion of experts, reduction in car fuel consumption in urban conditions are achieved by lowering frictional losses during cold start-ups which simultaneously result in less wear and a lowering of viscosity in constant throttle conditions. As Figure 9.6 shows, there is a plateau-like optimum for the correlative fuel savings in engines with HTHS values between 2.5 and 2.9 mPa s. The critical boundary to high wear conditions are seen by some to be different viscosities. Figure 9.5 shows piston ring wear in boundary to static friction conditions between 2.6 and 2.7 mPa s [33]. This threshold is viewed critically by different OEMs and is set individually. The European specification for fuel economy oils contains a span of 2.6–3.5 mPa s, whereby the fuel savings in the M 111 FE test must be at least 2.5% compared to the reference oil (SAE 15W-40). It has to be remembered that absolute fuel savings figures depend largely on the test method and the reference oil used. Standardized dynamometer tests, which more accurately reflect driving conditions, provide more realistic values than the established bench tests that cannot reproduce all operating conditions. The piston group was found to be one of the most important factors to reduce friction in the combustion engine. It was shown, that the viscosity of the engine oil can play a major role in saving fuel [34].

According to OEMs, HTHS values must not be lower than minimum 2.6 mPa s in all manufacturer's approvals and new engine oil developments because of possible wear between critical material pairings. In consequence of highest need to lower the fuel consumption, new materials and wear improved oils and the use of even lower viscous oils are under investigation.

9.1.3.8 Long Drain Intervals

Concerns about higher wear and thus shorter life caused by low-viscosity oils run contrary to the trends of the new generation of engines, which are designed to cope with even longer oil change intervals. As stated in manufacturer's specifications (see Section 9.1.3), oil change intervals for cars are currently 30 000 km for gasoline engines and 50 000 km for diesel engines. During the last years, evolution of the oil-drain intervals has stagnated because of increased demands for lower ash, phosphorus, and sulfur content of the oil at the same time as aggravated conditions in the engines.

The radionuclide technique (RNT), as a proven online tool, is experiencing a renaissance for examining the effects on wear in various operating conditions such as during running-in or to determine long-term stability. As can be seen in Figure 9.7 with the example of a new-generation DI turbodiesel, the rate of wear and total wear can be precisely and reproducibly selected for every critical material pairing, for example in valve trains, in bearings or in piston-cylinder geometries in every engine.

Apart from long-term wear, very high demands are made on oxidation stability and evaporation losses. This strengthens the trend towards synthetic and unconventional oil as the basis for such high performance engine oils. This is well illustrated by Figure 9.8 with the graphic correlation between evaporation tendency (Noack) and oil consumption. Evaporation losses, illustrated with the example of ILSAC thresholds for GF-2 and GF-3, serve as a generally recognized and reproducible value. A technically realized milestone for fully synthetic engine oils based on present synthetic base oils is a threshold of 5– 6%.

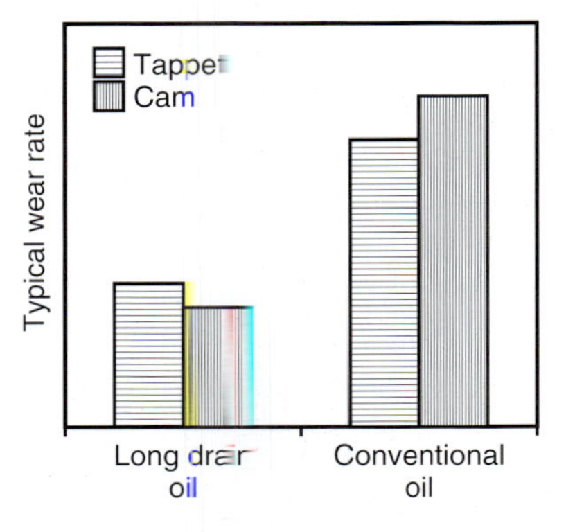

Figure 9.7 Wear characterization via radionuclide technique.

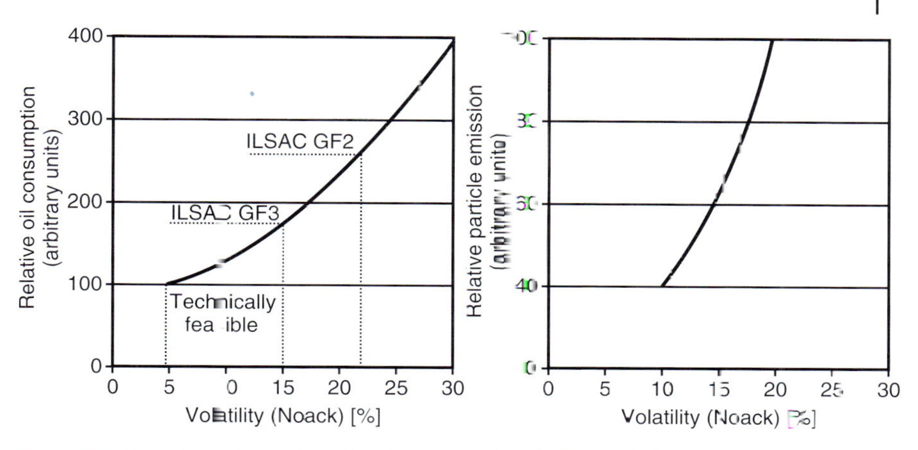

Figure 9.8 Oil consumption and relative oil generated particulate emissions versus evaporation loss.

The suitability of extended oil change interval oils is tested in thermally stressed test engines, which run hot and without oil top-up. The typical indicators of ageing like viscosity and TBN are measured. Based on the standardized VW T4 test which is used for current specifications, Figure 9.9 shows a comparison for modern ACEA oils of different viscosity grades demonstrating the influence of base stocks and evaporation loss.

9.1.3.9 Low Emission

Compared with car engine oils, heavy-duty engine oils already achieve drain intervals of 100 000 km and more. Because of the ever increasing number of trucks on roads, this is a useful contribution to improving environmental compatibility.

Apart from the CO, HC and SO_2 emissions, which are seen to be caused by the fuel, particulate emissions play a significant role in HD engines. These

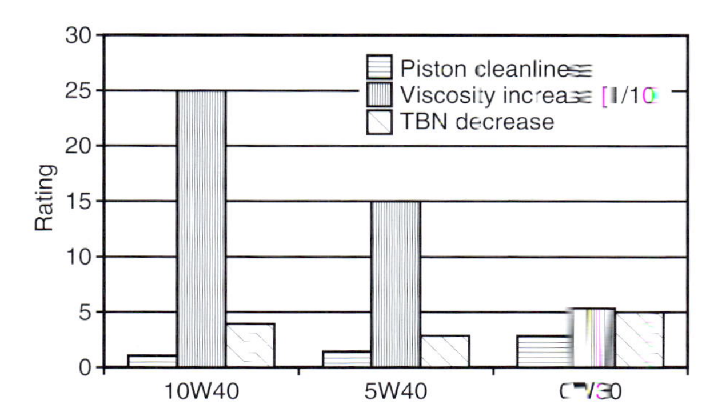

Figure 9.9 T4-engine tests for typical ACEA engine oils.

particulates are a result of incomplete combustion and are a mixture of fuel- and lubricant-based components. As the oil-based particles are largely caused by highly volatile elements in the formulation, evaporation losses have a direct effect on emitting pollutants (see Figure 9.8). Furthermore, it is assumed that sulfur compounds in diesel fuels will poison the catalytic converters of engines used for latest emission legislation. The introduction of 10 ppm sulfur in fuels places the sulfur content of HD engine oils in a new light. Low sulfur and most probably low phosphorus not just lead to some rethinking concerning additives but also to the rejection of Solvent Neutral oils which, as a rule, contain between 0.2 and 1.0% sulfur. As already with cars, unconventional and synthetic base oils are preferred so that the trend towards low viscosity, fuel economy oils will spread definitively to the heavy-duty sector.

As can be seen in Tables 9.7 and 9.8, during recent years engine oils have already been adapted to the new requirements of engine designs and exhaust gas-purification systems for cars and for trucks (e.g. ACEA C-categories, ACEA E6, MB 229.31, MB 229.51, MB 228.51, MAN 3477). Ash sulfur and phosphorus content had to be reduced by up to 50%. At the same time engine designs often needed increased oil performance with regard to engine cleanliness and oil ageing (higher temperatures), soot handling (higher soot content in the oil) and increased wear control (higher torques, lower weights, more soot). These new requirements called for much effort and innovation in the development of new formulations and additive technology. New and better anti-wear systems, more effective dispersant and detergent technology, and more active ageing-protection systems can satisfy the most challenging requirements of today's development of engine oils for low-emission vehicles.

9.2
Two-Stroke Oils

9.2.1
Application and Characteristics of Two-stroke Oils [35–37]

Two-stroke engines are mostly used when high specific power, low weight and low price are key parameters. These engines are thus often used in motorcycles, boats (outboard engines), jet-skis, lawn mowers, chain saws and small vans, whereby the vast majority are found in motorcycles and boats.

Almost all two-stroke engines use total-loss lubrication. The oil is not circulated as in the case of four-stroke engines but added to the fuel. A large part is burnt in the combustion process but about one quarter can be exhausted as unburned oil mist. Simple engines as found in older mopeds still use the premix method, which involves the operator manually adding a suitable two-stroke oil to the fuel tank at a ratio of about 1:20–1:100. More advanced designs use an automatic oil metering system. These either add a constant amount of oil to the fuel or add oil according to engine loading. Typical ratios in such are between 1:50 and 1:400.

In the majority of simple two-stroke, the engine breathes through classic carburetors. Contrary to four-stroke engines, the fresh fuel/air mixture in a two-stroke scavenges the cylinder after combustion. This simultaneous charging and emptying causes about 20–30% of the fresh mixture to be exhausted without burning.

This disadvantage, along with the only partial burning of the oil, causes many two-stroke engines to generate comparatively high emissions. In highly populated areas with a large number of small motorcycles, such as in many Asian cities, this leads to severe odour, smoke and noise pollution.

In recent years, these typical disadvantages have been countered by some advances in two-stroke technology. The development of direct or indirect fuel injection has led to significantly reduced emissions and improved fuel efficiency.

Today's engines require correspondingly high quality oils for reliable operation and long life. The principal criteria for the quality of two-stroke oils are as follows:

- Lubricity and anti-wear properties.
- Cleaning function (detergent/dispersant properties).
- Avoidance of deposits in the exhaust system.
- Low smoke.
- Spark plug cleanliness and the avoidance of pre-ignition.
- Good fuel miscibility even at low temperatures.
- Corrosion protection.
- Good flowing properties.

About 85–98% of two-stroke oils are base oils with the rest consisting of various additives that, similarly to four-stroke engine oils, provide a series of the above-mentioned characteristics. In principle, all common base oils can be used, ranging from Brightstocks, Solvent Neutral types to fully synthetic polyalphaolefins. As most two-stroke oils need not perform particularly well at low temperatures, Brightstocks are often used to achieve the desired viscosity. Apart from hydrocarbon types, higher quality two-stroke lubricants often contain various synthetic esters and this is particularly the case for biodegradable oils that were specially developed for marine outboards.

The additives in two-stroke oils are usually matched to the requirements of the engine. As in the case of four-stroke engines, anti-wear additives are included which chemically interact with metal surfaces to protect against wear particularly in boundary friction conditions. Along with the most commonly used zinc dialkyldithiophosphates, non-ash forming types such as dithiophosphoric acid esters of alkyl and aryl esters or phosphoric acids are used.

Dispersant and detergents (DD systems) are added to the oil to keep the engine clean and to avoid deposits in the combustion chamber and around the piston rings. Alkalis or earth alkalis of sulfonates and/or phenolic compounds are often used. Dispersants are often high-molecular-weight compounds which are capable of trapping and suspending contaminants. Examples of these types of substances are polybutene succinimides whose properties result from the chemical bonding of a polar succinimide with oil-soluble polybutenes.

Furthermore two-stroke oils contain small quantities of anti-oxidants, corrosion inhibitors, defoamers and flow improvers in addition to anti-wear and DD additives.

Low-smoke two-stroke oils contain a significant amount of polybutenes (about 10–50%). These are fully synthetic fluids, which are available in various viscosities. Apart from good lubricity compared to mineral oils, these fluids offer much cleaner combustion and significantly less coking [38,39].

9.2.2 Classification of Two-stroke Oils

As with four-stroke oils, two-stroke oils are allocated to certain performance groups that provide information about suitable applications. The basis for all the below-mentioned classification systems are a series of laboratory and functional tests, the latter being bench tests performed on the latest generation of two-stroke engines.

9.2.2.1 API Service Groups

The API currently lists three categories (Table 9.10) which cover all engines from low-power lawn mowers to high-performance motorcycles. Engine tests are no longer performed because the specified test engines are no longer manufactured. In future, it is planned to replace the API groups with Japanese JASO (Japanese Automotive Standards Organization) and global International Standards Organization (ISO) classifications. There are still a number of oils on the market with API classifications because this system was widely accepted in the past.

9.2.2.2 JASO Classification

JASO, to which all major Japanese vehicle manufacturers belong, today classifies two-stroke oils into three groups, FB, FC and FD,according to JASO M345 standard (Table 9.11).

All three categories use the same test engines and the corresponding performance categories allocated according to pre-determined thresholds. The test

Table 9.10 API groups.

API	Application	Test engine	Test criteria
TA	Mopeds, lawn mowers, electricity generators, pumps	Yamaha CE 50 S (50 cm^3)	Piston seizing, exhaust system deposits
TB	Scooters, small motorcycles	Vespa 125 TS (125 cm^3)	Pre-ignition, power loss due to combustion chamber deposits
TC	High-performance motorcycles, chain saws	Yamaha Y 350 M2 (350 cm^3) Yamaha CE 50 S	Pre-ignition, power loss due to combustion chamber deposits Piston seizing, ring sticking

Table 9.11 Engine test criteria for JASO classifications.

Test engine	Test criteria	Test parameters
Honda Dio AF 27	Lubricity	Piston ring wear, ring scuffing, piston seizing
Honda Dio AF 27	Detergent effect	Piston ring sticking as a result of lacquering, coking, deposits on piston and in combustion chamber
Suzuki SX 800 R	Exhaust smoke	Smoke particles
Suzuki SX 800 R	Exhaust deposits	Back pressure in exhaust system

results are determined in comparison to an exactly defined, high-performance reference oil (JATRE 1) and published as an index relative to JATRE 1 (Table 9.12). The key test criteria are the lubricity and detergent effect of the oil as well as its tendency to create smoke and deposits in the exhaust system. The first specification for a low-smoke oil was created with the laying-down of JASO FC. Oils fulfilling JASO FD performance were introduced in 2004 and have highest demands on engine cleanliness (detergent effect).

9.2.2.3 ISO Classification

In the mid-90 s, after JATRE 1 oils were tested in European engine tests, it became clear that JASO FC oils could no longer satisfy the latest demands of European two-stroke engines. A series of extended tests which satisfied all demands were thus developed in Europe. In addition to the testing of smoke, exhaust system deposits, lubricity and detergent effect according to JASO, a new category with a 3-h Honda Dio test to quantify improved piston cleanliness and detergent effect was added. The reference oil for all tests was JATRE 1. These new guidelines were created by the CEC working parties in which European engine and lubricant manufacturers are represented.

The ISO now classifies two-stroke oils into three categories, ISO L-EGB, -EGC and -EGD.

The categories ISO-L-EGB and -EGC mirror the requirements of the JASO categories FB and FC while requiring additional proof of piston cleanliness. ISO-L-EGC and -EGD require proof of low smoke similarly to JASO FC. Table 9.13 shows all engine-based evaluation criteria.

Table 9.12 JASO performance categories (reference oil: JATRE 1 = 100).

Test criteria	JASO FB	JASO FC	JASO FD
Lubricity	>95	>95	>95
Detergent effect	>85	>95	>125
Exhaust smoke	>45	>85	>85
Exhaust deposits	>45	>90	>90

Table 9.13 Summary of the ISO categories (reference oil: JATRE 1 = 100).

Test criteria	ISO-L-EGB (incl. JASO FB)	ISO-L-EGC (incl. JASO FC)	ISO-L-EGD
Lubricity	>95	>95	>95
Smoke	>45	>85	>85
Exhaust deposits	>45	>90	>90
Detergent effect	>85 (1-h test)	>95 (1-h test)	>125 (3-h test)*)
Piston cleanliness	>85 (1-h test)*)	>90 (1-h test)*)	>95 (3-h test)*)

*) New requirements in addition to JASO FC.

9.2.3 Oils for Two-Stroke Outboard Engines

Neither API nor JASO or ISO classifications contain quality guidelines for outboard engine oils. These are usually oils whose formulation and characteristics have been matched to the engine technologies, which have become established for powering boats. The main difference to other two-stroke oils lies in their additive chemistry. The additives in these oils are all non-ash-forming because these engines display a marked coking tendency in certain parts of the combustion chamber, such as the ring grooves. If the wrong additives were used, this would lead to major functional impairment and possibly to breakdowns. In principle, the additives have the same functions as in other two-stroke oils but are chemically different. No components are used which can form deposits or ash-like residues at high thermal loads or during combustion. Metal salts such as zinc (anti-wear) or calcium/magnesium (detergents) which are often found in engine oils cannot be used. However, all such oils use typical base oils. The performance categories of two-stroke oils for outboard engines were primarily developed by the American NMMA (National Marine Manufacturers Association). All important American outboard manufacturers belong to the NMMA. Back in 1975 the minimum requirements of such oils were incorporated into the TCW specification. In 1988, a far-reaching revision was issued with the title TCW 2. During the following years, problems were encountered with technologically advanced engines, which were run on TCW or TCW 2 oils. This initiated a further tightening-up of minimum requirements, which was released as TCW 3. In 1997, the standards for oil quality were again increased to keep pace with continuing developments in engine technology. The new specification, TCW 3-R (R = recertified) now includes laboratory tests and tests on five different engines, three of which are outboards. Table 9.14 shows the tests that have to be performed on a newly developed outboard engine oil to achieve TCW 3-R classification. The costs generated by testing a TCW 3-R product development have never been so high. The engine tests alone generate costs of about $ 150 000–200 000.

Table 9.14 Test criteria for NMMA TCW 3-R.

Engine tests	
Test engine	**Test criteria**
Yamaha CE 50 S	Lubricity (seizures)
Yamaha CE 50 S	Power loss due to pre-ignition
Mercury 15 HP (2 runs)	Compression losses Ring sticking Piston cleanliness
OMC 40 HP	Ring sticking Piston cleanliness
OMC 70 HP	Ring sticking Piston cleanliness
Laboratory tests	
Low-temperature viscosity	Limited viscosity at −25 °C
Miscibility	Mixing with fuel at −25 °C
Corrosion protection	Standard rust tests compared to reference oil
Compatibility	Stability after mixing with reference oil
Filterability	Flow rate compared to pure fuel

9.2.4 Environmentally Friendly Two-stroke Oils [40]

The ever increasing severity of environmental legislation is also effecting the development of two-stroke oils, especially outboard engine oils. Such ecologically optimized oils often have regionally differing classifications, which reflect local environmental legislation and their biodegradability depends on varying minimum requirements. At the international level, the ICOMIA (International Council of Marine Industry Associations) has specified harmonized requirements [38]. In 1997, the ICOMIA Standard 3–97 was passed for environmentally friendly outboard engine oils. From a technical point of view, products thus labelled must fulfil at least TCW 3-R as well as offering very low algae, daphnia and fish toxicity and rapid biodegradability as defined by ISO and OECD standards. These oils are based on fully synthetic components with the base oils usually being rapidly biodegradable synthetic esters. By using correspondingly high quality esters, these products are the very best two-stroke oils and can even be used for lubricating chain saws. The use of ester-based lubricants combines the highest technical performance with improved environmental compatibility.

9.3 Tractor Oils

Relatively newer generations of construction and agricultural machinery make differing demands on functional fluids. For reasons of simplified servicing but

Table 9.15 Application of tractor oil types.

UTTO	STOU
Hydraulic system	Hydraulic system
Gearbox	Gearbox
Wet brakes	Wet brakes
	Engine

also because of the general wish to rationalize stocks, universal oils were developed which satisfy the various functional requirements of such machines.

These oils should guarantee long machinery life in all manner of climatic conditions as well as extending service intervals and reducing down-time. And the oil manufacturers welcome their possible use in a wide variety of machinery.

These days, two different oil technologies are used which are characterized by their application area. They are universal tractor transmission oils (UTTO) and super tractor oils universal (STOU). Table 9.15 shows where they are used in tractors.

The demands made on tractor oils have increased sharply in line with advances in vehicle technology and ease of operation. Earlier generations of tractors had manual-shift gearboxes, as well as relatively simple rear axle designs. These days, the state-of-the-art is complex hydraulic systems, hydrodynamic drive units (retarders/power splitting) or wet (oil) braking systems. In the past, simple engine oils or low-additive gear oils were used for general lubrication as well as for the hydraulic circuits.

Technical advances in tractors required a significant improvement in operating fluids as they did in the whole automotive area. Apart from big improvements in additive technology to cope with greater mechanical demands, the ageing stability of the fluids has increased reflecting dramatically higher specific power outputs and lower oil volumes.

All-season use is now a standard requirement for tractor oils as it is in the automotive area. The result of this is that the viscosity grades (defined according to SAE J 300) and thus temperature ranges have been extended from SAE 15W-30 and 10W-30 to 15W-40, 10W-40 and 5W-40.

The performance of such universal oils as hydraulic fluids corresponds to, at least, HLP and HVLP levels because of the additives included to guarantee universal use.

The use of these products in vehicle gearboxes and wet brakes makes much greater demands on the fluid. Highly stressed mechanical drive units make great demands on the wear protection and life of the lubricant. As a rule, gearbox suitability is indicated by the automotive API category, normally at least GL-4. However, in-house specifications such as ZF TE-ML 06/07, Allison C-4 or Caterpillar TO-2 may also have to be met. A special challenge to oil formulations is the use of these products in wet brakes which began back in the 1960s. Not least because of safety considerations, such oils must offer high thermal stability and balanced and stable friction characteristics in brakes. Too high or too low

Table 9.16 Manufacturer specifications for tractor oils.

Manufacturer	Oil type	Specification
AGCO Massey Ferguson	STOU	M 1139 M 1144, M 1145
AGCO Massey Ferguson	UTTO	M 1135 M 1141, M 1143
Case New Holland	UTTO	CNH MAT 3505,3525,3526
Fendt	STOU	KDM 41.2011
Ford	STOU	M2C159-B/C
Ford	UTTO	M2C86B/M2C134D
J I. Case	UTTO	MS 1207, MS 1209, MS 1210
John Deere	STOU	J27
John Deere	UTTO	J20C, J20D
New Holland	UTTO	FNHA-2-C 201.00

friction values can easily lead to excess wear on pads and discs but also to uneven braking and unpleasant brake screeching. Fine-tuning these oils with special friction modifiers is one of the most difficult aspects of developing such oils.

While UTTO oils can satisfy the above-listed applications, their use as engine oils require vastly different additives. Apart from the UTTO additive objectives of friction and wear control, low-temperature stabilizers, oxidation, corrosion and foam inhibitors, engine oils additionally require significant quantities of detergents and dispersants. These ensure engine cleanliness and adequate sludge (in particular soot) transportation. As a rule, normally aspirated engines require, at least, an API CE oil and if a turbocharger is fitted the oil should be API CF or CF-4. For the latest engine generations oils of ACEA E7 or API CI-4 performance level is recommended. In some cases, tractor engine manufacturers issue approvals for oils after they have successfully completed tests in the corresponding engines. Examples of this are Deutz DQC II and DQC III.

The latest generation of tractor oils not only include products with significantly higher performance than in the past, but also oils which offer better environmental compatibility. Such products often contain rapeseed or sunflower base oils or synthetic, rapidly biodegradable esters.

Tractor manufacturers now issue their own in-house fluid specifications, which satisfy all specific requirements of the corresponding machinery. Table 9.16 lists several manufacturer specifications.

9.4 Gas Engine Oils [41–43]

The running of combustion engines on gas instead of liquid fuels is nothing new. Gas has long been used to power vehicles and stationary engines. Such engines require a variety of lubricating oils, depending on the type of engine and the operating conditions.

9.4.1
Use of Gas Engines: Gas as a Fuel

Natural gas engines generate significantly less emissions than gasoline powered units. This has led to increasing use, especially in the mobile sector. However, as gas is not as universally available as gasoline or diesel fuel at filling stations, gas power tends to be used for fleets which can be filled centrally, for example urban buses, school buses or other short-haul transport vehicles. Many car and truck engines can be adapted to run on gas at no great expense.

Gas engines for stationary applications is particularly interesting in areas where gas is cheaply available. For example, they are often used to power electricity generators or pipeline transmission compressors. Apart from natural gas, landfill gas is being increasingly used these days – rubbish tips and sewage treatment plants often use the gas to drive power generators. In the stationary field, both two- and four-stroke engines are used. Similarly to mobile gas engines, these are based on conventionally fuelled designs but are tailor-made for their particular application. The operating conditions of vehicle and stationary engines differ greatly. While vehicle engines last for about 5000 h at speeds of up to 6000 rpm, stationary engines can run for decades but at significantly lower speeds. This influences the selection of materials and operating fluids.

Several gases are used in gas engines. Most commonly used, and especially for cars, is natural gas, either under pressure as CNG (compressed natural gas, mostly methane) or as LNG (liquefied natural gas, mostly propane–butane). CNG is by far the most common. As to the use of gas as a fuel, a number of quality criteria have to be considered. Hydrocarbon structure, calorific value, the presence of water and above all, contamination with other gases such as hydrogen sulfide, halogen compounds or silanes all have a decisive impact. As opposed to the gas used for mobile applications, stationary engines often have to run on varying gas qualities that depend on local conditions. The design and lubricants for engines burning landfill gases have to be chosen carefully because these gases can contain a number of contaminants and can often be corrosive.

9.4.2
Lubricants for Gas Engines

There is currently no universal, harmonized specification for passenger car gas engine oils. The large variations in operating conditions between mobile and stationary engines generally require oils with different additive packages. A general differentiation is made between high-, medium- and low-ash types which are recommended by the manufacturers in line with the designed use of the engine.

As a rule, gas engine oils are subject to high oxidation and nitration that can accelerate the ageing of the oil. Gas-powered cars normally use the same conventional engine oils as are used in gasoline-powered engines using gas are ACEA A3/B4, A5/B5 or C2/C3 qualities as well as API SH/SJ/SL/SM performance types. Similarly to the automotive sector multigrade oils are used to cope with variable operating conditions and to guarantee reliable lubrication at low ambient temperatures. As the number of CNG-powered cars increases, there is increasing pressure to develop oils, which are especially formulated for these applications, having higher anti ageing configuration and lower ash forming tendency.

Special multigrade oils have already been developed for use in heavy diesel engines, with CNG-powered buses being the major application. These have been tested and approved by various engine manufacturers. Examples of these approvals are Mercedes–Benz Sheet 226.9 or MAN M 3271. These oils were tested in bench tests as well as in realistic field trials.

Stationary gas engines can make significantly more complex demands on the oil and this has an effect on their development. While the common ACEA or API bench tests suffice for CNG-powered car engines, laboratory tests on gas engine oils are limited to initial oil screening. The real development takes place in field trials in which engines often have to run for years before they are evaluated. While particular attention is paid to sludge formation, carbon deposits, valve train wear and low temperature flowing in car engines, other effects are important in stationary engines. Particularly important is controlling oil ageing caused by oxidation and nitration but also pre-ignition which is caused by high levels of ash-forming additives that can give deposits ending up in hot spots disturbing the compression phase. This problem is most prevalent in two-stroke engines, which generally need low-ash oils. Simply because of their long service lives, gas engine oils should protect against valve seat recession and spark plug fouling. These problems sometimes go unnoticed for thousands of hours in stationary engines. As regards running on corrosive gases (caused by sulfur or halogen contamination), special attention must be paid to adequate corrosion protection. Oils for stationary gas engines thus require significantly more development work and have to be better matched to the engine and operating conditions than normal automotive engine oils. The development and application of gas engine oils thus normally takes place in close cooperation between the oil and engine manufacturers who normally issue an approval after successful completion of trials.

9.5 Marine Diesel Engine Oils [44]

These lubricants are heavily influenced by the type of fuel used and the design of the engines themselves. A number of similar engines are also used in stationary applications to generate electricity with conventional fuels or with steam power.

9.5.1
Low-speed Crosshead Engines

Low-speed diesel engines generating up to 1000 kW per cylinder at 50–120 rpm use the crosshead principle (large engines with over 900 mm bores and 3000 mm strokes and 20 000 kW outputs). In crosshead designs the cylinder block and the crankcase are separate units. Sealing is provided by stuffing boxes and the liners are lubricated with cylinder oils by means of dosing devices. Depending on the bore and stroke of the cylinder, up to 16 dosing devices may be fitted.

The crankcase bearings are lubricated with crankcase oils, which are sometimes called system oils. These oils also lubricate the crosshead bearings and guides. As opposed to cylinder oiling which is a form of total-loss lubrication, the crankcase oil is recirculated.

9.5.2
Medium-Speed Engines

Medium-speed engines run at around 200–100 rpm. The overall design of these engines is roughly similar to that of vehicle internal combustion engines, they do not use the crosshead design and the crankshaft is connected to the pistons and thus the cylinders with connecting rods. In such designs, the same oil is used to lubricate the cylinder walls and the crankshaft bearings and forms part of a circulation system. These engines are sometimes called trunk piston engines. Figure 9.10 shows a simplified diagram of a low-speed crosshead engine.

These days, marine diesel engines are fuelled by the worst and heaviest crude oil fuel cuts. Residues are gathered for all areas of refining to make low cost fuels. These include vacuum residues, propane de-asphalting residues, heavy solvent extracts from lube refining and other by-products. Refining residues as well as the large share of high molecular weight substances provide poor combustion characteristics and create large amounts of deposits. High sulfur contents up to 3% lead to oil acidification. The formulation of such marine diesel lubricants is considerably influenced by these fuel characteristics. Although the fuel is filtered in centrifuges, high ash and asphalt levels in the fuels lead to significant quantities of solid impurities being formed during combustion.

9.5.3
Lubricants

Cylinder oils in crosshead engines are total-loss products that lubricate the sliding motion of the piston rings in the cylinder liner. To avoid the deposit of combustion residues, these oils must have good dispersant properties. In addition, they must be capable of neutralizing the corrosive acids, which result from the high sulfur levels in the fuels. To satisfy these requirements, cylinder oils have a large proportion of over-based components (up to 30% over-based calcium sulfonates or other over-based components). The alkalinity required for

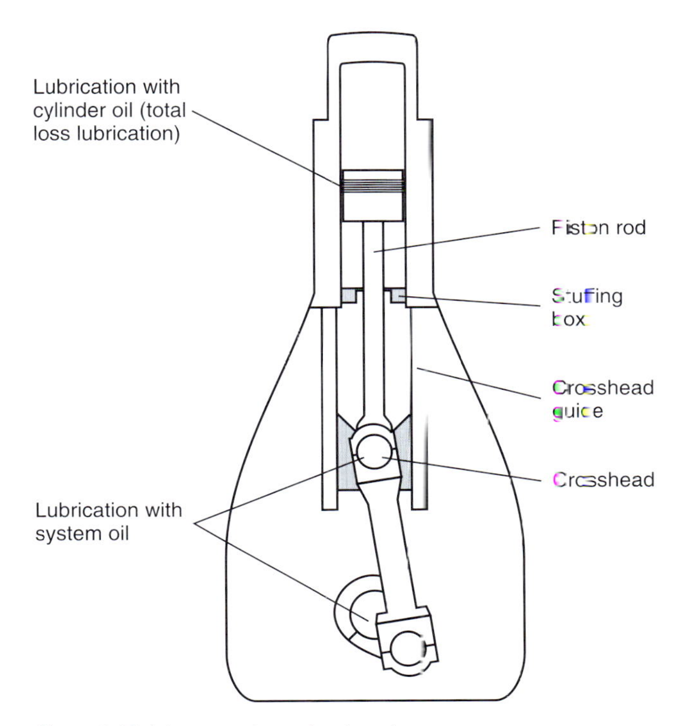

Figure 9.10 A low-speed crosshead engine

neutralization leads to total base numbers (TBNs) of up to 100. When additives for such oils are being selected, special attention must be given to the good colloidal solubility of the additives in the base oil to avoid precipitation. The large surface areas of cylinders in crosshead engines require the applied oil to disperse rapidly and reliably and this is achieved by good spreadability.

The stuffing boxes fitted to crosshead engines ensure that the crankcase oils are hardly contaminated by combustion chamber residues and these oils thus contain relatively few additives. Additives to combat thermal oxidation are essential and the neutralizing capacity can be relatively low with a TBN of five. The contamination of crankcase oil with cylinder oil leads to increased emulsifying capacity which is undesirable. Water separation is therefore another vital characteristic of crankcase oils.

An increase in dispersant and detergent properties caused by the ingress of cylinder oil can also lead to wear problems as the effect of typical anti-wear and EP additives such as zinc dialkyldithiophosphate is thereby diminished.

The type of oil used to lubricate trunk piston engines (combined crankcase and piston lubrication) is also largely determined by the sulphur content of the fuel. Oils with total base numbers of 12–40 are common and the neutralizing capacity of the oils is very soon exhausted in some cases so that the TBN of used oils can be much lower.

During the development of such oils, tests are performed on 1- or 3-cylinder Bolnes engines but final results are only available from very time-consuming on-board trials lasting at least 1 year.

It is certain that significant restrictions on the sulfur content of marine diesel fuels will be imposed in the coming years. This will result in a lowering of the high proportion of over-based components necessary for such oils at present.

References

1 Fuchs, M. (2000) The world lubricants market – current situation and outlook. *Proceedings of the 12th International Colloquium Tribology 2000 Plus*, January 11–13, Vol. I, TAE Esslingen, p. 9.

2 Hubmann, A., Reifer, K. and Baumann, W. (1994) Motorenöl mit optimierter Umweltrelevanz. *Mineralöltechnik*, 4.

3 Dearlove, J. and Cheng, W.K. (1995) Simultaneous Piston Ring Friction and Oil Film Thickness Measurements in a Reciprocating Test Rig, SAE Paper 952470.

4 Antusch, S. Dienwiebel, M., Nold, E., Albers, P., Spicher, U. and Scherge, M. (2010) On the tribochemical action of engine soot. *Wear*, **269**, 1–12.

5 Harenbrock, M., Kolczyk, M., Gray, C. and Robson, R. (2004) Ageing of Filter Media in Automotive Engine Oils. Proceedings of the 14th International Colloquium Tribology, Esslingen.

6 Schwarze, H., Brouwer, L., Müller-Frank, U. and Kopnarski, M. (2008) Ölalterung und Verschleiß im Ottomotor. *Motortech. Z.*, **69** (10), 878–886.

7 Schwarze, H., Knoll, G., Longo, C., Kopnarski, M. and Emrich, S. (2010) Auswirkung von Ethanol E85 auf Schmierstoffalterung und Verschleiß im Ottomotor. *Motortech. Z.*, **71** (4), 286–292.

8 Riga, A.T. and Roby, S.H. (1994) Low-temperature properties of crankcase motor oils – a fundamental approach to pumpability phenomena. *Lubr. Eng.*, **50**, 441–417.

9 Küpper, C., Artmann, C., Pischinger, S. and Rabl, H.-P. (2013) Schmierölverdünnung von direkteinspritzenden Ottomotoren unter Kaltstartrandbedingungen. *Motortech. Z.*, **9**, 710–715.

10 Shimokoji, D. and Okuyama, Y. (2009) Analysis of Engine Oil Deterioration under Bio Diesel Fuel Use, SAE 2009-01-1872.

11 Henderson, K.O. and Sticking, J.M. (1990) The Effect of Shear Rate and Shear Stress on Viscosity Determinations at Low Temperature for Engine Oils, SAE Paper 902091

12 Selby, T.W. (1985) Further Consideration of Low Temperature, Low Shear Rheology Related to Engine Oil Pumpability, SAE Paper 852115.

13 Kollmann, K., Gürtler, T., Land, H., Warnecke, W. and Müller, H.D. (1998) Extended Oil Drain Intervals – Conservation of Resources or Reduction of Engine Life (Part II), SAE Paper 981443.

14 Amann, C.A. (1997) The Stretch For Better Passenger Car Fuel Economy, SAE Paper 972658.

15 McGeehan, J.A. (1999) The Pivotal Role of Crankrease Oil in Preventing Soot Wear and Extending Filter Life in Low Emission Diesel Engines, SAE Paper 1999-01-1525.

16 Mang, T. (1997) International Conference on Tribology, Calcutta.

17 MacgMach, W. and Gervé, A. (2000) IAVF Information Brochure, Das CEC Referenzöl-System: Entwicklung, Erfolg, Herausforderungen, Karlsruhe.

18 Prince, R.J. (1992) Base oils from petroleum, in *Chemistry and Technology of Lubricants* (eds R.M. Mortier and S.T. Orszulik), Blackie, London.

19 Grossmann, W. (1991) Additives for lubricants. *Erdöl Erdgas Kohle*, **107** (10), 417–421.

20 Rasberger, M. (1988) *Chemistry and Technology of Lubricants*, John Wiley & Sons Inc., London, p. 83 ff.

21 Lauterwasser, F., Hutchinson, P., Wincierz, C., Ulzheimer, S. and Gray, D. (2012) The Role of VI Improvers in the Formulation of Fuel Efficient Engine Oils with Long Drain Intervals Proceedings of the 18th International Colloquium Tribology, Esslingen.

22 ASTM D4684 Determination of Yield Stress and Apparent Viscosity of Engine Oils at Low Temperature, ASTM Standard Test Method.

23 Davies, R.E., Draper, M.R., Lawrence, B.J., Park, D., Seeney, A.M. and Smith, G.C. (1995) Lubricant Formulation Effects on Oil Seal Degradation, SAE Paper 952340.

24 van Eberan-Eberhorst, C. and Cantab, M.A. (2000) *Mineralöltechnik*, **4**.

25 Choi, E., Akiyama, K., Astida, T., Kado, K., Ueda, F. and Otira, H. (1995) Engine Testing Comparison Of The Relative Oxidation Stability Performance Of Two Engine Oils, SAE Paper 952530.

26 Jahanmir, S. (1986) Examination of wear mechanisms in automotive camshafts. *Wear*, **108**, 235–254.

27 Coritch, M.J. How Polymer Architecture Affects Permanent Viscosity Loss of Multigrade Lubricants, SAE Paper 982638.

28 van Dam, W., Broderick, S.H., Freerks, R.L., Small, V.R. and Willis, W.W. (1997) TBN Retention – Are We Missing the Point? SAE Paper 972950.

29 Lanzerath, P. (2011) *Dissertation, Alterungsmechanismen von Abgaskatalysatoren für Nutzfahrzeug-Dieselmotoren*, TU Darmstadt.

30 Winkler, A., Ferri, D., Dimopoulos-Eggenschwiler, P. and Aguirre, M. (2010) Analyseverfahren zur Alterung von Dieseloxidationskatalysatoren. *Motortech. Z.*, **6**, 428–434.

31 Devlin, M.T., Lam, W.Y. and McDonnell, T.F. (1998) Comparison of the Physical and Chemical Changes Occuring in Oils During Aging in Vehicle and Engine Fuel Economy Tests, SAE Paper 982504.

32 Völtz, M. (1993) *Biologisch Schnell Abbaubare Schmierstoffe und Arbeitsflüssigkeiten*, Expert Verlag, pp. 136–157.

33 Nakada, M. (1996) Requirements to Lubricants from Engine Technology, 2nd Annual Fuels & Luber Conference, Manila, Philippines.

34 Carter, B.H. (1992) Marine lubricants, in *Chemistry and Technology of Lubricants* (eds R.M. Mortier and S.T. Orszulik), Wiley-VCH Verlag GmbH.

35 Bartz, W.J. (1983) *Handbuch der Betriebsstoffe für Kraftfahrzeuge – Teil 2*, Expert Verlag.

36 von Eberan-Eberhorst, C. (1993) Lubrication of Two-stroke and Four-stroke Engines, SAE Technical Paper 937051

37 Groff, E.G., Sheaffer, B.L. and Johnson, W.P. (1998) Two-stroke Engines and Emissions, SAE Special Publication SP 1327.

38 Kawabe, H. and Konishi, Y. (1993) Effect of Polybutene on Prevention of Clogging in the Exhaust System of Two-stroke Engines, SAE Technical Paper 93[illegible]A085

39 Souillard, G.J., van Quaethoven, L. and Dyer, F.E. (1971) Polyisobutylene, a New Synthetic Material for Lubrication, SAE Technical Paper 710730.

40 International Council of Marine Industry Associations (1997) Standard 27–97, Egham, UK.

41 Chamberlin, W.B., Curtis, T.T. and Smith, D.M. (1996) Crankcase Lubricants for Natural Gas Transportation Applications, SAE Technical Paper 961920.

42 Zareh, A. (1996) An overview of industrial gas engines and their lubrication – Part 1. *Lubr. Eng.*, **52** (10), 730–740.

43 Zareh, A. (1996) An overview of industrial gas engines and their lubrication – Part 2. *Lubr. Eng.*, **52** (11), 798–808.

44 Deuss, T., Ehnis, H., Freier, R. and Künzel, R. (2010) Reibleistungsmessungen am Befeuerten Dieselmotor. *Motortech. Z.*, **71** (5), 326–330.

45 Wilkinson, J. (1994) Biodegradable Oils – Design, Performance, Environmental Benefits and Applicability, SAE Technical Paper 941077.

10
Gear Lubrication Oils

Thorsten Bartels and Wolfgang Bock

The gear lubrication oil is a machine component of particular significance for gear and transmission. During operation, the lubricant comes into contact with most of the other inbuild machinery components. Apart from the important function of lubricating the sliding rolling contacts, the oil also fulfils the task of cooling and removing the friction heat generated in the sliding rolling contacts.

A comparison of the expenses and costs relating to the production or manufacture of the machine elements of a transmission, such as the roller bearings, toothed wheels, shafts, seals or gearboxes, shows that, in general, the lubricant is a transmission component that can be relatively simply produced at low cost. However, in order to ensure the reliable and long-term service life, the selection of the suited lubricant, in comparison to the selection of the other machine components, is of decisive significance during the construction and designing phase. For example, a new unlubricated ball-bearing used in a vehicle's alternator reached a service life of 4 min. A ball-bearing of the same type used under the same operating conditions reached a service life of several hours after the application of approx. of 2 mL oil on to the bearing's races.

Various types of lubricant are used in the lubrication of gear and transmissions, whereby a lubrication oil mainly consists of a base-oil and an additive adjusted to the base-oil and the application. The following base-oils are not only used for the lubrication of transmissions but also many other applications:

- Mineral oils
- Synthetic hydrocarbons (polyalphaolefins)
- Poly(alkylene glycols) (homopolymers)
- Esters (environmental friendly oils, mainly on synthetic basis)
- Naphthenic oils.

Apart from the various base-oils, the type and quantity of the additive, which depends heavily on the base-oil, has a significant influence on the function and service life of gear and transmission.

Lubricants and Lubrication, Third Edition. Edited by Theo Mang and Wilfried Dresel.

In principle, any type of lubricant, available on the market, including engine oils, can be used in any gear and transmission, thus ensuring for the moment their functionality.

However, at this point we would like to warn about the risks involved in this scenario, when some next best lubricant is used without knowing the operating and environmental conditions of the application. The operating conditions include the switching periods and the forces to be transferred, the so-called specific loads of each single machine component that result from operational speed and the transferred torque. The operation of a transmission causes friction losses in all sliding rolling contacts generating a heating-up. In this case, insufficient or inappropriate lubrication, often combined with inadequate cooling of the friction contact points, leads to a short-term failure of the system. The most frequent failure criteria for gears and transmissions are as follows:

- Extreme abrasive wear.
- Early endurance failure, fatigue of components.
- Scuffing and scoring of the friction contacts.

The selection of a lubricant, which is not adapted to the respective construction components and the operating and environmental conditions of gears and transmissions, can given an early failure of the machine, leading to maximum consecutive damages up to the breakdown of complete systems. The resulting repair and standstill times of the system lead to unforeseen costs. In most cases, this is the result of trying to save on the relatively low lubricant costs. This example highlights the particular significance of the lubricant as a machine element in gear and transmissions today.

10.1 Requirements of Gear Lubrication Oils

In many areas of machine designing, the torque transfer plays a decisive role. Figures 10.1 and 10.2 give a general survey over the various transmission types used today.

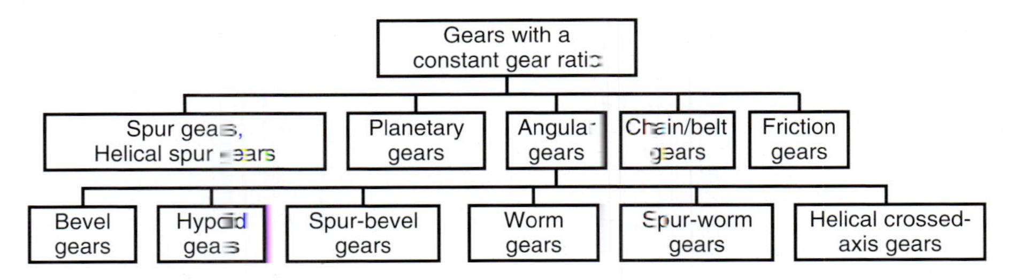

Figure 10.1 Gears with a constant gear ratio.

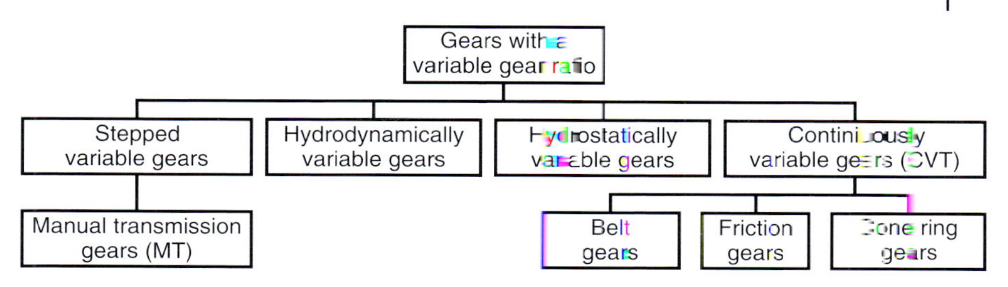

Figure 10.2 Gears with a variable gear ratio.

Each of the above-mentioned types for torque transfer makes specific requirements on the lubricant that must be met to ensure the reliable function of the machines and plants. Thus, the heavily loaded lubricants in hypoid gears require high oxidation stability, together with a very good scuffing and scoring, and wear load capacity, due to the high load of the tooth contacts. At the same time, the formation of a load-capable and separating film thickness for sufficient lubrication and cooling of the sliding rolling contacts in hypoid gears requires a lubricant with an adequately high viscosity at operating temperature.

On the other hand, lubricants used in hydrodynamic gears, such as torque converters, hydrodynamic wet clutches or retarders, do not need to have a good scuffing and scoring load capacity. However, they must have a high oxidation stability. Due to the viscosity-dependent losses, lubricants that are used in hydrodynamic gears therefore have a clearly lower viscosity at operating temperature in comparison to lubricants used in hypoid gears.

The above-mentioned gear types are used in machine and plant construction with variable exposure time. The requirements made on these gears, closely connected to the requirement made on the gear lubrication oils, should be viewed in an industry-wide manner where the service life and the oil drain intervals are concerned. Figure 10.3 presents the service lives currently required for machines according to the most important industries using gears for torque transfer making obvious the large differences in the required service lives, from 100 to 500 operating hours for gears in household machines to more than 100 000 hours service life for gears used in paper machines.

The heat development in a gear generates heat and raises the temperature of the oil sump and the temperature in the oil tank. This is of a major significance for the lubricant's service life, since it accelerates the oil's ageing process and, therefore, can cause a reduction of the oil's service life. The heat development and, thus, the oil temperature, is determined by the type of gear, the transferred torque, the specific load, as well as from the switching periods – permanent or intermittent operation–and from the environmental conditions – use of the gear in warm or cold climates, or in a mobile or stationary application.

Gears that require extended service lives, such as paper machines or presses, the lubricant has to be exchanged according to the mechanical and

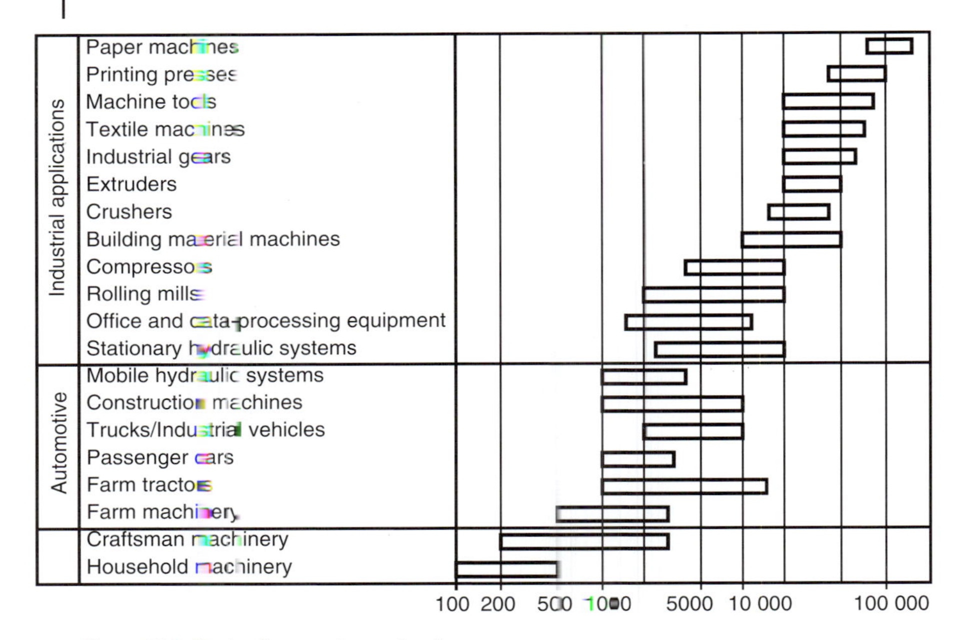

Figure 10.3 Service live requirements of gears.

thermal claim, in compliance with the oil producers' recommendations. Assuming that the average oil sump temperature in such transmissions is approx. 90 °C, the oil should be exchanged in intervals of 2500 h of operation. An increase of the oil sump temperature by 10 K leads to a 50% reduction of the service life, reducing the temperature by the same value doubles the oil's service life in general.

Today, gears for which short or medium service lives are required, as for example in passenger cars or mobile hydraulic systems, have life-time fluids. The oil sump temperatures in these transmissions often rise up to 130 °C. Therefore, the use of synthetic lubricants is recommended in these applications. Suitable synthetic lubricants, such as poly(alkylene glycols) glycols or synthetic hydrocarbons, reduce the friction losses and thus the temperature of the transmissions. In general, the service life and, therefore, the oil drain interval of synthetic lubricants are three times longer than the service life of mineral oils. However, the current prices for synthetic lubricants are also three times higher than those for mineral oils.

According to the market share of mobile and stationary gears, one differentiates, as shown in Figure 10.3, between 'gears for industrial application' and 'gears for automotive application'. With respect to the gear lubricant, household and craftsman machinery only play a minor role, since only relatively short service lives are required here. In most cases, these machines are not lubricated with oil but with grease.

10.2
Gear Lubrication Oils for Motor Vehicles

Gear lubrication oils for motor vehicles have to meet the specific requirements of the gear drives. These gear drives transfer the torque from the engine to the drive gears and consist of a gearbox and the drive axle that use different all gears to transmit the power to the wheels. In Europe, mostly, the gearboxes that serve to change the gear ratio are manual transmissions with synchronization. In America and in Asia Pacific, they are mostly automatic or semiautomatic transmissions or so-called constantly variable transmissions. With respect to these automatics or CVTs, the lubricant is not only responsible for the gear drive lubrication but also for the function-related operation of wet clutches, wet brakes as well as torque converters, retarders and dual clutches.

Today, it is not possible to meet all requirements of the mentioned types of transmissions with only one fluid. Due to the increasing technological developments together with an increasingly specific component designing with new gear components, for example sensorics, the development trend is more and more orientated towards individualization. Currently, more and more tailor-made individual lubricant solutions are being developed for specific applications, special transmissions and transmission family-types. Like the development of new engine oils, the current development of gear lubrication oils for motor vehicles has not yet come to an end and is mainly driven by several factors:

- Vehicle operators welcome an improving reliability of the vehicles and increasing oil-drain intervals to reduce the operative costs incurred.
- The requirements and specifications issued by the motor, transmission and vehicle manufacturers intend to considerably extend the oil-drain intervals. Today's trend is to supply all transmissions with so-called fill-for-life fluids. Today, most of the manual transmissions and axle gears in passenger cars are operated with fill-for-life fluids (>300 000 km). With respect to automatic transmissions in passenger cars, these long oil-drain intervals will soon be achieved. As far as commercial vehicles are concerned, the required oil-drain intervals of the manufacturers are currently experiencing an extension. It is planned to achieve transmission service lives of 500 000 km, 750 000 km up to 1 000 000 km without drain.
- The legislation in the industrialized countries has an influence on the lubricant formulation mainly due to reasons of environmental friendly fluids and disposal. This means that the use of mineral-based oils is decreasing. Today, environmentally friendly, synthetic and ester-based base oils are often used for reasons of environmental protection. Due to the better thermal oxidation stability, the ester-based oils are used together with PAO and hydrocracked base oils.

Attempts are furthermore made to reduce lubricant toxicity, especially in conventional components of commonly used additives. In this respect, the reduction of the chlorine content is of a great importance due to the disposal of the used

Table 10.1 Gear oil viscosity classification and gear oil performance level.

ISO	ISO 3448
SAE	SAE J300
SAE	SAE J306
MIL	MIL-L-2105
AGMA	
MIL	MIL-L-2105E
SAE	J2360

lubricants. In order to reduce the particles emission, the discharge of oil into the environment and, thus, the oil consumption, future gear lubricants for vehicles are to be tested very exactly with respect to elastomers used in the seal production.

- Driver of motor vehicles expect modern car to ensure an improved performance, a reduced noise emission and a reduced fuel consumption. This trend is expressed in the increasing use of lubricants with a low viscosity together with a reduced fill volume. Furthermore, multigrade oils are currently used on a large scale to considerably reduce the fuel consumption (Table 10.1).

10.2.1
Driveline Lubricants for Commercial Vehicles

Commercial vehicles are divided into light trucks with a maximum capacity of 3–6 tons and trucks with a maximum capacity of more than 6 tons. In 2005, the worldwide annual demand for gear lubrication oils for commercial vehicles amounted to approximately 800 000 tons. In the same year, some 2.1 million commercial vehicles were produced in the Western and Eastern Europe, which corresponds to drive-line lubricant volume demand of 170 000 tons of gear oils for the commercial vehicles' gearboxes and axle gears for the European market. In 2005, the number of manufactured light trucks amounted to some 1.7 million vehicles. According to their market share, the most important manufacturers are PSA (22%), Ford (15%), VW (12%), Renault (12%), Daimler–Chrysler (8%), Fiat (8%), Daewoo (3%), Skoda (3%), Hyundai (2%), Mitsubishi (1%), Iveco (1%) and others (13%).

Because the fill volume per transmission unit is two to six times larger and amounts to 10 to 30 l, trucks are far more interesting for the oil industry. At the same time, the oil development also confronts a bigger challenge in meeting the requirements made on the fluids used in the commercial vehicles' transmissions. In 2005, the number of manufactured trucks amounted to approximately 0.45 million in all European countries. According to their market share, the most important manufacturers are Daimler–Chrysler (23%), Volvo (16%), Iveco (15%), Scania (13%), MAN (12%), Renault (10%), DAF (9%) and others (2%).

Table 10.2 Heavy-duty axle applications (commercial vehicles, trucks, buses).

Chrysler	MS-9020
DaimlerCrysler	DC 235.8
API	GL-5
GM	8 863 370
Ford	SQM-2C9002-AA
Volvo	STD 1273.12
Scania	STO 1:0
Renault	RVI TDL
MAN	MAN 342
MAN	M 3343
VW	TL 727
ZF	ZF Ecofluid X

Companies such as ZF and Getrag should also be mentioned, since they play an important role as transmission manufacturers for a large number of worldwide automotive companies and produce transmissions for both commercial vehicles and passenger cars and deliver these transmissions to the mentioned automotive companies. The special oil requirements of these companies, above all of ZF, have a major influence on the current trend in the development of new fluids for the transmissions of commercial vehicles.

Tables 10.2 and 10.3 list the most important current (2005) specifications issued by the transmission manufacturers and automotive companies split into manual transmission and rear axle gears. With respect to the specifications, the modified new SAE classes are to be taken into account. Above all, the new classification SAE J2360 is to be mentioned that will replace the old MIL-PRF-2105E. In addition to this, the new SAE 'Automotive Gear Lubricant Viscosity Classification' J306 is also to be taken into account. In this specification, the shear loss of the fluids after shear (stay-in-grade) attached much importance.

Table 10.3 Light-duty axle applications (passenger cars).

Chrysler	MS-9763
API	GL4
Ford	M2C-119A
GM	9 985 476
Clak	MS-8 Rev. 1
Volvo	STD 1273.10
Volvo	STD 1273.13

This trend in the development of lubricants for commercial vehicles leads to a significant change in the current ratio between service fill on the one hand and the factory fill on the other hand. The share in service fill and drain of currently 75% – with a factory fill stake of 25% –will be reduced to 20% for the service fill – with a factory fill stake of 80% during the next 10 years.

From today's point of view, it can be said that all mentioned specifications constitute a challenge for the development of gear lubrication oils for commercial vehicles. Apart from the very high chemical and physical requirements included in these specifications concerning the oxidation stability (test at 150–160 °C), corrosion, filterability, foam, and so on, compared in Sections 10.2.1–10.2.4, the mentioned specifications include a large number of mechanical–dynamic tests and require very high safety levels with respect to the scuffing and scoring, pitting and wear resistance of the toothings using different, standardized, non-standardized and company-internal test procedures. This decelerates the development of new oils and makes it increasingly complicated and expensive. A particular obstacle in the development of oils for commercial vehicles is the improvement and optimization of the synchronization behaviour during shifting operations.

Although many commercial vehicles of American manufacturers still do not use synchronized transmissions, almost all commercial vehicles in Europe are equipped with these transmission systems. The SSP 180 test bench (see also Section 19.3.7) has proven very succesful in the oil development for a sufficient synchronization behaviour (Table 19.17). Tables 10.4 and 10.5 show fluid specification referring to heavy-duty transmissions with and without synchronization.

Table 10.4 Heavy-duty synchronized transmissions and transaxle applications (commercial vehicles, trucks, buses).

DaimlerCrysler	DC 235.11
API	GL-1
API	GL-3
MACK	GO-J
MACK	GO-J PLUS
MAN	MAN 341
Volvo	STD 1273.07
ZF	ZF Ecofluid M

Table 10.5 Heavy-duty non-synchronized manual transmissions (commercial vehicles, trucks).

API	MT-1
Eaton	PS-164
Eaton Bulletin	2053

One of the benefits for the oil development on the FZG SSP180 is the fact that the development is carried out with regular components at real operating conditions. However, a disadvantage is that almost all automotive companies use different friction material pairs, for example steel–molybdenum, steel–brass, and also often coatings such as carbon, sinter or paper. This is aggravated by the use of different component geometries with different conical angles, single or biconical, as well as different operational parameters. According to the type of synchronization, real operating conditions with axial forces, load-increase speeds, surface pressures and inertia are created for the oil development. Often, automotive companies use two or more friction material pairs in the synchronization of a gear family at the same time and expect the fluid developer to fully meet the given operating conditions with the same lubricant or lubricant technology apart from all the other requirements.

The main problem in the fluid development is to adjust the same friction or friction constancy during a required gear service life of approximately 100 000 gearshift operations with different material–lubricant combination without significantly changing the other chemical–physical and dynamic–mechanical properties.

The following example, shown in Figures 10.4–10.6, gives an overview of a lubricant development which complies with most of the required chemical–physical and dynamic–mechanical limits, including an excellent gear-shifting behaviour in a molybdenum–steel synchronization. This is clarified by the shown individual gearshift operations, which show a constant friction coefficient during the entire synchronization time (shift time). The trend shown in Figure 10.4a, in which the average friction coefficient has been applied during the required service life of 100 000 gearshift operations, remains almost constant and sufficiently high during all gearshift operations. The gear-shifting behaviour of the oil with this molybdenum–steel synchronization is to be considered excellent.

Different material combinations such as sinter–steel, which is used by the manufacturer in the same gear family display an inferior synchronization behaviour in the bench test. The individual gearshift operation at the beginning of the test in Figure 10.5 (gearshift 10 000) still shows the desired stable friction coefficient during the entire synchronization time (shift time). However, this behaviour changes after a while and experiences a sharp increase in the friction towards the end of the gearshift operations. The big difference between the static friction coefficient and sliding friction coefficient leads to a jolting during the gear-shifting operation, which is perceived as an undesirable phenomenon. This effect will increase during the following gearshift operations and results in a clashing of the synchronization, as shown in Figure 10.6 (gearshift 48 500). Thus, as shows the trend figure in comparison to the molybdenum–steel synchronization, the required number of gearshift operations of 100 000 is not achieved. The test has to be stopped after 48 500 gearshift operations. A second test under the same operating conditions confirms the obtained results. The lubricant technology is rejected by the gear manufacturer as unsatisfactory.

(a)

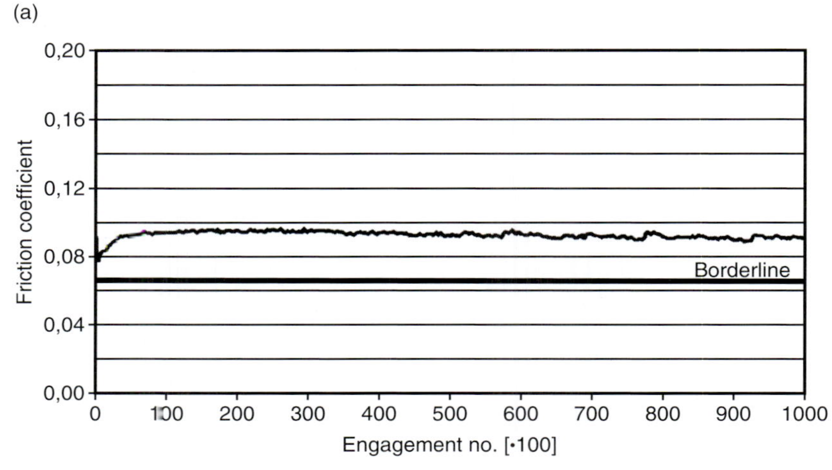

(b)

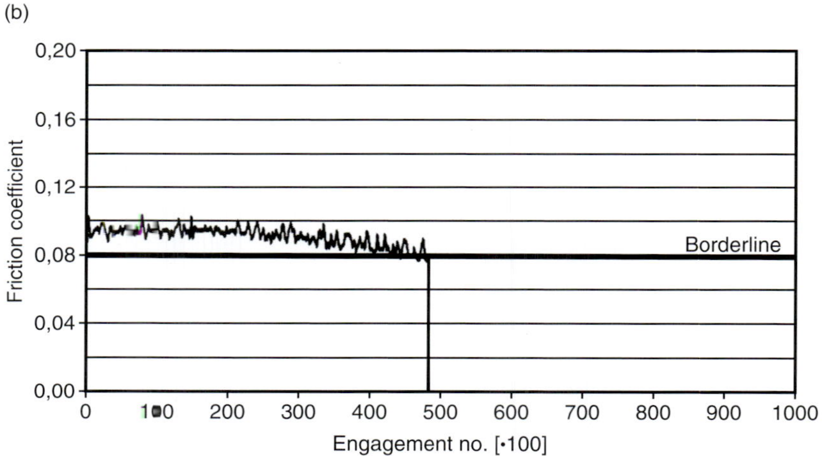

Figure 10.4 ZF SSP 180 test run – same fluid with different synchronization materials. (a) Upshift, Mo-synchronizer. (b) Upshift, sinter-synchronizer.

10.2.2
Driveline Lubricants for Passenger Cars

The development of passenger cars is mainly influenced by the worldwide fierce competition of numerous producers. Especially against this background, the competitiveness urges many automotive companies to enter co-operation and mergers according to which identical components and drivelines are manufactured at different locations and for different vehicles. For the oil industry, this means to supply approved lubricants that have to be available in the same quality worldwide or at least in the production locations for gear-filling purposes.

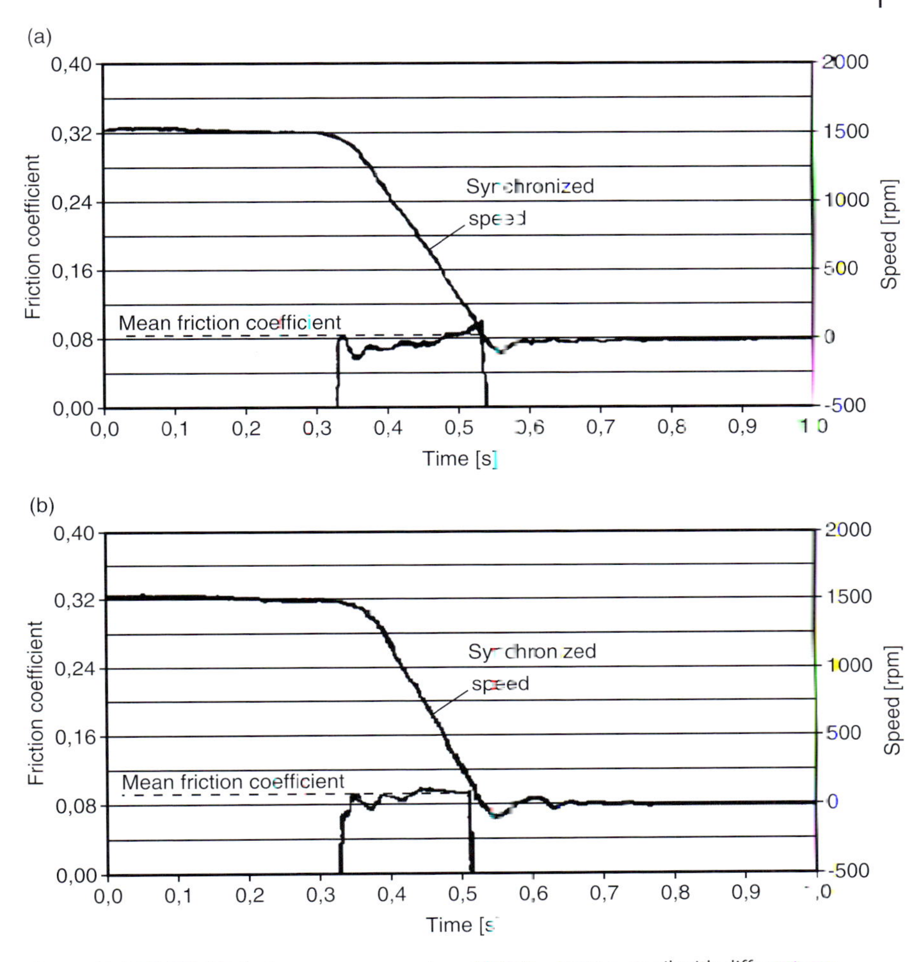

Figure 10.5 ZF SSP 180 test run, engagement no. 10 000 – same gear oil with different synchronization materials. (a) Upshift, Mo-synchronizer. (b) Upshift, sinter-synchronizer.

As far as the customers are concerned, the competition's main criteria are the passenger car's driving comfort, design, economy and sportiness, which serve as trendsetters to ensure high sales figures. In particular with respect to passenger cars, the mentioned vehicle trends have a certain influence on the engine and total drive-line and thus on the fluids used in these transmissions.

A passenger car's sportiness requires a high performance level and a high torque transfer in identical or even smaller spaces for gearboxes. Due to the economy required at the same time, the lubricants are expected to optimize the total efficiency of the engine and total drive-line without having to be changed during the entire service life of more than 300 000 km.

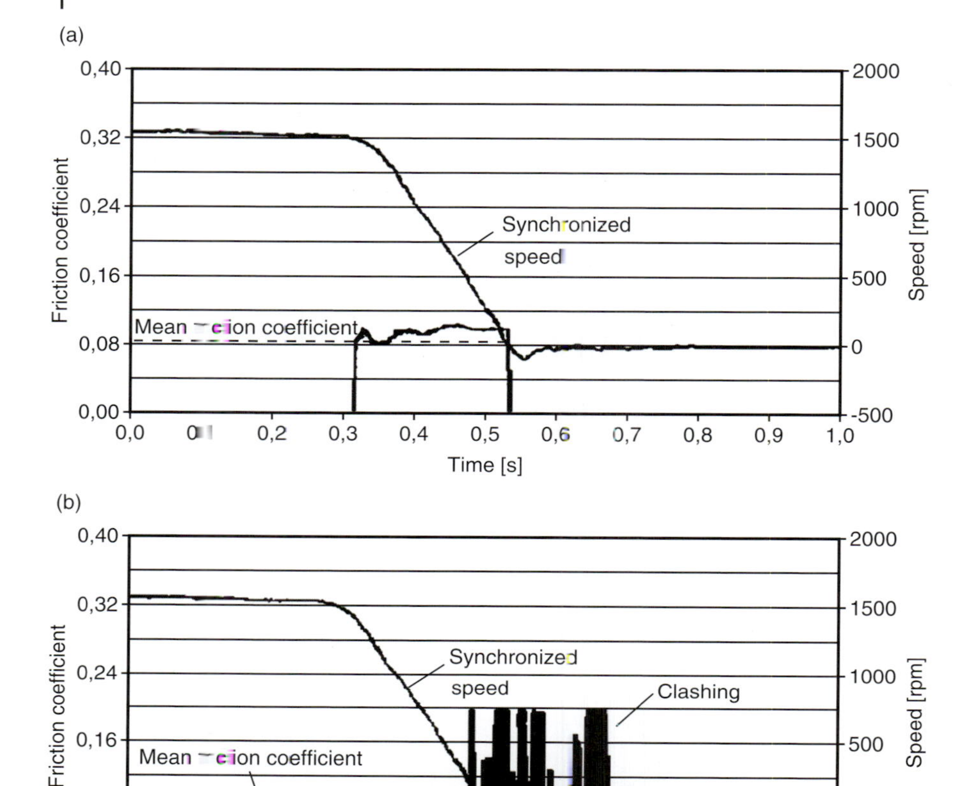

Figure 10.6 ZF SSP 180 test run, last engagement same fluid with different synchronization materials. (a) Upshift, Mo-synchronizer, engagement no. 100 000. (b) Upshift, sinter-synchronizer, engagement no. 48 500.

The best performance costs an increased heat generation in the transmissions that accelerates the ageing of the gear lubrication oils. The design of new vehicles leads to the development of carefully selected chassis, often with a low air resistance. This is often connected with an inferior air circulation and cooling of the transmissions. Additional capsules facilitate the reduction of driving-related noise to advance the driving comfort. At the same time, the poor heat dissipation leads to a heat stowing in the capsular spaces. Especially, these factors increase the temperature in the vehicle's transmission and enhance the operational conditions, thus accelerates the ageing of the fluid.

New passenger cars include transmissions with a higher number of gear stages so that today passenger cars using gear drives with six stages are not unusual.

Here, the constructors are more and more required to exploit all technical possibilities of dimensioning and material, always with an effect on the corresponding fluid when new lighter or more solid materials have to be used in the transmission as new friction partners with a friction and wear characteristics unknown so far.

An increasing number of vehicles utilize automatic transmissions and all wheel drive systems. Today, the necessary control elements and sensor systems consisting of electric components such as meters and sensors and so on, are installed in the vehicles' drive-lines. Here, the compatibility of these sensible components with the fluid, for example with regard to copper corrosion or at high oil sump temperatures leads to enhanced requirements made on the fluid and its development.

Today, transmissions for passenger cars with a synchronized manual transmission has a market share in Europe of almost 90%. As already shown in Section 10.4.1 for commercial vehicles, all manufacturers of passenger cars make very high comfort-related requirements on the synchronization. The variety of the friction material partners used in the synchronization of passenger cars is constantly increasing so that here more and more individual solutions are also required with respect to the fluid. Another requirement on the gear lubrication oils for manual transmission systems, as well as for rear axle gears, is the very high scuffing capacity to ensure the protection of the spur gears' which have significantly a higher sliding ratio than the planetary gears' toothings normally found in automatic transmission, or continuously variable transmissions.

All mentioned development trends, together with the requirements on the gear lubrication oils for passenger cars connected with these, are reflected in the specifications issued by leading automotive companies which continuously determine an increasing number of requirements for the screening, and approval tests. Today, some vehicle and gear manufacturers have even decided to synthetically age or oxidize fresh oils in order to additionally test them afterwards in a synthetically quickly aged condition using the conventional test methods (see Section 19.3.1.7).

Several important specifications for fluids for manual transmissions and axle fluids for gears of passenger cars are listed in Tables 10.6 and 10.7.

Worldwide, there are further specifications for off-highway construction machines and commercial vehicles with importance for the fluid development (Table 10.8).

Table 10.6 Synchronized manual transmission and transaxles (passenger cars).

Volvo	STD 1273.08
DaimlerCrysler	DC 235.10
VW	TL 726

Table 10.7 Light duty synchronized manual transmission (light trucks, passenger cars).

API	PM-1
Ford	SM-2C-1011 A
Ford	M2C 200C
VW	TL 52512
VW	TL 52171
VW	TL 52178
BMW	602.00.0

Table 10.8 Heavy duty automatic transmissions (commercial vehicles, trucks, buses).

Allison	C-4
Allison	TES-295
Chrysler	MS-9602
ZF	TE-ML 02
ZF	ZFN 13015

10.2.3
Lubricants for Automatic Transmissions and CVTs

The most important consumers of fluids for automatic transmission and drive systems, the so-called automatic transmission fluids (ATFs), are off-highway vehicles and machines, as well as commercial vehicles such as buses (city buses, intercity buses, coaches and mini/midi buses). In addition to this, ATFs are also used to fill power steering systems in trucks, commercial vehicles and passenger cars.

A total of 90% of all passenger cars on the North American and Asian vehicle market utilize automatic transmission systems. The total estimated ATF volume worldwide for the year 2006 amounts to approximately 1.2 million tons. The total volume breaks down as follows: North America (61%), Asia Pacific (15.5%), Europe (12.5%), Latin America (8.8%) and Middle East (2.2%).

Most manufacturers of automatic transmission systems require fluids for the application, which meet the listed specifications (Table 10.9).

These specifications refer to the Asian, North American and European markets and to the business for service fill and re-lubrication of vehicles. Currently, the factory fill is subject to enhanced specifications (Table 10.10).

European manufacturers of vehicles and, especially, transmission systems have issued particular specifications mainly for off-highway construction machines and commercial vehicles.

Table 10.9 Light duty automatic transmissions (light trucks, passenger cars).

SAE	SAE J311
TASA	TASA ATF
Chrysler	MS-7176
DaimlerCrysler	DC 236.12
DaimlerCrysler	DC 236.20
DaimlerCrysler	Mopar 4+
Ford	M2C 202 B
Ford	Mercon V
Ford	Mercon SP
GM	GM 6418 M
GM	Dexron III H
GM	Dexron VI
GM Opel	B 040 1068
GM Opel	B 040 1073
GM Opel	B 040 2060
VW	TL 52162
VW	TL 52182
Porsche	040204
ZF	ZFN 13026
ZF	ZFN 13014
ZF	ZFN 904

Automatic transmissions (wet clutches) require hydrodynamic clutches, torque converters and wet brakes. Furthermore, a series of wet friction clutches and brakes as well as their shifting and friction characteristics play a major role during the automatic gearshift operation. This fact is taken into account in the mentioned specifications. With respect to the friction characteristics, the

Table 10.10 Off-highway vehicles and construction machines (railway, excavators, cranes).

Caterpillar	TO-4
Terex	EMS 19003
Komatsu Dresser	B22-0003
Komatsu Dresser	B22-0005
Voith	G607
Voith	G1363
ALLISON	TES-353
ZF	ZF Powerfluid
ZF	ZFN 130031

lubricant, apart from the toothing requirements not dealt with here in detail, is nearly the most significant element.

10.2.3.1 Fluid Requirements for Hydrodynamic Transmissions

The hydrodynamic clutches facilitate an improved and smooth driving and operating behaviour, especially with respect to heavy vehicles with a high inertia mass. Here, so-called Föttinger hydrodynamic clutches are predominantly used. A centrifugal pump wheel (rotor) as the working machine and a turbine wheel (stator) as the engine are arranged opposite each other in a confined space in a conical housing. Thus, the exchange of energy between rotor and stator takes place over a short distance via the filled fluid. A hydrodynamic transmission displays a constantly variable operation, whereby the resulting torque independently adapts to the respective load status by changing the rotational speed. When the driving speed remains constant, the rotational speed's slip will increase as the load or output torque increase. The more slip between rotor and stator is generated, the more inefficient operates the torque converter and the more loss and heat will be produced.

Apart from the hydrodynamic start-up clutches, hydrodynamic brakes, the so-called retarders and intarders are used, especially in commercial vehicles. They serve to limit the output speed when driving down long hills in order to reduce the load on the vehicles. Thus, the retarders protect the vehicle's brakes against overheating and failure. A retarder will also convert the fluid losses generated due to the slip between rotor and stator for the application of the braking torque into heat.

The fluid machines ensure the uncoupling of the driveline and output torque and, thus, a constantly smooth and comfortable force transfer in the drive shaft. They will, however, only work adequately with slip and are thus characterized by significant fluid losses. Apart from the larger weight, this constitutes another reason for the generally higher fuel consumption of vehicles with automatic transmissions in comparison to vehicles with manual transmissions.

During start-up procedures and permanent operation as well as during braking, the lubricant as a force transfer medium in the hydrodynamic transmissions is subject to an extremely quick generation of heat despite installed radiator systems in comparison to gear drives. Due to the function-related fluid losses, hydrodynamic clutches, torque converters and brakes always require low-viscosity, mildly additivated gear lubrication oils with a high oxidation stability. Short-term oil temperatures of more than 160 °C during the operation of a vehicle are not unusual. Especially for these fluid machines, a good viscosity–temperature behaviour, corrosion protection, optimal foaming behaviour and air release properties are a must. With respect to the viscosity and viscosity–temperature behaviour fluids for automatic transmissions are very similar to engine oils.

Therefore, engine oils are frequently used for light transmission applications, which often become problem if the transmission is also utilizing wet clutches. On one hand, the high-additivated engine oils have a low viscosity and extreme oxidation resistance. On the other hand, they cause problems in the wet clutches and brakes with respect to their friction characteristics. To facilitate the distinction between ATFs and the engine oils, ATFs are always dyed with red colour.

10.2.3.2 Fluid Requirements for Wet Clutches and Brakes

Like in the synchronization, the friction characteristics are of great importance for a comfortable and regular gearshift behaviour during the transmission's entire service life in wet clutches and brakes as well. A wet clutch or brake consists of steel disks and separator discs. Such a separator disc or friction partner consists of a disk with an organic or sintered coating, paired with a steel disc. The minimum amount required is a single disc but wet clutches with six and more disc-plates are not unusual. The clutches and brakes are opened and closed hydraulically. Apart from the desired friction characteristic, the oil predominantly serves to ensure the cooling of the friction partners. Often the oil is injected from the interior into the disc-plate via the rotating shaft. Apart from the oil, the coating's compression, the spline and the material are also of great importance in the friction contact, which will not be dealt with in detail.

To adjust a fluid to a required friction characteristic, a test in the wet clutch or, at least, a test of the disc-plates used in the wet clutch is required. Therefore, test benches have been developed especially for this purpose that enable a more precise adjustment of the gearshift behaviour to select the best combination of material and fluid.

A tester which demonstrates the different sliding–friction behaviour of oils is the low-velocity friction apparatus (LVFA). Figure 10.7 gives an example of measured friction behaviour for some fluids as a function of the sliding speed.

Tests of original disk-plates can also be done using an SAE 2 rig or a DKA friction test bench. These machines work with given centrifugal masses that

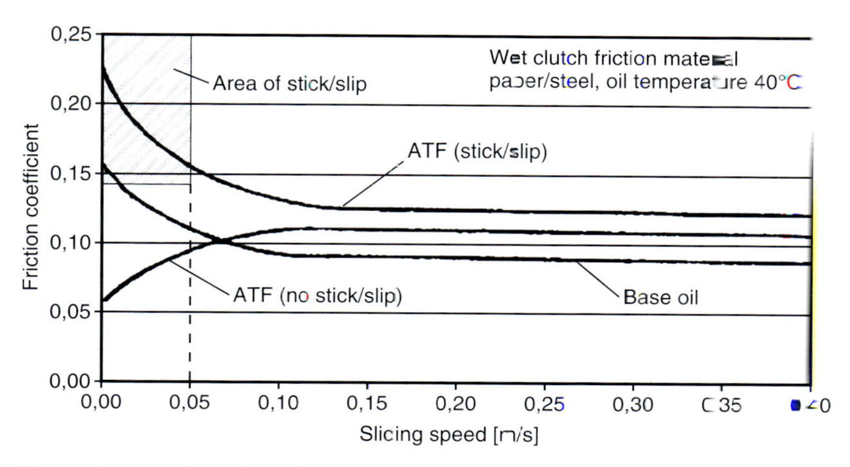

Figure 10.7 LVFA (low velocity friction apparatus). Determination of the dependence of friction on sliding speed to avoid stick-slip behaviour.

reduce the speed of the disk-plate blocks under defined test conditions. The static and dynamic friction coefficient and friction losses can be determined by using a high-resolution measuring technology.

According to the definition, the static friction is measured during the tests at low rotational speed differences (<2 min^{-1}) or towards the end of the gear-shifting operation. In this case, a load-collective-capable variation of the DKA-1B test bench is used which allows determining the load dependence of the static and dynamic friction and friction stability during a certain period of time. Figure 10.8 shows typical friction coefficients of a four-stage load-collective test.

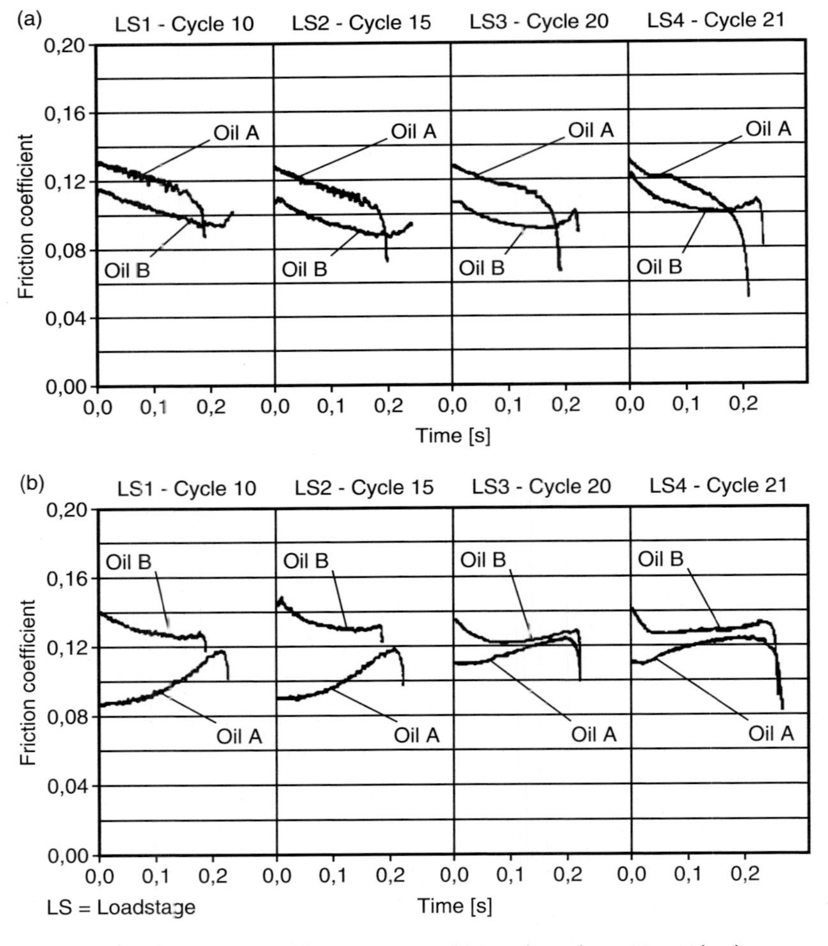

Figure 10.8 DKA 1B test run, different ATF technologies. Wet clutch – loadstage test. (a) Run-through no. 1 at 4 loadstages (LS 1–LS 4 = 66 cycles) analogous to 1st run. (b) Run-through no. 50, at 4 loadstages (LS 1–LS 4 = 3300 cycles) analogous to 50th run multiplied with 66 cycles each.

Figure 10.8 compares the friction characteristics at the 1st and at the 50th cycle of two different ATF technologies. In order to achieve a smooth and soft gear-shifting, it is tried to reach a relatively high dynamic friction coefficient at the beginning of the gear-shifting operation which will decrease during braking and then fade to an even slightly lower static friction coefficient. During the test, oil A at first shows such a desired friction curve in the 1st cycle. It is, however, not able to constantly maintain this behaviour, as shows the 50th cycle. The long-term friction coefficient of oil A towards the 50th cycle sinks as well. In this case, the user will prefer oil B which even after the 50th cycle does not display such a high increase in the static friction and the long-term friction of which shows a relatively constant behaviour over all.

To test the gearshift and friction behaviour, especially in case of permanent slip of the friction partners, the gear manufacturer ZF has developed another test bench. The so-called GK test bench or CSTCC (continuously slipping torque converter clutch) enables the simulation of almost all real operating conditions. Even slip-controlled clutches of hydrodynamic clutches and torque converters can be simulated using the GK test bench (see also Section 19.6.3.3).

From the lubricant manufacturers' point of view, only the use of the described, very expensive rigs, especially the DKA and GK test benches, will enable a target-orientated ATF development for the adjustment and improvement of the friction properties of wet clutches, slip-controlled torque-converters brakes for a certain lubricant–material pair. Here, the additives as well as the selection and adjustment of the right base oils have a decisive influence. The use of these highly technical testers increases the development expense significantly, thus making the fluid development more expensive.

10.2.3.3 **Fluid Requirements for CVT Applications**

Constantly variable transmissions in motor vehicles enable the operation of a combustion engine along certain preferred characteristic curves in the engine operating map (ignition map). In contrast to all other vehicle transmissions, the CVT enables an ideal alignment of the supply torque of a combustion engine with the request torque of the vehicle. The resulting benefits of the CVT in comparison to all other vehicle transmission system include the following:

- The exploitation of the engine's at any speed, through the operation along the engine's characteristic curve of the maximum torque (sporty driving).
- Achieving an economical performance through the operation along the engine's characteristic curve of minimum fuel consumption (economic driving).

Figure 10.9 shows the ignition map of a passenger car's 101-kW Otto engine. It displays the two characteristic curves 1 as the graph to describe the curve of maximum torque available and 2 as the curve along the minimum fuel consumption.

Only a correspondingly adjusted and optimally controlled CVT enables the driving operation along the shown characteristic curves 1 or 2. Such transmissions require the highest values possible with respect to the operating range of the final control element and the efficiency. Till today, mainly three constantly

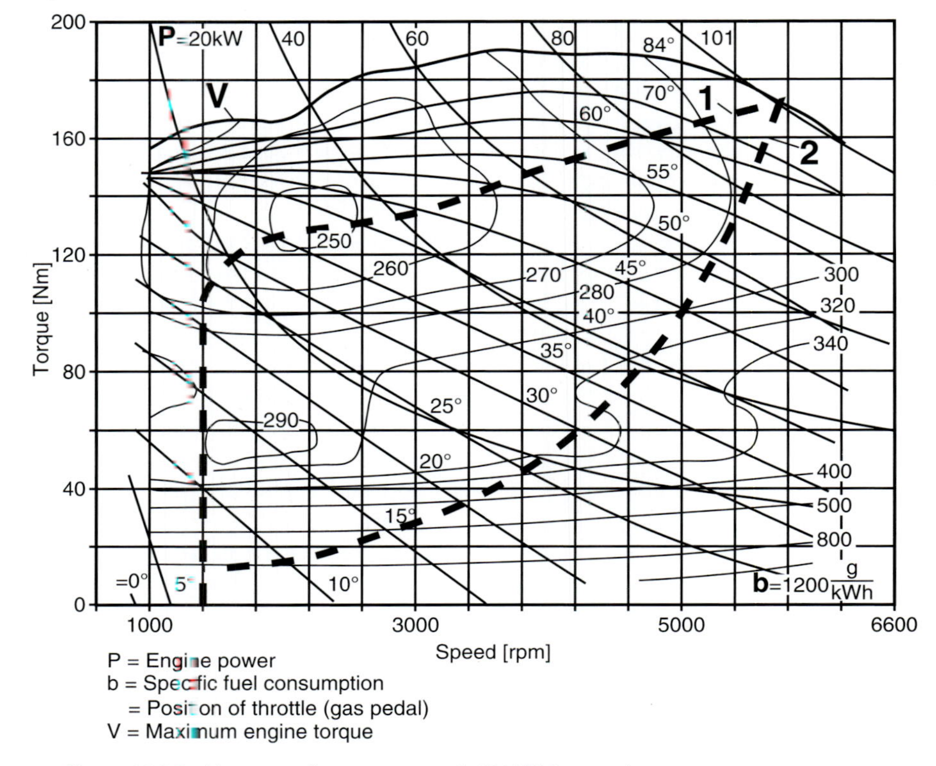

Figure 10.9 Ignition map of a passenger car's 101 kW Otto engine.

variable transmission concepts have proven to be successful to meet these requirements.

10.2.3.4 B-CVT Push Belt and Link Chain Drives

For passenger cars with a low or medium engine power of up to 150 kW, the belt drive, mostly a 'Van Doorne' push belt, has been proven successful. Figure 10.10 shows the work principle. The belt wraps the cone pulleys that have been hydrostatically and axially applied to the drive and output shafts. Thus, the radii of the belt drive's course as well as the gear ratio can be varied.

For the instationary slide–roll contacts between the belt drive and the cone's pulley surface, the fluid is an extremely important component. The contact points are often subject to very high contact temperatures and mixed-lubrication regime. Here, a very good wear protection of the oil against the extremely hardened surface of the cone pulleys is required. In addition, the pulleys grinded surfaces require from the fluid a very high pitting capacity. On one hand, a slip of the belt drive due to insufficient contact pressure forces should be avoided by means of control engineering. On the other hand, however, it cannot be avoided completely. Accordingly, the fluid also has to ensure a sufficient scuffing load capacity for these operating conditions.

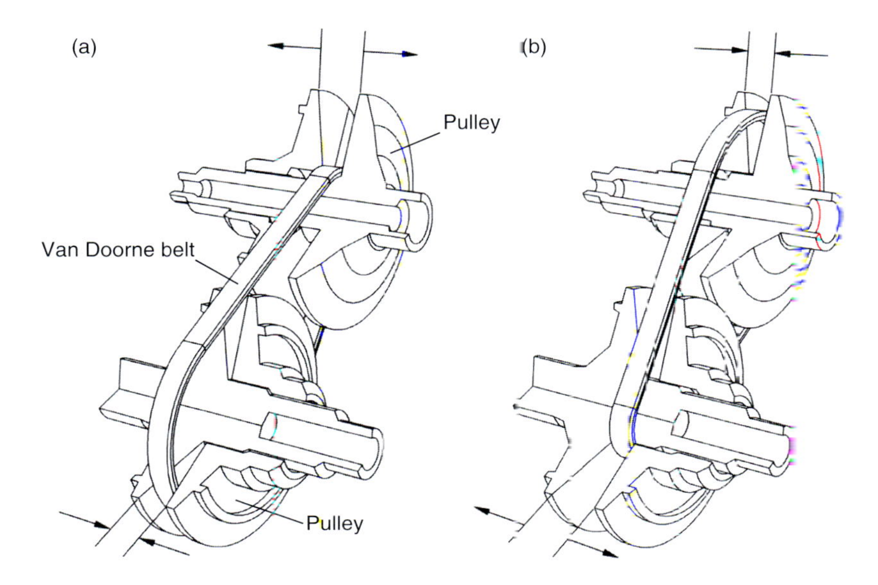

Figure 10.10 CVT variator, principle of a B-CVT (Van Doorne push belt). (a) Gear ratio in position low. (b) Gear ratio in position top.

Nearly all belt drives of vehicles have a toothed wheel gear stage as well as a hydrodynamic start-up clutch or torque converter. In order to increase the driving comfort and reduce the losses, slip-controlled clutches are used. Thus, the same requirements apply for both CVT fluids and ATFs. Currently, CVTs are filled with ATF oils, which are slightly modified lubricant variations or have been individually adjusted to the respective CVT. The viscosity, additives and base oils are very similar to ATFs, however, the so-called friction modifiers have great importance.

10.2.3.5 T-CVT Traction Drives

As of a certain vehicle performance, the belt drive system's mechanics has reached its limits in power transmission. This limit is approximately reached at an input power of slightly more than 150 kW (420 Nm), that is, in mid-range passenger cars or limousines. Currently, continuously variable transmission concepts are tested in this respect, which are based on so-called traction drives. Figure 10.11 gives a schematic overview of the operation of a traction drive.

The continuously variable adjustment of a traction drive takes place by tipping the transmission or idle wheels which, axially pressed together, roll on the half or full-toroids' races in a force-conclusive set-up. In traction drives as well, the lubricant is to be considered an important construction element and is as significant as the material, the surface treatment and the hardening of the rollers and toroids. Of all transmission types, half and full-toroid gears stand out due to the highest surface pressure and circumferential speeds in the slide–roll point

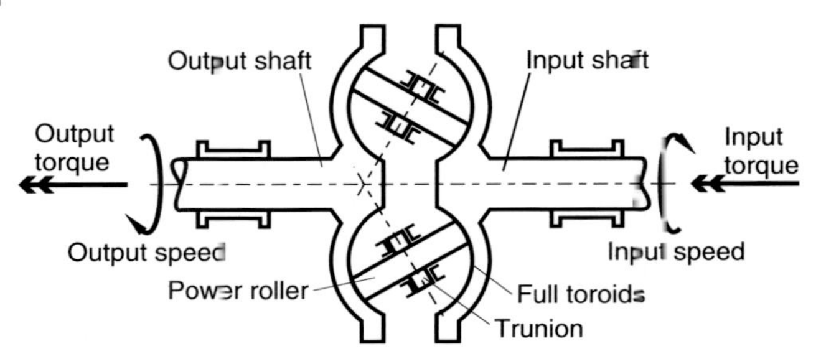

Figure 10.11 CVT variator, principle of a T-CVT (traction drive).

contacts. Surface pressures of 4500 MPa and circumferential speeds of approximately 50 m s^{-1} in traction drives are not unusual.

The transferable torque is a function of the normal force in the contact point and the friction coefficient in the slide–roll contact point. In traction drives, this friction depends heavily on the fluid, the material and the slip. According to the material partners, mainly those fluids are used which enable a high torque transfer performance at the lowest slip possible, thus having a high friction coefficient. In addition to this, the wear and corrosion protection must be ensured.

Adequately added, naphthene-based hydraulic fluids have proven very successful for these applications. However, synthetic oils of the cycloaliphatic hydrocarbon type with a particularly high friction are even more suitable for these purposes. Such oils are also called traction fluids. When contacting traction fluids in transmissions, for example in roller bearings and toothed wheels, the higher friction coefficient however, can lead to excessive and undesired overheating.

Another disadvantage of the traction fluids is a relatively low flash point of 130–150 °C. Therefore, the very high contact temperatures in the slide–roll contacts cause undesired evaporation losses. For this reason, with the lubricant technologies known today the demand for a fill for life for traction fluids can hardly be met.

10.2.3.6 H-CVT Hydrostatic Dynamic Powershift Drives

Hydrostatic dynamic powershift drives are used in agriculture and in tracked vehicles with a usually very high drive power of more than 300 kW. In these transmission systems, planetary gear stages branch the drive power into a closed hydrostatic circuit consisting of a controllable adjustment pump, mostly an axial piston pump, and a hydrostatic constant engine, mostly of bent axis design. The branching of the power and the control of the output speed depending on the volume flow rate takes place through the adjustment of the axial piston pump.

The requirements made on the fluids for this application are limited to the gears, roller bearings and hydraulic systems. For reasons of pumpability, the very good viscosity–temperature behaviour of the low-viscosity hydraulic fluids used is of great importance. In addition, these transmissions are to be protected against wear and corrosion using a suitable additive technology. ATFs and engine oils are also often used in these applications.

10.2.4 Multifunctional Fluids in Vehicle Gears

Special gear lubricants, so-called multipurpose fluids, are used in agricultural and working machines such as tractors, harvesters, and so on.

In these vehicles, the long-term and perfect function of the wet clutches and wet brakes is to be ensured. The scuffing load capacity of the hypoid gears is guaranteed using a suitable fluid. Friction coefficient, hydraulic performance and wear requirements. In order to ensure the driving and working operation even at low temperatures, torque converters have not least to work adequately and safely, even under conditions of permanent slip of the wet clutches. Therefore, multipurpose oils have almost always a low viscosity and stand out due to a very good viscosity–temperature behaviour. The presence of water and dirt has a significant impact on these oils, especially in respect to foaming and air release properties (Table 10.10).

This is aggravated by the fact that the mentioned requirements often have to manage within only one system. Apart from these complex requirements made on the multipurpose oils, which are called UTTOs (universal tractor transmission oils–not for tractor engines), another engine-related performance is usually required in addition (Table 10.11). In this case, these oils can also be used as engine oils and are then called STOUs (super tractor oils universal) (Table 10.12).

Table 10.11 UTTO multi-functional farm and tractor, agricultural machines (hypoid gears, synchronizers, wet clutches, hydraulics).

John Deere	JDM J11 D
John Deere	JDM J11 E
John Deere	JDM J20 C
John Deere	JDM J20 D
Massey Ferguson	CMS M1127
Massey Ferguson	CMS M1135
Massey Ferguson	CMS M1143
Ford	ESN-M2C-86-C
Case	JI Case 1316
New Holland	STD 200 HYD OIL
New Holland	NHA-2-C-200
New Holland	NHA-2-C-201
New Holland	M2C134-D

Table 10.12 STOU multi-functional farm and tractor, agricultural machines (engine, hypoid gears, sychronizers, wet clutches, hydraulics).

Massey Ferguson	CMS M1139
Massey Ferguson	CMS M1143
Ford	ESN-M2C-159-C
John Deere	JDM J27
Renk	530 BW
ZF	TE-ML 06

Against the background of these requirements it is easy to understand that the major manufacturers of tractors and agricultural machines, such as Ford, John Deere and Massey Ferguson, have developed their own specifications for UTTOs and STOUs. The most important of these specifications are listed in Table 10.13.

10.3 Gear Lubricants for Industrial Gears

Wolfgang Bock

10.3.1 Introduction

The gear oil is an important engineering element in power transmission engineering and is used in almost all application areas, components and finished products. In recent years, German power transmission and fluid technology has repeatedly achieved record results. With a world market share of around 24%, it represents a major engine of the economy and driver of innovation [1]. The innovative products are highly regarded everywhere in the world, and in many areas are the benchmark for the competition. The future of German power transmission and fluid technology will undoubtedly lie in the development of even more intelligent power transmission solutions, energy-efficient systems, and even more compact and efficient power transmission solutions. The coordination of the components that come together in the drive train is particularly important for this. A key role is played by the lubricants used. The increasing scarcity and rising prices of raw materials, in particular the base oils for lubricants, is currently leading to a situation in which the end customer is increasingly coming to regard the lubricant as a high-grade, value-retaining machine element. Designers and project engineers, too, are increasingly focusing on the

Table 10.13 Minimum requirements – industrial gear oils, type CLP/CKC – according to DIN 51517, part 3 – extract – dated 2011.

	Required typical parameters of CLP industrial gear oils according to DIN 51 517, part 3 (2011) – extract										
Property	**Unit**		**Minimum requirements**								**Method**
Type of lubricant according to DIN 51502/ISO 6743-6 and ISO 12925-1			CLP 100/ CKC 100	CLP 150/ CKC 150	CLP 220/ CKC 220	CLP 320/ CKC 320	CLP 460/ CKC 460	CLP 680/ CKC 680	CLP 1000/ CKC 1000	CLP 1500/ CKC 1500	
Kinematic viscosity at 40 °C	$mm^2 s^{-1}$	(min.–max.)	90–110	135–165	198–242	288–352	414–506	612–748	900–1100	1350–1650	DIN EN ISO 3104
Viscosity index (VI)		(min.)	90	90	90	90	90	85	85	85	DIN ISO 2909
Flashpoint (COC)	°C	(min.)	200								DIN EN ISO 2592
Pour point	°C	(max.)	−12	−9	−9	−9	−9	−3	−3	−3	DIN ISO 3016
Water content	%	(max.)	< 0.1								DIN 51777-2
Foaming behaviour (immediately/after 10 min.) Seq. I, II, III	ml	(max.)	100/10	100/10	100/10	100/10	150/60	150/60	150/60	150/60	ISO 6247
Demulsibility	min	(max.)	30	30	30	30	45	60	60	60	DIN ISO 6614
Copper corrosion, 3 h/100 °C degree of corrosion		(max.)	1								DIN EN ISO 2160
Steel corrosion, method A		(min.)	pass								DIN ISO 7120
Ageing properties, 312 h/95 °C increase of viscosity at 100 °C	%	(max.)	+6 (increase of precipitation number max. 0.1 ml)								DIN EN ISO 4263-4

(continued)

Table 10.13 (Continued)

	Required typical parameters of CLP industrial gear oils according to DIN 51 517, part 3 (2011) – extract			
Property	Unit		Minimum requirements	Method
FZG mechanical gear test rig A/8.3/90	failure load stage	(min.)	12	DIN ISO 14635-1
FE8 dynamic-mechanical roller bearing lubricant test rig D 7.5/80-80 rolling element wear	mg	(max.)	30	DIN 51819-3
cage wear	mg	—	to be reported	
Behaviour towards the SRE-NBR 28/SX sealant after 7 d ± 2 h @ (100 ± 1) °C				
Relative change in volume	%	(max.)	0/+10	DIN ISO 1817
Change in shore A hardness	%	(max.)	−10/+5	
Reduction of tensile strength	%	(max.)	30	
Reduction of elongation	%	(max.)	40	

lubricant as a design element. Lubricants are also an important issue in a large number of research works carried out, for example in the scope of the German machinery manufacturer association (VDMA[a]) and the German drive train research association (FVA[b]). Industrial gear oils – particularly synthetic oils – nowadays have a significant market share in the lubricants market.

Depending on the application, gear oils can be split into two main groups:

- *Lubricating oils in circulating industrial systems* (stationary gear oils as defined in DIN 51517 [2], ISO 6743-6 [3] and ISO 12925 [4]).
- *Lubricating and gear oils for automotive applications* (mobile gear oils, car and truck gear oils. Automatic transmission fluids as defined by API GL4, GL5, etc. [5]).

Optimization and further development nearly always lead to an increase in component performance. In the future, more and more work will be performed in an increasingly short time.[1] Components and gearboxes will get smaller and more compact. From the very beginning, design and component selection will focus on costs and cost-effective engineering. The gear oil as one of the most important and complex machine element will have to keep pace with these changing application conditions and criteria. As oil volumes continue to get smaller, oil circulation more rapid and more energy is transferred to the oil, the thermal and oxidative loads will increase. As a rule of thumb, every 10 °C increase in temperature causes the reactive speed, that is, ageing speed, to double and this shortens the life of the lubricant (by up to half in threshold zones). Compared to mineral oils, synthetic oils together with optimized additives have enormous potential to reduce friction and thus lower sump temperatures (by up to 30 °C). This ultimately can lead to a significant increase in the life of the lubricant.

a) *VDMA*: The VDMA (Verband Deutscher Maschinen- und Anlagenbau e.V., German Machinery Association) represents over 3000 companies and is one of the most influential trade associations in Europe.
b) *FVA*: The FVA is the leading innovation network of electrical, mechanical and mechatronic drive technology. In cooperation between economy and science the foundations for tomorrow's success will be created.

10.3.2 Industrial Lubricants: Industrial Gear Oils – Statistics

10.3.2.1 Market Shares/Market Situation

In 2015, approximately 30 000 tons of industrial gear oils were consumed in Germany. These oils are primarily used in stationary units and gears. The biggest part of the demand for industrial gear oils (around 70–80%) continues to be met

1) Broschüre des Forschungsfonds der Fachgemeinschaft Fluidtechnik im VDMA, 60498 Frankfurt, Delphi-Studie.

Based on mineral oil, CLP-M:	ca. 21.000 to	ca. 70 %
Based on synthetic oils:	ca. 9.000 to	ca. 30 %
→	Polyglycols (CLP-PG)	ca. 6 %
→	Polyalphaolefins (CLP-PAO)	ca. 20 %
→	Synthetic esters (rapidly biodegradable), CLP-E	ca. 4 %
→	Others (e.g. USDA-H1)	ca. <1%

Tendency in Germany + Europe : 'Pro Synthetic Lubricants'
(High technical performance and for extreme operating conditions)

Figure 10.12 Market shares of synthetic industrial gear oils.

by mineral oil-based gear oils. Frequently, however, the performance of CLP standard industrial gear oils, based on mineral oil (in accordance with DIN 51517, part 3) is no longer sufficient to meet the more demanding requirements of customers and gear manufacturers. Even highly developed, mineral oil-based special gear oils come up against their limits and are no longer able to meet the required lubrication tasks. This automatically leads to the use of synthetic gear oils, which have higher performance characteristics in comparison with mineral oil-based products. Around 10–15% of synthetic gear oils are based on polyalphaolefins, around 6% on polyglycols and around 4% on synthetic esters. Most of all newly designed gear sets and applications with high energy density are equipped with synthetic gear oils to meet the requirements of these more and more compact and lower weight units with increasing power density. The enormous growth in the use of wind power installations is a particular area that has resulted in a boom in new developments and research activities in the field of power transmission engineering. The trend to ever larger turbines (5 MW installations in offshore projects) is encouraging plant engineers to specify increasingly compact and lower-weight units with increasing efficiency, power transmission and reliability, and a simultaneous reduction of maintenance intervals. The only way of meeting these requirements in the future will be through the use of synthetic oils (Figure 10.12).

10.3.3
Composition of Industrial Gear Oils

As a rule, industrial gear oils consist of a base fluid, normally known as a base oil, and chemical substances known as additives. The quality and performance of a gear oil depends on the quality of the base fluid and the combination of additives or additive systems. The classification and performance of a gear oil is ultimately decided by the type of base fluid and additives used along with technical and ecological aspects.

Industrial gear oils mostly comprise up to 85–95% of base oils of different viscosities. Depending on the type of gear oil, additives account for between 5 and 15% (special oils contain significantly more). Gear oils are primarily produced from paraffin-based mineral oils, synthetic polyalkylene glycols, synthetic esters and synthetic polyalphaolefins.

10.3.3.1 Mineral Oils (CLP-M)

Mineral Oil-Based Gear Oils

In the case of mineral oils, the crude oil is passed through various distilling and refining processes to extract the base oils required for the production of lubricants. The composition of mineral oils is always a mixture of different types of hydrocarbons. In the case of industrial gear oils, by far the biggest part of the hydrocarbons is generally in a paraffinic bond. These have good solubility and good response to additives, and have been developed by decades of experience to be harmless to seals, coating materials and other standard materials used in gear construction. The different viscosities are adjusted by mixing solvent-neutral (SN) grades with high-viscosity brightstock grades and/or cylinder oils. The provenance (origin) of the base oils plays a key role in their classification.

Mineral oils are mixtures of different hydrocarbon compounds. Depending on the number of paraffinic (XP), aromatic (XA) and naphthenic (XN) compounds, oils are divided up between two main naphthenic and paraffinic groups.

If the majority of carbon atoms are in naphthenic compounds, such base oils are referred to as naphthenic.

The largest percentage of industrial gear base oils are paraffinic. They offer good viscosity–temperature behaviour (viscosity index is >90). Additives generally performing well, are also easily soluble and have a good additive response in paraffinic base oils. Such mineral oils are also compatible with and generate no problems for seals and machine finishes.

Mineral oil-based gear oils have been used as standard lubricants for decades. As a result of their chemical composition, mineral oil-based gear oils should only be used up to sump temperatures of 85–90 °C. Short-term peak temperatures of around 100 °C and over are tolerable but can lower the life of the lubricant. Constant high temperatures cause premature oil ageing. Exceeding threshold temperatures chemically changes the oil molecules by oxidation and thermal decomposition (ageing) and this condition is signalled by odours, deposits, residues and by the formation of acidic reaction products in the system. The lower permissible temperature is defined by the pour point or the low-temperature flow characteristics of the gear lubricant.

10.3.3.2 Synthetic Hydrocarbons: Polyalphaolefins (CLP-PAO)

Polyalphaolefins (PAO): Synthetic Base Oils

Currently, the most important group of synthetic lubricants is the group of synthetic hydrocarbons, particularly the polyalphaolefins (PAO). These are paraffin-like base oils, obtained by polymerization of olefins. Pre-definable chain lengths (the degree of polymerization) allow the production of differing viscosities, a high natural viscosity index and a low pour point. The four most important base fluids in polyalphaolefins are PAO 5 (V40 ≈ 32 $mm^2 s^{-1}$), PAO 8 (V40 ≈ 46 $mm^2 s^{-1}$), PAO 40 (V40 ≈ 400 $mm^2 s^{-1}$) and PAO 100 (V40 ≈ 1200 $mm^2 s^{-1}$). The appropriate ISO viscosity classes are obtained by mixing these base components of differing viscosities. Polyalphaolefins do not contain any sulphur or nitrogen compounds. The high natural viscosity index (VI >135), which is also shear stable over the entire life, makes the viscosity of these oils much less temperature-dependent than is the case with mineral oils (VI approx. 100). The essential advantages of using PAO-based base oils in a lubricant in conjunction with suitable additives are the high oxidation and temperature stability and the extremely good low-temperature properties. Even at high temperatures, polyalphaolefins exhibit lower oxidation and ageing properties as a result of the defined structure of the polyalphaolefins and the saturated compounds. Studies carried out using special oxidation tests (RPVOT, ASTM D 2272 [6]) have shown the life of polyalphaolefins to be up to four times that of mineral oil-based gear oils.

Polyalphaolefins, or so-called synthetic hydrocarbons are petrochemical products and are miscible with mineral oils in all ratios. They are noteworthy for their excellent viscosity–temperature behaviour. The viscosity index of finished PAOs is about 140–160. PAOs have a high flashpoint and an extraordinarily low pour point.

As a rule, gear oils based on polyalphaolefins contain about 2–10% (or a higher content of) ester oils. This ester oil content (which causes elastomers to swell) compensates for the shrinking effect of PAO molecules. As a result, PAO-based gear oils are neutral to elastomers and are comparable to mineral oils. Their compatibility with paints and plastics is comparable to that of mineral oils and thus unproblematic.

Polyalphaolefins are used in heavily stressed industrial gearboxes or in gearboxes in the foodstuff industry. They can be used at high temperatures (comparable with polyglycols) and in high oxidation conditions and offer significantly longer life than mineral oils. Generally speaking, residual mineral oil contents of 10–20% can be tolerated when changing-over to polyalphaolefins.

10.3.3.3 Polyglycols (CLP-PG)

Polyalkylene Glycols (PAG): Synthetic Base Oils

Polyalkylene glycols are polymers of ethylene oxide (EO) or propylene oxide (PO) or mixtures of these components. Depending on the mixing ratio of EO: PO, key properties are produced, such as water solubility. Water-soluble polyglycols are mostly made up on an EO basis or EO/PO basis. Non-water-soluble polyglycols are obtained by PO polymerization.

- *Polyglycols EO:PO 1:0, PEG, polyethylene glycol*: Completely soluble in water, highly polar, very low coefficient of friction, highly surface active.
- *Polyglycols EO:PO 1:1, mixed polyglycol*: Soluble in water and alcohols, highly polar, low coefficient of friction.
- *Polyglycols EO:PO 0:1, PPG, propylene glycol*: Insoluble in water, limited solubility in hydrocarbons, average polarity, low coefficient of friction.

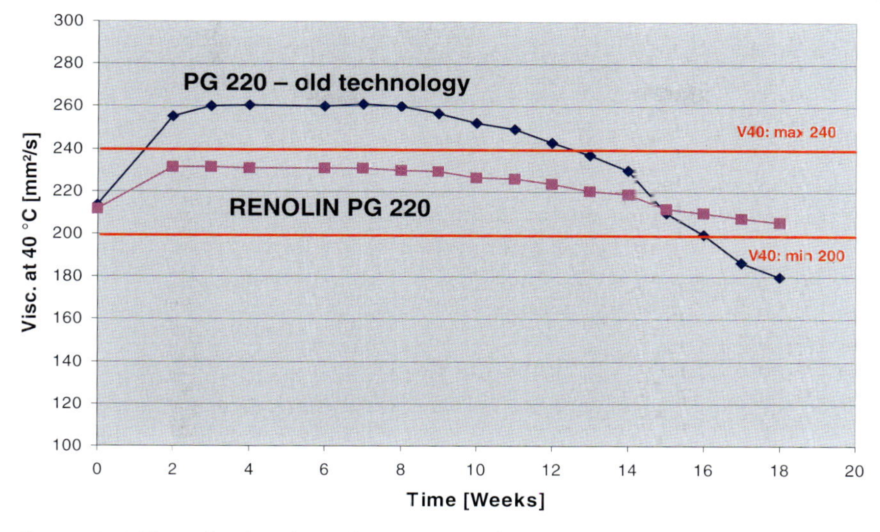

Figure 10.13 Thermal and oxidative decomposition of PAG gear oils.

Compared to mineral oils, the viscosity–temperature graphs of polyglycols are much flatter. The natural, shear-stable viscosity index of polyglycols is between 180 and 250. This means that polyglycol-based lubricants have very good viscosity–temperature behaviour. As a result of their oxygen compounds, polyglycols are highly polar. Due to their high polarity, polyglycol-based lubricants have a high load bearing capacity and outstanding lubricity.

Polyglycol-based lubricants have extremely low friction coefficients, particularly in applications with high levels of sliding friction in steel/bronze contacts. For this reason, they are particularly used in worm gears or in unfavourable friction conditions, and at very high temperatures. Such oils are recommended for extreme applications, for example embossing rollers, paper-making machine oils.

There are generally no solid deposits and residues produced in the thermal decomposition of polyglycols. The decomposition products of polyglycols are mostly low-molecular cleavage breakdown products, which are volatile only at high temperatures (Figure 10.13).

A big disadvantage of polyglycols is their restricted compatibility with paints, elastomers and aluminium alloys. Polyglycol-based gear oils can only be used with certain materials and components. This not withstanding, compatibility should be tested before every application. Another disadvantage of polyglycols is their incompatibility with mineral oils. Systems, which were previously running on mineral oils or hydrocarbons must be thoroughly cleaned and flushed. If polyglycols absorb too much water during use, material compatibility, corrosion and metal problems may occur.

Some polyglycols can be used continuously at temperatures between −20 and 100 °C and even up to 110–120 °C for short periods. PAG industrial gear oils are remarkable for their polar character, low coefficient of friction, good adhesion and large load-bearing capacity.

10.3.3.4 **Synthetic Esters (CLP-E)**

Synthetic esters include a broad range of products with large structural variety. Esters are produced by the chemical modification of alcohols and acids. Both alcohols and acids can be combined to effectively manipulate the characteristics of the product. The raw materials for synthetic esters originate from petrochemicals as well as from natural substances and their chemically modified products. Because of the great diversity of raw materials used and the different control of the production process, the group of materials that can be obtained is extensive and has great structural variety. The most widespread synthetic esters are the dicarboxylic acid esters, polyol esters and complex esters.

Esters have a good viscosity–temperature behaviour (viscosity index 160–180), extremely good fluidity at low temperatures, with a low pour point and a low evaporation tendency. The very high polarity of synthetic esters is shown in very good wear protection and load carrying capacity of the base oil without additives. They therefore achieve outstandingly low friction values and show an excellent cleaning effect. In view of the risk of hydrolysis (reverse reaction of the esterification), moisture/water presents the greatest risk for synthetic ester-based lubricants. These disadvantages can be offset in final-formulated gear oils by appropriate selection of high-grade, saturated esters and special additives. Esters can be mixed with and are compatible with mineral oils and polyalphaolefins, and they are rapidly biodegradable. Due to their environmental compatibility, synthetic ester-based lubricants are frequently used in environmentally sensitive areas, for example in water engineering and/or in wind power plants.

They also tend to swell elastomers and strip paint finishes. As a result, ester-compatible paint, elastomer and plastics must be selected when these products are used. Esters offer excellent lubricity, particularly in boundary friction conditions. Synthetic esters are nearly always used as base fluids for environmentally harmless, rapidly biodegradable industrial gear oils that are primarily used in ecologically sensitive areas such as in the water industry or in wind energy applications.

The influence on the coefficient of friction of gearbox brake-rings and synchromesh mechanisms must not be ignored.

Figures 10.14–10.17 illustrate the chemical structure and most important characteristics of the different base fluids used in industrial gear oils [7,8].

10.3.3.5 **Gear Oil Additives**

Additives are chemical compounds which are added to base fluids with the intention of supplementing, improving or giving some special characteristics to the base oils. In most cases, a number of additives or additive packages are used to achieve the desired effect. Additive components can either complement each

o **Mineral Oil**

H_3C CH_3 CH_3

- 'conventional base oils' (VI ca. 100)
- universally applicable
- good additive response, good additive solubility
- good compatibility with materials

Figure 10.14 Base fluids for industrial gear oils: mineral oil.

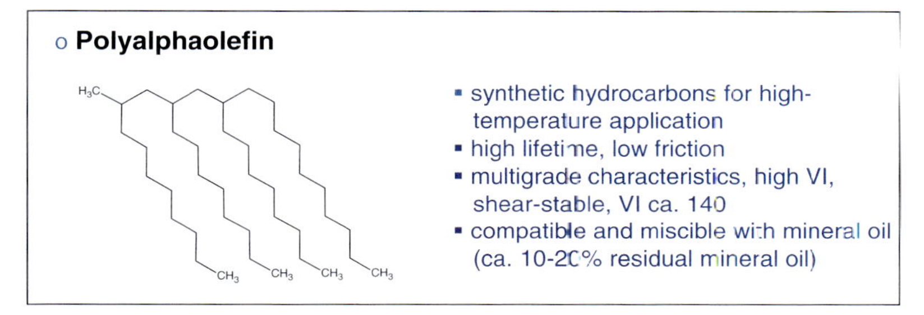

Figure 10.15 Base fluids for industrial gear oils: polyalphaolefin.

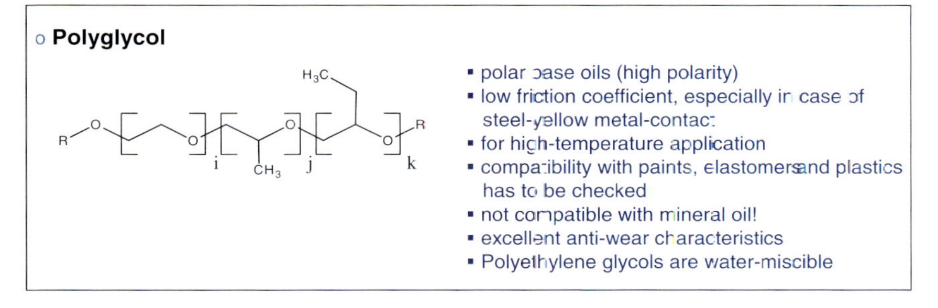

Figure 10.16 Base fluids for industrial gear oils: polyglycol.

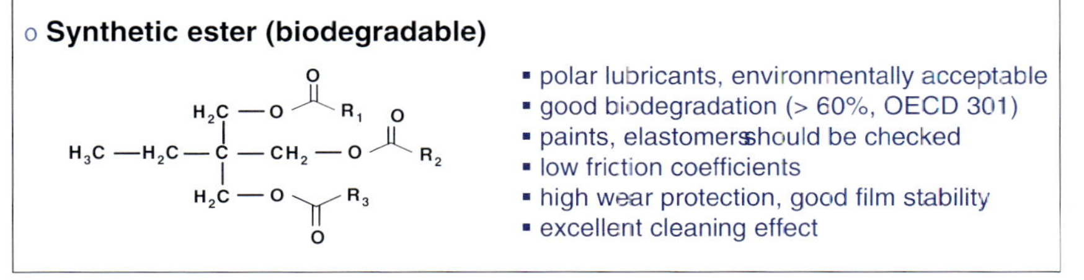

Figure 10.17 Base fluids for industrial gear oils: synthetic ester (biodegradable).

other or be mutually destructive. The following characteristics can be improved with additives: ageing behaviour, oxidation protection, corrosion protection, wear protection, EP capacity, viscosity–temperature behaviour, foaming, detergent capacity, water release, coefficient of friction and much more. Top-performance lubricants are developed by achieving a good balance of synergistically acting additives in optimum ratios.

The most important additives for general lubricating oils and gear oils are as follows:

- Surface-active additives such as corrosion inhibitors, metal deactivators, wear inhibitors friction modifiers, detergents, dispersants, extreme-pressure additives.
- Additives that influence the base oil such as anti-oxidants, defoamers, VI improvers and pour point depressants.

By differentiating between zinc-free and ash-free wear protection additive systems (phosphorous–sulphur compounds) and those containing zinc and ash (zinc dialkyldithiophosphates etc.), a rough classification of the additive systems used in general and gear oils can be made. However, mixtures of these two additive groups are also used.

Mainly phosphorus sulphur and/or phosphorus-based additives are used in industrial gear oil formulations.

10.3.4 Classification of Gear Oils

10.3.4.1 Requirements and Specifications

DIN 51502 together with DIN 51517 (ISO 6743-6) classifies the following general lubricating and gear oils: C, CL, CLP, CLPF, CLP-HC, CLP-PG and CLP-E lubricating oils.

The letter L indicates the presence of substances that increase corrosion protection and ageing resistance. The letter P indicates the presence of substances that reduce wear in boundary friction conditions. Lubricants containing graphite or molybdenum disulphide are indicated with the letter F. The suffix HC indicates lubricants based on synthetic components (polyalphaolefins – PAO), polyglycols bear the suffix PG and ester oils the suffix E.

Parallel, the ISO standard ISO 12925-1 and the new AGMA 9005/E02 are the most important international standards for industrial gear oils.

For application in wind turbine gear sets, a new specification – IEC/ISO 61400-4 – was published as draft version.

10.3.4.2 Requirements for Industrial Gear Oils According to DIN 51517, Part 3

The requirements for extreme-pressure industrial gear oils are laid down in DIN 51517, part 3. In these requirements, especially the wear protection with regard to scuffing (FZG A/8,3/90: failure load stage min. 12) and bearing wear protection (FE8 D-7,5/80 kN/80 °C/80 h) are fixed. These requirements define the minimum standards for industrial gear oils that contain extreme-pressure/anti-wear additives, including anti-oxidants and corrosion protection properties.

The requirements placed on lubricants in modern gear units are continuously increasing. The revisions of DIN 51517 in the last years (ISO 12925 – under discussion) take account of these new requirements [2,4], particularly those of roller bearing manufacturers. The introduction of the mechanical-dynamic test on the FE8 roller bearing lubricant tester into DIN 51517, part 3, in the year 2004 underlines these increasing requirements.

Table 10.13 shows the typical properties of CLP industrial gear oils required in accordance with DIN 51517, part 3 – extract from the essential relevant viscosity classes. DIN 51517 has last been revised in 2011. The specification and test standards have been adapted to the latest international standard. At the same time, relevant requirement criteria have been more tightly defined. It should be emphasized here that the specified water content of <1000 ppm applies to mineral oil-based oils (not hygroscopic oils). The ageing behaviour is tested in the TOST test for gear oils (95 °C, 312 h). The cage wear in the FE8 test has to be reported. The limits with regard to the behaviour of gear oils in relation to seal materials have been revised.

DIN 51517 actually presents only the minimum requirements for the industrial gear oils to be used. Efforts are currently being made to harmonize ISO 12925 and DIN 51517. The aim is to create the basis for a uniform standard of requirements. Figure 10.18 shows the general classification of gear oils in accordance with DIN, ISO and AGMA. Gear unit manufacturers and OEMs place much higher requirements on qualified industrial gear oils.

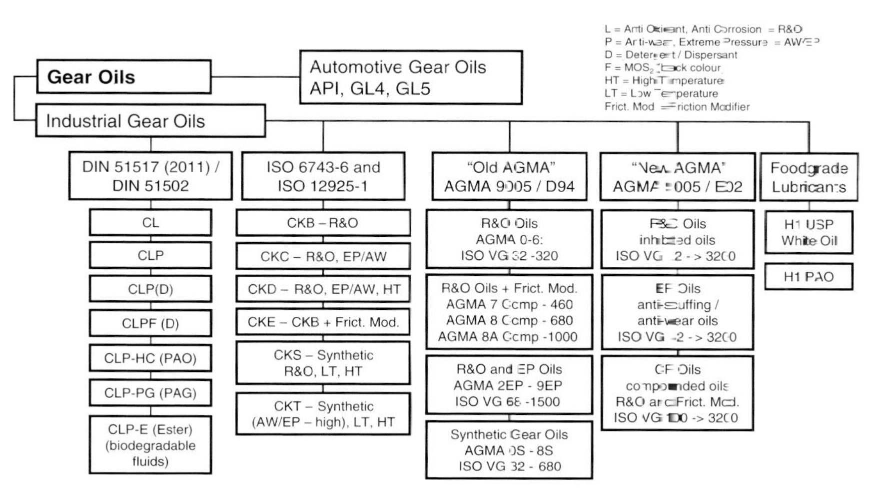

Figure 10.18 General classification of gear oils.

10.3.5 Temperature Ranges and Lifetime of Industrial Gear Oils

During use, lubricants are subject to thermal and oxidative stresses. Figure 10.19 shows the temperature ranges recommended by a leading manufacturer of industrial gearboxes. It also shows an overview of the expected lifetime of different industrial gear oils at a mean temperature of 80 °C. Further information concerning the application temperatures of gear lubricants is contained in the GFT (German Association of Tribology) sheet no. 5. This also details the temperature thresholds for various types of lubricants [9–12].[2)]

Compared to mineral oil, synthetic lubricants can be used at about 10–20 °C lower temperatures and can withstand about 10–20 °C higher temperatures (at double the life of the lubricant). This can be demonstrated in the FZG test procedure FZG A/8.3/90 [13], which shows the temperature reduction by using synthetic gear oils (Figure 10.20, Table 10.14). This is of particular interest to applications in extreme conditions (very low and very high ambient temperatures).

Synthetic industrial gear oils based on polyalphaolefin and on polyalkylene glycol have better viscosity–temperature characteristics compared with mineral oil. This means that these types of products have a higher viscosity index, the viscosity does not much depend on temperature compared with mineral oil.

10.3.5.1 Lifetime of Gear Oils: Oxidation Stability (RPVOT Test)

Thermal and oxidative stresses cause lubricants to age and change. Of course, even altered and aged lubricants have to provide the necessary friction and

According to:	Flender	GFT* Sheet No. 5
Mineral oils, CLP-M:	-10°C to 90°C (short periods: 100°C)	"-15°C to 130°C"
Polyglycol, CLP-PG:	-20°C to 100°C (short periods: 110°C)	"-30°C to 180°C"
Polyalphaolefins, CLP-PAO:	-20°C to 100°C (short periods: 110°C)	"-45°C to 150°C"
Synth. Ester („Bio"), CLP-E:	-15°C to 90°C	"-30°C to 180°C"

* GFT = German Association of Tribology: General Temperature Range

Lifetime at an average oil temperature of 80°C (according to Flender)

Oil	Lifetime
Mineral oils, CLP-M:	2 years or 10.000 h
Synth. Ester („Bio"), CLP-E:	2 years or 10.000 h
Polyglycol, CLP-PG:	4 years or 20.000 h
Polyalphaolefins, CLP-PAO:	4 years or 20.000 h

Contaminations (e.g. water, dirt, abrasion / wear products, acid gases, etc.) have a negative influence on the lifetime.

Figure 10.19 Temperature ranges and lifetime of industrial gear oils.

2) GfT-Arbeitsblatt 5, Zahnradschmierung, Gesellschaft für Tribologie, D-47443 Moers.

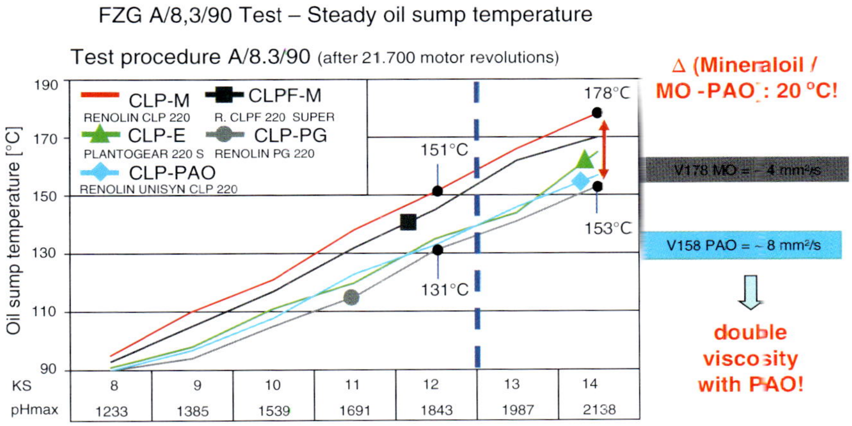

Figure 10.20 Advantages of PAO-based synthetic gear oils.

wear-protection properties, wetting properties and other lubricant properties associated with gear lubricants. However, many conventional, mineral base oils are often unable to satisfy extreme applications. In order to satisfy higher thermal and oxidative demands, the commonly-used paraffinic solvent cuts sometimes have to be replaced by synthetic base oils such as polyglycols, polyalphaolefins and esters. Optimum lubricity can, of course, only be achieved in conjunction with selected additives. The expected life of industrial gear oils,

Table 10.14 Load stages in the FZG test.

Load stage no.	Tooth normal force F_n, N	Hertzian contact pressure at contact point P_c pH-A, N mm^{-2}
1	99	146
2	407	295
3	1044	474
4	1800	621
5	2786	773
6	4007	927
7	5435	1080
8	7080	1232
9	8949	1386
10	11029	1538
11	13342	1691
12	15826	1841
13	18365	1987
14	21272	2138

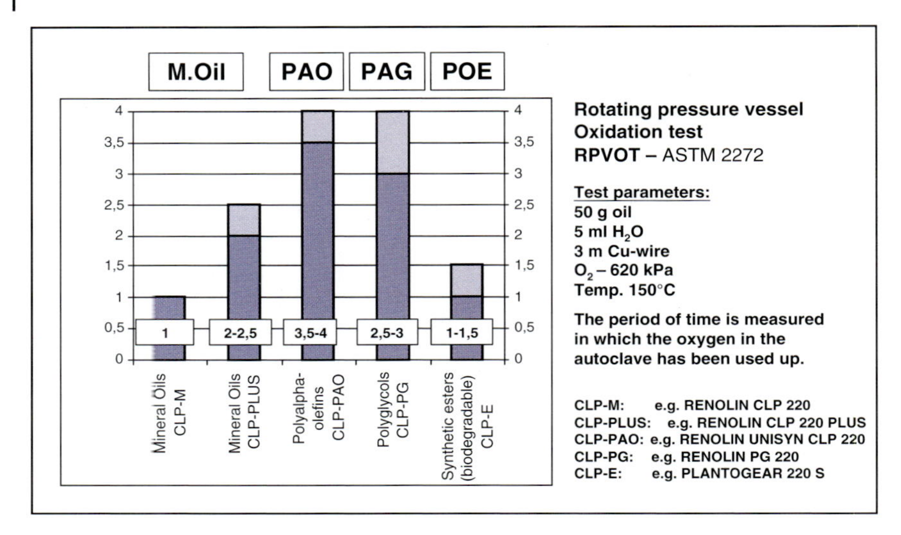

Figure 10.21 RPVOT test results of different industrial gear oils.

with particular respect to oxidation stability, can be examined by comparative rotating pressure vessel oxidation tests. This test measures the time in which the oxygen in a gear oil is used-up in a pressurized autoclave. Long life (minutes ∧ lifetime factor) is equivalent to good oxidation stability [14]. Comparative tests have shown that synthetic industrial gear oils offer much greater oxidation stability than mineral oil-based products (Figure 10.21).

10.3.6 Cost-Benefit Ratio of Industrial Gear Oils

Compared to standard mineral oils, the price of synthetic base fluids is about 3–5 times higher. The longer lifetime of synthetic base oil compensates for the higher product price by a factor of 2 as well as reducing sump temperatures by 10–30 °C and reducing servicing requirements. All these aspects must be considered when the overall cost-benefit of a product is being evaluated. The higher cost of synthetic gear oils must therefore be set against the savings generated by longer life (compared to mineral oils) and the reduction in servicing costs. If all parameters are considered, the use of synthetic base fluids can significantly reduce total lubricant-related system costs.

10.3.7 Filtration Behaviour: Electrical Conductivity of Gear and Lubricating Oils

The reduction of particles, contaminants, dirt, dust and water, is of great interest in power transmissions. The lifetime of bearings is influenced by the purity. Therefore, more and more fine-meshed filters are used in circulating systems.

Table 10.15 Electrical conductivity of industrial gear oils.

Fine mesh filtration: electrostatic phenomena occur with high viscosity, special filter material and low conductivity of lubricants!

			Electrical conductivity (pS m^{-1})		
Product	**Base oil type**	**ISO VG**	**Test date**	**23 °C**	**50 °C**
Renolin CLP 68	MO	68	Jan 10	11	88
Renolin CLP 220	MO	220	Jan 10	6	39
Renolin CLP 68 PLUS	MO	68	Jan 10	123	860
Renolin CLP 220 PLUS	MO	220	Jan 10	47	290
Renolin highgear 220	MO	220	Jan 10	1509	9300
Renolin Unisyn CLP 68	PAO	68	Jan 10	76	296
Renolin Unisyn CLP 220	PAO	220	Jan 10	24	154
Renolin PG 32	PAG	32	Jan 10	400 000	914 000
Renolin PG 46	PAG	46	Jan 10	370 000	695 000
Renolin PG 100	PAG	100	Jan 10	200 000	480 000
Renolin PG 220	PAG	220	Jan 10	195 000	400 000
Plantohyd 68 S	Ester	68	Jan 10	123	465
Plantogear 320 S	Ester	320	Jan 10	48	267

If a fluid with a high viscosity and a low electrical conductivity is passing these specific filter materials, sometimes electrostatic phenomena can be seen. These phenomena depend on the filter/filter material and on the electrical conductivity, the temperature and the viscosity of the gear oil. Hydrocarbon-based products like mineral oil and polyalphaolefin show low/medium electrical conductivity whereas products with higher polarity and a higher water content have a better electrical conductivity. Additives can influence the electrical conductivity (Table 10.15).

10.3.8
Oil and Water: Saturation Values of Dissolved Water in Oil

According to the minimum requirements for industrial gear oils, the water content should be lower than 0.1%, which is lower than 1000 ppm for mineral oils and polyalphaolefins. In general, the more polar polyalkylene glycols can have 2000–2500 ppm water dissolved in the oil. In general, gear and lubricating oils should be used with a low water content. A high water content, especially free water reduces the lifetime expectations of bearings, will create corrosion problems and other undesired phenomena. Therefore, it is of great interest to know about the saturation level of gear and lubricating oils depending on temperature, viscosity and type of gear oil. The saturation values of dissolved water in oil

Table 10.16 Saturation level of water/humidity (industrial gear oils, ISO VG 220).

	Saturation level (mg kg^{-1})				
	25 °C	**40 °C**	**60 °C**	**80 °C**	
Renolin CLP 220	—	—	420	640	Mineral Oil
Renolin Unisyn CLP 220	—	—	370	550	Polyalphaolefin -PAO
Renolin PG 220	—	32 700	30 600	32 300	Polyglycol -PAG
Plantogear 220 S	—	2100	2800	3800	Ester
Renolin highgear 220	—	—	820	930	Mineral oil, Mo-containing

(no free water) of different types of CLP industrial gear oils can be seen in Table 10.16 and Figure 10.22.

It can be seen that the saturation level is depending on temperature (the higher the temperature, the more water can be dissolved in the oil). Mineral oil-based products and polyalphaolefins have a saturation level between 550 and 700 ppm water at 80 °C. Polyglycols of high polarity can absorb water up to a level of 32 000–33 000 ppm water at 80 °C. The saturation level of polar, ester-based products is around 4000 ppm water at 80 °C (Figure 10.22). According to our experience, the water content can only be measured according to the Karl Fischer method, DIN 51777, part 2 (indirect method).

The influence of additives on the water content is depending on the type of additive. There are additives on the market that contain dissolved crystallized water linked directly to the additive molecules, which increase the water content.

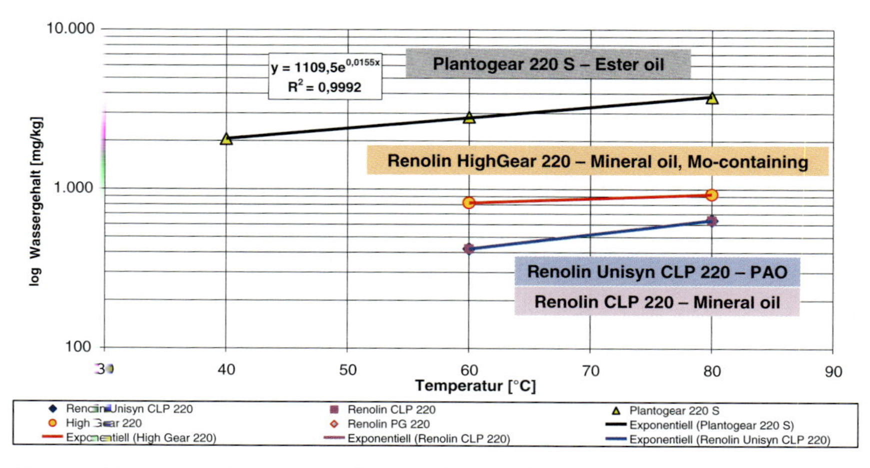

Figure 10.22 Saturation level of water/humidity (industrial gear oils, ISO VG 220).

10.3.9 Special Industrial Gear oil Formulations

10.3.9.1 Special Corrosion Protection Gear Oils

Often gear sets and components are stored over a long period of time. Due to high humidity, condensation effects, corrosion problems often occur in gear sets, bearings and components on the inner surface layers. Remaining oil films are washed off, condensation water can penetrate the inner gear and corrosion can occur, which pre-damages gear sets and bearings. This can be avoided by using so-called VCI industrial gear oils which guarantee performance according to DIN 51517, part 3, CLP in combination with excellent corrosion protection in the oil- and vapour phase for steel and yellow metal materials (VCI = vapour corrosion inhibitors).

10.3.9.2 Detergent/Dispersant Types of Gear Oils

Due to oxidation effects, thermal stress, prolonged service intervals, lubricants tend to build sludge, oxidation products and deposits. These oxidation products can block lubricant supply pipes. Because of this the lubricant supply to bearings can be interrupted which can lead to damage and break-down of gear sets. Detergent gear oil technology can be described as 'Clean Gear technology'. Detergent/dispersant components in the gear oil lead to a high sludge carrying capacity, avoid the formation of sludge and deposits.

Together with a high anti-oxidant level, the inner lubricant supply in gear sets, to the gears and the bearings, is guaranteed.

Also under the influence of a high amount of free water, protection against scuffing and scoring is guaranteed. 'Clean Gear technology' with 4.8% free water has been tested in the FZG test. Without cooling the oil in the test, a failure load stage >12 in the FZG A/8,3/90 test can be reached.

10.3.9.3 'Plastic Deformation' Additive Technology

Pre-damages on the surface of gear sets can cause big problems, especially before reaching the service intervals of the units. By using 'plastic deformation' (PD) additives in special types of gear oils, the pre-damaged surface can be smoothed, the specific load in the contact zone can be reduced, and therefore the service intervals can be reached. The PD additive technology is based on liquid molybdenum and special EP/AW additives.

10.3.9.4 'Reducing Friction' by Special Industrial Gear Oils

Synthetic gear oils have great potential regarding the lowering of internal and external friction and thus can lower overall sump temperatures. Different types of industrial gear oils have an influence on frictional torque and persistent sump temperatures. Lubricant test results including those from the 'newer' tests provide vital information about the oil-related reduction in friction in gearboxes and roller bearings.

A breakthrough in the development of low-friction gear oils was the establishing of the FZG efficiency test based on a modified FZG gear test rig. Results of the tests performed show that careful oil selection can reduce sump temperatures in test conditions by up to 30 °C. The transferability of these test results to practical day-to-day conditions can be demonstrated.

The lowering of sump temperatures significantly cuts the thermal and oxidation stresses on a gear oil. At the same time, viscosity remains stable and the strength of the oil film is not compromised. Oil change intervals can be extended while operating costs are correspondingly lower. In real-life applications, it is thus worthwhile to conduct a cost-benefit analysis while considering the excellent tribological characteristics of type CLP-PAO, CLP-E and CLP-PG synthetic industrial gear oils.

10.3.10
Gear Oils for Wind Turbines: Demands and Characteristics

As a rule, specially approved industrial gear oils are used for the main gears (ISO VG 320) – mostly multistage planetary spur gears, and for the azimuth gears (ISO VG 220). Viscosity selection is based on requirements of the gear tooth meshing (slow on the rotor side and fast on the generator side) and the requirements of the corresponding bearings. ISO VG 320 gear oils thus represent a compromise between the specific 'needs' of the planetary gears, the spur gears and the roller bearings. In addition, such gear oils have to satisfy special demands regarding compatibility with seals, paints, filters, corrosion preventives, running-in oils, and so on.

10.3.10.1 Demands on Wind Turbine Gear Oils

Of all the demands on gear oils, the Siemens Flender specifications are viewed as fundamental requirements and are among the most important together with DIN 51517, part 3 (Table 10.17). Prominent among the physical/chemical demands are compatibility with internal gearbox paints, static elastomer tests (NBR – 100 °C, 1008 h, FKM – 110 °C, 1008 h) and the Freudenberg dynamic elastomer test. Experience has shown that compatibility with gearbox paints under specific test conditions is not an unimportant factor. Paint compatibility is affected by the base oil, as well as by the additives and their components. As regards seals, FKM materials are relatively uncritical. NBR compatibility is a problem so far, as the corresponding tests are carried-out at very high temperatures for 1008 h (a drop in test temperature to 95 °C is planned). The dynamic tests are performed in line with Freudenberg in-house procedures. At present, a working group is examining the demands, test temperatures and test durations and a change to the current demands and test parameters is currently under discussion. Apart from Flender and Freudenberg, a number of leading lubricant and additive manufacturers are represented in this working group to discuss and perhaps practically modify the present specifications.

Table 10.17 Siemens Flender specifications for industrial gear oils (revision 13) – excerpt.

Property	Method	Unit	Requirement
Internal coating compatibility	Mäder		
P22-8050 anthracene brown			Pass
Nuvopur aqua primer 510.1.6.1400			Pass
Liquid sealing compatibility	Henkel		
Loctite 128068			Pass
Static elastomer 72 NBR902 (100 °C, 1008 h)	ISO 1817		
Δ Hardness shore A		pts	−5 to +5 (+7)
Δ Volume		%	−2 to +5 (+6)
Δ Tensile strength		%	−50 to +20
Δ Rupture elongation		%	(−65) −60 to +20
Pass/Fail comment			Pass
Static elastomer 75 FKM585 (110 °C, 1008 h)	ISO 1817		
Δ Hardness shore A		pts	−5 to +5 (+7)
Δ Volume		%	−2 to +5 (+6)
Δ Tensile strength		%	−50 to −20
Δ Rupture elongation		%	(−65) −60 to +20
Pass/Fail comment			Pass
Dynamic elastomer 72 NBR902 (80 °C, 1000 h)	DIN 3761		
Leakage	Freudenberg	ml	0
Sealing time		h	1008
Wear band width on sealing edge		mm	≤0.5
Depth of shaft wear		μm	≤5
Radial load with spring		%	+10 to −45
Interference with spring		mm	≤0.6
Interference without spring		mm	≤0.7
Visual assessment of sealing edge			
Dynamic elastomer 75 FKM 585 (90 °C, 1000 h)	DIN 3761		
Leakage	Freudenberg	ml	0
Sealing time		h	1008
Wear band width on sealing edge		mm	≤0.4
Depth of shaft wear		μm	≤10
Radial load with spring		%	+10 to −35
Interference with spring		mm	≤0.5
Interference without spring		mm	≤0.6
Visual assessment of sealing edge			

(continued)

Table 10.17 (*Continued*)

Property	Method	Unit	Requirement
Micropitting test (Grey staining)	FVA 54 (I-IV)		
Load stage fail		Fail load	≥10
Load stage GFT rating		GSC	High
Endurance stage GFT rating		GSC	High
FZG scuffing (A/8.3/90)	ISO 14635-1		
Load stage		Fail load	>12
FZG scuffing (A/16.6/90)	ISO 14635-1		
Load stage		Fail load	By agreement >12
FE-8 bearing	DIN 51819-3		
Rollers weight loss		mg	<30
Cage weight loss		mg	Report
Flender foam test	GG-V 425		
Original oil			
Volume at T1		%	≤15
Air–oil dispersion at T5		%	≤10
Original oil + 2% castrol alpha SP220 S			
Volume at T1		%	≤15
Air–oil dispersion at T5		%	≤10
Original oil + 4% castrol alpha SP220 S			
Volume at T1		%	≤15
Air–oil dispersion at T5		%	≤10

In addition, important mechanical-dynamic tests – demands on wind turbine gear oils with regard to micro-pitting and scuffing avoidance (standard test and test with higher speeds), roller bearing wear test FE8, as well as diverse Flender foam tests, also with corrosion preventive contamination should be mentioned.

Of particular significance is good micro-pitting protection as well as scuffing prevention/protection. Scuffing prevention influences gearbox design, that is, gears are designed considering the Hertzian stress in rolling contact (pitch point) according to FZG scuffing tests (DIN-ISO 14635-1). The same applies to roller bearing selection. The corresponding FE8 test parameters are defined in the area of extreme mixed friction at very high axial loads and high oil temperatures.

An important selection criterion for industrial gear oils is the protection of the teeth against micro-pitting. Any gear oils used must offer a high degree of

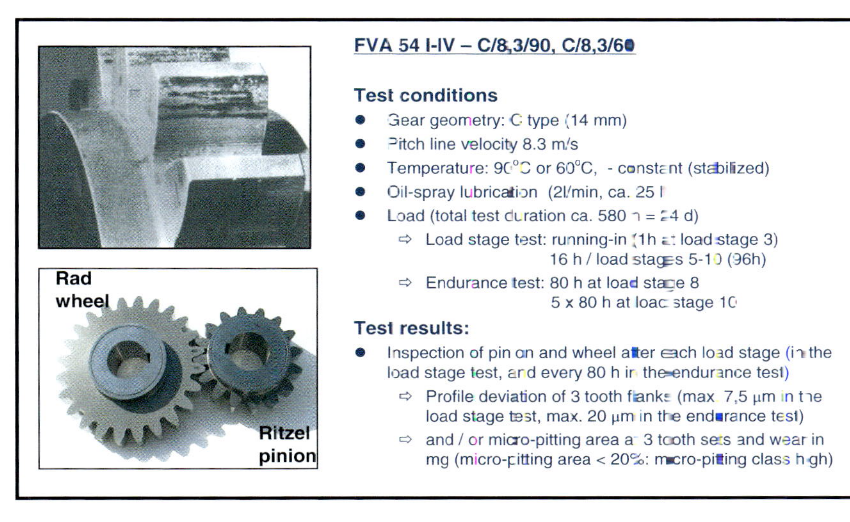

Figure 10.23 Micro-pitting test according to FVA 54 I-IV – test parameters.

protection against micro-pitting. This property is tested according to FVA 54 I-IV with the therein specified C-meshing, pitch line velocity of $8.3\,\mathrm{m\,s^{-1}}$ and oil injection temperatures of 90 and 60 °C. Micro-pitting is tested in both a load stage test as well as an endurance test. In the load stage test, the median tooth profile deviation must be less than 7.5 µm and the maximum threshold for the endurance test is 20 µm. Also, the tooth profile deviation must not increase under load. Figures 10.23 and 10.24 show an overview of the demands regarding micro-pitting for industrial gear oils for wind turbines.

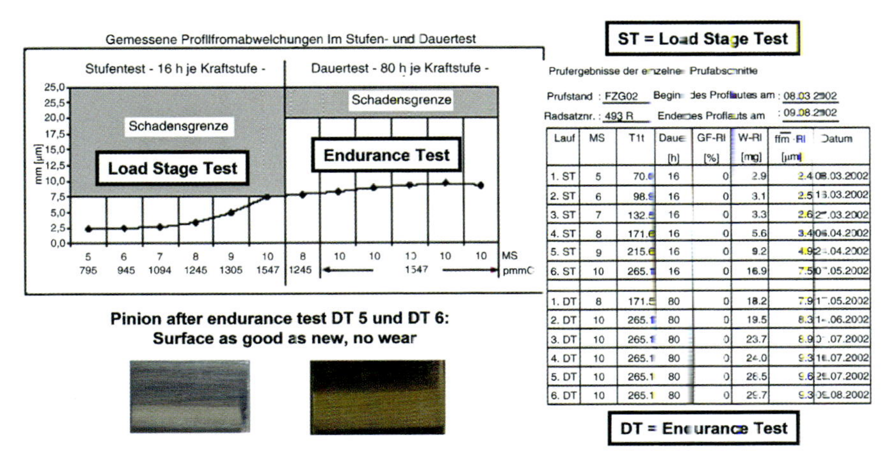

Prufergebnisse der einzelnen Prufabschnitte

Prufstand : FZG02 Beginn des Proflautes am : 08.03.2002
Radsatznr. : 493 R Endes des Proflauts am : 09.08.2002

Lauf	MS	T1t	Dauer [h]	GF-RI [%]	W-RI [mg]	ffm-RI [µm]	Datum
1. ST	5	70.6	16	0	2.9	2.4	08.03.2002
2. ST	6	98.9	16	0	3.1	2.5	13.03.2002
3. ST	7	132.5	16	0	3.3	2.6	2[illegible].03.2002
4. ST	8	171.6	16	0	5.6	3.4	06.04.2002
5. ST	9	215.6	16	0	9.2	4.9	2[illegible].04.2002
6. ST	10	265.1	16	0	16.9	7.5	0[illegible].05.2002
1. DT	8	171.5	80	0	18.2	7.9	1[illegible].05.2002
2. DT	10	265.1	80	0	19.5	8.3	1[illegible].06.2002
3. DT	10	265.1	80	0	23.7	8.9	0[illegible].07.2002
4. DT	10	265.1	80	0	24.0	9.3	16.07.2002
5. DT	10	265.1	80	0	28.5	9.6	25.07.2002
6. DT	10	265.1	80	0	29.7	9.3	09.08.2002

Figure 10.24 Micro-pitting according to FVA 54 I-IV – example: test results with Renolin Unisyn CLP 320.

Scuffing is generally tested with the FZG test according to DIN-ISO 14635-1. The standard scuffing test, performed at a pitch line velocity of 8.3 m s^{-1} and an initial oil sump temperature of 90 °C, can be run at up to load stage 14 which is equivalent to a Hertzian stress of 2138 N mm^{-2}. This test highlights the highly positive influence of fully synthetic gear oils based on polyalphaolefins. Low coefficients of friction, a high viscosity index and optimum mechanical efficiency can lead to a possible lowering of oil sump temperatures by about 20 °C at correspondingly high loads when compared to mineral oils. This means that fully synthetic, polyalphaolefin-based gear oils offer lower power losses and better efficiency under load in comparison to mineral oils. This leads to an increase in the load-bearing viscosity (see Figure 10.20).

Lower oil sump temperatures are also recorded in practice. A drop in oil sump temperature in the gearbox housing of about 5–7 °C was achieved in a wind farm when a CLP 320 mineral oil was replaced with a fully synthetic, polyalphaolefin-based lubricant: RENOLIN UNISYN CLP 320. Other advantages included high viscosity at equivalent operating conditions, greater lube film stability, less oxidation and longer service life of the fully synthetic gear oil.

10.3.10.2 Special Tests for Wind Turbine Gear Oils

FAG Wind Turbine Four-Stage Test (Schaeffler-FAG)

The FAG FE8 four-stage test was specially developed for wind turbine gear oils. In the past (about 10–15 years ago), gear oils contained highly active phosphorous–sulphur compounds. In the FE8 test rig, these generated roller wear rates of 200–300 mg. These days, industrial gear oils are formulated with mild phosphorous–sulphur compounds to meet the roller wear specifications of less than 30.

The FAG FE8 four-stage test attempts to replicate different load and mixed friction conditions at different speeds, temperatures and test parameters.

Stage 1 can be described as a short-term test and is performed on the FE8 test rig according to DIN 51819, parts 1–3 at 80 KN axial load, 80 °C for a duration of 80 h.

Stage 2 describes a fatigue test with moderate mixed friction and is performed on the FE8 test rig at 75 rpm, 100 KN axial load, 70 °C for a duration of 800 h.

Stage 3 is a so-called fatigue test under EHL conditions (10 bearings). The test is performed in the FAG test rig L11 at 9000 rpm, an axial load of 8.5 KN, about 80 °C and for a duration of 700 h.

Stage 4 involves a deposit test at higher temperatures in the presence of water. This modified (PM) paper-making machine oil test from FAG performed on a special FAG test rig at 750 rpm, an axial load of 60 KN, at up to 140 °C for a duration of 600 h.

A wind turbine gear oil must pass all these different test procedures with good results. Figure 10.25 shows a summary of the FAG wind turbine four-stage test (with an example of results of a tested wind turbine gear oil).

Testing of the suitability of oils for roller bearings

- Stage 1: Short-term wear test under extreme mixed friction
- Stage 2: Fatigue test under moderate mixed friction
- Stage 3: Fatigue test under EHL conditions
- Stage 4: Residue formation test at increased temperature with water added

Example: RENOLIN UNISYN CLP 320

Summary result wind turbines - 4 stage test

	criterion	test	result	
Stage 1*	wear at boundary lubrication	FE8-80h	1,0	passed
Stage 2**	fatigue beh. at mixed friction cond.	FE8-800h	1,0	passed
Stage 3***	fatigue behavior at EHL-cond.	L11-700h	1,0	passed
Stage 4***	fatigue behavior and residues with water added	FE8-WKA	1,0	passed
		summery:	1,0	passed

Figure 10.25 FAG wind-turbine four-stage test.

10.3.10.3 SKF Specifications for Wind Turbine Gear Oils

In its WTGU specification for wind turbine gear oils, the company SKF placed an emphasis on the high chemical and thermal stability of the lubricant. To summarize, it can be noted that SKF focuses on chemical stability (SKF roller test, SKF oil film ageing test) along with FE8 performance (wear protection), filtration and corrosion protection

10.3.10.4 Low-Speed Wear Behaviour of Industrial Gear Oils

Gearboxes in wind turbines usually have slow- and fast-running stages. The low-speed wear behaviour of industrial gear oils must be considered. This feature can be examined using the DGMK 377 method and/or the FZG method. At a peripheral speed of 0.05–0.57 $m\,s^{-1}$ and a high Hertzian stress (load stage 12), this involves determining total wear per test stage in milligrams Figure 10.26 shows a comparison of industrial gear oils based on mineral oils, polyalphaolefins, esters and polyglycols. The high viscosity index of synthetic oils and the correspondingly thicker lubricating film during the test can significantly lower the low-speed wear when compared to mineral oils.

10.3.10.5 Low-Temperature Viscosity of Industrial Gear Oils

Industrial gear oils in wind turbines have to perform under a number of different conditions and temperatures. The question of which maximum viscosity is permissible is still vigorously discussed in various specialist journals. Following the automotive sector, a low temperature specification threshold of 150 000 mPa s has been accepted in some circles. However, the behaviour of gear oil formulations with regard to the pour point should also be considered.

The pour point and the boundary viscosity define the lowest temperature at which an oil can be used normally. We have tested the low-temperature viscosity

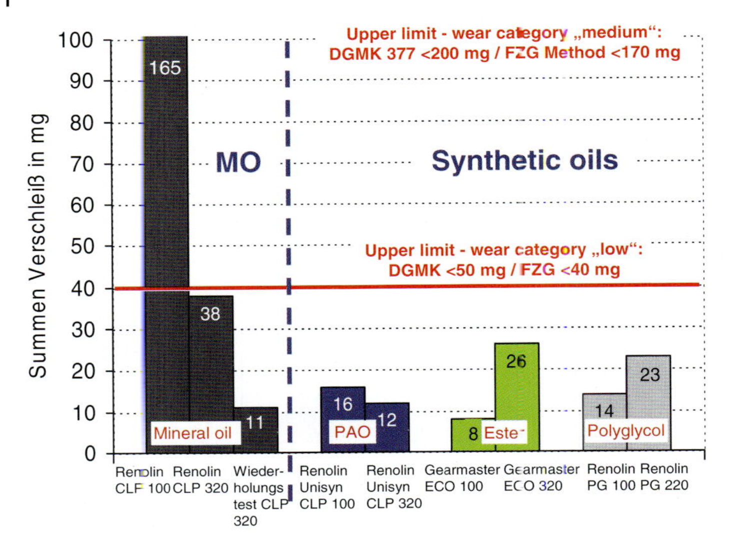

Figure 10.26 Slow-running wear behaviour of industrial gear oils.

in a rotational viscometer (Brookfield viscometer) and we have defined the maximum dynamic viscosity in mPa s with regard to the flowability limit in automotive gear oil specifications which is defined at 150 000 mPa s. Especially polyalphaolefin-based products show excellent low-temperature flowability and viscosity properties. The flowability limit is approximately 5–10 °C over the pour point (Figures 10.27–10.29).

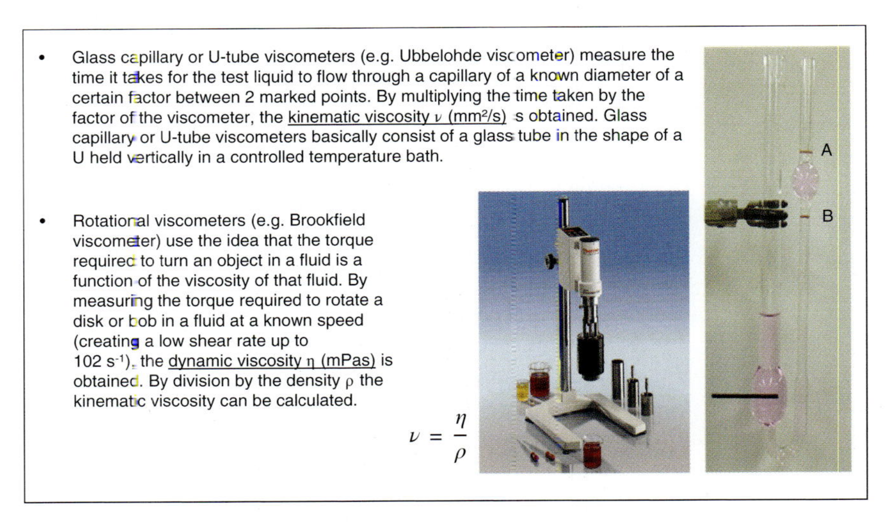

Figure 10.27 Low temperature viscosity measurement.

AMERICAN NATIONAL STANDARD ANSI/AGMA 9005-E02

Table 2 - Minimum performance requirements for antiscuff/antiwear (EP) oils

Property	Text method: ISO/ASTM	Requirements										
Viscosity grade	3448/D2422	32	46	68	100	150	220	320	460	680	1000-3200	>3200
Viscosity @ 40°C, mm²/s	3104/D445	See table 4										Report[1])
Viscosity @ 100°C, mm²/s	3104/D445	Report[1])										
Viscosity index [2]), min.	2909/D2270	90								85		Report[1])
Bulk fluid dynamic viscosity @ cold start-up [2]), mPa · s, max.	None/D2983	150 000										

A.14 Viscosity - dynamic (Brookfield)

Based on the experience of automotive gear manufactures, the same critical Brookfield viscosity limit of 150,000 cP has been proposed for industrial gear applications. Because of the wide variety of field conditions, the temperature for the maximum Brokfield viscosity is not specified in this standard. Rather, the temperature should be specified by the end user and it should relate to the lowest actual lubricant temperature at cold startup.

A.12 Pour point

Pour point is an indicator of the lowest temperature at which an oil flows under the influence of gravity. Pour point should not be used as the only indicator of the low temperature limit at which a lubricant may function acceptably. Initial agitation by gears, ...

[10] is used to determine pour point. It is recommended that the pour point of the oil used should be at least 5°C lower than the minimum ambient temperature expected.

Figure 10.28 Low temperature viscosity – threshold values according to AGMA 9005-E02.

Method	Standard	RENOLIN UNISYN CLP 220	RENOLIN UNISYN CLP 320
V-40 Brookfield [mPas]	DIN 51398 / DIN ISO 9262 / ASTM 2893 Repeatability: 8-10 % depending on level !	290.000	530.000
V-35 Brookfield [mPas]		115.000	220.000
V-30 Brookfield [mPas]		55.000	100.000
V-25 Brookfield [mPas]		31.000	55.000
V-20 Brookfield [mPas]		16.000	30.000
V-15 Brookfield [mPas]		10.000	17.000
Pourpoint [°C]	DIN ISO 3016	-42	-42
V 40 [mm²/s]	DIN EN ISO 3104	220	320
V 100 [mm²/s]		26,7	35,0
VI	DIN ISO 2909	155	155

flowability limit in autmotive gear oil specs: max. 150.000 mPas!

Flowability limit of RENOLIN UNISYN CLP 220: ~ -35 °C
Flowability limit of RENOLIN UNISYN CLP 320: ~ -30 °C

low pourpoint, relatively high viscosity at low temperatures

Figure 10.29 Low temperature viscosity of RENOLIN UNISYN CLP 220/320.

As far as the low-temperature viscosity of mineral oils is concerned, it should be remembered that very large viscosity deviations (measured values compared to calculated values) can occur near to the pour point. In principle, the given viscosity values for temperatures under 0 °C should be those actually measured.

In the case of fully synthetic, polyalphaolefin-based gear oils, a relatively good correlation exists between the calculated and the measured viscosity values even for temperatures of −10, −20 and −30 °C.

10.3.10.6 Conclusion

The specifications of wind turbine gear oils from a lubricant manufacturer's and component manufacturer's point of view are discussed. The highest demands are made on the technical performance of such gear oils with regard to mechanical-dynamic wear protection, chemical stability and long-term stability. Comprehensive manufacturer, DIN and ISO tests must be passed, and compliance with roller bearing and gearbox manufacturers' specifications must be given before a wind turbine gear oil is approved. Such testing activity generates high testing and laboratory costs. Apart from laboratory trials in mechanical-dynamic test rigs, qualifying oils must pass comprehensive field trials in various ambient conditions. Wind turbine gear oils are mostly based on fully synthetic base fluids with polyalphaolefins predominating. Wind turbine gear oils are specialties with the highest quality demands. Specific testing and specification criteria are currently being critically discussed, especially with regard to elastomer and chemical compatibility.

10.3.11
Summary

Industrial gear oils represent an important element in engineering design. Apart from standard, mineral oil-based gear oils, synthetic-based products are gaining acceptance.

Synthetic gear oils offer outstanding wear protection properties in gearboxes and roller bearings. They out-perform the previously used standard lubricants based on mineral oils.

The outstanding features of synthetic gear oils are their high-temperature stability, excellent oxidation resistance and the extending of the oil change intervals. They have excellent wear protection properties and can therefore safely protect industrial gear sets and roller bearings in operation. The requirement criteria placed on modern gear oils, particularly those for wind power applications, have increased enormously in recent years, and can be met only with high-performance formulations. A judgement always has to be made of the extent to which reactive or less reactive additives/additive systems must be used. Synthetic base oils and modern additive technologies, optimally matched to the requirements of the drive chain, increase the economy, availability and reliability of geared installations. The higher initial price of synthetic products can be countered by savings generated by significantly longer oil life when compared to mineral oils as well as the overall reduction in total operating costs.

References

1 FVA (2006) Annual Report, Frankfurt.
2 DIN 51517 (August 2011) Teil 1–3: Schmierstoffe, Schmieröle C, CL, CLP, Mindestanforderungen.
3 ISO 6743-6 (1990) Lubricants, Industrial Oils and Related Products (Class L) – Part 6: Family C (Gears) (E).
4 ISO 12925 (1996) Lubricants, industrial oils and related products (Class L) – Family C (gears) – Specifications for lubricants for enclosed gear systems.
5 DIN 51502 (August 1990) Schmierstoffe und verwandte Stoffe, Kurzbezeichnungen der Schmierstoffe und Kennzeichnung der Schmierstoffbehälter, Schmiergeräte und Schmierstellen.
6 ASTM D 2272 (2014) Standard Test Method For Oxidation Stability Of Steam Turbine Oils By Rotating Pressure Vessel.
7 Wollhofen, Gerhard P. u.a. (1990) *Getriebeschmierung in der Anlagentechnik*, Expert Verlag, Renningen, Germany.
8 Mang, Theo and Dresel, Wilfried (Hrsg.): (2001) *Lubricants and Lubrications*, Wiley-VCH Verlag, Weinheim, Germany.
9 Lohmann+Stolterfoth GmbH, D-58408 Witten, LSN 9351400, 1990: Getriebeschmierung, Schmierempfehlung, Übersicht.
10 Spilker, M., Ferch, J. and FUCHS Mineralölwerke GmbH, Mannheim (Juli/August 1995) Reibungsmindernde Getriebeöle im Langzeitversuch, In: Tribologie und Schmierungstechnik, 42. Jahrgang.
11 Friedrich Flender AG, 46393 Bocholt, Betriebsanleitung BA 7300 DE 07.98: Schmierstoffempfehlung.
12 Siemens (Flender) requirements – Eignungsnachweie für Öle, die in FLENDER-Stirnrad-, Kegelrad- und

Planetengetrieben sowie Getriebemotoren eingesetzt werden – Rev. 13 (28.02.2011).
13 DIN ISO 14635-1 (May 2006) Gears – FZG test procedure A/8.3/90 for relative scuffing load carrying capacity of oils.
14 ASTM D 2112 – 00 (April 2000) Standard Test Method for Oxidation Stability of Inhibited Mineral Isolating Oil by Pressure Vessel.

11
Hydraulic Oils

Wolfgang Bock

11.1
Introduction

Hydraulics describe the transfer of energy and signals through fluids. In this transmission process, power is transferred to drive, control and move. Hydraulic fluids based on mineral oils, synthetic fluids and fire-resistant fluids are used in all types of machinery and equipment. Hydraulics are a part of everyday life. There is hardly a machine or aircraft that operates without hydraulics. Hydraulic-component manufacturers supply nearly all industries including the agriculture and construction machinery sectors, conveyor technology, foodstuffs and packaging industries, woodworking and machine tools, shipbuilding, mining and steel industries, aviation and aerospace, medicine, environmental technology and chemicals. Many of these industries are leading players in the global market. Fluid technology makes a significant contribution to the competitiveness of these industries. The innovative development of hydraulic components and systems using the very latest materials, lubricants and electronics gives new impulses to technical developments [1]:

- Fluid technology is an essential technology – many applications are economical only if fluid technology is used.
- Fluid technology is omnipresent – whether in stationary or mobile applications – throughout the world.
- Fluid technology benefits the environment – as an environmentally harmless technology and used in environmentally sensitive plants fluid technology contributes to our quality of life.
- Fluid technology promotes an orderly future – wherever something happens, forces and torque are needed, as are hydraulics.
- Fluid technology, in general, fulfils the needs of a variety of end users.

The field of fluid technology and therefore hydraulics is divided into hydrostatics and hydrodynamics. In hydrostatic systems, the transfer of energy requires

Lubricants and Lubrication, Third Edition. Edited by Theo Mang and Wilfried Dresel.

static pressure, and so pressures are high but flow rates are low [2]. In hydrodynamic systems, the kinetic energy of the flowing fluid is used, so pressures are low but flow rates are high. Fluids designed for hydrodynamic applications are known as power-transmission oils and fluids designed for hydrostatic applications are known as hydraulic oils [2–4]. The fluid is the most important element in hydrostatic and hydrodynamic systems and must be treated like a machine element in the planning, realization and commissioning of hydraulic systems.

After engine oils, hydraulic oils are the second most important group of lubricants. They account for approximately 13–14% of the total lubricant consumption (124 400 of the consumption in 2012) [2,5,6]. In 2013, mineral-based hydraulic oils accounted for approximately 80–85% of all hydraulic oils in Germany. Fire-resistant fluids had a market share of about 6%, rapidly biodegradable hydraulic fluids about 7% and synthetic polyalphaolefins (PAO)- or HC-based fluids about 1% [2,5,6]. Amongst the mineral-based fluids, the products based on hydrotreated base oils (HC base oils and group III base oils) increased significantly in the past few years.

Since the early 1950s, hydraulics have been growing rapidly. The German VDMA (Verein Deutscher Maschinen- und Anlagenbau e.V., German Machinery and Plant Manufacturers' Association) founded an 'Oil Hydraulics and Pneumatics' consultative group in 1959. According to the VDMA [1], in 2003 sales of hydraulic equipment and machinery in Germany totalled approximately €4.1 billion with exports totalling ~50%. Fluid technology is thus a rapidly growing industry and its growth rate is significantly higher than that of engineering.

Modern hydraulics can be divided into three principal areas: stationary, mobile and aviation hydraulics. Each of these areas makes special demands on its components and the hydraulic medium. In recent years, the performance of hydraulic systems has increased significantly. This is reflected in higher pressures, higher system temperatures and lower system volumes that increase circulation and thus the stress on the medium.

Hydraulic fluid developments, both to date and in the future, and their correct applications are of enormous economic significance. Optimum applications save energy, reduce maintenance intervals, reduce wear, increase machine life and thus enable worthwhile savings.

11.2
Hydraulic Principle: Pascal's Law

The principle of the hydrostatic displacement machine is based on Pascal's law from the seventeenth century, which states 'Pressure applied anywhere to a body of fluid causes a force to be transmitted equally in all directions. This force acts at right angles to any surface within, or in contact with, the fluid'. The static pressure in a fluid thus enables force to be transferred. Figure 11.1 illustrates Pascal's hydrostatic principle. Figure 11.2 shows the principle of a hydraulic press [6].

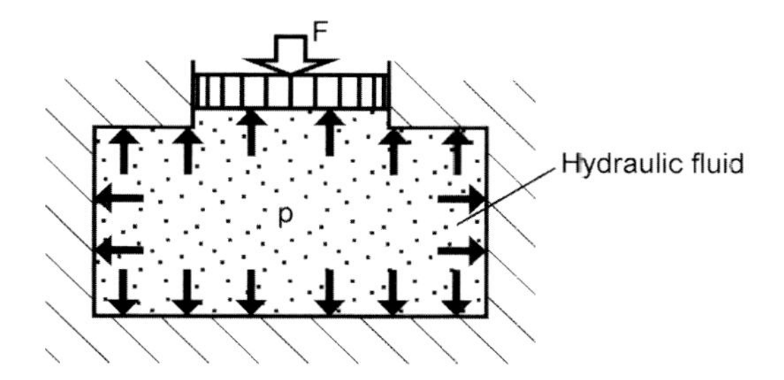

Figure 11.1 Pascal's hydrostatic principle.

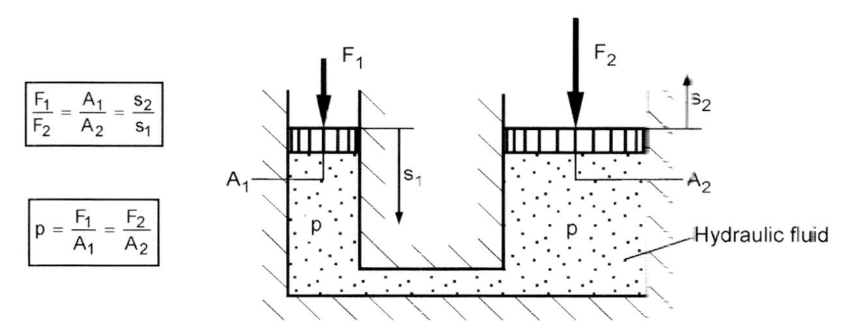

Figure 11.2 Principle of a hydraulic press.

11.3
Hydraulic Systems, Circuits and Components

The hydraulic transfer of power is characterized by the simplicity of its elements, long life, high performance and economic factors. The variety of hydraulic applications is largely determined by the behaviour of the hydraulic fluid.

11.3.1
Elements of a Hydraulic System

The most important elements of a hydraulic system are as follow:

- Pumps and motors (e.g. gears, rotary vanes and piston pumps)
- Hydraulic cylinders (e.g. single- and double-action cylinders)
- Valves (e.g. pressure limiters and control valves)
- Circuit components (e.g. fluid tanks, filter systems, pressure tanks, pipework etc.)
- Seals, gaskets and elastomers

Figure 11.3 shows a schematic illustration of a simple hydraulic circuit [6].

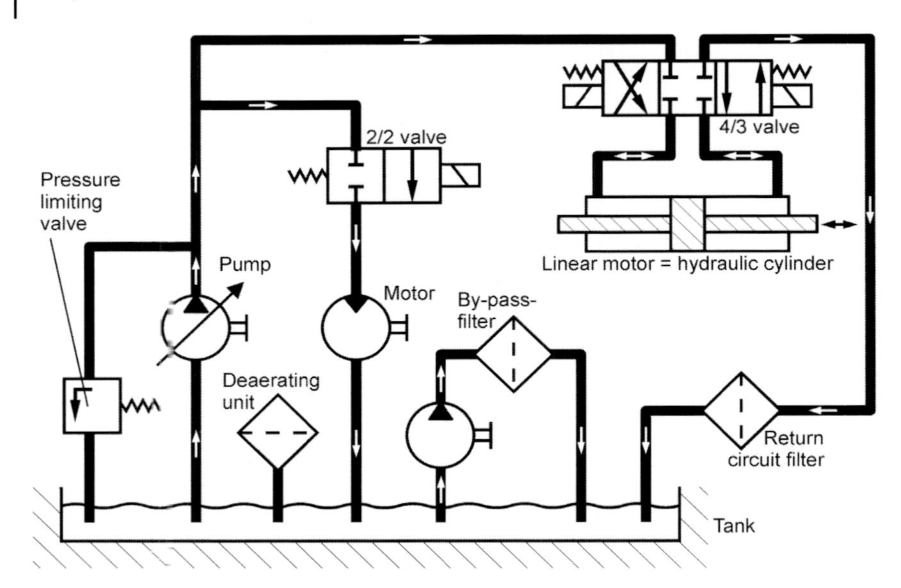

Figure 11.3 Schematic illustration of a simple hydraulic circuit.

11.3.1.1 Pumps and Motors

Pumps and motors are used in hydraulic systems to transfer energy. Electrical or mechanical energy is transformed into hydraulic energy by hydrostatic machines such as displacement pumps. The most important types of pumps are gears, rotary vanes and axial and radial piston pumps. Gear pumps are used for flow rates of 0.4–200 cm^3 $revolution^{-1}$ and pressures of 160–250 bar (internal gear pumps up to 350 bar) [2–4,6].

Rotary vane pumps can generally create pressures up to 160 bar but new developments enable use of pressures between 210 and 290 bar. Flow rates are about 30–800 cm^3 $revolution^{-1}$ [2–4,6]. Radial piston pumps cover the range up to 700 bar but these types of pumps are used for pressures of about 480 bar [2–4,6]. Axial piston pumps are divided into swashplate and bent-axis pumps. These can create pressures of up to 450 bar and higher. Flow rates vary greatly depending on pump dimensions [2–4,6].

Pumps and motors are subjected to great hydraulic stress. The main functions of hydraulic fluids are to transfer the energy, protect drive components and bearings from wear and corrosion and reduce friction, and thus to reduce the accumulation of deposits.

Figures 11.4–11.7 illustrate the most important types of hydraulic pumps and motors: gear pumps, rotary vane pumps, radial piston pumps and axial piston pumps.

11.3.1.2 Hydraulic Cylinders

Hydraulic cylinders transform hydraulic pressure and hydraulic energy into linear movement. This can then perform mechanical work. Hydraulic cylinders can be single- or double-action cylinders. In these, the primary functions of the fluid

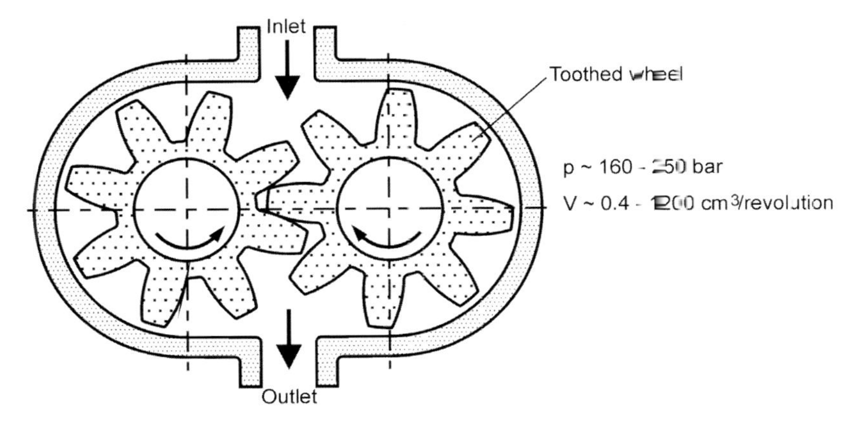

Figure 11.4 Gear pump.

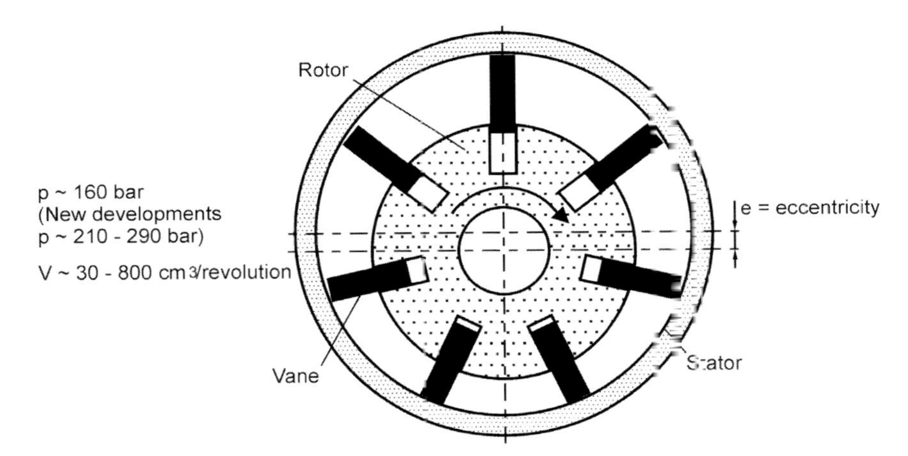

Figure 11.5 Rotary vane pump.

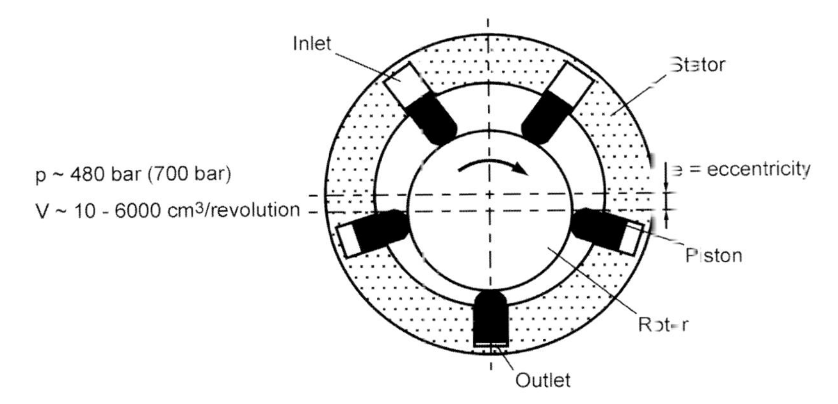

Figure 11.6 Radial piston pump.

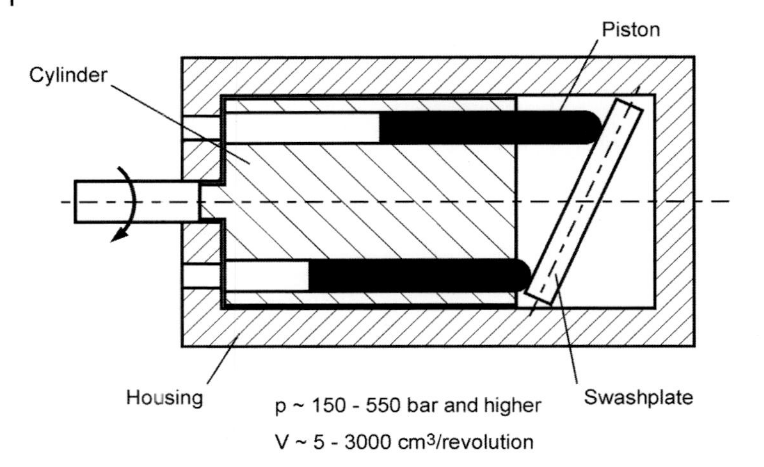

Figure 11.7 Axial piston pump.

are to seal and lubricate the pistons and guides, avoid stick–slip, minimize wear and avoid corrosion [2,4,6].

11.3.1.3 **Valves**

Valves are mechanisms that control the start, stop, direction or flow of a hydraulic medium from a pump or a pressure vessel. Flow valves have preset switching points. Proportional and servo valves are electro-hydraulic, that is their movement is proportional to the electrical input signal. The differences between these valves are their mechanical design, static and dynamic properties, avoidance of sludge and their price. The hydraulic fluid in a valve should dissipate heat, reduce wear, minimize friction and avoid corrosion. It is equally important that no deposits should form in the narrow tolerances found in valves. Long fluid-change intervals and high thermal loads (e.g. caused by solenoid magnets) must not lead to deposits in, or the gumming of, flow valves [2,4,6].

11.3.1.4 **Circuit Components**

Circuit components include fluid tanks, filter systems, pressure tanks and pipework. The hydraulic fluid must be compatible with all of the materials used in these elements including all coatings and paint finishes [6].

11.3.1.5 **Seals, Gaskets and Elastomers**

Every seal or elastomer in a hydraulic system is fully or partially exposed to the hydraulic fluid when the system is in operation. Interaction between the sealing material and the hydraulic medium is thus unavoidable.

The hydraulic medium can influence the sealing material insofar as it causes shrinkage or swelling. This, in turn, affects the volume of the seal and also alters mechanical properties such as hardness, elasticity, tensile strength and elongation behaviour.

Elastomeric seals are influenced chemically by temperature, oxygen, water, additives and the oxidation by-products of the hydraulic fluid. It is, therefore, vital that the seals and the hydraulic fluid are chemically compatible.

A seal is mechanically stressed by the pressure and pulsation of the fluid. In addition, dynamically stressed seals such as piston and rod seals are subjected to sliding friction [6–8].

Physical and chemical factors directly influence the mechanical wear of a seal. Swelling softens the sealing material, leading to higher friction and thus greater wear and power consumption. Normally, it is accepted that seals can swell within defined limits to avoid leakage.

Ideally, a hydraulic fluid should have no effect on sealing materials and elastomers while at the same time should protect them from wear, dissipate heat, reduce friction and avoid the accumulation of deposits in crevices. Seal manufacturers usually test the compatibility of their products with different hydraulic fluids and include them in compatibility lists. Lubricant manufacturers are primarily concerned about the behaviour of seals when in contact with hydraulic fluids as detailed in DIN 51 524 (dated 2006) and the effect of mineral-oil-based fluids on standard reference elastomers (SRE) NBR 1 sealing material as detailed in DIN 53 521 and DIN 53 502. These standards establish thresholds for volume and hardness changes to sealing materials. Fundamental and comparison tests are performed with reference fluids. These reference fluids are specified and classified according to an ASTM standard. ASTM Fluid 2 and Fluid 3 and their successor products, IRM 2 and IRM 3, are used for testing purposes [7–9]. Basically, a seal must not shrink when in contact with a hydraulic fluid because of the risk of leakage, although slight swelling is permissible. The seal must not harden, but slight softening is allowed. The duration of compatibility tests on seals with hydraulic media is 7 days at 100 °C [10].

Rapidly biodegradable hydraulic fluids based on vegetable oils, esters and polyglycols (PAG) are tested for longer periods to obtain more practically relevant results. The behaviour of rapidly biodegradable hydraulic fluids in contact with SRE is tested for 1000 h at temperatures ranging from 80 to 100 °C, as reported in CETOP R81 H, ISO 6072. The sealing materials listed in this test include HNBR, FPM AC 6, NBR 1 and AU grades. The properties tested now include hardness changes, volume changes, tensile strength and elongation breakage point. According to this test, the tensile strength and elongation breakage point must not exceed 30% in the presence of rapidly biodegradable hydraulic fluids [9,11–13]. Table 11.1 shows the general compatibility of elastomers with different hydraulic fluids [7,8,13].

Table 11.1 Compatibility of elastomers with hydraulic fluids.

	NBR	HNBR	AU	FPM	EPDM mineral oil free
	Average temperature range of the elastomers in °C (permanent operating range)				
	−30/(−40)/+100	−20(−30)/+140	−30/+80 (+100)	−20/+200[a]	−50/+150[a]
HL/HLP/HLPD mineral oils	+	+	+	+	−
HFD	−	−	−	+[b]	+[b]
HFC	+	+	+[b]	−	+
HFB	+	+	−	+	−
HFA	+	+	+[b]	+	−
HETG	+[b]	+[b]	+[b]	+	−
HEES	+[b]	+[b]	+[b]	+	−
HEPG	+[b]	+	−	+	+

a) Maximum temperature in air.
b) Check the application in the case of dynamic stress of elastomer material.

Symbol according to DIN/ISO 1629 resp. ASTM D 1418	Chemical name	Trade name
NBR	Acrylonitrile–butadiene–caoutchouc	Perbunan, Nipol, Europrene
HNBR	Hydrotreated acrylonitrile–butadiene–caoutchouc	Zetpol, Theiban
AU	Polyurethane–caoutchouc (polyesterurethane–caoutchouc)	Desmopan/Urepan
FPM	Fluoro–caoutchouc (FKM)	Viton, Fluorel, Tecnoflon
EPDM	Ethylene–propylenediene–caoutchouc	Vistalon, Buna EPG, Keltan
SBR	Styrene–butadiene–caoutchouc	Buna SB
CR	Chlorobutadiene–caoutchouc	Neoprene/Chloroprene
PTFE	Polytetrafluoroethylene	Hostaflon/Teflon

In principle, the conception 'like dissolves like' applies. To illustrate this, non-polar elastomers such as ethylene propylene diene monomer (EPDM) rubber swell to the point of dissolving in non-polar hydrocarbons contained in mineral-oil-based hydraulic fluids. Conversely, polar fluids such as HEPG fluids (polyglycols) function perfectly with these non-polar elastomer rubbers [7,8,13].

Elastomer tests with fire-resistant hydraulic fluids are described in the 7th Luxembourg Report. As food-grade lubricants are based on white oils and polyalphaolefins, they are used in the DIN 51 524, ISO 12922 and ISO 6743/4 elastomer test procedures [10,14].

General seal compatibility statements must consider the different manufacturing processes and the different composition of elastomers in practice. For example, nitrile butadiene rubber (NBR) materials can contain different amounts of acrylic nitrile and this has to be kept in mind when testing and evaluating elastomers and lubricants. The new ISO working paper, 'Hydraulic fluid power – compatibility between fluids and standard elastomeric materials' (ISO TC 131/SC7N343 Draft), dated 1999, lists methods that describe the influence of fluids on standard elastomeric materials [9].

11.4 Hydraulic Fluids

11.4.1 Composition of Hydraulic Fluids: Base Fluids and Additives

11.4.1.1 Base Oil or Base Fluid

In general, a hydraulic fluid consists of a base fluid, usually called a base oil, and chemical substances, usually called additives. The quality and performance of a hydraulic fluid depend on the quality of the base fluid and the combination of the additives or additive systems used. Additives improve certain characteristics that the base fluid cannot provide to a certain extent. Keeping in mind technical and ecological aspects, the types of base fluids and additives ultimately decide the classification of a hydraulic oil [5].

Mineral-oil-based fluids (paraffinic oils, naphthenic oils and white oils) and/or mixtures thereof are used as base fluids or base oils. Synthetic fluids based on hydrocracked oils (HC oils or so-called group III oils), polyalphaolefins, ester oils (POE) and PAG are mainly used in fire-resistant, rapidly biodegradable fluids or special hydraulic fluids. Natural vegetable oils such as rapeseed oil are often found in rapidly biodegradable fluids. Food-grade hydraulic oils are generally based on special white oils, polyalphaolefins and polyglycols (see Chapters 4 and 5) [2,5,6].

Mineral oils account for ~88% (mainly paraffinic group I oils); synthetic oils account for 12% (80% ester, 15% polyglycols etc.). Hydrocracked/hydrotreated oils increase market share due to their excellent technical properties and good price/performance ratio.

11.4.1.2 Hydraulic Fluid Additives

The required lubricity of hydraulic fluids is normally provided by special additives. The additives are usually included in the form of additive packages (mixtures) to achieve the desired results. The additive components can either complement each other or counter each other. The characteristics that can be improved by the use of additives include ageing stability, corrosion protection, wear protection (AW – anti-wear properties), EP behaviour (EP – extreme pressure properties), viscosity–temperature behaviour, foaming, detergency, water separation, friction coefficient and many more.

The most important additives for hydraulic fluids are as follows:

- 'Surface-active additives' such as rust inhibitors, metal deactivators, wear inhibitors, friction modifiers, detergents/dispersants and so on.
- 'Base oil active additives' such as antioxidants, defoamers, viscosity index (VI) improvers, pour point improvers and so on.

A rough classification of hydraulic fluid additive systems can be made by differentiating between additive systems containing zinc and ash and zinc- and ash-free (ZAF) systems (see Chapter 6) [5,15]. On average, zinc-containing hydraulic oils account for 70–80% of the total volume, and zinc- and ash-free fluids account for 20–30% of the total volume.

11.4.2 Primary, Secondary and Tertiary Characteristics of a Hydraulic Fluid

The primary functions and properties of a hydraulic fluid are as follows [2,5,6]:

- Transferring pressure and motion energy
- Transferring forces and moments when used as a lubricant
- Minimization of wear to sliding surfaces under boundary friction conditions
- Minimization of friction
- Protection of components against corrosion (ferrous and non-ferrous metals)
- Dissipation of heat
- Suitability for a wide range of temperatures and good viscosity–temperature behaviour
- Prolonging the life of machinery and so on

A hydraulic fluid must satisfy the following characteristics:

Secondary characteristics: High ageing stability, good thermal stability, inactive to materials, compatibility with metals and elastomers, good air separation, low foaming, good filterability, good water release, good shear stability in the case of non-Newtonian fluids and many more [2,5,6].

Tertiary characteristics: Low evaporation as a result of low vapour pressure, toxicologically harmless, ecologically safe, low flammability (fire resistance) and so on [2,5,6].

The wide variety of different characteristics required for hydraulic fluids necessitates special performance that cannot be satisfied by just one base oil. Special chemical substances (additives) improve and complement the technical performance of hydraulic fluids. Synthetic base fluids can satisfy specific performance requirements such as environmental compatibility, high thermal stability, fire resistance and use in food-grade applications [5].

11.4.3
Selection Criteria for Hydraulic Fluids

Fluid selection depends on the application, such as the working temperature range, design of the hydraulic system, type of pump, working pressure and environmental conditions. The required fluid life, availability and economic and ecological factors also determine the type of hydraulic oil used. From a rheological standpoint, the viscosity of the fluid selected should be as low as possible. This guarantees instant hydraulic response when the system is activated. On the other hand, a minimum viscosity is required to reduce leakages and to guarantee adequate lubrication of the pump and other moving parts [2–6]. Any change in hydraulic fluid temperature has a direct effect on viscosity. For this reason, the working temperature range of a hydraulic system should be kept relatively narrow to maintain viscosity fluctuations as small as possible. For the accurate fluid selection, it is assumed that the working and ambient temperatures are known. In sealed systems, it is the circuit temperature; in open systems, it is the tank temperature. The viscosity of the fluid selected should be in the optimum range, between 16 and 36 $mm^2 s^{-1}$ ($V_{optimum}$ = optimum operating viscosity = 16–36 $mm^2 s^{-1}$). Under threshold conditions (during cold starts and short-term overloading), the viscosities shown in Table 11.2 could be applied, depending on the type of pump used [2,4,16–21].

Normal operating temperature depends on ambient temperature, pressure and other factors. In low- and medium-pressure stationary hydraulic systems, the operating temperature should be about 40–50 °C (tank temperature). For systems that operate at higher pressures (over 400 bar), the average system temperature can be about 10–20 °C higher [2,6,22]. It must be remembered that the fluid temperature at the pump outlet and downstream of motors or valves is higher than that in the tank or the system's average temperature. The leakage temperature that is influenced by pressure and pump speed is always more than the system or tank temperature. The fluid temperature should never exceed 90 °C (max. 100–120 °C) in any part of the system [22,23]. If these conditions cannot be fulfilled, due to extreme circumstances, at lower ambient temperatures it is recommended that pumps and motors are flushed [4,6]. Start-up and working viscosity (operating viscosity of hydraulic oils) are set by the various ISO

Table 11.2 Viscosities by the use of different type of pumps.

Type of pump	Maximum permissible viscosity in $mm^2 s^{-1}$ (during cold starts)	Minimum permissible viscosity (during load conditions, maximum permissible oil temperature of 90 °C) in $mm^2 s^{-1}$
Gear pumps	About 1.000	10–25
Piston pumps	1.000–2.000	10–16
Rotary vane pumps	200–800	16–25

viscosity grades. Most applications are satisfied by the viscosity grades 15, 22 (at low ambient temperatures), 32, 46 and 68 [17]. Normally, oils with a viscosity index of about 100 are used. Hydraulic oils with higher VIs (better viscosity–temperature range) are recommended for special hydraulic systems at higher temperatures, at lower ambient temperatures and in mobile applications [2,5,6]. If viscosity index improvers (VI improvers) are used, they must be shear-stable (over the lifetime of the fluid) to ensure that the oil retains its mechanical properties throughout its life. High-viscosity oils can be used in older systems to reduce leakages and wear. High-viscosity-index hydraulic oils can enable grade rationalization in industrial applications. For example, an HVLP 46 can replace up to four/five viscosity grades (ISO VG 15-68) in industrial applications. The selection diagram (Figure 11.8) shows recommended viscosity grades in relation to ambient temperature [17].

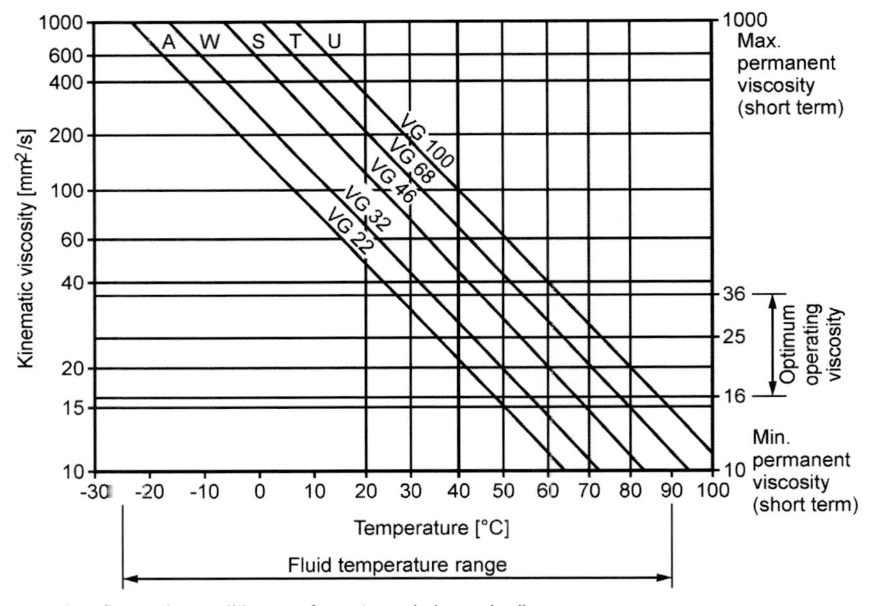

A = for arctic conditions or for extremely long pipelines
W = for winter conditions in Central Europe
S = for summer conditions in Central Europe or for enclosed areas
T = for tropical conditions or for areas with high temperatures
U = for excessively high temperatures (e.g. due to internal combustion engines)

Depending on the construction of the pumps (axial piston units) and the operating conditions the following viscosity ranges are valid:

10 mm2/s (t_{max} = +90°C) ..... 1000 mm2/s (t_{min} = -25°C)

[5 mm2/s (t_{max} = +115°C) ..... 1600 mm2/s (t_{min} = -40°C)] special components

Figure 11.8 Selection diagram for hydraulic fluids.

The average temperature of mineral-based hydraulic oils in stationary systems should not exceed 50–60 and 80–90 °C in mobile systems. Fluids containing water (e.g. HFC fluids) should be kept below 35–50 °C (attention: vapour pressure of water) [6].

The volume of fluid in stationary systems should be three to five times (depending on air release of the fluid) the volume pumped per minute. In mobile systems, the tank volume should be one to two times the pumped volume but in special circumstances it can be less [2,4,6].

11.4.4 Classification of Hydraulic Fluids: Standardization of Hydraulic Fluids

11.4.4.1 Classification of Hydraulic Fluids

Depending on their ultimate use, hydraulic oils can be allocated to one of the two main groups – fluids for hydrostatic applications and fluids for hydrodynamic applications.

Hydrostatic applications can be divided into sub-groups with regard to ISO, CETOP and national (e.g. DIN) classifications [24]:

- DIN 51 524 or ISO 6743/4 hydraulic oils [10,24].
- ISO/CD 12922, VDMA 24317, CETOP RP97 H and DIN 51 502 fire-resistant hydraulic fluids according to the 7th Luxembourg Report or Factory Mutual USA Factory Mutual System-Insurance Company USA (FM Global) [14,24–28].
- ISO 15 380 (VDMA 24568) and ISO 6743/4 rapidly biodegradable hydraulic fluids [11,12,25]: ISO 15 380, Lubricants, industrial oils and related products (class L) – Family H – Specification for environmentally acceptable fluids – HETG, HEES, HEPG and HEPR.
- NSF H1, H2 and FDA food-grade hydraulic oils [29]: NSF International – The Public Health and Safety Company (a not-for-profit, non-governmental organization – USA).
- STOU and UTTO universal mobile hydraulic oils [30].

The hydraulic oils used in hydrodynamic applications can be allocated to automatic transmission fluids (ATF), coupling and converter fluids (see Chapter 10).

Figure 11.9 shows the different categories of hydraulic fluids and their principal applications.

11.4.5 Mineral-Oil-Based Hydraulic Fluids

According to DIN 51 524 and ISO 6743/4, these fluids can be categorized as shown in Table 11.3 [10,25]:

Category L: Lubricants, industrial oils and related products.

Category H: Hydrostatic hydraulic systems.

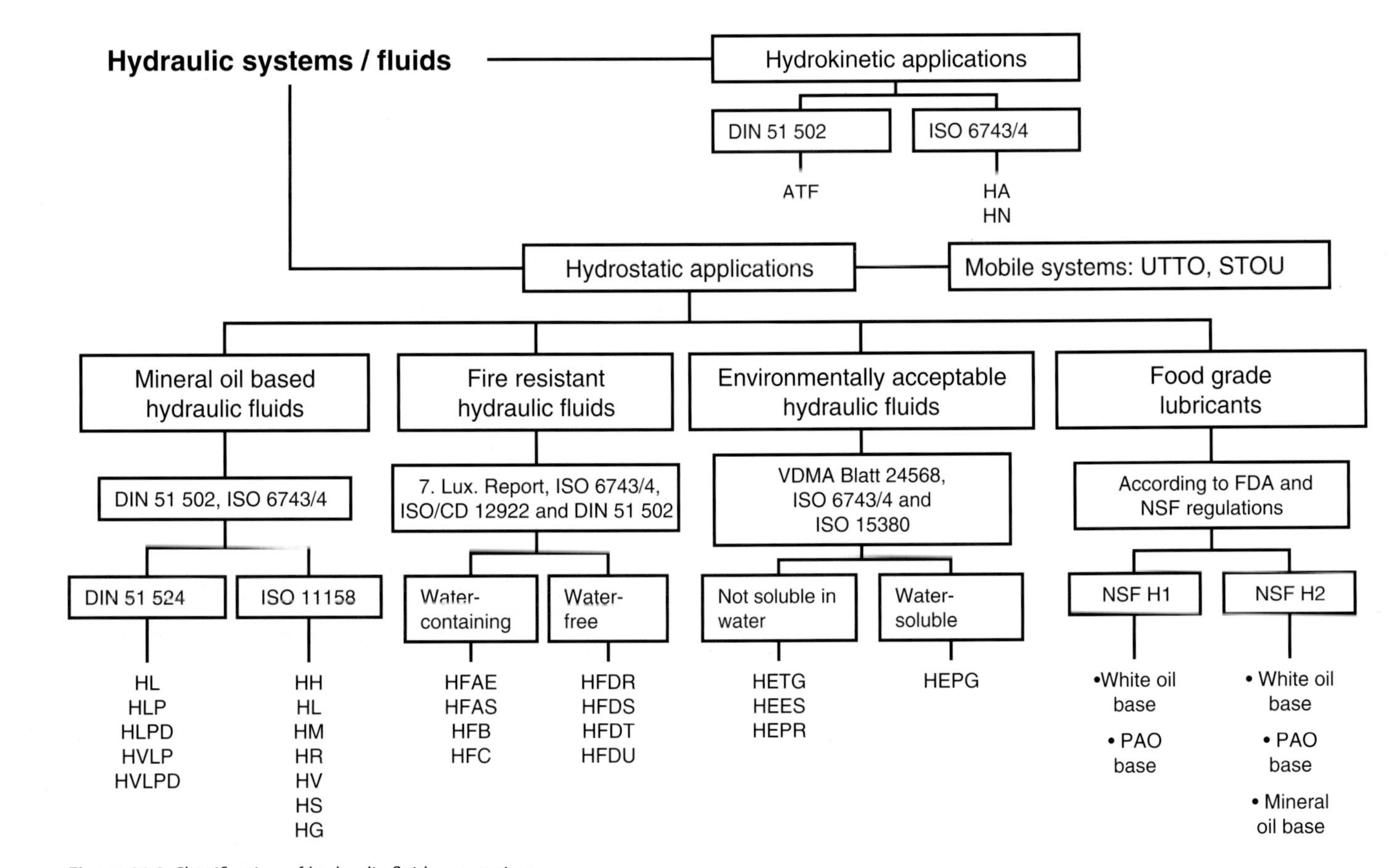

Figure 11.9 Classification of hydraulic fluids – overview.

Table 11.3 Classification of mineral-oil-based hydraulic fluids (categories according to DIN 51 502 and ISO 6743/4).

Category (symbol)		Composition typical characteristics	Field of application operating temperatures
DIN	ISO-L		
–	HH	Non-inhibited refined mineral oils	Hydraulic systems without specific requirements (rarely used nowadays)/−10 to 90 °C
HL	HL	Refined mineral oils with improved anti-rust and anti-oxidation properties	Hydrostatic drive systems with high thermal stress; need for good water separation/−10 to 90 °C
HLP	HM	Oils of HL type with improved anti-wear properties	General hydraulic systems that include highly loaded components, need for good water separation/−20 to 90 °C
–	HR	Oils of HL type with additives to improve viscosity–temperature behaviour	Enlarged range of operating temperatures compared with HL oils/−35 to 120 °C
HVLP	HV	Oils of HM type with additives to improve viscosity–temperature behaviour	For example, hydrostatic power units in construction and marine equipment −35 to 120 °C
–	HS	Synthetic fluids with no specific inflammability characteristics and no specific fire resistance properties	Special applications in hydrostatic systems, special properties/−35 to 120 °C
–	HG	Oils of HM type with additives to improve stick–slip-behaviour, and anti-stick–slip properties	Machines with combined hydraulic and plain bearing way lubrication systems where vibration or intermittent sliding (stick/slip) at low speed must be minimized/−30 to 120 °C
HLPD		Oils of HM type with DD additives reduce friction	Hydrostatic drive units with high thermal stress, which require EP/AW additives, DD additives keep contaminants in suspension, for example machine tools and mobile hydraulic equipment

11.4.5.1 H Hydraulic Oils

Type H, mineral-based hydraulic oils are generally base oils without additives. Accordingly, they are hardly used in the western Europe.

Classification: ISO 6743/4–HH [25].

11.4.5.2 HL Hydraulic Oils

Hydraulic fluids with additives are used to improve ageing stability and corrosion protection.

These oils are used in hydraulic systems that need not be protected from wear such as in steel and rolling mills where the prevailing conditions cause the fluids to be contaminated with water. Thus, these types of fluids can separate water well, usually also release air rapidly and are compatible with special – white metal and Morgan – bearings (Spec. Section 1: HL/CL; Spec. Section 2: HLP/CLP). If higher viscosities are used as general lubricating oils, they must fulfil the corresponding CL and CKB specifications according to DIN 51 517-2 [31] and ISO 6743/6 [32], respectively.

HL hydraulic oils according to DIN 51 524-1 [10].

HL hydraulic oils according to ISO 6743/4 [25].

Table 11.4 shows the specification profile of hydraulic oils according to DIN 51 524-1 (2006).

11.4.5.3 HLP Hydraulic Oils

Compared with HL fluids, these oils contain additional agents to reduce wear and/or improve EP behaviour. This is the dominant group of hydraulic oils in Europe and the rest of the world. They are universal hydraulic oils for a wide range of applications and highly loaded components and are used for applications that require good ageing stability, corrosion protection and wear protection. At the same time, these lubricants offer good demulsifying properties. These fluids are used as universal products in hydraulic presses, die-casting machines and steel mills when a fluid with wear protection is specified.

HLP according to DIN 51 524-2 [10].

HM according to ISO 6743/4 [25].

Table 11.5 illustrates the specifications of hydraulic oils according to DIN 51 524-2 (April 2006). Table 11.6 shows the specifications of HVI hydraulic oils according to DIN 51 524-3.

11.4.5.4 HVLP Hydraulic Oils

Compared with HLP grades, these fluids have a high viscosity index (VI > 140; HLP oils: VI ≈ 100). They therefore show good viscosity–temperature behaviour. The high VI is achieved by using additives and/or an appropriate base oil. The natural high VI of the base oil is preferred because shear losses do not occur. According to DIN 51 524-3 in conjunction with DIN 51 350-6 (VKA – determination of shear stability of lubricating oils containing polymers), the shear loss has to be reported (recommended maximum 15–20%; OEMs < 15%). A low start-up viscosity at low temperatures and a high operating viscosity at higher

Table 11.4 Minimum requirements of hydraulic fluids – DIN 51 524, Part 1, HL (Revised Version 2006 – April).

DIN 51 524 Part 1 – Hydraulic fluids with improved anti-rust and anti-oxidant properties									
Grade (DIN 51 502)		HL 10	HL 15	HL 22	HL 32	HL 46	HL 68	HL 100	HL 150
ISO viscosity grade (DIN 51 519)		VG 10	VG 15	VG 22	VG 32	VG 46	VG 68	VG 100	VG 150
Viscosity at 0 °C (−20 °C), $mm^2\,s^{-1}$, maximum (DIN 51 562-1)		90 (600)	150	300	420	780	1400	2560	4500
Viscosity at 40 °C, $mm^2\,s^{-1}$, minimum–maximum (DIN 51 562-1)		9.0–11.0	13.5–16.5	19.8–24.2	28.8–35.2	41.4–50.6	61.2–74.8	90.0–110	135–165
Viscosity at 100 °C, $mm^2\,s^{-1}$, minimum (DIN 51 562-1)		2.5	3.2	4.1	5.0	6.1	7.8	9.9	14.0
Pour point, °C, maximum (DIN ISO 3016)		−30	−27	−21	−18	−15	−12	−12	−12
Flashpoint (COC), °C, minimum (DIN EN ISO 2592)		125	140	165	175	185	195	205	215
Purity class (ISO 4406:1999)		21/19/16 (defined requirements depend on the system)							
Contamination with solid particles, mg kg^{-1}, maximum (DIN ISO 5884 or ISO 4405:1991)		50							
Filterability without H_2O, Stage I F_I/Stage II F_{II}, minimum % (E DIN ISO 13357-2)		80/60							
Filterability with H_2O, Stage I F_I/Stage II F_{II}, minimum % (E DIN ISO 13357-1)		70/50							
Demulsibility, 54 °C, minutes, maximum	(DIN ISO 6614)	20			30			–	
Demulsibility, 82 °C, minutes, maximum		–			–			30	

(continued)

Table 11.4 (*Continued*)

Water content in % m/m, maximum (DIN EN ISO 12937 – method A)		0.05							
Steel corrosion, method A (DIN ISO 7120)		Pass							
Copper corrosion, 100 °C, 3 h, degree of corrosion, maximum (DIN EN ISO 2160)		2							
Ageing properties (maximum increase in neutralization number after 1000 h) in mg KOH g^{-1}, maximum (DIN 51 587 or DIN EN ISO 4263-1)		≤ 2.0							
Behaviour towards the SRE-NBR 1 sealant after 7 d ± 2 h at (100 ± 1) °C (DIN 53 538-1, DIN ISO 1817 and DIN 53 505)	Relative change in volume, %	0 to 18	0 to 15		0 to 12		0 to 10		
	Change in Shore A hardness	0 to −10	0 to −8	0 to −7	0 to −6				
Air release, 50 °C, minutes, maximum (DIN ISO 9120)		5	5	5	5	10	13	21	28
Foam volume in ml, maximum (ISO 6247:1998, incl. Corr. 1: 1999) at 24 °C at 93.5 °C at 24 °C after 93,5 °C		150/0 75/0 150/0							

Table 11.5 Minimum requirements of hydraulic fluids – DIN 51 524, Part 2, HLP – April 2006, DIN 51 524 Part 3 HVLP – April 2006.

DIN 51 524 Part 2 – Anti-wear hydraulic oils (2006)								
Grade (DIN 51 502)	HLP 10	HLP 15	HLP 22	HLP 32	HLP 46	HLP 68	HLP 100	HLP 150
ISO viscosity grade (DIN 51 519)	VG 10	VG 15	VG 22	VG 32	VG 46	VG 68	VG 100	VG 150
Viscosity at 0 °C/(−20 °C), $mm^2 s^{-1}$, maximum (DIN 51 562-1)	90 (600)	150	300	420	780	1400	2560	4500
Viscosity at 40 °C, $mm^2 s^{-1}$, minimum–maximum (DIN 51 562-1)	9.0–11.0	13.5–16.5	19.8–24.2	28.8–35.2	41.4–50.6	61.2–74.8	90.0–110	135–165
Viscosity at 100 °C, $mm^2 s^{-1}$, minimum (DIN 51 562-1)	2.5	3.2	4.1	5.0	6.1	7.8	9.9	14.0
Pour point, °C, maximum (DIN ISO 3016)	−30	−27	−21	−18	−15	12	12	−12
Flashpoint (COC), °C, minimum (DIN EN ISO 2592)	125	140	165	175	185	195	205	215
Purity class (ISO 4406:1999)	21/19/16 (defined requirements depend on the system)							
Contamination with solid particles, mg kg^{-1}, maximum (DIN ISO 5884 or ISO 4405:1991)	50							
Filterability without H_2O, Stage I F_I/Stage II F_{II}, minimum % (E DIN ISO 13357-2)	80/60							
Filterability with H_2O, Stage I F_I/Stage II F_{II}, minimum % (E DIN ISO 13357-1)	70/50							
Demulsibility, 54 °C, minutes, maximum (DIN ISO 6614)	20			30		–		
Demulsibility, 82 °C, minutes, maximum	–			–		30		
Water content in % *m*/*m*, maximum (DIN EN ISO 12937 – method A)	0.05							

(*continued*)

Table 11.5 *(Continued)*

Steel corrosion, method A (DIN ISO 7120)		Pass							
Copper corrosion, 100 °C, 3 h, degree of corrosion, maximum (DIN EN ISO 2160)		2							
Ageing properties (maximum increase in neutralization number after 1000 h) in mg KOH g^{-1}, maximum (DIN 51 587 or DIN EN ISO 4263-1)		≤ 2.0							
Behaviour toward the SRE-NBR 1 sealant after 7 d ± 2 h at (100 ± 1) °C (DIN 53 538-1, DIN ISO 1817, and DIN 53 505)	Relative change in volume, %	0 to 18	0 to 15		0 to 12		0 to 10		
	Change in Shore A hardness	0 to −10	0 to −8		0 to −7		0 to −6		
Air release, 50 °C, minutes, maximum (DIN ISO 9120)		5	5	5	5	10	13	21	32
Foam volume, in ml, maximum (ISO 6247:1998, incl. Corr. 1: 1999) at 24 °C at 93.5 °C at 24 °C after 95 °C		150/0 75/0 150/0							
FZG mechanical gear test rig, A/8.3/90: failure load stage, minimum (DIN 51 354-2 or DIN ISO 14635-1)		–			10				
Vane pump wear, mg (DIN EN ISO 20763)	Ring, maximum	–	–	–	120			–	–
	Vanes, maximum	–	–	–	30			–	–

DIN 51 524 Part 3 – High-VI hydraulic oils (2006)

	Grade (DIN 51 502)	HVLP 10	HVLP 15	HVLP 22	HVLP 32	HVLP 46	HVLP 68	HVLP 100
HVLP 150	ISO viscosity grade (DIN 51 519)	VG 10	VG 15	VG 22	VG 32	VG 46	VG 68	VG 100
VG 150	Viscosity at 0 °C/(−20 °C), $mm^2 s^{-1}$, maximum (DIN 51 562-1)	To be reported by the supplier, except −20 °C of HVLP 100 and HVLP 150.						
Viscosity at 40 °C, $mm^2 s^{-1}$, minimum–maximum (DIN 51 562-1)	9.0–11.0	13.5–16.5	19.8–24.2	28.8–35.2	41.4–50.6	61.2–74.8	90.0–110	135–165
Viscosity at 100 °C, $mm^2 s^{-1}$, minimum (DIN 51 562-1)	To be reported by the supplier							
Viscosity index, minimum (DIN ISO 2909)	140							120
Pour point, °C, maximum (DIN ISO 3016)	−39	−39	−39	−30	−27	−24	−21	−18
Flashpoint (COC), °C, minimum (DIN EN ISO 2592)	125	125	175	175	180	180	190	200
Purity class (ISO 4406:1999)	21/19/16 (defined requirements depend on the system)							
Contamination with solid particles, mg kg^{-1}, maximum (DIN ISO 5884 or ISO 4405:1991)	50							
Filterability without H_2O, Stage I F_I/Stage II F_{II}, minimum % (E DIN ISO 13357-2)	80/60							
Filterability with H_2O, Stage I F_I/Stage II F_{II}, minimum % (E DIN ISO 13357-1)	70/50							
Demulsibility, 54 °C, minutes, maximum (DIN ISO 6614)	20			30		–		
Demulsibility, 82 °C, minutes, maximum	–			–		30		
Water content in % *m/m*, maximum (DIN EN ISO 12937 – method A)	0.05							
Steel corrosion, method A (DIN ISO 7120)	Pass							
Copper corrosion, 100 °C, 3 h, degree of corrosion, maximum (DIN EN ISO 2160)	2							

(*continued*)

Table 11.5 (*Continued*)

Ageing properties (maximum increase in neutralization number after 1000 h), mg KOH g^{-1}, maximum (DIN 51 587 or DIN EN ISO 4263-1)		≤ 2.0							
Behaviour toward the SRE-NBR 1 sealant after 7 d ± 2 h at (100 ± 1) °C (DIN 53 521, DIN ISO 1817 and DIN 53 505)	Relative change in volume, %	0 to 18	0 to 15		0 to 12	0 to 10			
	Change in Shore A hardness	0 to −10	0 to −8		0 to −7	0 to −6			
Air release, 50 °C, minutes, maximum (DIN ISO 9120)		5			13	21			32
Foam volume, in ml, maximum (ISO 6247:1998, incl. Corr. 1: 1999) at 24 °C at 93.5 °C at 24 °C after 95 °C		150/0 75/0 150/0							
FZG mechanical gear test rig, A/8.3/90. failure load stage, minimum (DIN 51 354-2 or DIN ISO 14635-1)		–		–	10				
Vane pump wear, mg (DIN 51 389-2)	Ring, maximum	–	–	–	120	–	–	–	
	Vanes, maximum	–	–	–	30	–	–	–	
Relative viscosity reduction due to shearing after 20 h (DIN 51 350-6)	at 40 °C	To be reported by the supplier							
	at 100 °C	To be reported by the supplier							

Table 11.6 Minimum requirements of hydraulic fluids – ISO 6743. Part 4 – HM, ISO 11158.

ISO 6743/4, ISO 11158: Specification for type HM mineral oil hydraulic fluids oils with improved anti-rust and anti-oxidant and anti-wear properties (a typical application is for general hydraulics)

Characteristics or test	Units	Specifications								Standard or test method
Viscosity grade		VG 10	VG 15	VG 22	VG 32	VG 46	VG 68	VG 100	VG 150	ISO 3448
Density at 15 °C	kg dm^{-3}	a)	a)	a)	a)	a)	a)	a)	a)	ISO 3675
Colour[b)]		a)	a)	a)	a)	a)	a)	a)	a)	ISO 2049
Appearance at 25 °C[c)]		Clbr	Clbr	Clbr	Clbr	Clbr	Clbr	Clbr	Clbr	Visual
Flashpoint, Cleveland open cup, minimum	°C	125	140	165	175	185	195	205	215	ISO 2592
Viscosity at 0 °C (−20 °C), $mm^2 s^{-1}$, maximum (DIN 51 562-1)	$mm^2 s^{-1}$ [d)]	90 (600)	150	300	420	780	1400	2560	4500	ISO 3104 and ISO 3105
Kinematic viscosity at 40 °C, minimum–maximum[d)]	$mm^2 s^{-1}$ [d)]	9,0–11,0	13,5–16,5	19,8–24,2	28,8–35,2	41,4–50,6	61,2–74,8	90,0–110	135–165	ISO 3104 and ISO 3105
Viscosity at 100 °C, $mm^2 s^{-1}$, minimum (DIN 51 562-1)	$mm^2 s^{-1}$ [d)]	2,5	3,2	4,1	5,0	6,1	7,8	9,9	14,0	ISO 3104 and ISO 3105
Viscosity index		a)	a)	a)	a)	a)	a)	a)	a)	ISO 2909
Pour point, maximum	°C	−30	−27	−21	−18	−15	−12	−12	−12	ISO 3016
Neutralization value[e)]	mg KOH/g	a)	a)	a)	a)	a)	a)	a)	a)	ISO 6618

(continued)

Table 11.6 *(Continued)*

Characteristics or test	Units	Specifications								Standard or test method
ISO 6743/4, ISO 11158: Specification for type HM mineral oil hydraulic fluids oils with improved anti-rust and anti-oxidant and anti-wear properties (a typical application is for general hydraulics)										
Water content	% (m/m), maximum	0025	0025	0025	0025	0025	0025	0025	0025	ISO 6296 or ISO 12937 or ISO 20764
Copper corrosion, 100 °C, 3 h, maximum	Class	2	2	2	2	2	2	2	2	ISO 2160
Rust prevention, procedure A procedure B		Pass[a)]	Pass[a)]	Pass[a)]	Pass Pass	Pass Pass	Pass Pass	Pass Pass	Pass Pass	ISO 7120
Foam volume in ml, maximum at 24 °C at 93.5 °C at 24 °C after 95 °C	ml ml ml	150/0 80/0 150/0	150/0 80/0 150/0	150/0 80/0 150/0	150/0 80/0 150/0	150/0 80/0 150/0	150/0 80/0 150/0	150/0 80/0 150/0	150/0 80/0 150/0	ISO 6247
Air release at 50 °C, maximum 75 °C, maximum	minimum minimum	5 -	5 -	5 -	5 -	10 -	13 -	- a)	- a)	ISO 9120
Water separation time to 3 ml emulsion at 54 °C time to 3 ml emulsion at 82 °C	minimum	30 –	30 –	30 –	30 –	30 –	30 –	– 30	– 30	DIN ISO 6614
Elastomer compatibility, NBR1, 100 °C, 168 h. relative increase in volume change in shore A hardness		0 to 18 0 to -10	0 to 15 0 to -8	0 to 15 0 to -8	0 to 12 0 to -7	0 to 12 0 to -7	0 to 10 0 to -6	0 to 10 0 to -6	0 to 10 0 to -6	ISO 6072

Oxidation stability, 1 000 h: delta neut. number, maximum insoluble sludge	mg KOH/g mg	2,0[a]	2,0[a]	2,0[a]	2,0[a]	2,0[a]	2,0[a]	2,0[a]	2,0[a]	ISO 42631
Wear protection, FZG A/8.3/90[f]	Fail stage	–	–	–	10	10	10	10	10	ISO 14635-1
Vane pump[g)] weight loss cam ring	mg, maximum	–	–	–	120 30	120 30	120 30	–	–	ISO 20763, procedure A
weight loss vanes	mg, maximum	–	–	–				–	–	

a) Report only.
b) For purposes of identification, dye may be used by agreement between supplier and end user.
c) Clear-bright is abbreviated as Clbr. Cleanliness level expressed according to ISO 4406 may be used by agreement between supplier and end user.
d) $mm^2 s^{-1}$ is equivalent to cSt.
e) Initial neutralization number is influenced by the presence of functional moieties in the total additive package.
f) Applicable from ISO VG 32 to ISO VG 150.
g) Applicable from ISO VG 32 to ISO VG 68.

temperatures are significant technical advantages over equi-viscous HLP oils. The absorption of energy is low, preheating of the system is usually not necessary and operating temperature is achieved rapidly. The use of HVLP oils affords significant rationalization potential. Generally speaking, HVLP oils have a multigrade character [2,4,5]. The use of these oils has grown rapidly over the past few years. They are perfect for fluctuating operating temperatures such as those found in mobile hydraulic systems, canal locks and cable-car hydraulics.

HVLP according to DIN 51 524-3 [10].

HV according to ISO 6743/4 (HR = HVL, without EP additives according to ISO 6743/4) [25].

The hydraulic oil grades described above, which fulfil DIN 51 524 and ISO 6743/4, all show demulsifying properties. They fulfil the demulsifying thresholds reported in DIN 51 599 and DIN ISO 6614.

In contrast with this, detergent and dispersing lubricants cannot fulfil the requirements relating to water separation.

The classification standard DIN 51 502 identifies detergent/dispersant (DD) oils with the letter D. Because tangible requirements are not specified, a direct comparison of ISO classifications is not possible. In a broader sense, HG hydraulic oils according to ISO 6743/4 can be allocated to this group [24,25].

11.4.5.5 HLPD Hydraulic Oils

Compared with HLP products, these hydraulic oils contain DD additives that finely disperse, suspend and emulsify water, dirt, ageing products and contaminants and thus hinder the accumulation of deposits on hydraulic components. The most polar DD additives also reduce friction and wear. Their polarity also affords good wetting, reduces stick–slip and leads to a lower coefficient of friction. These oils are often used in machine tools and other systems with sensitive control valves because they combat deposits. For machine tools, where cutting fluids can contaminate the hydraulic system, HLPD oils combat valve and cylinder gumming by emulsifying ingressed cutting fluid. HLPD oils hinder the precipitation of solid and fluid impurities. These are kept in suspension and are removed at the filtering stage. This is why filter capacities must often be increased if HLPD oils are used [2,4,6].

11.4.6
Fire-Resistant Hydraulic Fluids

Fire-resistant fluids have been developed for mining, die-casting, steel mill and aviation applications. These fluids have significantly higher ignition temperatures or fire-resistant properties and thus afford better fire protection than mineral oils. The use of these fluids is compulsory for some applications, for example underground coal mining.

Fire-resistant hydraulic fluids are classified according to the Luxembourg Report [14], ISO 6743/4 [25], VDMA sheets [26,27], CETOP RP97 H [28], DIN 51 502, Factory Mutual (FM-USA) [10] and new ISO 12922. Evaluated factors include the physical characteristics of the different types of fluids and their fire-resistant and technical specifications. The grading of the different types of fluids is identical in DIN 51 502, ISO 6743/4, ISO 12922 and the Luxembourg Report [10,25].

- HFAE: Oil-in-water emulsions (mineral-oil-based)
- HFAS: Synthetic, water-based solutions (ester- or/and polyglycol-based)
- HFB: Water-in-oil emulsions (mineral-oil-based)
- HFC: Water-based monomer and polymer solutions (polyglycol-based)
- HFD: Water-free fluids (ester-based)

11.4.6.1 HFA Fluids

HFA fluids are seldom used in industrial applications because of their poor anti-wear characteristics and very low viscosity. The additives used in HFA fluids usually improve corrosion protection for steels and non-ferrous metals, guarantee the bio-stability of the fluid and ensure compatibility with sealing materials (seal wear reduction). HFA fluids also contain friction-reducing additives but they only provide limited protection against wear. They are principally used in mining hydraulics and some steel mills [33]. The practical concentration of mineral-oil-based HFAE emulsions or synthetic HFAS solutions is between 1 and 5% in water (depending on water quality). Demands in Germany and other parts of the world for better HFA fluid biodegradability and less water pollution caused by HFA fluids, because large industrial leakages of these fluids pollute the soil and groundwater, have led to the development of new products. HFA fluids are gaining acceptance as hydraulic fluids in hydroforming machinery and in industrial robots used in the automobile industry (concentration: 5–7% in water). HFA concentrates are also recommended for plunger pump systems to protect the circuit against corrosion.

HFA fluids have to be carefully matched to the hardness of the water used to ensure adequate stability. This is particularly important in the mining industry [14,33]. HFA fluids should be used at their recommended concentrations to ensure optimum technical performance and bio-stability (e.g. crevice corrosion).

11.4.6.2 HFB Fluids

HFB fluids are water-in-oil emulsions with a (flammable) mineral oil content of about 60% (m/m). HFB fluids are almost used only in the UK mining industry and other UK-influenced countries (Commonwealth countries). Because of their high mineral oil content, these fluids do not pass prescribed spray-ignition tests in Germany and several other countries [14].

11.4.6.3 HFC Fluids

HFC fluids are normally based on a mixture of minimum 35–50% fully demineralized water with polyglycols (monomers or polymers) as thickeners. Low molecular weight ethylenes and/or propylene glycols are used to improve low-temperature behaviour of the fluid. Carefully matched additive systems that improve wear and corrosion protection and foaming behaviour ensure the required tribological properties. HFC fluids are normally somewhat alkaline (pH > 9) and contain fluid and gas-phase corrosion inhibitors. HFC hydraulic fluids can be used at pressures of up to and above 200–250 bar (new pumps > 350–400 bar). HFC fluids are used for a wide range of applications in the steel-making industry, foundries, forging plants, die-casting machines and hydraulic presses in offshore hydraulic systems and wherever pressurized hydraulic fluid leaks pose a fire hazard. The temperature range of HFC fluids is between −20 °C (start-up) and +60 °C (working temperature) because higher temperatures would lead to large fluid losses as a result of the high water content and the evaporative loss at high temperatures (vapour pressure) [14,33]. Water loss will increase the viscosity significantly.

11.4.6.4 HFD Fluids

HFD fluids are synthetic, water-free and fire-resistant hydraulic fluids. Usually, HFDR fluids based on phosphoric acid esters or HFDU fluids based on carboxylic acid esters or polyolesters are used. Until the 1980s, HFD fluids were the predominant hydraulic fluids. HFDS fluids based on chlorinated hydrocarbons then disappeared from the market because of PCB problems and replaced by HFC fluids for mining applications and by HFDU fluids for industrial applications. Currently, HFDR fluids containing phosphoric acid esters are used as control and regulator fluids in turbines and as hydraulic fluids in aviation systems [6]. HFDU fluids based on carboxylic acids and polyolesters cover the largest segment of HFD fluids [6]. They are normally used for hydrodynamic clutches and high-performance hydraulic systems at pressures of 250–350 bar and temperatures of 70–100 °C (and higher). HFDU fluids have tribological properties similar to those of mineral oils [2,5,6]. They show excellent characteristics in boundary lubrication and biodegradability. The ester-based HFDU fluids, currently available in the market, are only roughly comparable with HFC fluids in terms of fire protection. HFDU fluids (water-free) have flashpoints whereas HFC fluids (water-based) have no flashpoint due to the water content.

11.4.7 Biodegradable Hydraulic Fluids

Environmentally friendly, rapidly biodegradable hydraulic fluids were originally developed to ensure ecological compatibility. They are used in stationary and mobile systems. Their market share is growing rapidly and they are replacing mineral-oil-based hydraulic fluids in several areas. This trend has been underlined by the creation of VDMA guideline 24568 (old – not longer valid) [11] that specifies minimum requirements for HETG, HEES and HEPG fluids and the inclusion of these three fluid groups into ISO 6743/4 [25] in combination with the new ISO 15 380 specification (enlarged to include HEPR).

Rapidly biodegradable hydraulic fluids are allocated to several product families according to VDMA 24568 (minimum technical requirements – old) [11],

VDMA 24569 (changeover guidelines – old) [11], ISO 6743/4 [25] and ISO 15380; ISO 15380 is the actual well-respected classification and standard for biodegradable fluids:

HETG: Triglyceride (non-water-soluble and vegetable-oil types)

HEES: Synthetic ester types (non-water-soluble) – included partly saturated ester and saturated ester

HEPG: Polyglycol types (water-soluble)

HEPR: Polyalphaolefins and related hydrocarbon products

Table 11.7 shows the range of fire-resistant and environmentally friendly hydraulic fluids.

11.4.7.1 HETG: Triglyceride and Vegetable-Oil Types

Vegetable oils are also called 'natural esters'. The natural esters used for the HETG group of lubricants are primarily triglycerides. These are triple-valence glycerols and (fatty) acids. The most important esters are rapeseed oil but sunflower oil is also used. The physical and chemical properties of these base oils result from their fatty acid distribution. Natural fatty acids have even numbers and are unbranched. Fatty acids can be saturated but mono or multiple unsaturated fatty acids are also used. In general, one can say that a high proportion of unsaturated or short-chain fatty acids produces lubricants with a low pour point. A higher proportion of double bonds (unsaturated fatty acids) increases the sensitivity of a lubricant to oxygen and higher temperatures. Natural ester triglycerides are obtained from the seeds of oil plants such as rapeseed or sunflower by pressing or extraction. The oil thus obtained is subjected to a number of purification processes. Rapeseed oil and other natural fatty acids can be used as the raw material for HETG lubricants or as the raw material for synthetic esters.

The selection of the most suitable natural ester and the corresponding matched additive system is of fundamental importance to the quality of an environmentally friendly lubricant based on harvestable raw materials. Because HETG hydraulic oils withstand lower thermal and oxidative loads than mineral oils, they are used for medium-temperature and low-pressure applications (often in total-loss application). They have very low market share.

The minimum requirements of triglycerides are described in ISO 15380 (Table 11.1) [2,5,6,13,34].

11.4.7.2 HEES: Synthetic Ester Types

Synthetic esters are a group of substances with wide variation in structure. Esters are manufactured by chemically altering alcohols and acids. Alcohols and acids from a broad range of raw materials can be randomly combined to achieve the desired product characteristics such as thermal and hydrolytic stability, low-temperature properties, seal compatibility and so on. The chain length and the branching characteristics (linear or branched) of alcohols can be varied, as can

Table 11.7 Classification of fire-resistant fluids and rapidly biodegradable fluids.

Category according to the 7 th Luxembourg Report, DIN 51 502 and ISO 6743/4/ISO 12922 FM-USA	Composition typical characteristics	Field of application operating temperatures
Water-containing fire-resistant hydraulic fluids		
HFA E	Oil-in-water emulsions, mineral oil or synthetic ester Water content > 80%	Power transmissions, about 300 bar, high working pressures, powered roof support
HFA S	Mineral-oil-free aqueous synth. chemical solutions Water content > 80%	hydrostatic drives, about 160 bar, low working pressures 5 to < 55 °C
HFB	Water-in-oil-emulsions Mineral oil content about 60%	For example in the British mining industry, not approved in Germany 5–50 °C
HFC	Water polymer solutions water content > 35%	Hydrostatic drives, industry and mining applications −20 to 60 °C
Water-free, synthetic-fire resistant hydraulic fluids		
HFD R	Water-free synthetic fluids, consisting of phosphate esters, not soluble in water	Lubrication and control of turbines, industrial hydraulics −20 to 150 °C, in hydrostatic applications often 10–70 °C
HFD U	Water-free synthetic fluids of other compositions (e.g. carboxylic acid esters)	Hydrostatic drives, industrial hydraulic systems −35 to < 90 °C
Category according to ISO 15380 (VDMA 24568)	**Composition typical characteristics**	**Field of application operating temperatures**
Water-free rapidly biodegradable hydraulic fluids		
HEPG	Polyalkylene glycols soluble in water	Hydrostatic drives, e.g. locks, 'water hydraulics' −30 to < 90 °C (reservoir temperature −20 to 80 °C)
HETG	Triglycerides (vegetable oils) not soluble in water	Hydrostatic drives, e.g. mobile hydraulic systems −20 to < 70 °C (reservoir temperature −10 to 70 °C)
HEES	Synthetic esters not soluble in water	Hydrostatic drives, mobile and industrial hydraulic systems −30 to < 90 °C (reservoir temp. −20 to 80 °C)
HEPR	Polyalphaolefins and/or related hydrocarbons not soluble in water	Hydrostatic drives, mobile and industrial hydraulic systems −35 to < 80 °C (reservoir temperature −30 to 100 °C)

the number of OH groups (monoalcohols, diols and polyols). The acids can include monocarboxylic acids, dicarboxylic acids and polycarboxylic acids with aliphatic, aromatic, straight or branched structures. They can be saturated or partially unsaturated [6,13]. The most widely used synthetic esters are probably dicarboxylic acid esters, polyol esters (e.g. TMP esters) and complex esters (manufactured from more than one alcohol or acid component). The raw materials for synthetic esters originate from, on the one hand, petrochemicals and, on the other hand, from natural materials and their chemical transformation. Partially unsaturated ester oils form the largest group of rapidly biodegradable lubricants. Fully synthetic, saturated ester oils based on saturated TMP, polyol and complex esters show the best technical performance. Their advantages include outstanding oxidation stability, good material compatibility and excellent tribological performance. These products, which are much more expensive than mineral oils, are used in high-pressure, high-temperature and highly stressed hydraulic systems, for example mobile hydraulics and hydrostatic drives, under difficult conditions [35–39].

The minimum requirements for ester oils are specified by ISO 15380, Group HEES and synthetic esters [11]. These requirements do not differentiate between partially unsaturated and saturated systems. The VDMA guideline 24568 is substituted by the official edition of ISO 15380 (dated 2002) [12]. Table 11.8 shows the requirements of the types of HEES hydraulic fluids according to ISO 15380. It lists synthetic esters according to their ageing stability (unsaturated systems – ageing stability according to Baader: 72 h at 110 °C; synthetic saturated esters – ageing stability in the dry TOST test without the addition of water). Furthermore, special material compatibility tests (such as non-ferrous metal compatibility according to Linde) are included in the specifications [13].

11.4.7.3 HEPG: Polyglycol Types

As already mentioned, polyglycols differ from the hydraulic fluids because they are water soluble. Depending on the substances used to manufacture the polyglycol (polyethylene oxide and polypropylene oxide), their molecular mixture ratio and the alcohol start-up molecule used, polyglycols of different types can be synthesized. Ethylene-oxide-based polyglycols are highly water soluble, poorly miscible with mineral oil and have high polarity. Polyglycols with a high proportion of propylene oxide are not water soluble, or only slightly soluble, to some extent miscible with mineral oil and are significantly less polar than polyethylene glycols. As a result of their 'water solubility', glycol-based hydraulic fluids can, and may, contain water. The water solubility of polyglycols (and thus their mobility in the ground – groundwater contamination) and their incompatibility with mineral oils have limited their acceptance. Polyglycols are primarily used in the water-supply industry, canal lock hydraulics and offshore applications as rapidly biodegradable hydraulic fluids especially when the application leads to unavoidable contamination of the hydraulic fluid with ingressed water [13].

Table 11.8 Minimum requirements of hydraulic fluids – ISO 15380 – May 2004.

Characteristics of test	Units	Requirements					Test method or standard
Viscosity grade		22	32	46	68	100	ISO 3448
Density at 15 °C	$kg\ m^{-3}$	a)	a)	a)	a)	a)	ISO 12185
Colour[h)]		a)	a)	a)	a)	a)	ISO 2049
Appearance at 25 °C[c)]		Clbr	Clbr	Clbr	Clbr	Clbr	
Ash content, maximum	% (m/m)	d)	d)	d)	d)	d)	ISO 6245
Flashpoint Cleveland open cup, minimum	°C	165	175	185	195	205	ISO 2592
Kinematic viscosity[e)]	$mm^2\ s^{-1}$	d)	d)	d)	d)	d)	ISO 3104
at −20 °C, maximum	$mm^2\ s^{-1}$	300	420	780	1 400	1 500	ISO 3104
at 0 °C, maximum	$mm^2\ s^{-1}$	19,8–24,2	28,8–35,2	41,4–50,6	61,2–74,8	90,0–110	ISO 3104
at 40 °C, minimum–maximum	$mm^2\ s^{-1}$	4,1	5,0	6,1	7,8	10,0	ISO 3104
at 100 °C, minimum							
Pour point, maximum	°C	−21	−18	−15	−12	−9	ISO 3016
Low temperature fluidity after 7 d	°C	d)	d)	d)	d)	d)	ASTM D 2532
Acid number[f)], maximum	g KOH/g	d)	d)	d)	d)	d)	ISO 6618
Water content, maximum	$mg\ kg^{-1}$	1 000	1 000	1 000	1 000	1 000	ISO 12937 or ASTM D 1744 or DIN 51 777-1
Copper corrosion, 100 °C 3 h, maximum	rating	2	2	2	2	2	ISO 2160
Rust prevention, procedure A		Pass	Pass	Pass	Pass	Pass	ISO 7120
Foam at 24 °C, maximum	ml	150/0	150/0	150/0	150/0	150/0	ISO 6247
at 93 °C, maximum		75/0	75/0	75/0	75/0	75/0	
at 24 °C, maximum		150/0	150/0	150/0	150/0	150/0	
Air release, 50 °C, maximum	min	7	7	10	10	14	ISO 9120
Water separation, time to 3 ml emulsion at 54 °C, maximum	min	g)	g)	g)	g)	g)	DIN ISO 6614
Elastomer compatibility[g)] HNBR, FPM, NBR1, AU after 1.000 h at	°C	60	80	80	100 Except NBR1 and AU.	100 Except NBR1 and AU.	ISO 6072

Change in shore-A-hardness, maximum	IRHD	± 10	± 10	± 10	± 10	± 10	ISO 6072
Change in volume, maximum	%	−3 to +10	−3 to +10	−3 to +10	−3 to +10	−3 to +10	ISO 6072
Change in elongation, maximum	%	30	30	30	30	30	
Change in tensile strength, maximum	%	30	30	30	30	30	
Oxidation stability	h	a),d)	a),d)	a),d)	a),d)	a),d)	ASTM
Modified TOST, dry TOST	%	20	20	20	20	20	D943[h)]
Baader test, 110 °C, 72 h							DIN 51 554-3
Increase in viscosity at 40 °C, maximum							
Wear protection, FZG A/8.3/90, minimum	failure load stage	i)	10	10	10	10	DIN 51 354-2
Vane pump weight loss	mg	120	120	120	120	120	IP 281 or
Ring, maximum	mg	30	30	30	30	30	CETOP TP 67 H
Vane, maximum							DIN 51 389-2

a) Report.
b) For purpose of identification, dye may be used by agreement between supplier and end user.
c) Clear-bright is abbreviated as Clbr.
d) Criteria of performance or characteristics values to be negotiated between supplier and end user.
e) $mm^2 s^{-1}$ is equivalent to cSt.
f) Initial acid number is influenced by the presence of functional moieties in the total additive package.
g) The type of elastomer and definition of compatibility are to be agreed between supplier and end user.
h) Modification consisting in driving the oxidation stability test without water, and water is replaced by hydraulic fluid.
i) Not applicable to viscosity grade 22.

11.4.7.4 HEPR: Polyalphaolefin and Related Hydrocarbon Products

Because of their biodegradability, low molecular weight polyalphaolefins and the correspondingly derived hydrocarbons are classified as environmentally friendly hydraulic fluids. In ISO 15380 [12], these products were classified as HEPR. The inclusion of this new group in ISO 15380 is an extension compared with the 'old VDMA 24568' [11]. HEPR fluids are more rapidly biodegradable than mineral oils but significantly less biodegradable than most ester oils and natural oils such as rapeseed oil. The technical properties of these oils are similar to those of mineral oils.

11.4.8 Hydraulic Fluids for the Food and Beverage Industry

The classification of hydraulic oils for the food and beverage industry is based on Food and Drug Administration (FDA) and NSF guidelines [29].

The NSF issue of H1 or H2 lubricant registrations depends on the FDA classification of components such as base oils and additives (FDA code of federal regulations 21 CFR 178.3570). In 1998, the NSF organization has taken over the activities of the United States Department of Agriculture (USDA) which stopped their activities.

Since 2008, InS Services (UK) Ltd is offering lubricant registrations as an alternative to NSF.

The European hygiene guideline 93/43 EWG together with the hazard analysis and critical control points (HACCP) management system promotes the use of safe lubricants in the food-processing industries. Unfortunately, the European law does not provide any further guidelines about food-safety lubricants.

11.4.8.1 H2 Lubricants

This classification indicates that these lubricants are suitable for general use in the foodstuff industry if no direct contact with the foodstuffs can occur [29].

11.4.8.2 H1 Lubricants

The classification H1 stands for lubricants that can be used in the food-processing industries if occasional, technically unavoidable contact between the foods and the lubricants occurs. In other words, if there is a chance of lubricants coming into contact with foods, the lubricants must be NSF H1 registered or registered according to InS H1 [29].

The base oils for such products are principally special white oils, special polyalphaolefin grades and special polyglycols. H2 lubricants can be standard lubricants if special precautions are taken. As the USDA is not issuing any new registrations since 1998, all previously issued USDA registrations are not valid anymore. All formerly registered products had to become re-registered by NSF International or InS Service (UK) Ltd.

DIN and European activities for this group of lubricants have resulted in a new standard called ISO 21469: 'Safety of Machinery – Lubricants with Incidental Product Contact – Hygiene Requirements' [40].

ISO 21469 is not a replacement for H1 registrations but an additional certificate which proves to food manufactures that the lubricant manufacturing plant is applying a comprehensive risk assessment to its facility.

11.4.9
Automatic Transmission Fluids

ATF are functional fluids for automatic transmissions in vehicles and machinery. In special circumstances, ATF are used in manual synchromesh gearboxes and hydraulic gearboxes/transmissions. These products have excellent viscosity–temperature behaviour, shear stability, high oxidation stability, minimal foaming and outstanding air release. As a result of different manufacturers' demands on a fluid's friction characteristics (coefficient of friction on the brake and clutch bands), there are several types of AFT (see Chapter 9) [15,30].

11.4.10
Fluids in Tractors and Agricultural Machinery

Gear oils and hydraulic fluids are specified for the gearboxes and hydraulic systems of tractors and agricultural machinery fitted with and without 'wet' brakes. The most commonly used types of oils are universal tractor transmission oil (UTTO), super tractor universal oil (STUO) and multi-functional oil (MFO) (see Chapter 9) [15,30].

11.4.11
Hydraulic Fluids for Aircraft

Since the 1930s, the use of hydraulic systems in aviation has increased considerably. The principal functions of aircraft hydraulics are controlling the elevator, rudder, ailerons, flaps, landing gear, doors, brakes and so on. The most commonly used pumps are of the axial piston type with pressures between 100 and 280 bar. Fluid volumes, between 20 and 40 l, are relatively small. The hydraulic fluids used must be free from particle contaminants because aircraft hydraulics have to work under extreme climatic conditions with enormous fluctuations in temperature. The hydraulic fluids must have outstanding thermal stability and good viscosity–temperature behaviour. These fluids should also be fire-resistant. A wide variety of different hydraulic fluids are currently used in aircraft. These include those based on phosphoric acid esters, polyalphaolefins, silicate esters and special mineral oils. One standard they must fulfil is US military specifications, for example

Table 11.9 Performance requirements of US hydraulic specifications US Steel 126 and 127.

Performance requirements ASTM test	US Steel 126	US Steel 127
Viscosity D-88	Suitable for specified application	Suitable for specified application
Viscosity index D-567	Not less than 80	Not less than 90
COC Flash point D-92	375°F minimum	375°F minimum
Hydraulic pump test D-2882 (100 h at 2000 psig)	0.05% Total wear (by weight)	50 mg maximum
Four-ball wear test D-2266 (40 kg, 1800 rpm, 130°F, 1 h)	No more than 0.30 mm scar diameter	Not more than 0.50 mm scar diameter
Rotary bomb oxidation D-2272	120 min minimum	120 min minimum
Low-temperature cycling test (US steel method)	Must pass at 15°F	Must pass at 15°F
Water emulsion test D-1401 at 130 °F	ml ml ml min Oil water emulsion 40 37 3 ≤ 30	ml ml ml min Oil water emulsion 40 37 3 ≤ 30
Rust-prevention test D-665 A	No rust	No rust

MIL H 5606 or others. These specifications serve as basis for other specifications issued by manufacturers of aircraft and aircraft hydraulics [6].

11.4.12
International Requirements on Hydraulic Oils

The requirements of hydraulic fluids are normally set out in national standards or company-specific specifications. Apart from ISO 6743/4 and DIN 51 524, there are several specifications issued by leading component manufacturers. Although these specifications often overlap, there are differences in detail, normally in relation to thermal/oxidation stability, mechanical/dynamic tests, filterability tests and in-house tests performed by some manufacturers. Table 11.9 shows US Steel's 126 and 127 requirements. Table 11.10 shows some OEM hydraulic tests and part of specifications.

11.4.13
Physical Properties of Hydraulic Oils and Their Effect on Performance

11.4.13.1 Viscosity and Viscosity–Temperature Behaviour

Viscosity is the most important measure of the load-carrying properties of a hydraulic oil. A differentiation is made between dynamic and kinematic viscosities [25,41,42].

Table 11.10 Major OEM Hydraulic Oils Specifications–Extract

Performance (ASTM Test)	Denison HF-0 Piston and vane pump test (hybrid pump)	Denison HF-1 Piston pump test	Denison HF-2 Vane pump test	Bosch Rexroth[1)]/ Vickers[2)]	Cincinnati Milacron P-68,P-69, P-70	Rust and Oxidation Cincinnati Milacron P-38, P-54, P-55, P-57
SLUDGE & METAL CORROSION (1000 hrs, D943)						
Neutralisation No. (D974)	1.0 mg KOH/g max.	1.0 mg KOH/g max.	1.0 mg KOH/g max.			
Insoluble Sludge, max.	100 mg	100 mg	100 mg			
Total Copper, max.	200 mg	200 mg	200 mg			
C.M. THERMAL STABILITY TEST						
168 hrs. at 135 °C (Cincinnati P-70, ISO 46)						
Results After Test						
Viscosity (D445)					5% max. change	5% max. change
Neutralisation No. (D974)					± 50% max. change	0.15 max. increase
Sludge or Precipitate	100 mg/100 ml max.	100 mg/100 ml max.	100 mg/100 ml max.		25 mg/100 ml max.	25 mg/100 ml max.
Condition of Steel Rod						
Visual					No discoloration	No discoloration
Deposits (per 200 ml)					3.5 mg max.	3.5 mg max.
Metal Removed (per 200 ml)					1.0 mg max.	1.0 mg max.
Condition of Copper Rod						
Copper Rod Rating, Visual (CM)	Report	Report	Report		5 max.	5 max.
Copper Weight Loss	10.0 mg max.	10.0 mg max.	10.0 mg max.		10.0 mg max.	10.0 mg max.
HYDROLYTIC STABILITY (D2619)						
Copper Weight Loss (mg/cm^2)	0.2 max.	0.2 max.	0.2 max			
H_2O Acidity (mg KOH)	4.0 max.	4.0 max.	4.0 max.			
PUMP WEAR TEST (D2882)				Bosch Rexroth 1) Fluid Rating RD 90235, 2016, Germany	50 mg max., 100 hours	

(continued)

Table 11.10 *(Continued)*

Performance (ASTM Test)	Denison HF-0 Piston and vane pump test (hybrid pump)	Denison HF-1 Piston pump test	Denison HF-2 Vane pump test	Bosch Rexroth[1)]/ Vickers[2)]	Cincinnati Milacron P-68, P-69, P-70	Rust and Oxidation Cincinnati Milacron P-38, P-54, P-55, P-57
DENISON AXIAL PISTON & VANE PUMP WEAR TEST (in cooperation with Denison when deemed necessary)	T6 H20C according to TP30533	P-Pump	T-Pump			
VICKERS HP VANE PUMP TEST[1)]				Total-90 mg max.[2)]		
MANNESMANN REXROTH INHOUSE PUMP TEST[2)]				Inhouse Report[2)]		
FILTERABILITY (DENISON PROCEDURE)						
Filtration Time w/o Water	600 secs max.		600 secs max.			
Filtration Time with Water	Not to exceed two times the filtration rate w/o water		Not to exceed two times the filtration rate w/o water			
PALL Filtration			80 min			
RUST TEST (D665A, D665B) 24 hrs.	Procedure A&B	Procedure A&B	Procedure A&B		Procedure A	Procedure A
FOAM (D892) after 10 min.	No foam	No foam	No foam			
NEUTRALISATION NUMBER (D974) mg KOH/g					1.5 max.	0.2 max.

Industrial lubricants and hydraulic oils are allocated ISO viscosity grades on the basis of their kinematic viscosity that in turn is described by the ratio of dynamic viscosity and density. The reference temperature is 40 °C. The official SI unit for kinematic viscosity is $m^2 s^{-1}$. In the petroleum industry, the unit cSt or $mm^2 s^{-1}$ is used. The ISO viscosity classification, DIN 51 519, for liquid industrial lubricants defines 18 viscosity grades from 2 to 1500 $mm^2 s^{-1}$ at 40 °C. Every grade is defined by the average mean viscosity at 40 °C and the permissible deviation of ±10% of this value [43]. The viscosity–temperature behaviour (V–T) is of great importance to hydraulic oils. Viscosity increases sharply as temperature decreases and decreases as the temperature increases. In practical terms, the threshold viscosity of a fluid (permissible start-up viscosity, about 800–2000 $mm^2 s^{-1}$) must be observed for the different types of pumps [16–18,36,37,44]. The minimum permissible viscosity at higher temperatures is defined by the start of the boundary friction phase. The minimum viscosity should not be less than 7–10 $mm^2 s^{-1}$ to avoid unacceptable wear to pumps and motors [16–18,21,36,37,44]. In viscosity–temperature diagrams, the viscosity of a hydraulic fluid is plotted against temperature. In linear terms, V–T graphs are hyperbolic. By mathematical transformation, these V–T graphs can be shown as straight lines. These lines enable the exact viscosity to be determined over a wide range of temperature. The measure of V–T behaviour is the VI and the V–T line gradient in the diagram. The higher the VI of a hydraulic oil, the less its viscosity changes with temperature, that is the flatter the V–T line (see Chapter 3). Mineral-oil-based hydraulic oils usually have natural VI of 95–100. Synthetic, ester-based hydraulic oils have natural VI of 140–180 and polyglycols have natural VI of 180–200 (Figure 11.10) [2,5,6,44].

The viscosity index can also be increased by additives (polymeric additives that must be shear-stable) called VI improvers. Hydraulic oils of high VI enable easy pump start-up, less loss in performance at low ambient temperatures, and improved sealing and wear protection at high operating temperatures. High-VI oils improve the efficiency of a system and increase the life of components subjected to wear (higher viscosity at operating temperatures will result in better volume efficiency). High VI hydraulic fluid can save energy by 5–15% due to better volumetric efficiency.

11.4.13.2 Viscosity–Pressure Behaviour

The viscosity–pressure behaviour of a lubricant is principally responsible for the load-carrying properties of a lubricating film. The dynamic viscosity of fluid media increases with pressure. The following relationship governs the dependence of dynamic viscosity on pressure at a constant temperature.

Viscosity–pressure behaviour and the increase in viscosity at increasing pressures have a high positive effect on specific loading (such as on bearings) because the viscosity of the lubricating film increases when exposed to high partial pressure and this increases load-carrying capacity. Assuming an increase in pressure from 0 to 2000 bar, the viscosity would increase by factors of 2 for an HFC fluid,

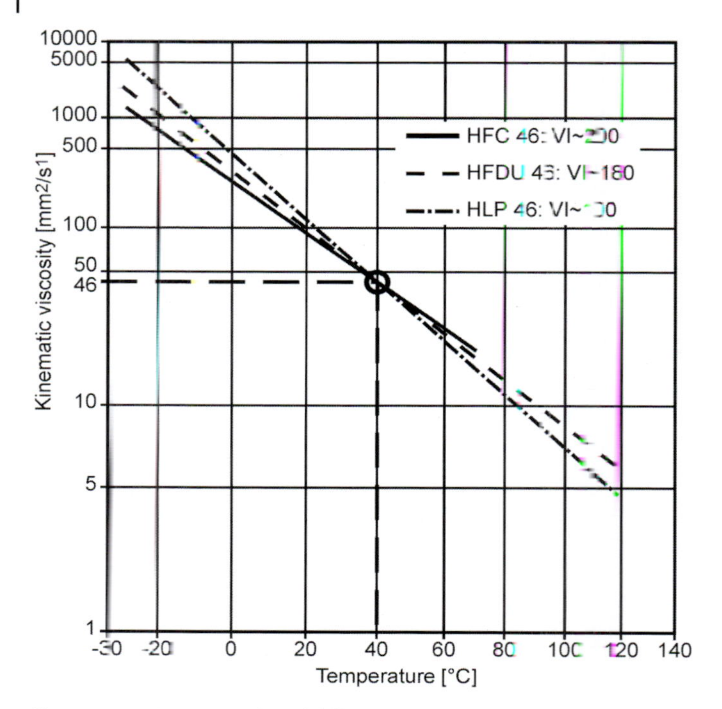

Figure 11.10 Viscosity index of different hydraulic fluids.

30 for a mineral oil and 60 for an HFD fluid [2,4]. This explains why roller bearings have a relatively short life if water-based (HFA, HFC) lubricants are used. Figures 11.11 and 11.12 show the dependence of viscosity on pressure for different hydraulic fluids. Viscosity–temperature behavior can also be described by an exponential statement (see Chapter 3) (Figure 11.13) [2,4,6,45,46]:

$$\eta = \eta_0 \times e^{\alpha P}$$

where η_0 is the dynamic viscosity at atmospheric pressure, α is the viscosity–pressure coefficient and P is the pressure. For HFC, $\alpha = 3.5 \times 10^{-4}\,\text{bar}^{-1}$; for HFD, $\alpha = 2.2 \times 10^{-3}\,\text{bar}^{-1}$; and for HLP, $\alpha = 1.7 \times 10^{-3}\,\text{bar}^{-1}$.

11.4.13.3 Density

Losses in the pipework and elements of a hydraulic system are directly proportional to the density of the fluid [2,42]. For example, pressure losses are directly proportional to density:

$$\Delta P = (\rho/2) \times \zeta \times c^2$$

where ρ is the density of the fluid, ζ is the resistance coefficient, C is the fluid flow velocity and ΔP is the pressure loss. The density ρ is the mass per unit

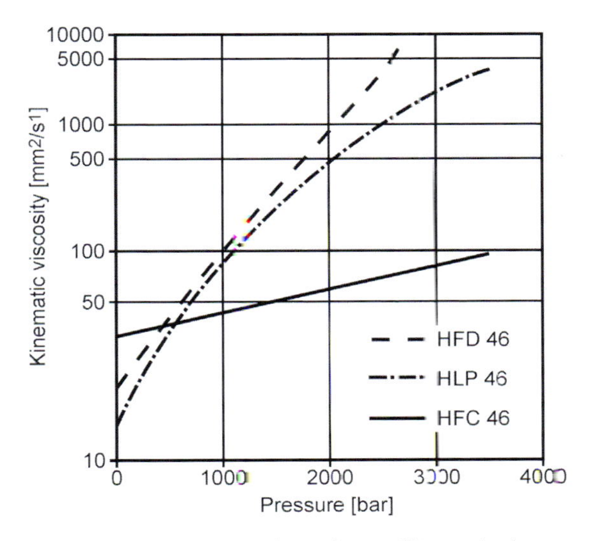

Figure 11.11 The pressure dependence of kinematic viscosity.

volume of fluid.

$$\rho = m/V [\text{kg m}^{-3}]$$

The density of hydraulic fluids is measured at 15 °C according to DIN 51 757. The density of a hydraulic fluid is dependent on temperature and pressure because the volume of a fluid expands as its temperature increases. The change in volume of a fluid as a result of heating is thus expressed as

$$\Delta V = V \times \beta_{(\text{Temp.})} \times \Delta T$$

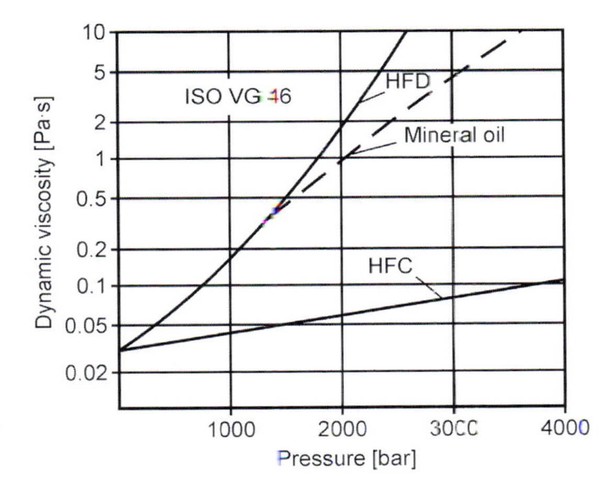

Figure 11.12 The pressure dependence of dynamic viscosity

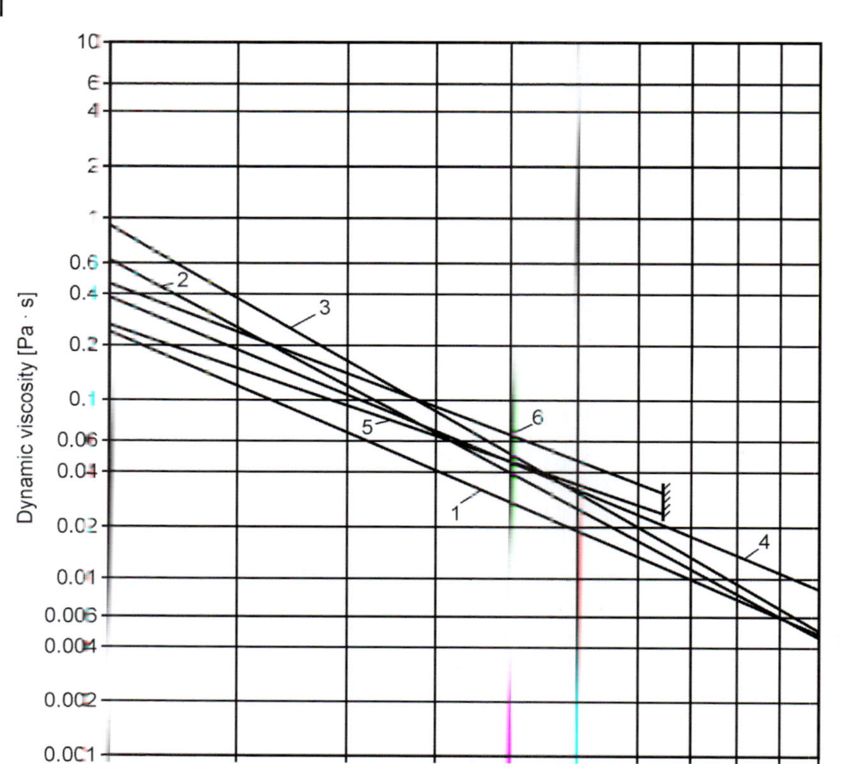

Figure 11.13 Viscosity–temperature characteristics of hydraulic fluids [2,4,5].

and this causes the change in density:

$$\Delta\rho = \rho \times \beta_{(\text{Temp.})} \times \Delta T$$

In hydrostatics, it is sufficient to apply the linear formula to the above equation at temperatures between −50 and +150 °C. The thermal volume-expansion coefficient, $\beta_{(\text{Temp.})}$, can be applied to all different types of hydraulic fluids [2,4–5,45,46].

Base oil	$\beta_{(Temp.)}$ (K^{-1}) Thermal volume expansion coefficient at atmospheric pressure
HLP 46 Mineral oil	$0.65–0.70 \times 10^{-3}$
HFC 46 Water, glycol	0.75×10^{-3}
HFD 46 Ester oil	0.75×10^{-3}

As the coefficient of expansion for mineral oils is approximately 7×10^{-4} K^{-1}, the volume of a hydraulic oil would increase by 0.7% (vol.) if its temperature increased by 10 °C. Figure 11.14 shows the temperature dependence of the density of common hydraulic fluids [2,4].

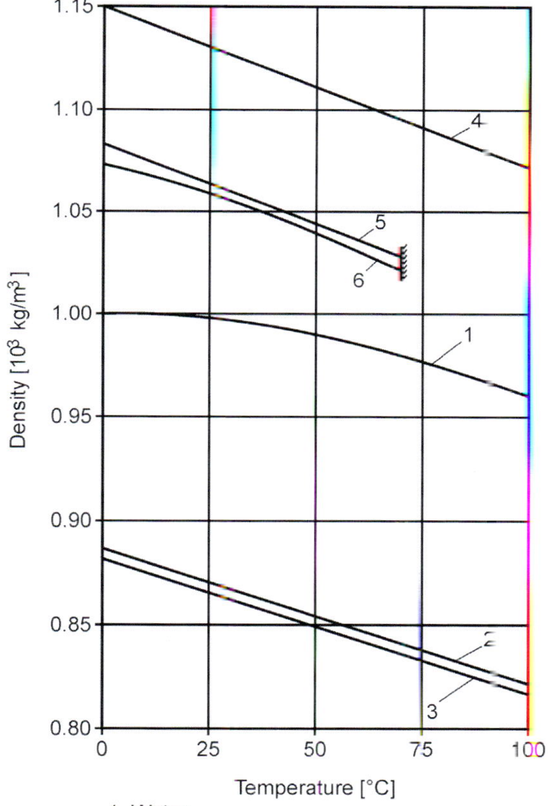

1. Water
2. Mineral oil, HLP 32
3. Mineral oil, HLP 46
4. Phosphoric-acid-ester, HFD R 46
5. Water-glycol-solution, HFC 46 (40% H_2O)
6. Water-glycol-solution, HFC 46 (45% H_2O)

Figure 11.14 Temperature dependence of the density for different hydraulic fluids.

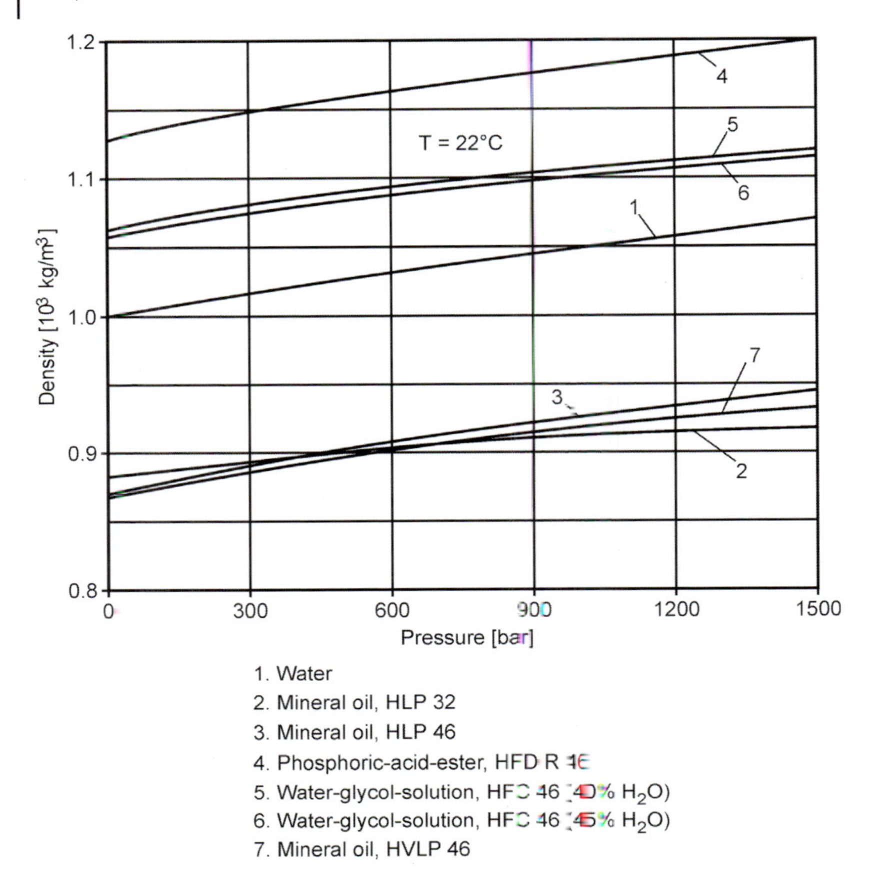

1. Water
2. Mineral oil, HLP 32
3. Mineral oil, HLP 46
4. Phosphoric-acid-ester, HFD R 46
5. Water-glycol-solution, HFC 46 (40% H_2O)
6. Water-glycol-solution, HFC 46 (45% H_2O)
7. Mineral oil, HVLP 46

Figure 11.15 Pressure dependence of the density for different hydraulic fluids.

The density–pressure behaviour of hydraulic fluids should also be included in a hydrostatic evaluation because the compressibility of fluids affects their dynamic performance. The dependence of density on pressure can simply be taken from corresponding diagrams (Figure 11.15) [2,4,45,46].

11.4.13.4 Compressibility

The compressibility of mineral-oil-based hydraulic fluids depends on temperature and pressure. At pressures up to 400 bar and temperatures up to 70 °C, which are top-end values in industrial systems, compressibility is irrelevant to the system. The hydraulic fluids used can be viewed as incompressible. At pressures from 1000 to 10 000 bar, however, changes in the compressibility of the medium can be observed. Compressibility is expressed in terms of the compressibility coefficient β or the compressibility modulus M (Figure 11.16; $M = K$) [2,45,46].

$$M = 1/\beta \text{ bar} = 1/\beta \times 10^5 \text{ N m}^2 = 1/\beta \times 10^5 \text{ Pa}$$

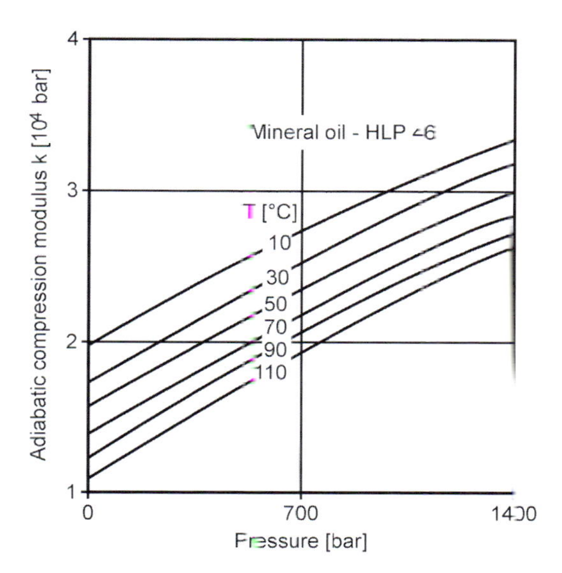

Figure 11.16 Adiabatic compression module of hydraulic oil.

The change in volume can be determined by using the following equation:

$$\Delta V = V \times \beta \times (P_x - P_{start})$$

where ΔV is the volume change, P_x is the maximum pressure and P_{Start} is the initial pressure.

11.4.13.5 Gas Solubility and Cavitation

Air and other gases can dissolve in fluids. A fluid can absorb a gas until a saturation point is reached: this need not negatively affect the characteristics of the fluid. The solubility of a gas in a fluid depends on the base fluid, type of gas, pressure and temperature. Up to about 300 bar, gas solubility is proportional to pressure, and Henry's law applies [2,6]:

$$V_G = V_F \times \alpha_V \times P/P_0$$

where V_G is the volume of gas dissolved, V_F is the volume of fluid, P_0 is the atmospheric pressure, P is the fluid pressure and α_V is the Bunsen coefficient (1.013 mbar, 20 °C).

The Bunsen coefficient is highly dependent on the base fluid and indicates how much (% v/v) gas is dissolved in a unit volume of the fluid under normal conditions. The dissolved gas can escape from the hydraulic fluid at low static pressure (high flow velocity and shearing stress) until a new saturation point is reached. The speed at which the gas escapes is usually higher than the speed at which the gas is absorbed. Gas that escapes as bubbles changes the compressibility of the fluid similarly to air bubbles. Even at low pressures, small quantities of air can drastically reduce the incompressibility of a fluid. Up to about 5% (v/v) undissolved air can be found in mobile systems with high fluid circulation. This undissolved air has a highly negative effect on the performance, the load rigidity and dynamics of a system (see Sections 11.4.13.6 and 11.4.13.7). As the pressurization of fluids in systems is usually very rapid, any air bubbles can suddenly heat up to a high temperature (adiabatic compression). In extreme cases, the ignition temperature of the fluid might be reached and micro-diesel effects can occur. The gas bubbles can also collapse in pumps as a result of pressurization and this can cause erosion damage (sometimes called cavitation or pseudo-cavitation) [2,6]. The situation can become worse if vapour bubbles are created in the fluid. Thus, cavitation occurs when the pressure falls below the gas solubility or vapour pressure of the fluid. Cavitation mainly occurs in open systems with a constant volume. These can include inlet and intake circuits and pumps. Therefore, the primary reasons of cavitation are very low absolute pressure as a result of flow losses in narrow cross sections, filters, manifolds and chokes, excessive inlet head and pressure losses resulting from excessive fluid viscosity. Cavitation can cause pump erosion, poor efficiency, pressure peaks and excessive noise. It can also detrimentally affect the stability of choke controls and cause foaming in tanks if the fluid–air mixture is returned to the tank at atmospheric pressure [2,4,6,47].

11.4.13.6 Air Release

The flow of hydraulic fluids back into tanks can drag air into the fluid. In addition, leaks in pipework at restrictions and partial vacuums can draw air bubbles into the system. Turbulence in the tank or local cavitation promotes the formation of air bubbles in the fluid.

This air separates the surface of the fluid or it will be drawn into the pump and this can cause other system or component damage. The speed at which the air bubbles rise to the surface depends on the diameter of the bubbles, viscosity of the fluid, density and quality of the base oils. The higher the quality and purity of the base oils, the faster is air release. Low-viscosity oils usually release air more rapidly than high-viscosity oils because of the speed at which the bubbles rise:

$$C = (\rho_{\mathrm{FL}} - \rho_{\mathrm{L}}) \times X/\eta$$

where ρ_{FL} is the density of the fluid, ρ_{L} is the density of the air, η is the dynamic viscosity and X is a constant that depends on the density and viscosity of the fluid [2,6].

Systems should be designed to hinder the ingress of air into the fluid and to assist the release of any trapped air bubbles. Critical areas include the tank, which should be fitted with baffles and deflectors, and the positioning of return pipes and circuits. Additives cannot positively influence air-release properties. Surface-active components (such as silicone-based defoamers) and contaminants (such as greases and corrosion preventives) detrimentally affect the air-release properties of hydraulic oils [5,48,49]. The air-release properties of mineral oils are generally better than that of fire-resistant fluids. The air-release properties of HPLD hydraulic fluid can be comparable with those of HLP hydraulic fluids.

Air-release properties are defined by DIN 51 381. The test involves air being blown into the oil. The air-release figure is the time the air (below 0.2% v/v) takes to escape from the oil at 50 °C under given conditions. The proportion of dispersed air is determined by measuring the density of the oil–air mixture [42] (see Chapter 18).

11.4.13.7 Foaming

Surface foaming occurs when the speed of air release is greater than the speed at which the bubbles at the surface of the fluid collapse, that is more bubbles are created than collapse. In the worst cases, this foam can be forced out of tank vents or openings or can be drawn into the pump. Additives, silicone-based or silicone-free defoamers can accelerate the collapse of bubbles by reducing the surface tension of the foam. They also negatively influence the air-release properties of the fluid, however, and this can cause compression and cavitation problems. Defoamers are thus used in very small concentrations (about 0.001%). The concentration of defoamer in a fluid can progressively decline as a result of ageing and settling on metal surfaces, and foaming problems often affect old, used fluids. Subsequent addition of defoamers should only be performed after consultation with the fluid manufacturer. Foaming behaviour is defined by ISO 6247/ASTMD 892. This involves defined air being passed through a porous stone that is submerged in the fluid. The volume of foam that gathers on the surface of the fluid is measured against time (at once, after 10 min) and at different temperatures (25 and 95 °C) [42]. Surface-active substances, detergent or dispersing additives, contamination in the form of grease, corrosion preventives, cleaners, cutting fluids, ageing by-products and so on can all have a negative effect on foaming behaviour [5,22,48,49] (see Chapter 13).

11.4.13.8 Demulsification

Demulsification is the capacity of a hydraulic fluid to repel ingressed water. Water ingress can result from a leak in a heat exchanger, the formation of condensation water in tanks as a result of considerable changes of the oil level, poor filtering, water contamination through faulty seals and extreme ambient conditions. Water in a hydraulic fluid can cause corrosion, cavitation in pumps, increases friction and wear and accelerate the decay of elastomers and plastics. Free water should be removed from hydraulic fluid tanks as soon as possible by means of draincocks. Contamination with water-miscible cutting fluids which is especially possible in machine tools can promote the formation of sticky residues after the evaporation of water. This can cause problems in pumps, valves and cylinders. A hydraulic fluid should rapidly and completely repel ingressed water. Demulsification is defined by DIN ISO 6614 but it cannot be used for hydraulic oils that contain detergent/dispersing additives. Demulsification is the time taken to separate an oil–water mixture. The parameters are as follows:

- Viscosity up to 95 $mm^2 s^{-1}$ at 40 °C; test temperature 54 °C.
- Viscosity > 95 $mm^2 s^{-1}$; test temperature 82 °C [42]

In DD hydraulic oils, fluid and solid contaminants and water are held in fine suspension and can be removed by using suitable filter systems without the hydraulic function of machine and the fluid is negatively affected. DD hydraulic oils are thus often used in hydrostatic machine tool equipment and mobile hydraulics. Such machinery, for which circulation times are rapid, which need to be constantly available and which are permanently subjected to the danger of water and other contamination, is the primary area of application of detergent hydraulic fluids. Hydraulic fluids with demulsifying properties are often recommended for application in steel and rolling mills where large amounts of water are present and low circulation times enable tank separation. Demulsifying properties in a modified form are used to determine the compatibility and demulsification of water-miscible cutting fluids with slideway oils and hydraulic oils. The ageing of a hydraulic fluid negatively affects its demulsifying properties [5,22,48,49] (see Chapter 18).

11.4.13.9 Pour Point

The pour point is the lowest temperature at which an oil still flows. A sample is systematically cooled and its flowing properties are tested every 3 °C according to DIN ISO 3016 [42]. The pour point and the boundary viscosity define the lowest temperature at which an oil can normally be used (see Chapter 18).

11.4.13.10 Copper Corrosion Behaviour: Copper Strip Test

Copper or copper-containing materials are often used in hydraulic components. Materials such as brass, cast bronze and sintered bronze are found in the bearing elements, guides or control units or sliding blocks of hydraulic pumps and motors. Copper pipes are used in cooling systems. Because copper corrosion can damage the entire hydraulic system, the copper strip test provides information about the corrosive effect of base fluids and additives on copper-containing materials. A procedure for testing the corrosive effect of mineral-oil-based and rapidly biodegradable hydraulic fluids on non-ferrous metals is known as the so-called 'Linde test' (a screening method for biodegradable oils with respect to copper-alloy corrosion; SAE Technical Paper 981516, April 1998), also known as VDMA 24570 (VDMA 24570 – Biologisch schnell abbaubare Druckflüssigkeiten – Einwirkung auf Legierungen aus Buntmetallen 03-1999) [11].

According to DIN EN ISO 2160, corrosion to a copper strip can take the form of discolouration or flaking. A ground copper strip is submerged in the fluid to be tested for a given time at a given temperature. Hydraulic and lubricating oils are normally tested for 3 h at 100 °C. The degrees of corrosion are categorized as follows: 1, slightly discoloured; 2, moderately discoloured; 3, heavily discoloured; and 4, corrosion (dark discolouration) [42] (see Chapter 18).

11.4.13.11 Water Content: Karl Fischer Method

If water enters a hydraulic system, a part might be so finely dispersed that it enters the oil phase and, depending on the density of the hydraulic oil, the water can also separate from the oil phase. This possibility must be considered when samples are being taken to determine the water content.

Determining the water content in mg kg^{-1} (mass) with the Karl Fischer method involves the addition of a Karl Fischer solution through direct or indirect titration [42] (see Chapter 18).

11.4.13.12 Ageing Stability: Baader Method

This is an attempt to replicate the influence of temperature and oxygen, in the form of air, on hydraulic fluids under laboratory conditions. By increasing the temperature above practical usage levels, increasing oxygen levels and including metal catalysts, an attempt is made to artificially accelerate the ageing of hydraulic oils. The increase in viscosity and the increase in the neutralization number (free acid) are recorded and evaluated. The laboratory test results are transferred to practical conditions. The Baader procedure is a practical means of testing the ageing of hydraulic and lubricating oils according to DIN 51 554-3.

For a given period of time, samples are aged at a predetermined temperature and in the presence of moving air while a copper coil which acts as an ageing accelerator is periodically submerged in the oil. According to DIN 51 554-3, the ageing stability of C, CL and CLP and hydraulic oils HL, HLP and HM is tested at 95 °C. The saponification number is given in mg KOH g^{-1} [42] (see Chapter 18).

11.4.13.13 Ageing Stability (TOST Test)

The ageing stability of steam turbine and hydraulic oils containing additives is determined in accordance with DIN 51 587.

The TOST test has been used for many years to test mineral-oil-based turbine and hydraulic oils. In a modified form (without water), the dry TOST test is used to determine the ageing stability of ester-based hydraulic oils according to DIN EN ISO 4263 – Part 1.

The ageing of a lubricant is characterized by an increase in the neutralization number when the oil is exposed to oxygen, water, steel and copper for a maximum of 1000 h at 95 °C (neutralization ageing curve). The maximum permissible increase in the neutralization number is 2 mg KOH after 1000 h [42] (see Chapter 18).

11.4.13.14 Neutralization Number

If the neutralization number of a hydraulic oil increases as a result of ageing, overheating or oxidation, the ageing products formed can damage the system components such as pumps and bearings. This, in turn, can cause serious system failures. The neutralization number is, therefore, an important criterion for evaluation of the condition of a hydraulic fluid [42].

The neutralization number according to ISO 6618 indicates the amount of acidic or alkaline substances in a lubricant. Acids in mineral oils can damage materials. A high acid content, which can be the result of oxidation, is, therefore, undesirable [42] (see Chapter 18).

11.4.13.15 Steel/Ferrous Corrosion Protection Properties

The steel/ferrous corrosion protection properties of steam turbine and hydraulic oils containing additives are determined in accordance with DIN ISO 7120.

Hydraulic fluids often contain dispersed, dissolved or free water and a hydraulic fluid must provide corrosion protection to all wetted components under all operating conditions including contamination with water. This test examines the performance of the corrosion-protection additives under a number of different operating conditions.

The oil to be tested is mixed with distilled water (Method A) or artificial seawater (Method B) and stirred continuously for 24 h at 60 °C while a steel rod is submerged in the mixture. After 24 h, the steel rod is examined for corrosion. The results enable assessment of the corrosion protection offered by the oil to steel components that are in contact with water or water vapour: degree of corrosion 0, no corrosion; degree of corrosion 1, little corrosion; degree of corrosion 2, moderate corrosion; and degree of corrosion 3, heavy corrosion [42] (see Chapter 18).

11.4.13.16 Wear Protection (Shell Four-Ball Apparatus; VKA, DIN 51 350)

The Shell four-ball apparatus is used to measure the anti-wear and EP properties of hydraulic fluids. The load-carrying capacity of hydraulic fluids is tested under boundary friction conditions. The procedure serves to determine values of lubricants with additives that withstand high pressures under boundary friction conditions between sliding surfaces. The lubricant is tested in a four-ball apparatus that consists of one (central) revolving ball and three stationary balls arranged as a cup. Under constant test conditions and for a predetermined duration, the contact scar on the three stationary balls is measured or the load on the revolving ball can be increased until it welds to the other three [42] (see Chapter 19).

11.4.13.17 Shear Stability of Polymer-Containing Lubricants

Polymer-containing lubricants, high molecular mass polymer molecules, are used as viscosity index improvers to improve the viscosity–temperature behaviour of oils. As their molecular mass increases, these substances become increasingly sensitive to mechanical stress such as that which exists between a piston and its cylinder. Several tests are used to evaluate shear stability under different conditions [42]: DIN 51 350-6, four-ball test; DIN 51 354-3, FZG test; and DIN 51 382, diesel fuel injector method.

The decrease in kinematic viscosity after shearing indicates the permanent decrease in viscosity that can be expected during operation (see Chapter 19).

The relative viscosity reduction due to shearing after 20 h according to DIN 51 350-6 (determination of shear stability of lubricating oils containing polymer-tapered roller bearing) is implemented in DIN 51 524-3 (dated 2006); recommended shear loss is below 15%.

11.4.13.18 Mechanical Testing of Hydraulic Fluids in Rotary Vane Pumps (DIN EN ISO 20763 [Q])

The Vickers pump test and a variety of other manufacturers' pump tests realistically evaluate the performance of a hydraulic fluid. At present, however, alternative tests (such as the DGMK 514 project, mechanical testing of hydraulic fluids) are being developed [50].

The Vickers test determines wear protection in a rotary vane pump. The oil to be tested is circulated through a rotary vane pump at a given temperature and pressure (the test conditions are 140 bar, 250 h, variable temperature and operating fluid viscosity 13 mm^2 s^{-1}). After completion of the test, the ring and vanes are examined for wear (Vickers V-104 C10 or Vickers V-105 C10). The maximum permissible wear values are <120 mg for the ring and <30 mg for the vanes [42] (see Chapter 19).

11.4.13.19 Wear Protection (FZG Gear Rig Test; DIN ISO 14635-1 [R])

Hydraulic fluids, particularly higher viscosity grades, are used as hydraulic and lubricating oils in combined systems. Dynamic viscosity is the key wear-protection factor in hydrodynamic lubrication. At low sliding speeds or high pressures under boundary–friction conditions, the wear protection offered by a fluid depends on the additives used (reactive layer formation). These boundary conditions are replicated by the FZG test.

The test is primarily used to determine the boundary performance of lubricants. Defined gear wheels turning at a defined speed are either splash- or spray-lubricated with an oil whose initial temperature is recorded. The tooth-flank load is increased in stages and the appearance of the tooth flanks is recorded. This is repeated until the final 12th load stage: load stage 10, Hertzian pressure at the pitch point 1.539 N mm^{-2}; load stage 11, Hertzian pressure at the pitch point 1.691 N mm^{-2}; and load stage 12, Hertzian pressure at the pitch point 1.841 N mm^{-2}. The starting temperature at load stage 4 is 90 °C, the peripheral speed is 8.3 m s^{-1}, the upper temperature is not defined; and gear geometry A is used.

The damage load stage as defined by DIN 51 524-2 is at least 10. ISO VG 46 hydraulic fluids that do not contain anti-wear EP additives normally achieve load stage 6 (about 929 N mm^{-2}) [42] (see Chapter 19). Zinc-containing hydraulic fluids normally achieve damage load stages 10–11 at least. Zinc-free, so-called ZAF, hydraulic fluids achieve damage load stage 12 or greater.

11.5 Hydraulic System Filters

Hydraulic oils are used for very many sensitive industrial manufacturing machines. Because of the use of these oils, hydraulic systems are reliable and designed to run for years. The minimum technical requirements of hydraulic fluids according to DIN, ISO and manufacturers' specifications are clearly defined and generally fulfilled by the fluids presently available in the market.

Table 11.11 Hydraulic component clearance.

Component	Typical critical component clearance (μ)
Gear pumps (under pressure)	
Gear to side plate	0.5–5
Gear tip to housing	0.5–5
Vane pumps	
Vane tip to stator	0.5–5 (1)
Vane to side plate	5.0–13
Piston pumps	
Piston to cylinder	5.0–40
Cylinder to valve plate	1.5 (0.5)–10 (5)
Servo valves	
Jets	130.0–450
Splash care	18.0–63
Piston valve (radial)	2.5–3
Control valves	
Jets	130.0–10 000
Piston valve (radial)	2.5–23
Dish valve	1.5–5
Plug valve	13.0–40
Component	**Film thickness (μm)**
Roller bearings	0.1–1.0
Hydrostatic slide bearings	0.5–100.0
Hydrodynamic slide bearings	1.0–25.0
Toothed wheels	0.1–1.0
Seals	0.05–0.5

These specifications do not, however, refer to 'good filterability' – requirements are not defined.

In the past, most hydraulic and lubricating oil systems in machine tools, presses and stationary and mobile systems were fitted with 25–50 μm filters. This mesh size was adequate to satisfy the requirements of critical system elements such as valves. Critical hydraulic system components include those with narrow passages and low flow rates. Table 11.11 summarizes the typical gaps and passage sizes in the selection of hydraulic components [51–54].

If dirt and contaminants are present in the oil, these critical gaps can influence the function of the system and wear rates. Experts differ on the size and amount of particles that constitute a critical situation.

11.5.1
Contaminants in Hydraulic Fluids

There are several types and causes of hydraulic fluid contamination. The first major differentiation is between primary and secondary contaminations. Primary contamination is that which exist in the hydraulic circuit before it was commissioned. This can include machining residues assembly residues and fresh-oil contaminants. Secondary contamination is that which formed after the system began to operate, for example mechanically abraded material, flow-related abrasion, corrosion, wear and dirt that enters the system via cylinder sealing materials or tank deaerating units [6,51–54].

After comprehensive trials by a leading manufacturer into the effect of contaminants on the life of roller bearings, now a great importance is given to the cleanliness and filtration of oils. Purity and the types of additives used have a significant influence on the life and likelihood of failure of roller bearings and thus a whole system [52,53,55]. Trials conducted within the framework of the FVA 179/1 research project 'Influence of Foreign Particles on Roller Bearings and Measures to Avoid Them' also investigated this subject. The causes of premature roller bearing failures are, above all, inadequate lubrication, particle contamination and overloading.

11.5.2
Oil Cleanliness Grades

Several methods can be used to classify oil cleanliness. The best known methods are ISO 4406 and NAS 1638. Determining oil cleanliness according to ISO 4406 involves examining the number and size of particles in a 100 ml sample of fluid. The number of particles in the categories >2, >5 and >15 μm are recorded. Normally only particles >5 and >15 μm are reported (old commonly used practice). The new ISO 4406 specification (December 1999) defines the particles in the categories >4, >6 and >14 μm. The particles can be counted using a microscope or by suitable automatic particle counters. ISO 4406 or NAS 1638 defines the maximum permissible contamination according to the type of hydraulic system, how sensitive it is and which critical components form part of the system. Depending on the operating conditions, the cleanliness categories in Table 11.12 are recommended [6,17,22,55] (Tables 11.13–11.15).

Table 11.12 Cleanliness categories.

Type of system/case of application/ filter size	Cleanliness category in accordance with ISO 4406 – old	Cleanliness category in accordance with NAS 1638 – old
Against fine soiling and mudding-up of sensitive systems; servo hydraulics	Minimum 13/10	3–4
Heavy duty servo systems, high-pressure systems with long service life	Minimum 15/11	4–6
Proportional valves, industrial hydraulics with high operating safety	Minimum 16/13	7–8
Mobile hydraulics, common mechanical engineering, medium pressure systems	Minimum 18/14	9–10
Heavy industry, low pressure systems, mobile hydraulic	Minimum 19/15	9–11

Table 11.13 Cleanliness requirements for hydraulic components – ISO 4406 (HYDAC, TAE Seminar, November 2003).

Type of system/case of application/component	Recommended cleanliness class
Against fine soiling and mudding-up of sensitive systems; servo hydraulics	15/13/10
Industrial hydraulic	17/15/12
• Proportional technology	
• High-pressure systems	
Industrial and mobile hydraulic	18/15/12
• Electromagnetic control valve technology	19/16/14
• Medium-pressure and low-pressure systems	
Industrial and mobile hydraulic with low requirements at wear protection	20/18/15
Forced lubrication in gears	18/16/13
Fresh oil (HL, HLP, HVLP, oil according to DIN 51 524)	21/19/16
Pumps/engines	18/16/13
• Axial piston pump	19/17/13
• Radial piston pump	20/18/15
• Gear pump	19/17/14
• Rotary vane pump	
Valves	20/18/15
• Directional valves	19/17/14
• Pressure valve	19/17/14
• Flow regulation valve	20/18/15
• Non-return valve	18/16/13
• Proportional valve	17/15/12
• Servo valve	
Cylinder	20/18/15

SAE Code 11A/8B/5C according to SAE AS 4059:2001 means

800 001–1 600 000 particles per 100 ml	> 4 µm
38 901–77 900 particles per 100 ml	> 6 µm
865–1730 particles per 100 ml	> 14 µm

11.5.3 Filtration

Filters intended to remove solid impurities from lubricants have been fitted to hydraulic systems for decades. The filters used are as follows:

- Tank vent filters to clean any drawn-in air.
- Pressure filters to clean the fluid entering the pump.

Table 11.14 Structure of the contamination classes according to ISO 4406 (HYDAC, TAE Seminar, November 2013).

ISO Code (acc. to ISO 4406)	>4 µm Particle count per 100 ml		ISO Code	>6 µm Particle count per 100 ml		ISO Code	>14 µm Particle count per 100 ml	
	From	To		From	To		From	To
10	500	1000	10	500	1000	10	500	1000
11	1000	2000	11	1000	2000	11	1000	2000
12	2000	4000	12	2000	4000	12	2000	4000
13	4000	8000	13	4000	8000	13	4000	8000
14	8000	16 000	14	8000	16 000	14	8000	16 000
15	16 000	32 000	15	16 000	32 000	15	16 000	32 000
16	32 000	64 000	16	32 000	64 000	16	32 000	64 000
17	64 000	130 000	17	64 000	130 000	17	64 000	130 000
18	130 000	260 000	18	130 000	260 000	18	130 000	260 000
19	260 000	500 000	19	260 000	500 000	19	260 000	500 000
20	500 000	1 000 000	20	500 000	1 000 000	20	500 000	1 000 000
21	1 000 000	2 000 000	21	1 000 000	2 000 000	21	1 000 000	2 000 000
22	2 000 000	4 000 000	22	2 000 000	4 000 000	22	2 000 000	4 000 000
23	4 000 000	8 000 000	23	4 000 000	8 000 000	23	4 000 000	8 000 000

Table 11.15 Contamination classes according to SAE AS 4059:2001 (HYDAC, TAE Seminar, November 2013).

		Maximum particle count per 100 ml					
		>4 µm	>6 µm	>14 µm	>21 µm	>38 µm	>70 µm
Class	SAE Code	A	B	C	D	E	F
	000	195	76	14	3	1	0
	00	390	152	27	5	1	0
	0	780	304	54	10	2	0
	1	1560	609	109	20	4	1
	2	3120	1220	217	39	7	1
	3	6250	2430	432	76	13	2
	4	12 500	4860	864	152	26	4
	5	25 000	9730	1730	306	53	8
	6	50 000	19 500	3460	612	106	16
	7	100 000	38 900	6920	1220	212	32
	8	200 000	77 900	13 900	2450	424	64
	9	400 000	156 000	27 700	4900	848	128
	10	800 000	311 000	55 400	9800	1700	256
	11	1 600 000	623 000	111 000	19 600	3390	512
	12	3 200 000	1 250 000	222 000	39 200	6780	1020

- Top-up filters that filter the hydraulic fluid as it is fed into the tank.
- By-pass filters in the tank circuit to improve the cleanliness levels.
- Return filters fitted to fluid return lines.

The filters can be of the cartridge or surface variety. Important data are the mesh and retention size of the filter (the designation $\beta 3 > 200$ describes a filter of 3 μm mesh size and a separation rate of 200, i.e. only one particle of 200 particles will pass the filter). In addition, the initial pressure difference (ΔP maximum. 0.1–0.2 bar) and the maximum output pressure difference (ΔP maximum 3–5 bar) in relation to the flow rate, viscosity and density are of importance. The primary filter materials are micro-fibreglass, metal meshes, cellulose paper and some other constructions. Hydraulic filters consist of an element, a housing, a contamination indicator and other components. In general, the fluid flows from the outside to the inside. The selection of mesh size is a matter of experience and depends on the specific requirements of critical components. As a rule, hydraulic systems use filter mesh sizes ranging from 3 to 40 μm [55,56]. When using filters with micron ratings of, for example, 1, 3 and 6 μm, attention must be paid. Especially, high molecular components of the fluids (e.g. VI improvers) and contaminations (e.g. grease, corrosion preventives) can block the filters. Filter blockage can occur if additive systems are incompatible (e.g. mixture of incompatible zinc-containing and zinc-free additives).

11.5.4
Requirements of Hydraulic Fluids

High-performance filtering systems make high demands on the filterability of hydraulic fluids. A hydraulic fluid should only generate a small pressure difference across the filter after long-term use. Base oils and additives should be easily filterable with filter mesh sizes of 1, 3, 6 and 10 μm. Nothing in the fresh fluid should cause the filter to block and thus reduce its life (this is examined by special laboratory tests). Naturally, the purity of the fresh fluid should be low.

According to ISO 4406, the cleanliness of drums should be 17/14 (19/15) and experience shows that the cleanliness of road tankers should be 15/12 (18/14), although transport, storage and environmental factors generally cause the cleanliness factor to deteriorate by 2–3 categories. In practice, poor filter life often results from contamination of the hydraulic fluid with water, dirt and other fluids, from inadequate maintenance of the system or from incorrect filter selection. Determining the exact cause normally requires expensive laboratory tests [55,56].

11.6
Machine Tool Lubrication

11.6.1
The Role of Machine Tools

Machine tools are the most important machines in the metalworking industry. With a share of approximately 20%, Germany is one of the world's leading manufacturers of machine tools. In terms of sales, Germany (DM 14 billion) is second to Japan (about DM 16 billion) but ahead of the United States (DM 9 billion), Italy (about DM 6 billion) and Switzerland (about 4 billion) [57]. Machine tools are used for a wide variety of operations including forming, cutting and bending; they are principally used for turning, milling, drilling, grinding and machining centre. They can combine any of these in a transfer system [58]. Machine tool construction is a major sector in engineering and their share of overall exports for Germany (about 60–70%) and Japan illustrate the importance of machine tools in national economies [59].

11.6.2
Machine Tool Lubrication

This section covers lubricating oils, hydraulic fluids and greases for machine tools. Apart from cutting fluids, hydraulic oils are volumetrically the most significant group of machine lubricants, followed by slideway oils and gear oils. Neat or water-miscible cutting fluids or metalworking fluids are described in Chapters 14 and 15.

The lubrication of machine tools is described in DIN 8659-1 and DIN 8659-2 and ISO 5169 and ISO 3498. These standards contain requirements that should be observed when manufacturers and users establish lubrication plans. These also satisfy the requirements specified in DIN/ISO 5170 (machine tool lubrication systems) [60].

Lubrication plans should cover all the components in a machine tool which need lubrication. These should describe

- the precise location of all lubrication points,
- the type of lubrication required,
- the lubricant itself according to DIN 8659-1 and DIN 8659-2 and ISO 3498 and the tank volume and
- the lubrication timetable.

The purpose of a lubricating plan as part of routine servicing is to ensure that a sufficient quantity of the correct lubricant is applied to the right point at the right time (VDI guideline 3009). Machine manufacturers normally supply lubricant recommendation tables with machines. These list the types of lubricants according to DIN 51 502, ISO 6743 and ISO 3498 for each viscosity grade by its brand name. On the basis of this information, a maintenance plan is created for every machine that shows the type of lubricant and the lubrication interval. For most machines, a maintenance plan and a lubrication chart are included in the service handbook.

Figure 11.17 shows an example of a lubrication plan for a centreless grinding machine. This shows lubricants conforming to DIN 51 502 and ISO 3498, lubrication intervals, tank volume and the location of all lube points.

Lubricant recommendations should be updated every 2 years to make use of new lube developments. Technically, similar lubricants can often be grouped to enable some lubricant rationalization [61].

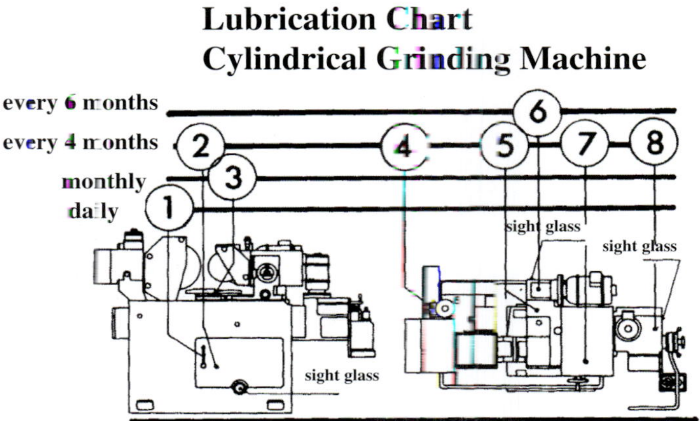

Lubrication Chart
Cylindrical Grinding Machine

every 6 months
every 4 months
monthly
daily
sight glass
sight glass
sight glass

	a	b	b	c	d	e	f	g	h
	1	2*	2**	3	4	5	6	7	8
Lubricant DIN 51502	CGLP 68	CLP 10	CLP 5	K2K	CGLP 68	HLP 32	CLP 150	CGLP 68	CGLP 68
ISO 3498	GCC 68	CC 10	CC 5	XM2	GCC 68	HM 32	CC 150	GCC 68	GCC 68
Tank volume ℓ	2,7	20	20	0,1	1	4	1	1	1

- a - centralized lubricating system
- b - grinding wheel spindle bearing
- c - slideway (table)
- d - slideway (dressing tool)
- e - spindle bearing
- f - worm gear
- g - slideway (dressing tool)
- h - slideway (grinding tool)
- 2* - speed 30-45 m/s
- 2 ** - speed 45-60 m/s

Figure 11.17 Lubrication chart of a machine tool.

Machine manufacturers often refer to the lubricant recommendations issued by component manufacturers. The recommendations issued by the manufacturers of hydraulic components, gearboxes, slideways and linear guides must be observed.

For lubrication, a machine tool can be divided into a number of major elements: hydraulic unit, gearbox, spindle, slideway, linear system, plain and roller bearings and, finally, cutting zone lubrication. In general, a different lubricant is recommended for every component, that is at least seven different types and viscosities of lubricants (excluding the cutting fluid) are required.

11.6.3 Machine Tool Components: Lubricants

11.6.3.1 Hydraulic Unit

Most hydraulic equipment is designed to use HLP (HM), HLPD (HG) fluids with an ISO viscosity between 32 and 46. Running temperatures range from 40 to

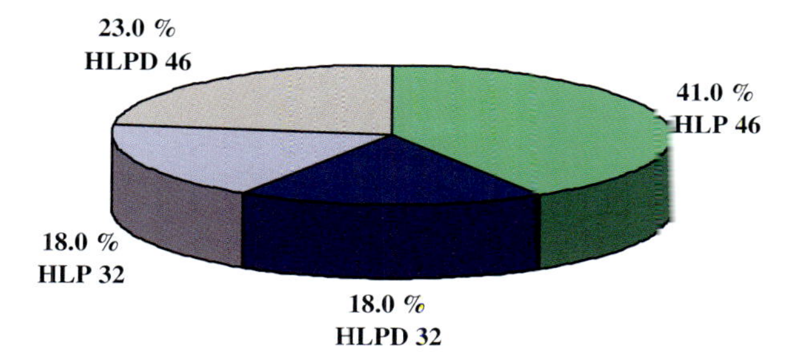

Figure 11.18 Hydraulic oils used in machine tools.

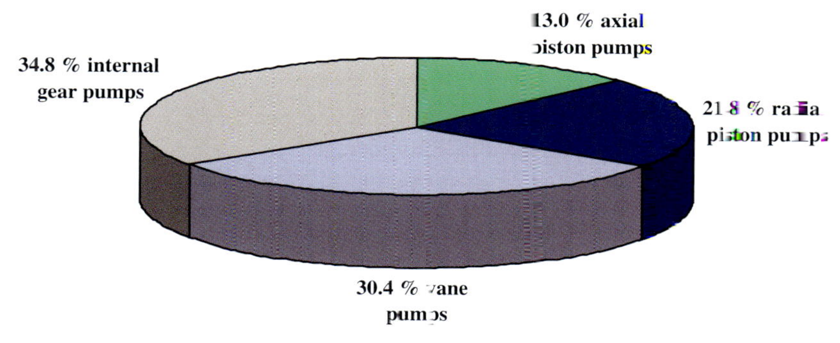

Figure 11.19 Hydraulic pumps used in machine tools.

60 °C and peak temperatures of 60–80 °C can occur [58,59]. Although operating pressures range from 50 to 100 bar (relatively low), pressures up to 400 bar are used in clamping fixtures. Generally, low system pressures are used to avoid chatter marks (compressibility of the fluid) that often occur at higher pressures. Moreover, higher pressures lead to more leakage and thus lower overall efficiency [58]. Figure 11.18 shows a list of hydraulic oils used in machine tools (survey of 12 German machine tool manufacturers, 1995) [59]. HLPD fluids are often used to solve friction and compatibility problems.

Rotary vane and internal gear pumps are used at pressures between 50 and 100 bar. Higher pressures are generated by radial and axial piston pumps. External gear pumps are seldom used because of the noise they generate. Figure 11.19 shows the types of pumps used in machine tools (survey of 12 German machine tool manufacturers, 1995) [59].

Figure 11.20 shows the pressures used in machine tools.

Actuator valves, sleeve valves, shut-off valves and throttle valves are used in machine tools. Many valves have hydrodynamic bearings that make them sensitive to stick–slip effects, contamination and deposits [58–62].

Machine tool hydraulic systems are normally equipped with mesh or fibre filters. Approximately 80% of machine tool manufacturers use filters in the 5–10 μm range; the remaining 20% use filters up to 25 μm [59].

Pressure range

80 % of all machine tools are working in a pressure range between 50 and 100 bar
13 % greater than 100 bar
7 % greater than 150 bar

Minimum viscosity of hydraulic fluids

Vane pumps:
Normally, a viscosity of min. 15 mm²/s at pressures up to 100 bar is required-
low viscosity fluids are currently being developed

Piston pumps:
Today, they are available for low viscosity fluids - but expensive

Gear pumps:
Need a lot of assembly volume

Figure 11.29 Working conditions used in machine tools.

Depending on the type of valves used, the pressure and the importance of the machine, the ISO 4406 cleanliness of the fluids should be between 15/11 and 17/13 or lower according to ISO 4406 [28,55,59].

11.6.3.2 Slideways

Machine tool slideways that guide supports and workpieces are amongst the most important elements of a machine tool. The special demands made on these slideways include precision, high performance, low manufacturing costs and low operating costs. The most important features of slideways are as follows:

- Low friction, no stick–slip at low feeds and high load-carrying capacity.
- Low wear and ultimate reliability against seizures.
- Torsional stiffness and minimal play.
- Good damping properties to reduce chatter marks on machined surfaces.

In general, hydrodynamic, hydrostatic and roller guides are used. Aerostatic and electromagnetic guides are seldom found in machine tools. Hydrostatic guides are losing popularity because of their price but can still be found in many machines. Currently, hydrodynamic and linear roller guides (linear systems) are often used. Hydrodynamic slideways are losing market share because they only enable relatively small feed velocities (maximum 0.5 m s^{-1}), often suffer from stick–slip and are more expensive to manufacture than linear roller guides. The most common material pairings used in hydrodynamic slideways are cast iron–cast iron, cast iron–plastic, cast iron–steel and steel–plastic. Slideway oils should conform to DIN 51 502, ISO 6743-13 and ISO 3498 [63]. Horizontal slideways are often lubricated with CGLP 68, HG 68 or G 68 slideway oils. Inclined or vertical slideways are lubricated with CGLP 220 HG 220 or G 220 oils. The oil is applied through central systems and lost after use. Slideway oils are general lubricating oils with additives to improve oxidation and corrosion protection. They also contain anti-wear agents, EP additives, surface-active substances and often adhesion improvers (tackifiers).

In recent years, roller or linear guides have been fitted increasingly to machine tools. In 1995, 8 of 12 German manufacturers surveyed used roller linear guides exclusively and four used hydrodynamic and roller guides. The lubricants used should separate the moving parts in the roller in the contact zone which counterrotate. The lubricant should also have damping characteristics in the contact zone (especially when the direction of movement changes) and reliably protect against wear and seizures. The lubricant should also form a stable and effective film in a very short time. Such total-loss lubricants are supplied to the linear guide zones via a central system. CGLP 68 and CGLP 220 grades are often used [60]. High-viscosity CGLP 220 slideway oils that contain surface-active components are often recommended. Alternatively, K2 K or similar greases can also be used (see Chapter 16).

Oils for hydrodynamic slideways and linear guides should have the following properties [58,59]:

- Chemical compatibility with all cutting fluids used.
- Good demulsification of emulsions and no sticky residues on slideways.
- Low coefficient of friction (static and dynamic).
- Avoidance of stick–slip (sliding and static friction alternates during stick–slip on slideways, which can cause chatter marks).
- Good pumpability in central lubrication systems.
- Good adhesion to slideways with tacky additives and/or without tacky additives.
- Good wear protection (EP and AW additives) with FZG ≥ 12.
- Good slideway material compatibility.
- Good corrosion protection (no black stains on slideways).
- Same additive systems as hydraulic oils (i.e. zinc-, ash- and silicone-oil-free hydraulic oils).
- Fulfil the specifications of hydraulic oil if hydraulic and slideway oils share a circuit.

11.6.3.3 Spindles: Main and Working Spindles

The function of spindles is to guide the tool and/or workpiece at the cutting zone. In addition, spindles should absorb external forces. The accuracy and surface quality of components made on machine tools depend on the static, dynamic and thermal behaviour of the spindle bearings. These are key elements of machine tools. Tool spindles can be supported on greased roller bearings, oil-lubricated roller bearings or hydrodynamic plain bearings. Roller bearings have almost completely replaced plain bearings. Oil-lubricated roller bearings are normally fed into total-loss oil from a central system or via an oil-mist system. Often, low-viscosity CL/CLP general lubricating oils according to DIN 51 517 or ISO VG 5-22 FC and FD spindle oils according to ISO 6743-2 are used. Spindle oils must lubricate and cool. They have to protect against steel and copper corrosion, and be oxidation stable. Depending on the application, lubricants with AW/EP additives are used. The spindle speed defined as the product of rpm (min^{-1}) and average bearing diameter (mm), determines whether a spindle should be lubricated with oil or grease [31,58,64].

11.6.3.4 Gearboxes and Bearings

Gearboxes are designed to convert and transfer movement and forces – they are units that transmit energy. Gearboxes in machine tools serve to reduce drive speed to the feed velocity of supports and so on. The gearboxes can have fixed or selectable ratios. Speed adjustments are often made with synchronized or non-synchronized motors. The different gearboxes include spur, worm, crown wheel and pinion or planetary types [58]. The stress on machine tool gearboxes is relatively small and ISO VG 68 to 320 CLP (DIN 51 517 – dated January 2004), CKC or CKD (ISO 6743/6) gear and general lubricating oils are often used. Worm drives are often lubricated with polyglycol-based CLP PG or CKE gear oils. Synthetic, polyalphaolefin-based CLP HC or CKT oils are used for thermally stressed gearboxes [31,58,60].

The bearings most often found in gearboxes are plain and roller bearings, although plain bearings are seldom used in machine tools. The most popular types are ball and cylindrical roller bearings. The corresponding lubricants are general lubricating oils or specific gear oils (see Chapter 10).

11.6.4 Machine Tool Lubrication Problems

Many different oils and greases are used in a machine tool. Hydraulic, gear, slideway and spindle oils and greases are the most important groups. An important point is the required compatibility of the lubricants with each other. Leaks can cause large amounts of hydraulic oils to contaminate cutting fluids (every year three to four times the volume of the hydraulic system enters the cutting fluid circuit [5?]). At the same time, metalworking fluids can enter the hydraulic oil circuit via cylinders and so on. Slideway oils as total-loss lubricants and metalworking fluids are in close contact and must, therefore, be compatible. Developments in the area of universal oils (chemically related lubricants) that eliminate the problem of poor compatibility are currently a high priority. Initial developments of fluid families, that is hydraulic, gear, slideway, spindle and metalworking oils that share common additive packages but are available in different viscosities, have already been tested in practice [59,65]. Low-viscosity neat oils have already overtaken water-miscible emulsions (the trend is oil instead of emulsion) [5,65]. Future developments will concentrate on unifluid systems that consist of one low-viscosity oil that serves as cutting fluid, slideway oil and hydraulic fluid. This concept, however, requires the redesign of components such as pumps to handle such low-viscosity fluids. The development of compatible systems could save large amounts of money now spent on the expensive monitoring and maintenance of currently used water-based cutting emulsions (see Chapter 14).

11.6.5
Hydraulic Fluids: New Trends and New Developments

11.6.5.1 Applications

In Germany, 80–85% of hydraulic fluids are mineral-oil-based. Of these, approximately 40% are used in mobile hydraulics and 60% in industrial systems. Approximately 7% of the hydraulic fluids are fire-resistant fluids used in underground mining, steel mills, foundries and power stations. Approximately 7% of the total hydraulic fluids are biodegradable hydraulic fluids used in mobile and stationary equipment and approximately 1% are formulated for special uses, for example in the food and beverage industries.

11.6.5.2 Chemistry

Base Fluids

Typical hydraulic fluids are composed of 95–98% base fluids and 2–5% additives (or 5–10%). As already mentioned above, the largest contributors to base fluids are mineral oils refined from crude oil (mainly paraffinic and naphthenic compounds and hydrocracked base oils). Other base fluids used are, basically, fully and partially saturated esters, polyglycols, polyalphaolefins and alkylates.

Additives

The most important additives for hydraulic fluids are listed in Table 11.16.

Approximately 70% of mineral-oil-based hydraulic fluids in Europe are zinc-containing fluids, although an increasing number of zinc-free hydraulic fluids are available. These fluids are formulated without zinc dialkyldithiophosphate (ZnDTP) as a multipurpose additive. Although this has a large effect on the performance of hydraulic oils, it is important to state that the element content itself has no decisive effect on the quality of hydraulic fluids. Zinc-free hydraulic fluids usually contain similar or even lower amounts of phosphorus and sulfur than Zn-containing fluids (Table 11.17). Traditional group I mineral base oils have an incorporated sulfur content that positively contributes to anti-wear performance and synergistic functions, for example as a radical scavenger for oxidation stability. Additionally, a certain aromatic content that is typical for group I base oils improves the solubility of additives and ageing products. New hydraulic fluids using modern base oils (hydrotreated, hydrocracked or PAO) have a much lower or zero sulfur and aromatic content and therefore need consequently higher additive treat rates to balance the shortfall of base-fluid-related sulfur and aromatic content.

Selected characteristics of hydraulic fluids are discussed in the subsequent sections. The use of Zn-free or Zn-containing additives usually makes a remarkable difference to characteristics and performance and can also crucially affect whether a hydraulic fluid is formulated with demulsifying or detergent/dispersant characteristics.

Table 11.16 Most important types of additives in hydraulic fluids and their chemistry.

Type	Chemistry
Antioxidants (AO)	Phenolic and aminic AO, zinc dialkyldithiophosphate (ZnDTP)
Copper deactivators	Nitrogen compounds (triazoles), dimercaptothiadiazoles
Steel/iron corrosion inhibitors	Carboxylic acid derivatives, sulfonates, succinic acid compounds
Anti-wear additives (AW)	Esters, ZnDTP
Extreme pressure additives (EP)	Phosphorus and sulfurous compounds, thiophosphates, sulfurized hydrocarbons (active and inactive)
Friction modifiers	Fatty acids, polar compounds, esters
Detergents/dispersants	Ca and Mg phosphates, sulfonates, phenates, polyisobutylensuccinimide
Antifoam agents	Silicone oil, silicone-oil-free, polymethylsiloxane
Viscosity index improvers (VI)	Polymethacrylates
Pour point depressants (PP)	Polymethacrylates
Dyes	Azo dyes, fluorescent dyes
Tackifiers	Polyisobutylenes, polar compounds

Table 11.17 Variation of element content of zinc-containing and zinc-free hydraulic fluids.

Element	Zinc-containing hydraulic fluids (ppm)	Zinc-free hydraulic fluids (ppm)
Zinc	200–500	0
Phosphorus	200–500	60–500
Sulfur	400–1000	0–1000
Sulfur (base oil)	300–1000 (or higher)	300–1000 (or higher)

11.6.5.3 Extreme Pressure and Anti-Wear Properties

Although the Vickers vane pump test is a traditional and important test for hydraulic fluids, the test results obtained are not sufficiently differentiating for state-of-the-art hydraulic fluids. Other commonly used EP/AW test methods include use of the Shell four-ball tester (DIN 51 530-1 and DIN 51 530-2), the Brugger test machine (DIN 51 347-2) and the FZG test rig (DIN ISO 14635-1). Tests on hydraulic fluids containing different additives (all ISO VG 46) have been conducted with these methods. The results showed that Zn-free hydraulic oils of the detergent/dispersant type usually have better EP performance than traditional Zn-containing demulsifying hydraulic oils and other hydraulic oils (Table 11.18).

A relatively new test rig for hydraulic fluids is the so-called FE 8 of FAG for the determination of roller bearing wear and friction coefficients. Although this procedure is normally used for greases and gear oils (DIN 51 819-3), some

Table 11.18 Typical EP/AW test results for different types of hydraulic fluids.

Test method	Conventional EP/AW	Medium EP/AW	High EP/AW
Four-ball	EP performance 1800–2000 N • Zn-containing, DD • Zn-free, demulsifying Zn-containing, demulsifying		> 2200 N • Zn-free, DD
Brugger EP performance	20–25 N mm^{-2} • Zn-containing, demulsifying • Zn-containing, DD, synthetic base oil • Zn-free, demulsifying	30–40 N mm^{-2} • Zn-containing, DD • Zn-free, demulsifying	> 50 N mm^{-2} • Zn-free, DD
FZG EP performance	11–12 • Zn-containing, DD • Zn-containing, DD, synthetic base oil • Zn-free, synthetic base oil, demulsifying	> 12 • Zn-free, DD • Zn-containing, DD	
Vickers EP/AW Performance	< 120 mg ring < 30 mg vane • All pass (generally)		

OEMs request a 'pass' for their in-house approval of hydraulic oils. This test sequence is mainly passed by Zn-free formulated hydraulic fluids. One can, in general, state that phosphorus/sulfur chemistry without Zn is more suitable for high-EP requirements (e.g. hydraulic systems with incorporated gear drives). Fluids formulated with Zn-based additives perform well in the mixed friction applications where only medium EP but good AW is required. Their multi-functional anti-wear performance is, therefore, usually better than that of Zn- and ash-free fluids.

11.6.5.4 Detergent/Dispersant Properties

Some OEMs set minimum limits for the dirt-carrying properties of hydraulic oils. A Fuchs in-house test reported by Chrysler [66] entails the use of a paper chromatographic method in which a colloidal graphite dispersion is applied to a paper strip that is dipped in the lubricant. Migration of the graphite under specified conditions is reported. The in-house limit of the OEM stipulates a minimum migration distance of 40 mm. Only hydraulic fluids with selected dispersant additives and elevated treat rates pass this test (Figure 11.21).

11.6.5.5 Air Release

It seems logical that hydraulic fluids with demulsifying properties would have much better air-release behaviour than fluids with detergent/dispersant

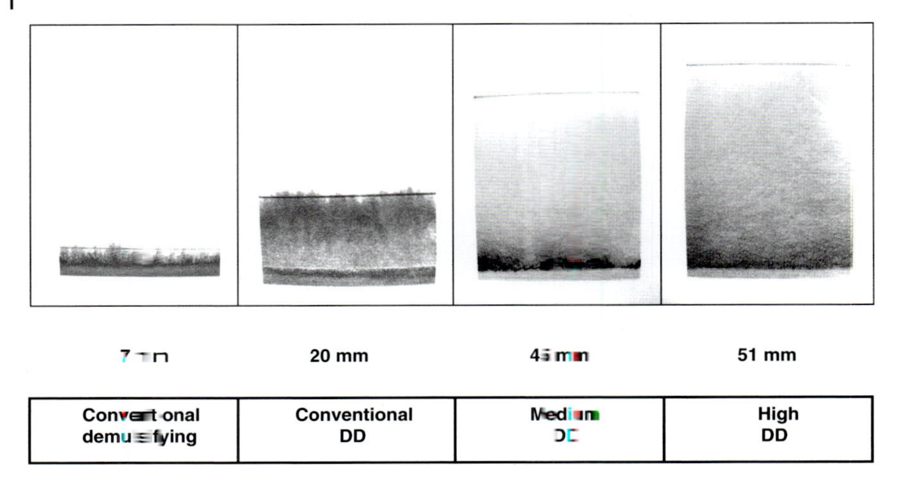

Figure 11.2 Detergent/dispersant properties: test results for different types of hydraulic fluids.

characteristics. Despite this, a careful DD formulation with state-of-the-art additive technology can have air-release characteristics similar to or better than those of demulsifying hydraulic fluids. This has been proven using DIN ISO 9120. All values were less than 10 min and therefore fulfilled DIN 51 524-2 that describes HLP hydraulic oils with limits increased from 10 to 13 min.

11.6.5.6 Static Coefficient of Friction

A series of tests using an inclination tribometer have been performed at SKC Gleittechnik in Coburg (Germany). The SKC test procedure was originally developed for slideway oils and determines the coefficients of friction of the fluids for a combination of plastic and steel. Static coefficients of friction for detergent/dispersant hydraulic fluids were between 0.130 and 0.177, lower than those of demulsifying fluids (0.214). Excellent values of 0.038–0.11 were achieved with synthetic esters because of their polar characteristics.

11.6.5.7 Oxidation Stability

The additives and selected base oil have a substantial effect on oxidation stability. Both hydrolytic and thermal stability are increased by changing from demulsifying to DD formulations. The type and treat rate of DD formulations affect the results as does the type of base oil. Results from the TOST test method according to ASTM 943 (DIN 51 587) are shown in Table 11.19. Choosing the appropriate hydraulic fluid can lead to a several-fold increase in fluid lifetime.

11.6.5.8 Shear Stability

Improved viscosity–temperature behaviour is desirable when hydraulic equipment operates over a wide range of temperature. The viscosity index is traditionally increased by addition of VI improvers that are usually polymers. These

Table 11.19 Examples of TOST test results.

Hydraulic fluid	TAN >2 mg KOH g⁻¹ (h)
Minimum test results to fulfil DIN 51 524	1000
Mineral oil, demulsifying, low treat rate	1200
Mineral oil, DD, medium treat rate	2300
Mineral oil, DD, high treat rate	3900
Hydrocracked base oil DD, high treat rate	4500

Table 11.20 Test results of high-VI hydraulic oils: tapered roller bearing shear-stability test DIN 51 350-6 (20 h, 60 °C, 40 ml).

	PAO HVLP 46	HEES saturated HVLP 46	Mineral oil, conventional HVLP 46	Hydrocracked HVLP 68
V_{-20}	2200	1970	1700	5900
V_{40}	46	46	46	68
V_{100}	7.8	9.2	8.9	10.7
VI	140	155	180	150
Shear loss, ΔV_{100} (%)	<1	<1	<40–50	<10

polymers are, unfortunately, eventually sheared, resulting in a decrease in viscosity. This behaviour is now considered in the revised version of DIN 51 524-3 that describes high-VI hydraulic oils (HVLP). Taper roller bearing shear-stability test is also being introduced in DIN 51 350-6. Some OEMs require a maximum shear loss of 15–20% in this test. Because the viscosity of conventional HVLP oils based on standard VI improvers decreases by 30–50% at 100 °C, conventional HVLP oils must be reformulated by using different additive chemistry, which increases the cost. An alternative method of formulation is to use base oils that already have in-built natural, shear-stable, high VI. Test results are presented in Table 11.20.

In many stationary hydraulic applications, multi-grade engine oils are used instead of regular hydraulic fluids. SAE 10W40 and SAE 15W40 are very common. These products have viscosities >95 cSt at 40 °C and are formulated with polymers subjected to high shear losses that result in critical decreases in viscosity.

11.6.5.9 Filtration of Zn- and Ash-Free Hydraulic Fluids

The filterability of Zn-free hydraulic oils very much depends on the type of chemistry used. A special multi-pass test rig has been designed for dynamic testing. Different pressures in the presence of 1% water (for better differentiation) are recorded over time. Depending on the formulation strategy,

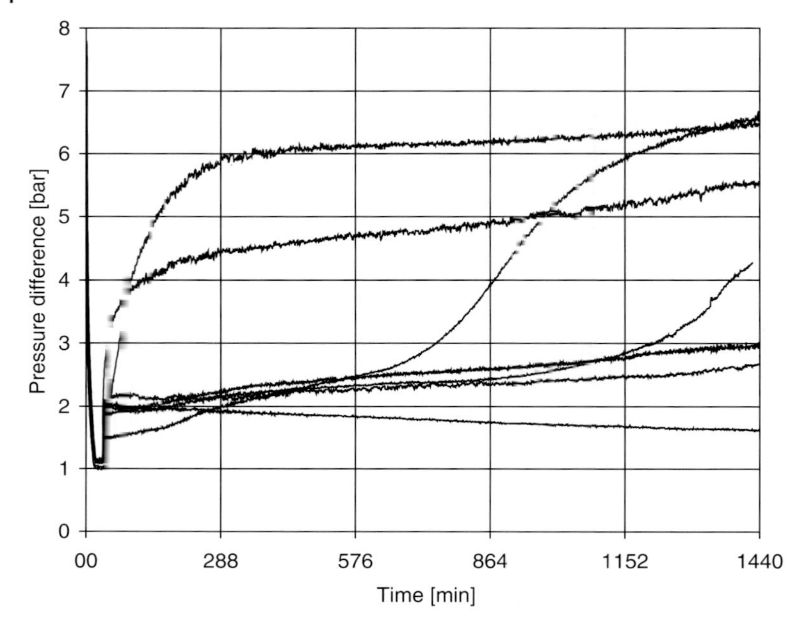

Figure 11.22 Multi-pass filterability test results for different Zn-free hydraulic fluids (8.7 l min^{-1}, 65 °C, 100 bar, 15 l volume, 3 µm/6 µm).

extremely different filtration behaviour can be observed. Under the practical conditions used with this test rig, long-lasting and filterable hydraulic fluids have been developed. Test results for a variety of Zn- and ash-free hydraulic fluid formulations are shown in Figure 11.22. In practice, the compatibility of Zn-containing and Zn-free hydraulic fluids should usually be tested before they are mixed.

11.6.5.10 **Electrostatic Charges**

Electrostatic phenomena usually depend on the fluids used and on the filter materials used for deep filtration. Because electrostatic charges in fluids mainly depend on the conductivity of the fluids, the risk potential of electrostatic charges in hydraulic fluids very much depends on the additives used (Table 11.21). Metal-containing fluids with zinc, calcium or magnesium compounds as additives are believed to be free from problems because their conductivity is $> 300\,\text{pS m}^{-1}$. The additives conduct electrostatic charges to the equipment housing. Although Zn-free and metal-free hydraulic fluids behave differently and do not conduct electrostatic charges, proper selection of Zn-free additives and advanced treatment procedures enable adequate conductivity to be achieved. Needless to say, for pure HL oils without any EP/AW additives, conductivity is very low and therefore must be considered as critical. A VDMA working group is actively investigating this phenomenon [67].

Table 11.21 Typical conductivity data for different types of hydraulic oil.

Type of hydraulic oil	23 °C (pS m^{-1})	50 °C (pS m^{-1})
Mineral oil, demulsifying, Zn-containing	200	800
Mineral oil, DD, Zn-containing	470	2000
Mineral oil, DD, Zn-containing	8.000	40 000
Mineral oil, demulsifying, Zn- and ash-free	4	17
Mineral oil, DD, Zn- and ash-free	140	690
Mineral oil, DD, Zn- and ash-free	360	1100

11.6.5.11 Micro-Scratching

When low-performance zinc and ash-free hydraulic oils are used instead of zinc-containing fluids, a new phenomenon, micro-scratching, very small scratches equally distributed all around the piston in the axial direction, is occasionally observed. Some discolouration also occurs and surface roughness reaches peaks of 2 μm compared with 0.1 μm for undamaged surfaces. Examples of damaged and undamaged pistons are shown in Figure 11.23. The fluid-related reason for this micro-scratching is the use of zinc- and ash-free hydraulic oils with low EP/AW performance. A special piston test rig at Busak + Shamban in Stuttgart (Germany) has been used to characterize different characteristics of hydraulic oils with regard to micro-scratching [68]. The conclusion was that Zn-containing hydraulic fluids and Zn-free hydraulic fluids containing high levels of selected EP/AW additives do not usually lead to micro-scratching. Film-forming properties also affect the behaviour positively and detergent/dispersant-type hydraulic fluids are, therefore, recommended in combination with zinc- and ash-free chemistry.

11.6.5.12 Updated Standards

DIN 51 524 for hydraulic fluids was updated in 2006 (new draft in 2012). Major changes were the declaration of purity levels, changes of limits, for example air release from 10 to 13 min, and introduction of a tapered roller bearing shear-stability test in DIN 51 524-3 (result to be reported only).

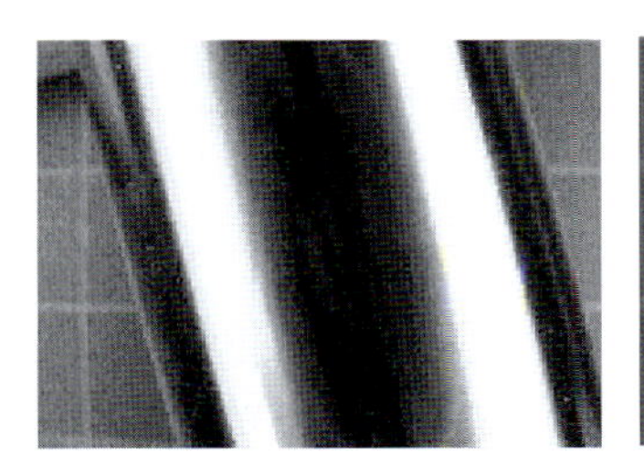

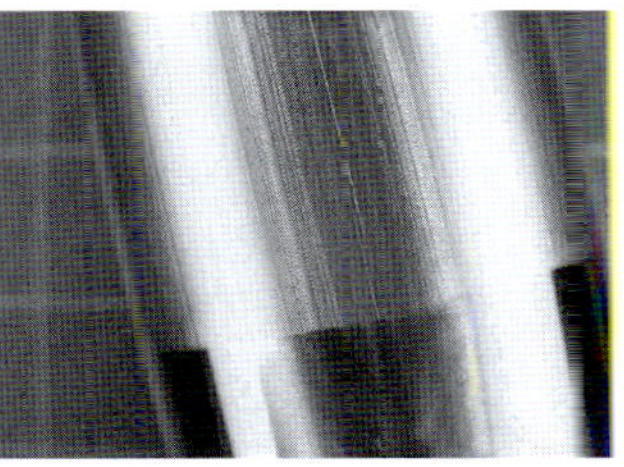

Figure 11.23 Pistons without and with micro-scratches.

Table 11.22 Important properties of hydraulic fluids (viscosity class ISO VG 46).

	Mineral oil	Emulsion/ solution	Polymer solution	Synthetic ester	Synthetic ester	Triglycerides rapeseed oil
DIN 51 502	HLP	HFA	HFC	HFDR	HFDU/HEES[a]	HETG
Density at 15 °C (g cm^{-3})	0.89	1.0	1.04–1.09	1.1–1.2	0.93–0.94	0.92–0.93
Kinematic viscosity at 40 °C (mm^2 s^{-1})	46	1–2	46	46	46	46
Viscosity index (VI)	100 (good)	–	150–200 (very good)	About 0–80 (low) 140–190	150 – 180 (very good)	200–220 (very good)
Average compression modulus, M (N m^{-2})	2×10^9	2.5×10^9	3.5×10^9	$2.3–2.8 \times 10^9$	$2.3–2.8 \times 10^9$	$1.8–2.5 \times 10^9$
Specific heat at 20 °C (kJ kg^{-1} K^{-1})	1.9–2.2	4.2	3.3	1.3–1.5	1.3–1.5	2.1
Thermal conductivity, λ, at 20 °C (W m^{-1} K^{-1})	0.13–0.14	0.6	0.3–0.43	0.11–0.12	0.11–0.13	0.15–0.18
Expansion coefficient, β_{Tr} (K^{-1})	7×10^{-4}	1.8×10^{-4}	7×10^{-4}	6.8×10^{-4}	7×10^{-4}	7.5×10^{-4}
Recommended temperature range/operating temperatures (°C)	−10 to 100	5[a] to < 55	−20 to 60	−10 to 100	−10 to 100[a]	0–70
Maximum temperature range/operating temperatures (°C)	35 to 120	0–55	25 to 65	20 to 150	−35 to 130[a]	−20 to 90
Flashpoint (°C)	About 220	n.a.[b]	n.a.[b]	240–250	250–300	About 315
Ignition temperature (°C)	310–360	None	None	500–550	450–500	350–500
Pour point (°C)	< −18	≈0	< −30	< −18	< −30	About −25
Bunsen coefficient, α_V, at 20 °C for air	0.08–0.09	0.02	0.01–0.02	0.012–0.02	0.012–0.02	0.05–0.06

Acoustic velocity at 20 °C (m s^{-1})	1.300	1.400	About 1.400	–	–	–
Vapour pressure at 50 °C (mbar)	10^{-4}/ 10^{-5}	120	About 50–80	$< 10^{-5}$	$< 10^{-5}$	$< 10^{-5}$
Risk of cavitation	Low	Very high	Middle	Low	Low	Low
Relative costs of the fluid (%)	100	10–15 dependent on concentration	200–300	800–900	300–600	200–250
Market share (%)	About [illegible]	< 1	About 1 [illegible]	< 1	< 1/[illegible]	About 1 [illegible]

a) Saturated esters.
b) Not applicable, no flashpoint.

European Eco-Label 'Margerite' with EU Directive 2005/360/EC was introduced in 2005. For biodegradable hydraulic fluids, it is based on technical specification of ISO 15 380. Besides requirements for biodegradability, sustainability is also requested – 50% of the raw materials must be derived from renewable materials.

ISO 6743-4 was also recently updated with respect to technical requirements for mineral-oil-based hydraulic fluids.

11.6.5.13 Conclusion

The variety of available hydraulic fluids are vast as are their individual technical requirements. Each application, with its ambient conditions, therefore must be considered in detail when selecting the optimum hydraulic fluid. The optimum depends on many different fluid characteristics. It may be the choice between Zn-containing and zinc- and ash-free hydraulic fluids, synthetic base oil or mineral base oil, high additive performance or not and demulsifying properties or detergent/dispersant fluid characteristics. The proper choice affects the performance, lifetime, availability and cost effectiveness of machinery and its hydraulic equipment. Hydraulic fluids are important liquid machine tools.

11.7 Summary

In technical and economic terms, widely varying specifications and application conditions require different hydraulic fluids. They should also satisfy a broad performance span. Table 11.22 shows the most important characteristics of hydraulic fluids and the most important fluid groups [2,4,6,33].

Hydraulic oils are key elements in machines and machine tools and therefore must be included in the planning of plant and equipment to reflect the different properties offered by the various types of fluids.

References

1 VDMA – Verband Deutscher Maschinen- und Anlagenbau (2005) Broschüre des Forschungsfond der Fachgemeinschaft Fluidtechnik im VDMA, Frankfurt, Statistik.

2 Murrenhoff, H. and der Fluidtechnik, Grundlagen (1997) *Teil 1: Hydraulik Umdruck zur Vorlesung*, Institut für fluidtechnische Antriebe und Steuerungen, Aachen.

3 Backe, W. Grundlagen der Ölhydraulik – Umdruck zur Vorlesung (1992) Institut für hydraulische und pneumatische Antriebe und Steuerungen der RWTH Aachen, Aachen.

4 Jaroslav und Monika Ivantysyn (1993) *Hydrostatische Pumpen und Motoren, Konstruktion und Berechnung*, Vogel Buchverlag und Druck KG, Würzburg.

5 Mang, T. (1998) *Hydrauliköle, Umdruck zur Vorlesung*, Institut für fluidtechnische Antriebe und Steuerungen, RWTH Aachen.

6 Bartz, W.J. und 18 Mitautoren (1995) *Hydraulikflüssigkeiten, Band 475*, Expert Verlag, Renningen.

7 Prazisionsdichtungen für die Pneumatik Firmendruckschrift, Katalog 3351, 1997, Parker Hannifin GmbH, Pradifa, Postfach 1641, 74306 Bietigheim/Bissingen.

8 Simrit Handbuch, Firmenschrift (1994) Fa. Karl Freudenberg, Sparte Dichtungs- und Schwingungstechnik, Weinheim.

9 ISO/DIS 6072, ISO TC 131/SC7 N343 (1999) Hydraulic Fluid Power – Compatibility Between Fluids and Standard Elastomeric Materials (Draft).

10 DIN 51 524 (April 2006) Teil 1–3, Druckflüssigkeiten, Hydrauliköle HL, HLP, HVLP, Mindestanforderungen.

11 VDMA Blatt 24568, VDMA Blatt 24569, Fluidtechnik biologisch schnell abbaubarer Druckflüssigkeiten, Technische Mindestanforderungen, Umstellungsrichtlinien, VDMA, Verband Deutscher Maschinen- und Anlagen- bau e.V., Beuth Verlag GmbH, 10772 Berlin.

12 ISO 15380 (2002) Lubricants, Industrial Oils and Related Products (Class L) – Family H (Hydraulic Systems) – Specifications for Categories HETG, HEES, HEPG and HEPR, (E).

13 Bartz, W., u.a. (1993) *Biologisch schnell abbaubare Schmierstoffe und Arbeitsflüssigkeiten*, Expert Verlag GmbH, Goethestraße, Ehningen.

14 Luxemburger Bericht, 7. (1994) Auflage, Anforderungen und Prüfung schwerentflammbarer Hydraulikflüssigkeiten zur hydrostatischen und hydrokinetischen Kraftübertragung und Steuer- ung, Dokument Nr. 4746-10-91 DE- Europaische Kommission, Generaldirektion V, Beschaftigung, Arbeitsbeziehungen und soziale Angelegenheiten, Luxembourg, April 1994 in conjunction with Factory Mutual (FM-USA).

15 Ethyl Handbuch (1995) *Lubricant Specification Handbook*, Ethyl Mineralöladditiv GmbH, Hamburg.

16 Umweltschonende Hydraulikflüssigkeiten – Einsatzrichtlinien, Anwendungstechnische Informationen ATI 9101, Status 10/1994. Sauer Sundstrand GmbH & Co., Krokamp, Neumunster.

17 *Hydromotoren, Axialkolbenmotoren, Radialkolbenmotoren, Zahnradmotoren, Langsamlaufer*, Firmenschrift RD14000/04.92, *Mannesmann Rexroth GmbH, Jahnstr, Lohr am Main, Firmendruckschrift*.

18 (1995) *Allgemeine Betriebs- und Wartungsanleitung für hydraulische Anlagen, Firmendruckschrift* Bosch GmbH, Geschaftsbereich Hydraulik, Pneumatik, Postfach, Stuttgart.

19 Tips für Hydraulik Firmenschrift HZ.00.A1.03. Dannfoss Antriebs- und Regeltechnik GmbH, Geschaftsbereich Hydraulik, Offenbach.

20 (1992) *Handbuch der Hydraulik Firmendruckschrift*, Vickers Systems GmbH, Bad Homburg, Firmendruckschrift.

21 Radialkolbenpumpen, Firmendruckschrift, Robert Bosch GmbH, Geschaftsbereich Automatisierungstechnik, Postfach, Stuttgart.

22 Hodges, P.K.B. (1996) *Hydraulic Fluids, Petroleum Consult. Norway*, John Wiley & Sons, New York, Toronto.

23 Klamann, D. (1984) *Lubricants and Related Products*, Verlag Chemie, Weinheim, Basel.

24 DIN 51 502 (1990) Schmierstoffe und verwandte Produkte, Kurzbezeichnung der Schmierstoffe und Kennzeichnung der Schmierstoffbehalter, Schmiergerate und Schmierstellen, August.

25 ISO 6743/4 (1997) Lubricants, Industrial Oils and Related Products (Class L) – Classification – Part 4: Family H (Hydraulic systems), with ISO 11158.

26 DIN 24 320 (1986) Schwerentflammbare Hydraulikflüssigkeiten – Eigenschaften, Anforderungen, August, Beuth GmbH, Burggrafenstr, Berlin.

27 VDMA Blatt 24317 Schwerentflammbare Druckflüssigkeiten, Richtlinien, VDMA, Verband Deutscher Maschinen- und Anlagenbau e.V., Beuth GmbH, Berlin.

28 CETOP RP77H, Flüssigkeiten für hydraulische Kraftübertragung, Einteilung, Europaisches Komitee Ölhydraulik und Pneumatik (CETOP).

29 United States Department of Agriculture (USDA) (1996) Miscellaneous Publication

No. 1419, 01/1996, List of Proprietary Substances and Nonfood Compounds together with NSF International, USA, Standard for "food grade Lubricants".

30 (1996) *Betriebsstoffe für Getriebe in Fahrzeugen und Arbeitsmaschinen*, FUCHS Technische Mitteilung 221, Fuchs Mineralölwerke GmbH, Mannheim, Firmendruckschrift.

31 DIN 51 517 (2004) Teil 1–3: Schmierstoffe, Schmieröle C, CL, CLP, Mindestanforderungen, Januar.

32 ISO 6743/6 (1990) Lubricants, Industrial Oils and Related Products (Class L) – Classification, Part 6: Family C (Gears).

33 Reichel, J. Druckflüssigkeiten für die Wasserhydraulik, DMT Gesellschaft für Forschung und Prüfung mbH, Antriebstechnik und Hydraulik, 45307 Essen. O+P, Ölhydraulik und Pneumatik, Nr. 43/1999

34 Umweltfreundliche Druckflüssigkeiten für Axialkolbenmaschinen, Firmenschrift RD 90221/05.93, Mannesmann Rexroth Hydromatik GmbH, Elchingen.

35 Omeis, J., Bock, W. and Harperscheid, M. (1998) Fuchs Mineralölwerke GmbH, Friesenheimer Str., Mannheim, the Development of a New generation of High Performance Biofluids, 49. SAE International Earth Moving Industry Conference, Peoria, IL, USA.

36 *Druckflüssigkeiten auf Mineralölbasis fur Flugelzellenpumpen, Radialkolbenpumpen und Zahnradpumpen sowie für Motoren*, Firmenschrift RD, Mannesmann Rexroth GmbH, Lohr am Main.

37 Druckflüssigkeiten und Schmierstoffe, *Firmenschrift SDF 08/92.697581 B*, Sauer Sundstrand GmbH & Co., Krokamp.

38 Feldmann, D.G. *Abschlußbericht zum Forschungsvorhaben Umweltschonende biologisch abbaubare Druckflüssigkeit auf Basis synthetischer Ester im Auftrag des VDMA*, TU Hamburg Harburg, Denickestr, Hamburg.

39 Reichel, J. (1998) DMT Gesellschaft für Forschung und Prüfung mbH, Franz-Fischer-Weg 61, 45307 Essen, Performance Testing of Biodegradable Hydraulic Fluids, 49. SAE International Earth Moving Industry Conference, Peoria, IL USA.

40 DIN-Arbeitskreis Lebensmittelschmierstoffe des Arbeitsausschusses Lebensmittelhygiene im NAL, Deutsches Institut für Normung e.V., 10722 Berlin together with NSF International, USA, Standard for "food grade Lubricants".

41 Moller, U.-J. and Boor, U. (1986) *Schmierstoffe im Betrieb*, VDI Verlag GmbH, Düsseldorf.

42 DIN-Taschenbucher: Mineralöle und Brennstoffe 1–5, Beuth GmbH, Berlin, Koln.

43 DIN 51 519, ISO 3448 (1976) ISO-Viskositätsklassifikation für flüssige Industrieschmierstoffe, Juli.

44 *Hydroaggregate für den Betrieb mit HFC-Druckflüssigkeiten, Firmenschrift AB 01-02.17, 3.2.1997*, Mannesmann Rexroth GmbH, Jahnstr, Lohr am Main, Firmendruckschrift.

45 Peeken, H. and Spilker, M. Druck- und Temperaturverhalten der Viskositat von Hydraulikflüssigkeiten (Teil 1), Dichte und Kompressibilitat von Hydraulikflüssigkeiten (Teil 2), O+P, Ölhydraulik und Pneumatik, Nr. 25+26/1981.

46 Peeken, H. and Spilker, M. (1981) *Untersuchungen zum Dichte- und Kompressionsverhalten von Schmierstoffen*, VDI Verlag GmbH, Düsseldorf.

47 Murrenhoff, H. (1998) *Servohydraulik, zur Vorlesung, Umdruck*, Institut für fluidtechnische Antriebe und Steuerungen, Steinbachstr, Aachen.

48 Mang, T. and Junemann, H. (1972) Beurteilung der Gebrauchseigenschaften von Hydraulikflüssigkeiten auf Mineralölbasis. *Erdol und Kohle*, 9–10.

49 Mang, T. and Junemann, H. (1973) Gebrauchseigenschaften von Hydraulikflüssigkeiten auf Mineralölbasis. *Mineralöltechnik*, 7–9.

50 DGMK Vorhaben 514 – Mechanische Prüfung von Hydraulikflüssigkeiten, Forschungsfond der Fachgemeinschaft Fluidtechnik im VDMA, Postfach, Frankfurt/Main.

51 Kirsch, W. and Mann, W.H. (1998) Hydac Technology, Sulzbach: Fluidcontrolling zur Erhöhung der Betriebssicherheit

fluidtechnischer Anlagen, IFK, 1. Internationales Fluidtechnisches Kolloquium, Aachen.
52 Essers, H. Senkung der Instandhaltungskosten durch eine konsequente Olpflege mit System, Firmendruckschrift 10/96, Pall Industriehydraulik GmbH, Dreieich.
53 Walzlager, Firmendruckschrift (1987) Fa. FAG, 8720 Schweinfurt.
54 Natscher, J. and Eisenbacher, E. (1995) Verunreinigungen in Hydraulikflüssigkeiten und ihr Einfluß auf die Konstruktion, in *Hydraulikflüssigkeiten* (ed. W.J. Bartz), Band 475, Expert Verlag, Renningen.
55 Olpflege mit System (1997) Technische Information, Pall GmbH, Industriehydraulik, Dreieich.
56 Bierwirth, A. (1998) Untersuchung der Filtrierbarkeit von Hydraulikölen in Abhangigkeit der Grundolzusammensetzung und der chemischen Additivierung und Aufbau eines dynamischen Filterprufstandes. Diplomarbeit, RWTH Aachen, Institut für fluidtechnische Antriebe und Steuerungen, H. Murrenhoff.
57 Werkzeugmaschinen – Obersicht, Zeitschrift Produktion, Nr. 4/99, Verlag Moderne Industrie, Landsberg.
58 Weck, M. (1997) *Werkzeugmaschinen – Konstruktion und Berechnung, 6. überarbeitete Auflage*, Springer, Berlin.
59 Gruneklee, A. (1996) Heutige und zukünftige Aufgaben eines Hydraulikfluids für die Werkzeugmaschine mit dem besonderen Blickpunkt auf extrem niedrigviskose Flüssigkeiten. Diplomarbeit, RWTH Aachen, Institut für fluidtechnische Antriebe und Steuerungen, H. Murrenhoff.
60 DIN 8659 (1980) Teil 1: Werkzeugmaschinen, Schmierung von Werkzeugmaschinen, Schmieranleitungen, April 1980, Teil 2: Werkzeugmaschinen Schmierung von Werkzeugmaschinen, Schmierstoffauswahl für spanende Werkzeugmaschinen, April.
61 (1986) *Schmieröle und Schmierfette für Walzlage* Firmendruckschrift, INA KG, Postfach, Herzogenaurach.
62 Witte, H. (1988) *Werkzeugmaschinen 6. überarbeitete Auflage*, Vogel Buchverlag, Würzburg.
63 ISO 6743/13 (1989) Lubricants, Industrial Oils and Related Products (Class L – Classification, Part 13 Family G (Slideways).
64 ISO 6743/2 (1981) Lubricants, Industrial Oils and Related Products (Class L – Classification, Part 2 Family F (Spindle Bearings, Bearings and Associated Clutches), (E).
65 Kiechle, A. (1996) Cost Aspects of Cutting Fluid Use, Mineralöltechnik No. 10, Dept. ZWT/PS, Mercedes Benz AG, Stuttgart, Germany, October, ISSN 0341–1893.
66 DaimlerChrysler (1993) DBL 6571 No. 4: Determination of Dirt Carrying Behavior.
67 VDMA, Working Group: Electrostatic Charges of Hydraulic Fluids, Frankfurt.
68 Trelleborg Busak+Shamban. In-house Test Procedure for Microscratching Stuttgart./BAF05/BAFA – Bundesamt für Wirtschaft- und Ausfuhrkontrolle, Amtliche Mineralöldaten, Wiesbaden 2005./DUF05/Dufour J.-C., Europalub 2004, Europalub Rueil-Malmaison 2005./SKC/SKC Gleittechnik. Determination of Static Coefficient of Friction Using an Inclination Tribometer Coburg.

Further Reading

Books

- Grundlagen der Fluidtechnik, H. Murrenholt, IFAS, Aachen Germany
- Hydrostatische Pumpen und Motoren, J. u. M. Ivantysyn, Vogel Verlag, Germany
- Properties according to VDMA 24317, August 1982
- TAE: 'Hydraulikflüssigkeiten', Nr. 24218/68.479, Esslingen, Germany

Standards

DIN 51 554-3, Testing of mineral oils; Test of susceptibility to ageing according to Baader; Testing at 95°C September 1973.

DIN 51 757, Testing of mineral oils and related materials – Determination of density, January 2011.

DIN EN ISO 20763, Petroleum and related products – Determination of anti-wear properties of hydraulic fluids – Vane pump method, October 2004.

DIN EN ISO 2160, Anodizing of aluminium and its alloys – Determination of mass per unit area (surface density) of anodic oxidation coatings – Gravimetric method, June 2011.

DIN EN ISO 4263, Part1, Petroleum and related products – Determination of the ageing behaviour of inhibited oils and fluids – TOST test, Procedure for mineral oils, March 2005.

DIN ISO 14635-1, Gears – FZG test procedures – Part 1: FZG test method A/8,3/90 for relative scuffing load-carrying capacity of oils, May 2006.

DIN ISO 3016, Petroleum oils; determination of pour point, October 1982.

DIN ISO 51524, Part 1, Pressure fluids – Hydraulic oils, HL hydraulic oils, Minimum requirements, Revised Version April 2006.

DIN ISO 51524, Part 2, Pressure fluids – Hydraulic oils, HLP hydraulic oils, Minimum requirements, Revised Version April 2006.

DIN ISO 51524, Part 3, Pressure fluids – Hydraulic oils, HVLP hydraulic oils, Minimum requirements, Revised Version April 2006.

DIN ISO 6614, Petroleum products – Determination of water separability of petroleum oils and synthetic fluids, April 2004.

DIN ISO 7120, Petroleum products and lubricants – Petroleum oils and other fluids – Determination of rust-preventing characteristics in the presence of water, May 2000.

DIN ISO 9120, Petroleum and related products – Determination of air-release properties of steam turbine and other oils – Impinger method, August 2008.

ISO 11158, Lubricants, industrial oils and related products (class L) – Family H (hydraulic systems) – Specifications for categories HH, HL, HM, HV and HG, September 2009.

ISO 15380, Lubricants, industrial oils and related products (class L) – Family H (Hydraulic systems) – Specifications for categories HETG, HEPG, HEES and HEPR, May 2002.

ISO 6247, Petroleum products – Determination of foaming characteristics of lubricating oils, June 1998/ASTMD 892, Standard Test Method for Foaming Characteristics of Lubricating Oils, 2013.

ISO 6618, Petroleum products and lubricants – Determination of acid or base number – Colour-indicator titration method, February 1997.

US Steel 126 and 127, US Steel hydraulic standards.

Vortrag 'Filter und Filtersysteme hydraulischer Arbeitsmaschinen', V. Lauer, HYDAC International GmbH, November 2013.

12 Compressor Oils

Wolfgang Bock and Christian Puhl

12.1 Air Compressor Oils

Compressors increase the pressure of air or any gaseous medium in one or more stages and thus transfer energy to the medium. As the volume of the medium is reduced, its temperature and density increase.

There are two types of compressor: displacement compressors and dynamic compressors. In displacement compressors, the gaseous medium is drawn into a chamber, compressed and expelled by a reciprocating piston. The principle of dynamic compressors is that turbine wheels accelerate a medium which is then abruptly de-accelerated [1]. In the past 25 years, traditional piston compressors have increasingly been replaced by rotary compressors and in particular screw compressors, so the market share of screw and rotary vane compressors is presently about >60%. The reasons for this are the lightweight and compact dimensions of rotary compressors, their low-noise and vibration-free operation and their reliability. Rotary compressors are generally characterized by the constant, relatively lower pressurization of larger volumes of air while piston compressors provide pulsating higher pressurization of smaller volumes.

Compared to piston compressors in which the oil primarily lubricates the bearings, pistons, cylinders and valves, the oil in oil-flooded screw and rotary vane compressors also has the additional function of cooling and sealing. It is possible to differentiate between air and gas compressors, vacuum pumps and refrigerant compressors by analysing the function of the oil. Figures 12.1 and 12.2 show a breakdown of compressors according to their construction and operative range (Bartz, W.J., Betrieb, W. and von Kompressoren S. Lehrgang Nr. 19293/68.358, Technische Akademie Esslingen, Ostfildern, Germany; Fuchs DEA Schmierstoffe GmbH & Co. KG, Luftverdichteröle – Technische Information Produktentwicklung und Anwendungstechnik, Herr Tinger, Friesenheimer Str. 15, Mannheim, Germany).

Lubricants and Lubrication, Third Edition. Edited by Theo Mang and Wilfried Dresel.

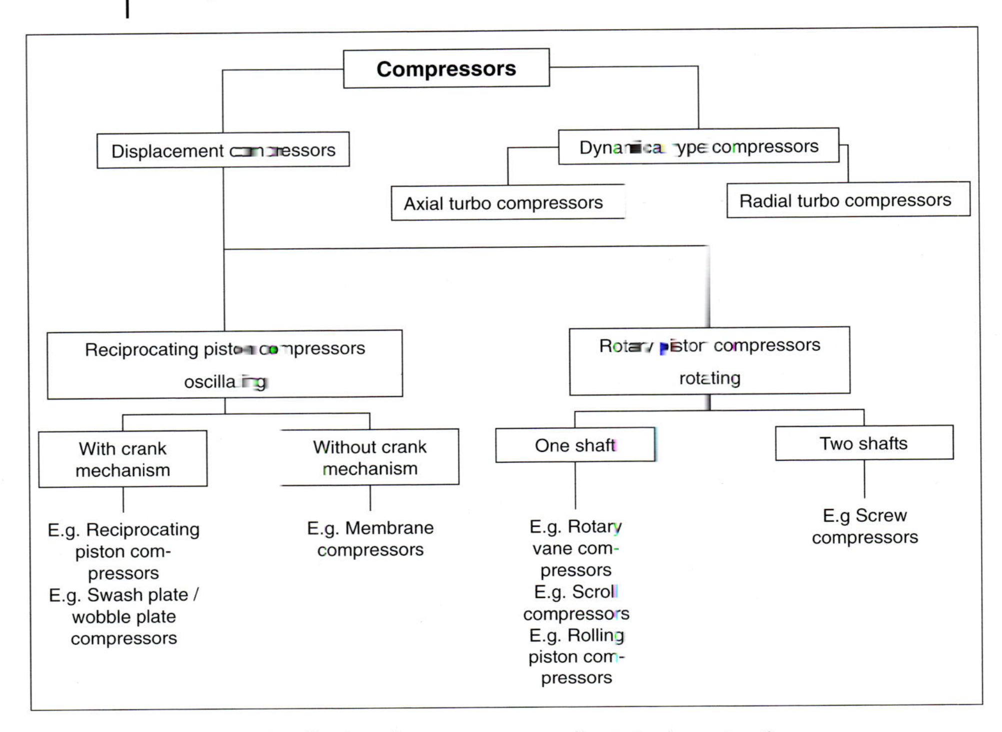

Figure 12.1 Classification of compressors according to their construction.

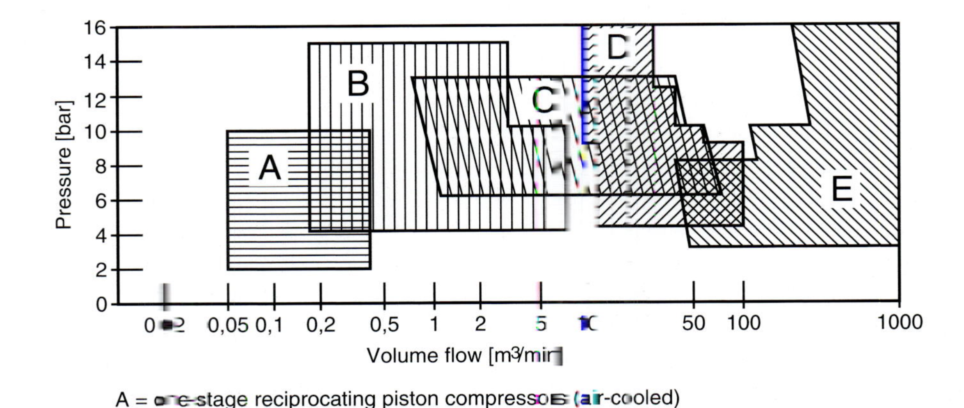

Figure 12.2 Classification of compressors according to their operative range.

12.1.1 Displacement Compressors

12.1.1.1 Reciprocating Piston Compressors

Reciprocating piston compressors increase the pressure of the medium by decreasing the volume of its chambers. The air or gas is compressed and displaced by the rising and falling of a piston in a sealed cylinder. The compression and pumping of the medium result from the periodic (oscillating) change in volume. Each stage of the process is controlled by inlet and outlet valves. The compressor is cooled either by air circulating around the ribs attached to the cylinder head or by water which flows through a jacket around the cylinder (Bartz, W.J., Betrieb, W. and von Kompressoren, S. Lehrgang Nr. 19293/68.358, Technische Akademie Esslingen, Ostfildern, Germany; Fuchs DEA Schmierstoffe GmbH & Co. KG, Luftverdichteröle – Technische Information Produktentwicklung und Anwendungstechnik, Herr Tinger, Friesenheimer Str. 15, Mannheim, Germany).

12.1.1.2 Lubrication of Reciprocating Piston Compressors

In reciprocating piston compressors, the piston is connected to the crankshaft by a connecting rod. In general, the cylinder and the drive share the same splash lubrication from oil in the crankcase. In larger compressors, the pistons are driven by cross-head rods (cross-head compressors, single or double action). In such compressors, the drive is lubricated by crankcase splash and separately from the cylinders (Bartz, W.J., Betrieb, W. and von Kompressoren, S. Lehrgang Nr. 19293/68.358, Technische Akademie Esslingen, Ostfildern, Germany).

The cylinders in a piston compressor represent the most difficult task for the lubricant and ultimately decide the choice of lubricants. The lubricant's primary tasks are the reduction of friction and wear, sealing the compression chambers and protection against corrosion. The peak stress occurs at the TDC and the BDC (top and bottom dead centre). At these points, there is a danger of the lubricant film tearing and allowing metal-to-metal contact. The oil is also subject to enormous stress resulting from the high temperatures created when the medium is compressed (which can cause oxidation and lead to deposits) and in the case of air, the oxygen enrichment. The cleanest possible air or gas should be compressed because contaminants can accelerate oxidation and wear (especially the water content of the air/gas and other contamination – for example, aggressive gases from the surroundings can influence the performance of the used lubricant in an extremely negative way). In the case of drive unit lubrication, the lubrication of the bearings is of primary importance.

Piston compressors are available in lubricated-with-oil and oil-free versions. Normally, lubricants based on mineral oil according to DIN 51 506 – VCL, VDL (or PAO- or diester-based lubricants) are used with viscosity grades of ISO VG 68 to ISO VG 150. Mobile compressors are often lubricated with monograde engine oil (SAE 20, SAE 30, SAE 40) [2,3]. Small- to medium-sized piston compressors are used for pressures up to 10 bar.

12.1.1.3 Rotary Piston Compressors: Single Shaft, Rotary Vane Compressors

In these compressors, the volume of the pressure chambers varies periodically between two extremes. The eccentric cylindrical rotor which is located in the cylindrical sleeve of the housing has sliders (normally steel or PTFE) fitted in grooves which alter the star-shaped chamber between the rotor and the housing. When the rotor turns, centrifugal force presses the sliders against the wall of the housing. During rotation, the enlargement of the chamber on the inlet side draws air in and as the volume gets smaller, the medium is compressed until it is expelled through the outlet. The inlet and compression steps are controlled by slots in the housing.

In single-action operation, pressures of up to 10 bar are possible and up to 16 bar in double-action operation. Volumes can reach 80 $m^3 min^{-1}$. The advantages of rotary piston compressors are compactness, continuous flow and vibration-free operation compared to piston compressors [1] (Bartz, W.J., Betrieb, W. and von Kompressoren, S. Lehrgang Nr. 19293-68358, Technische Akademie Esslingen, Ostfildern, Germany; Mannesmann Demag Verdichter, Druckluft Rotationsverdichter, Firmenschrift, Johann-Sutter-Straße 6–8, Schopfheim, Germany).

12.1.1.4 Lubrication of Rotary Piston Compressors

The pressure chambers of rotary piston compressors are cooled and lubricated by total-loss systems or by direct oil injection. The lubrication of rotary piston compressors is similar to the lubrication of the cylinders in reciprocating piston compressors insofar as the lubricant is subject to high outlet temperatures in both cases. In the case of oil-injected and oil-cooled rotary piston compressors, a quantity of oil is continuously injected into the compressor chambers. The quantity of the oil is such that the outlet temperature does not exceed 100–110 °C. At the same time, it seals the pistons against the housing and protects against wear. The cooling of the medium results in an increase in compression performance. Cooling and sealing increase the volumetric efficiency and thus the overall efficiency of the compressor. The oils used are normally VCL or VDL according to DIN 51 506 with an ISO VG between 68 and 150 or monograde SAE 20, SAE 30, SAE 40 engine oils. Rotary piston compressors are mostly used for vehicle and railroad applications (including road tankers). Outlet pressures are mostly less than 10 bar [1] (Mannesmann Demag Verdichter, Druckluft Rotationsverdichter, Firmenschrift, Johann-Sutter-Straße 6–8, Schopfheim, Germany). Figure 12.3 shows the lubrication circuit of a rotary vane compressor (oil cooled).

12.1.1.5 Screw Compressors

Screw compressors have two counter-rotating axial shafts (screws) and use the displacement principle. One of the two screws is the compressor and the other is an idler and both revolve in the same housing. The inlet area has a large cross section and volume. As the two shafts revolve, the volume gets smaller and compression takes place. The compressed medium then leaves the housing the

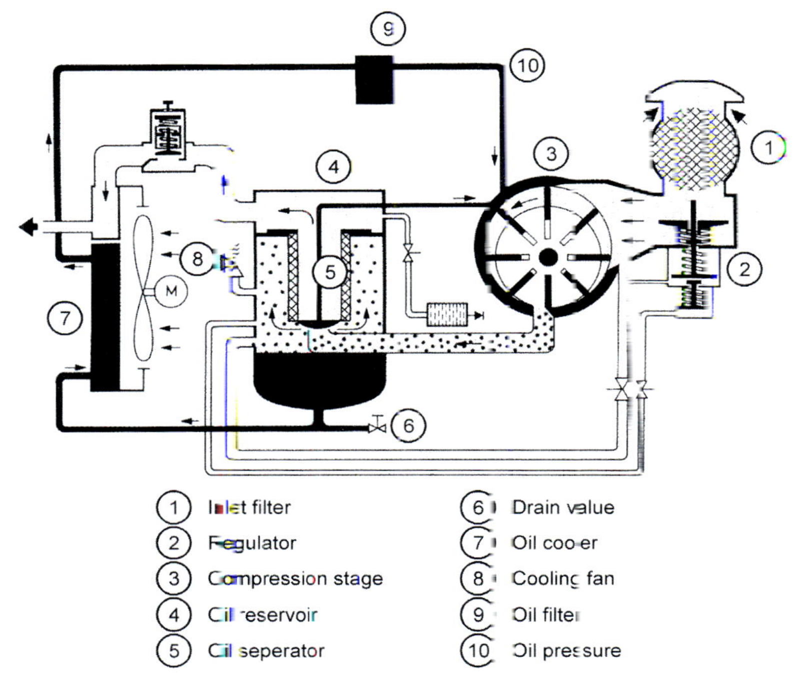

Figure 12.3 Lubrication circuit of a rotary vane compressor (oil cooled).

outlet. The compressor screw has a helical convex cross section while the idler has a helical concave cross section. The two rotors of oil-free screw compressors are geared to each other so that their surfaces never touch. On the other hand, the rotors in oil-flooded screw compressors contact each other and thus do not require the geared coupling.

The advantages of screw compressors are compact size, minimum vibration and continuous flow [1] (Bartz, W.J., Betrieb, W. and von Kompressoren, S. Lehrgang Nr. 19293/68.358, Technische Akademie Esslingen, Ostfildern, Germany; Comp Air Demag, Kolben- und Schraubenverdichter, Firmenschrift, Argenthaler Straße 11, Simmern/Hunsruck, Germany; Boge Kompressoren, Schraubenkompressoren, Firmenschrift, Lechtermannhof 26, Bielefeld, Germany).

12.1.1.6 Lubrication of Screw Compressors

In oil-injected screw compressors, the oil has a lubricating, sealing and cooling function. The lubricant is injected into the pressure chamber between the rotors at about 3–4 bar. It then forms a hydrostatic and a hydrodynamic lubricating film. The oil therefore lubricates the meshing rotors and the plain and roller bearings which are part of the geared coupling. Furthermore, it seals the gaps between the rotor and the housing. It also helps absorb heat and dissipate this via radiators. The temperature of the compressed air of about 80–100 °C is adjusted by the quantity of oil injected. Downstream oil separators (normally cartridge filters) remove the oil from the air. Residual oil quantities of 1–3 mg m^{-3} of air can be achieved. The separated oil de-aerates in tanks, is then filtered and cooled from about 80 to 50 °C. As the oil is on the pressure side of the screw compressor (e.g. up to 10 bar), this pressure can be used to re-inject the oil.

As the viscosity of the oil is of primary importance to elastohydrodynamic lubrication and thus for the mechanical stability of the film, it must be matched to start-up and normal running conditions. As a rule, ISO VG 46 lubricants cover most manufacturer's recommended viscosity thresholds of about >10 mm^2 s^{-1} at operating temperature to about 500 mm^2 s^{-1} when starting up. This range also satisfies most applications in central Europe. Higher viscosity ISO VG 68 oils or synthetic ester-PAG- or PAO-based lubricants are used in countries with high ambient temperatures. In recent years, lubricants based on hydrocracked oils (so-called group III oils) have found increasing acceptance. Screw compressor oils have mild EP/AW performance, FZG failure load stage ≥10 are typically required. In relation to their size and weight, the volumes achievable with screw compressors are excellent. Screw compressors are primarily used for mobile applications as well as for industrial applications such as in the glass and paper industries and general industry. Pressures of up to about 10 bar and higher are possible with screw compressors [2] (Comp Air Demag, Kolben- und Schraubenverdichter, Firmenschrift, Argenthaler Straße 11, Simmern/Hunsruck, Germany; Boge Kompressoren, Schraubenkompressoren, Firmenschrift, Lechtermannhof 26, Bielefeld, Germany).

12.1.1.7 Roots Compressors

Roots compressors normally consist of two symmetrical, figure-of-eight-shaped rotors in a housing. The counter-rotating rotors are driven by external gears and do not touch. The oil's sole task is to lubricate the rotors' gears and bearings. The benefits of this type of compressor are oil-free air, large volumes and low vibration (Bartz, W.J., Betrieb, W. and von Kompressoren, S. Lehrgang Nr. 19293/68.358, Technische Akademie Esslingen, Ostfildern, Germany).

12.1.1.8 Lubrication of Roots Compressors

Recommended lubricants include DIN 51 517 CL and CLP or HD SAE oils in the viscosity grades ISO VG 68 and ISO VG 100 [2].

12.1.2 Dynamic Compressors

12.1.2.1 Turbo Compressors

Turbo compressors are dynamic machines which convert dynamic energy into compression energy. The medium is accelerated by one or more rotors and the dynamic energy is converted into compression energy at the medium's fixed outlet point. Radial and axial turbo compressors differ in that the inlets to the rotors are either radial or axial.

The advantages of turbo compressors are high volumes, minimum vibration and oil-free air [1] (Bartz, W.J., Betrieb, W. and von Kompressoren, S. Lehrgang Nr. 19293/68.358, Technische Akademie Esslingen, Ostfildern, Germany)

12.1.2.2 Lubrication of Turbo Compressors

The oils for this type of compressors lubricate bearings, radial shaft seals and possibly gears via a positive-feed circuit. In some cases, the bearings are lubricated with grease. Ideally, the same lubricant should be used for the compressor and its drive. Most often, DIN 51 515 TDL 32, TDL 46 and TDL 68 turbine oils or TDL-EP grades (EP – extreme pressure additives) are used.

Turbo compressors are principally used for creating compressed air in mines and industrial manufacturing plants [2] (Fuchs DEA Schmierstoffe GmbH & Co. KG, Luftverdichteröle – Technische Information Produktentwicklung und Anwendungstechnik, Herr Tinger, Friesenheimer Str. 15, Mannheim, Germany).

12.1.2.3 Preparation of Compressed Air

The oil injected into oil-cooled screw and rotary piston compressors is always removed from the compressed air. The oil which is mixed with the highly compressed air is removed and collected in single or multistage downstream oil separators. Before the oil is recirculated, it is filtered and cooled. Depending on the specific requirements, the compressed air may then pass a number of subsequent treatment stages such as refrigerant dryers or absorption dryers (to reduce the water content coming from the humidity of the air/gas). Very low residual oil quantities in the air can be achieved by the fitting of a series of in-line oil separators (Mannesmann Demag Verdichter, Druckluft Rotationsverdichter, Firmenschrift, Johann-Sutter-Straße 6–8, Schopfheim, Germany; Comp Air Demag, Kolben- und Schraubenverdichter, Firmenschrift, Argenthaler Straße 11, Simmern/Hunsruck, Germany; Boge Kompressoren, Schraubenkompressoren, Firmenschrift, Lechtermannhof 26, Bielefeld, Germany).

12.1.2.4 Lubrication of Gas Compressors

Oxygen Compressors

Because of explosion hazards when oxygen is compressed, pressure chamber lubricants must be mineral oil-free. Water and water-based solutions such as glycerine can be used for cylinder lubrication. Mineral oil-based products may be used for compressor drives if it does not come into contact with the pressure chambers [2].

Oxygen compressors can be lubricated with inert lubricants based on perfluoroether oils (which are extremely expensive).

Acid Gas Compressors

Gases often contain acidic components such as SO_2 or NO_x. If standard compressor oils were used for such applications, the lubricating oil would soon become over-acidified. To counter this, lubricants are used for such applications which contain highly alkaline additives. These components can neutralize the acidic components in the gas. In these cases, it is recommended that monograde engine oils (20W-20, 30W, 40W) with high alkaline reserves (high TBN) are used [2].

Inert Gas Compressors

When inert gases are compressed, the same rules as for air compressors should be used [2] (Bartz, W.J., Betrieb, W. and von Kompressoren, S. Lehrgang Nr. 19293/68.358, Technische Akademie Esslingen, Ostfildern, Germany).

Hydrocarbon Compressors

Hydrocarbons such as ethane, propane, and so on are easily soluble in mineral oil. This causes the viscosity of the lubricating oil to fall if mineral oil-based products are used. For this reason, higher viscosity mineral oils such as ISO VG 100 and ISO VG 150 must be used in piston compressors whose crankcases are subject to low inlet pressures (1–3 bar). In the case of screw compressors (high pressure; 10–15 bar), ISO VG 68, 100, 150 and 220 ester- or polyglycol-based lubricants with lower hydrocarbon solubility are recommended. In some applications PAO based compressor oils are in use [2] (Bartz, W.J., Betrieb, W. and von Kompressoren, S. Lehrgang Nr. 19293/68.358, Technische Akademie Esslingen, Ostfildern, Germany).

Vacuum Pump Lubrication

Vacuum pumps are compressors whose inlet is connected to the chamber where the vacuum is created. VDL compressor oils can be used for low vacuums. Greater vacuums require synthetic oils with low vapour pressures (mostly synthetic ester oils). The lubricant selection must consider if the medium to be extracted is not air, but for example, a refrigerant. In such cases, a compatible refrigeration oil can be used [2].

12.1.2.5 Characteristics of Compressor Oils

Compressors whose chambers are lubricated pose particular safety problems if air or aggressive gases contact the lubricant. The selection of the most suitable lubricant depends on the type of compressor in question, the pressures involved, the outlet temperatures and the type of air/gas being compressed. Piston compressors which generate the highest pressures are particularly problematic. Turbo compressors which only have lubricated bearings and non-lubricated pressure chambers pose the least problems. Rotary and screw compressors with outlet pressures under 10 bar and correspondingly low outlet temperatures are examples of average compressor lubrication application. Table 12.1 shows an overview of normally used compressor oils.

In general, reciprocating piston compressors need lubricants with higher viscosity (ISO VG 100 or ISO VG 150), extremely low carbon residues, and no or mild EP/AW performance additives. Screw compressors need lubricants of lower viscosity (ISO VG 46 or 68) with excellent oxidation stability and mild/high AW/EP performance additives.

Table 12.1 Overview of normally used compressor oils.

Viscosity classification	Type piston compressors[a)]	Screw compressors oil-injected	Sliding vane compressors[a),b)]	Turbo compressors (axial and radial)[c)]
ISO VG 32		MO HC-oils PAO		TDL 32 TDL 32 EP synth. oils
ISO VG 46		MO HC-Oils PAO POE		TDL 46 TDL 46 EP Synth. oils
ISO VG 68 (SAE 20 W-20)	MO PAO Diester	MO HC-Oils PAO POE	MO Diester HC-Oils	TDL 68 TDL 68 EP Synth. oils
ISO VG 100 (SAE 30)	MO PAO Dieste		MO Diester HC-Oils	
ISO VG 150 (SAE 40)	MO PAO Diester			

The viscosity and the quality recommendations of the compressor manufacturers must be taken into consideration.
MO: mineral oil; PAO: polyalphaolefin; HC: hydrocrack oil (so-called group III oils); POE: biologically degradable polyolesters.

- *Diester, polyolester and PAO:* For very hard working conditions, increase of service intervals is possible.
- *HC oils (so-called group III oils):* For medium and hard working conditions.
- *MO:* For normal and medium working conditions.
- *Lubricants for roots-compressors:* HL, CL, CLP; ISO VG 100-150, DIN 51 524, DIN 51 517.
- *Lubricants for vacuum pumps:* ISO VG 68-150.

a) Total loss lubrication: HD-monograde motor oils HD 20W-20, HD 30, HD 40.

b) For oil-injected compressors in mobile equipment (e.g. railways, buses): multigrade motor oils e.g. 10W 40, 15W 40).

- For oil-injected compressors in stationary units: turbine oils according to DIN 51 515 TDL, air compressor oils according to DIN 51 506 VCL, VDL.
- For hard working conditions: monograde motor oils HD 20W-20, HD 30, MIL 2104 E.

c) Turbine oils according to DIN 51 515 TDL or TDL-EP with extreme pressure additives.

12.1.2.6 Standards and Specifications of Compressor Oils

DIN 51 506 describes the classification and requirements of lubricating oils which are used in reciprocating piston compressors with oil-lubricated pressure chambers (also for vacuum pumps). Lubricants for screw and oil-injected rotary vane and screw compressors are not included in DIN 51 506 [3]. Table 12.2 shows the classification of air compressor oils according to DIN 51 506. Tables 12.3 and 12.4 contain the minimum requirements of air compressor oils according to DIN 51 506.

Table 12.2 Classification of air compressor oils according to DIN 51 506 – Table 12.1 December 2013.

	Maximum compressed air temperature (°C)	
Lubricating oil category	**For air compressors and compressors, including storage tanks and pipe networks, stationary and on moving (mobile) equipment, for brakes, tippers, signals and conveyors**	
VDL	Up to 220	Up to 220
VB VBL	Up to 140	Up to 140

Rotary multi-vane compressors designed for a once-through lubrication can be operated at compressor end temperatures of up to 180 °C using lubricating oils doped in the same manner as motor lubricants or doped compressor oils, provided that the requirements specified for VCL lubricating oils in Table 12.2 are complied with.

According to this standard, such lubricants are pure mineral oils or mineral oils with additives to increase oxidation stability, ageing resistance and corrosion protection. The classification of the lubricants depends on the expected outlet temperatures and the general application. DIN 51 506 differentiates between lubricants for mobile applications and stationary applications with reservoirs. Principal differences between the listed groups – VB/VBL, VC/VCL and VDL – are the use of oxidation and corrosion inhibitors, ageing stability and residue formation and the quality of the base oils (cuts). The difference between group VB/VBL and group VC/VCL lies in the ageing behaviour (formation of residues coke after air-induced ageing). Group VDL oils have to pass a more difficult ageing test (formation of Conradson coke after air-induced ageing in the presence of ferrous oxide). DIN 51 506 VDL oils display the best thermal and oxidation stability and form the least residues. The selection criteria in DIN 51 506 were adopted into the 1983 ISO/DP 6521 draft [3,4]. Table 12.5 defines the requirements of air compressor oils for reciprocating piston compressors. Instead of mineral oil-based lubricants, synthetic compressor oils based on hydrotreated (group II base oils) or polyalphaolefin or ester oils have found increasing commercial acceptance (long-life oils).

The selection criteria for screw compressor and piston compressor oils differ greatly. In oil-flooded rotary vane and screw compressors, the injected oil is constantly in contact with the 80–100 °C hot medium gas or air being compressed. The compressed medium and the oil are well mixed and the oil has to be separated by downstream separators and filters. This places special demands on the lubricant. On the whole, these include low foaming, excellent air release and good demulsification (separation) of condensed water. In addition, of course, the specifications regarding wear protection (FZG ≥ 10, DIN 51 354), minimum formation of deposits, good corrosion protection and so on also apply. Most

Table 12.3 The minimum requirements of air compressor oils according to DIN 51 506 – VB, VBL, VDL.

Grades VB are pure mineral oils, Grades VBL and VDL contain additives L to increase ageing resistance and corrosion protection														
Lube oil group	**VB and VBL**									**Group VDL**				
Viscosity grade	ISO VG 22	ISO VG 32	ISO VG 46	ISO VG 68	ISO VG100	ISO VG150	ISO VG220	ISO VG320	ISO VG460	ISO VG 32	ISO VG 46	ISO VG 68	ISO VG100	ISO VG150
Kinematic viscosity (DIN 51562 part 1)														
Min. cST @ 40 °C max.	19.8–24.2	28.8–35.2	41.4–50.6	61.2–74.8	90–110	135–165	198–242	288–352	414–506	28.8–35.2	41.4–50.6	61.2–74.8	90–110	135–165
cST @ 100 °C	To be stated by the supplier													
Flash point, °C (COC) min. (DIN EN ISO 2592)	175		195		205	210	255			175	195		205	210
Viscosity index	To be stated by the supplier													
Pour point, °C, max. (DIN ISO 3016)	−9					−3	0			−9				−3
Oxide ash (DIN EN ISO 6245)	VB: max. 0.02% (m/m) oxide ash													
Sulfated ash, wt% max. (DIN 51575)	VBL and VDL: To be stated by the supplier													
Water-soluble acids (DIN 51559 part 1)	Neutral													
Neutralization number (acid), mg KOH g^{-1} max. (DIN 51558 part 1)	VB: max. 0.15 VBL and VDL: To be stated by the supplier													
Water, % (DIN 51777 part2)	0.05 max.													

(continued)

Table 12.3 (*Continued*)

Grades VB are pure mineral oils, Grades VBL and VDL contain additives L to increase ageing resistance and corrosion protection					
Lube oil group	VB and VBL		Group VDL		
Ageing characteristics % CRC[a] max. after air ageing (DIN ISO 6617)	2.0	2.5	Not required		
% CRC max. after air/Fe_2O_3 ageing (DIN 51352 part 2)	Not required		2,5	3,0	
Distillation residue % CRC max. of 20% distillation residue (DIN 51551 part1)	Not required		0.3		0.6
Kinematic viscosity at 40 °C, max. of 20% dist. residue (DIN EN ISO 3104 combined with DIN 51562 part1)	Not required		Maximum of five times the value of the new oil		

a) CRC: Conradson carbon residue.

Table 12.4 Minimum requirements of air compressor oils according to DIN 51 506 – VDL.

Lube oil group	VDL				
Viscosity grade	**ISO VG 32**	**ISO VG 46**	**ISO VG 68**	**ISO VG 100**	**ISO VG 150**
Kinematic viscosity (DIN 51 561/51 562-1) cST @ 40 °C	28.8–35.2	41.4–50.6	61.2–74.8	90–110	135–165
cST @ 100 °C	5.4	6.6	8.8	11	15
Flash point, °C (COC), min (DIN ISO 2592)	175	195		205	210
Pour point, °C, max. (DIN ISO 3016)	−9				−3
Ash, % wt., max. (DIN 51 575)	Sulfur ash to be stated by the supplier				
Water-soluble acids (DIN 51 558 part 1)	Neutral				
Neutralization number (acid), mg KOH g^{-1}, max. (DIN 51 558 part 1)	To be stated by the supplier				
Water, % (DIN ISO 3733)	0.1 max				
Ageing characteristics % CRC max. after air ageing (DIN 51 352 part 1)	Not required				
% CRC max after air/Fe_2O_3 ageing (DIN 51352 part 2)	2.5			3.0	
Distillation residue % CRC max of 20% distillation residue (DIN 51 356/51 551)	0.3				0.6
Kinematic viscosity at 40 °C max of 20% distillation residue (DIN 51 536/51 561/51 562-1)	Maximum of five times the value of the new oil				

manufacturers of oil-injected rotary and screw compressors issue their own lubricant specifications. A draft of ISO/DP 6521 for oil-injected screw compressors has existed since 1983. Table 12.6 describes the requirements of air compressor oils for screw compressors.

Where this does not apply, oils with dispersant additives, which tend to have poor water-separating properties, may be used satisfactorily.

The selection criteria of DIN 51 506 and ISO/DP 6521, along with the recommendation of the compressor manufacturers, should be observed when choosing an air compressor oil. At the same time, attention must be paid to the safety procedures issued by industrial and professional associations and to the specifications of the compressor manufacturers.

ISO 6743-3A (1987) divides oil-lubricated air compressors into six mechanical type groups but does not list any physical or chemical characteristics which the lubricants must fulfil [3,4].

Table 12.5 The requirements of air compressor oils for reciprocating piston compressors – ISO/DP 6521 – Draft 1983.

Category	Mineral oil based lubricants for reciprocating piston compressors										Test method
	ISO L DAA					ISO-L-DAB					
Viscosity grades	32	46	68	100	150	32	46	68	100	150	ISO 3448
Viscosity @ 40 °C, cSt ± 10%	32	46	68	100	150	32	46	68	100	150	ISO 3104 (IP71)
@ 100 °C cSt	To be stated					To be stated					
Pour point[a)], °C max	−9					−9					ISO 3016 (IP15)
Copper corrosion, max	1b					1b					ISO 2160 (IP154)
Rust	No Rust					No Rust					ISO/DP 7120A (IP135A)
Emulsion characteristics Temperature, °C Time (min) to 3 ml Emulsion, max	No requirement					54			82		ISO/DP 6614 (ASTM D1401)
						30				60	
Oxidation Stability after ageing @ 200 °C Evaporation Loss, %, max.	15					No requirement				ISO/DP 6617 Part 1 (DIN 51 352)	
Increase in Conradson Carbon residue, %, max.	1.5			2.0							
after ageing @ 200 °C Evaporation. loss, %, max. Increase in Conradson Carbon residue, %, max.	Not applicable					20		ISO/DP 6617 Part 2 (DIN 51 352)			
						2.5		3.0			
Distil. residue (20 % vol.) Conradson Carbon Residue, %, max.	Not applicable					0.3		0.6			ISO/DP 6616 with ISO/DP 6615 and ISO 3104
Ratio of viscosity of residue to that of new oil, max.						5					

a) When VG32 or VG46 oils are used in cold climate, pour points lower than −9 °C are required.

Table 12.6 The requirements of air compressor oils for screw compressors – ISO/DP 6521 – Draft 1983.

	Mineral oil-based lubricants for rotary screw compressors										
Category	**ISO-L-DAH**					**ISO-L-DAG**					**Test method**
Viscosity grades	32	46	68	100	150	32	46	68	100	150	ISO 3448
Viscosity @ 40 °C, cSt ± 10%	32	46	68	100	150	32	46	68	100	150	ISO 3104 (IP71)
Pour point[a], °C max.	−9					−9					ISO 3016 (IP15)
Copper corrosion, max.	1b					1b					ISO 2160 (IP154)
Rust	No Rust					No Rust					ISO/DP 7120A (IP135A)
Emulsion characteristics[b]	54				82	54				82	ISO/DP 6614 (ASTM D1401)
Temperature °C											
Time (min) to 3 ml											
Emulsion, max											
	30					30					
Foaming characteristics											
Sequence I @ 24 °C	300	300	ISO/DP 6247 (IP146)								
Tendency, ml, max	Nil	Nil									
Stability, ml, max											
Oxidation stability	To be decided					To be decided					To be established
Evapor. Loss, %, max.											
Increase in viscosity, %											
Increase in acidity, %											
Sludge, % weight											

a) When VG32 or VG46 oils are used in cold climate, pour points lower than −9 °C are required
b) Required only in those applications where condensation of atmospheric moisture is a problem.

12.2 Refrigeration Oils

12.2.1 Introduction

The lubrication of refrigeration compressors occupies a special position in lubrication technology. The longevity expected of refrigeration compressors is closely connected to the high quality which is required of refrigeration oils. The interaction with other substances which the refrigeration oil comes into contact with, and especially the extremely high and low temperatures, makes very specific demands on refrigeration oils.

The principal function of a compressor oil is to lubricate the moving parts of the compressor, for example the pistons or rotors, the valves and, in some cases, the slip-ring seals. Furthermore, the refrigeration oil must dissipate heat away from hot compressor components and assist in sealing the compression chambers and valves. The refrigeration oil serves as a hydraulic control and functional fluid in refrigeration compressors [5,6]. It is vital that refrigeration oil which reaches in colder sections of the circuit in the form of oil vapour or oil mist or as a result of splashing must be returned to the compressor by mechanical means (oil separator) or via the refrigerant flow (refrigerant solubility) in all operating conditions. Figure 12.4 shows the principle of a vapour compression refrigeration cycle and Figure 12.5 describes the schematic structure of a refrigeration system.

12.2.2 Minimum Requirements of Refrigeration Oils

The basic requirements of refrigeration oils are laid down in DIN 51 503-1. This standard defines the basic requirements of refrigeration oils according to the

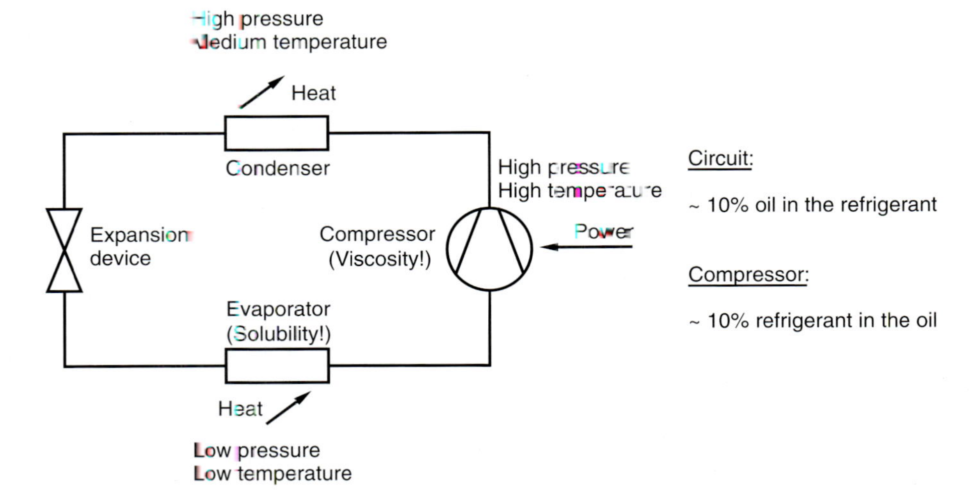

Figure 12.4 Principle of a vapour compression refrigeration cycle.

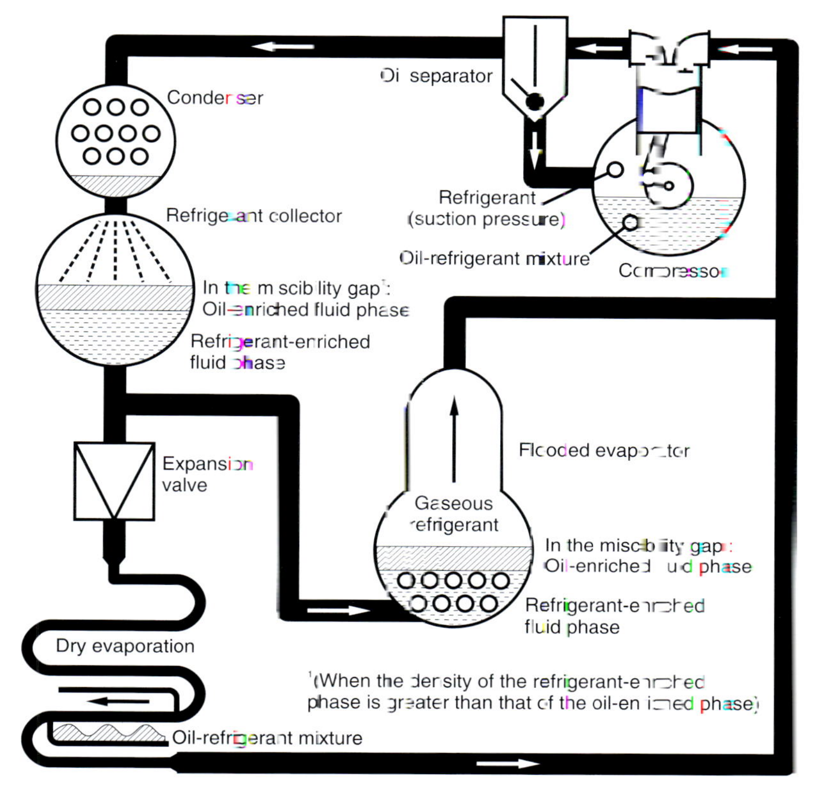

Figure 12.5 Schematic structure of a refrigeration system.

medium being compressed. The introduction of chlorine-free (partly) fluorinated refrigerants such as HFC R134a (to replace CFC R12) necessitated a revision of DIN 51 503 which appeared as DIN 51 503-1 in November 1997 (latest revision 2011-01).

DIN 51 503-1: Refrigeration Oils, Minimum Requirements (2011-01) [7]
Refrigeration oils are classified in alphabetical groups according to the refrigerant being compressed:

KAA	Refrigeration oils for ammonia (NH_3): non-miscible ammonia oils
KAB	Refrigeration oils for ammonia (NH_3): miscible ammonia oils
KB	Refrigeration oils for the refrigerant carbon dioxide (CO_2)
KC	Refrigeration oils for partly fluorinated and chlorinated hydrocarbons (HCFC)
KD	Refrigeration oils for (partly) fluorinated hydrocarbons (HFC, FC)
KE	Refrigeration oils for hydrocarbon refrigerants such as propane or isobutane

The various refrigerants available are described in DIN 8960 and in ANSI/ASHRAE 34-2010 (ASHRAE = American Society of Heating, Refrigerating and Air-Conditioning Engineers) [8].

In addition to appearance (visual inspection), density (DIN EN ISO 3675) and kinematic viscosity (DIN EN ISO 3104), several other properties for refrigeration oils are defined and determined:

Flashpoint	DIN EN ISO 2592
Total acid No	DIN 51 558-1
Total base No	DIN ISO 3771
Oxide ash	DIN EN ISO 6245
Water content	DIN 51 777-1 and 51 777-2
Pour point	DIN ISO 3016
Refrigerant miscibility ('miscibility gap')	DIN 51 514

These properties alone do not always provide sufficient information to judge the suitability of a particular refrigeration oil. However, data sheets on oil–refrigerant mixtures – so-called 'Daniel Plots' – contain information about the influence of dissolved refrigerant in the lubricant: the resulting composition and viscosity of the mixture in relation to temperature and vapour pressure can be extracted [9–11]. The thermochemical stability of oil/refrigerant mixtures is determined by the specific result in an oil ageing test according to ASHRAE 97/07 ('Sealed Glass Tube Test'/'Spauchus test').

DIN 51503-2 describes criteria for evaluating used refrigeration oils [8].

Table 12.7 gives an overview of important refrigerants and suitable refrigeration oils in the market.

The use of environmentally acceptable refrigerants – that is refrigerants with a reduced contribution to the global warming potential, so-called low-GWP refrigerants (GWP = global warming potential) – is becoming even more important.

In the meantime, the legal framework with an EU regulation to reduce the impact of HFC refrigerants (GWP of R134a: 1300) to the greenhouse effect is given: EU No. 517/2014.

To fulfil the valid emission limits during the next years (reducing step by step the emission of HFC refrigerants until 2030 to 21% of the initial value in 2015), the application of refrigerants with a high GWP value will become more and more difficult. Besides natural refrigerants such as carbon dioxide, ammonia and hydrocarbons (all: GWP<5), the use of partly fluorinated olefins, so-called HFO refrigerants, which also have low GWP values, will increase. Today HFO-1234 yf is already used in A/C systems with PAG based refrigeration oils. In stationary equipment HFO refrigerants will be used with polyolester oils (POE).

12.2.3
Classifications of Refrigeration Oils

12.2.3.1 Mineral Oils (MO) – Dewaxed Naphthenic Refrigeration Oils

Naphthenic mineral oils are still the most significant group of oils for refrigeration compressors using ammonia (NH_3) refrigerants along with HCFC (e.g. R22).

Table 12.7 Classification of important refrigerants and refrigeration oils.

Refrigerant type	ASHRAE name	Trade name	Chemical name/ formula	Refrigeration oil type[a]
(Partly) fluorinated refrigerants	R134a	Diverse	CH_2FCF_3	POE, PAG[b]
	R507	Solkane 507, AZ 50	R125/R143a	POE
	R404A	Diverse	R125/R143a/R134a	POE
	R407C	Diverse	R32/R125/R134a	POE
	R410A	Solkane 410, AZ 20	R32/R125	POE
Natural refrigerants	R717	Ammonia	NH_3	MO/PAO/AB[c]
	R744	Carbon dioxide	CO_2	POE/PAG[b]
	R600a	Isobutane	C_4H_{10}	PAO/POE/ MO/AB
	R290	Propane	C_3H_8	PAO/POE/ MO/AB
HFO refrigerants	HFO-1234yf	Solstice™ yf Opteon® yf	CH_2CFCF_3	POE/PAG[b]
	HFO-1234yf blends with HFC	Opteon® XP 10/Diverse	Diverse	POE
	Other HFOs/ HFC blends	Diverse	Diverse	POE
Partly chlorinated refrigerants	R22	Diverse	$CHClF_2$	MO/AB
	R401A	MP 39	R22/R152a/R124	MO/AB
	R401B	MP 66	R22/R152a/R124	MO/AB
	R402A/B	HP 80/81	R22/R125/R290	MO/AB
	R403A/B	69 S/L	R22/R218/R290	MO/AB
	R408A	FX 10	R22/R143a/R125	MO/AB

a) AB: alkylbenzene oil; MO: mineral oil; PAG: polyalkylene glycol; PAO: polyalphaolefins; POE: polyol ester oil.
b) PAG for vehicle A/C systems and special applications.
c) Without AW additives.

Naphthenic mineral oils are those which have more than 38% carbon in naphthenic X(N) bonds. Naphthenic refrigeration oils generally display very low pour points, good cold flowing and high thermal and chemical stability. Selected cuts are normally used [5.12].

12.2.3.2 Mineral Oils (MO) – Paraffinic Refrigeration Oils

Paraffinic mineral oils are those which contain less than about 33% carbon in naphthenic X(N) bonds.

Paraffinic refrigeration oils have been used in R11 and R12 turbo compressors ('old-type') (ISO VG 68 and 100) because of their good viscosity–temperature behaviour. These oils are not recommended for other applications because of their generally inadequate solubility in refrigerants (miscibility gap, e.g. R22). The boundaries between paraffinic and naphthenic oils are not rigid [5,6].

12.2.3.3 Semi-synthetic Refrigeration Oils: Mixtures of Alkylbenzenes (AB) and Mineral Oils (MO)

Semi-synthetic refrigeration oils are mixtures of highly stable alkylbenzenes and highly refined naphthenic mineral oils. The presence of alkylbenzenes greatly improves the miscibility and thermal stability of the naphthenic components. The proportion of synthetic components is usually between 30 and 60%. Semi-synthetic oils are recommended for HCFC systems, medium/low-temperature R22 systems and drop-in refrigerants such as R401A/B, R402A/B and R22 mixtures [9,13]. Together with the decline of those chlorine-containing, ozone-depleting refrigerants, the importance of MO–AB mixtures is also decreasing.

12.2.3.4 Fully Synthetic Refrigeration Oils: Alkylbenzenes

Fully synthetic refrigeration oils are based on chemically and thermally highly stable alkylbenzenes. Alkylbenzenes have been used as refrigeration oils for a number of years. Carefully selected and specially treated alkylaromatics are used. A number of complex manufacturing stages ensure that the products are free of difficult-to-dissolve waxy substances and other contaminants (e.g. sulfur). Alkylbenzene-based lubricants display excellent miscibility in HCFC refrigerants (e.g. R22) and mixtures thereof at evaporation temperatures down to −60 °C.

ISO VG 46 and VG 68 alkylbenzenes have proved particularly suitable for use in heavy-duty ammonia compressors with very high outlet temperatures. Compared to mineral oil-based refrigeration oils, alkylbenzenes form minimal amounts of coke and other ageing products.

Alkylbenzenes are also being used in hermetically sealed and semi-sealed compressors.

Due to the phase-out of chlorinated refrigerants – mainly in Europe but also in other regions (e.g. China), the use of R22 and correspondingly the use of alkylbenzene oils are declining.

12.2.3.5 Fully Synthetic Refrigeration Oils: Polyalphaolefins (PAO)

Polyalphaolefins are recommended for NH_3 compressors because of their good thermal stability. The formation of oxidation products (coke) is avoided even at high compressor outlet temperatures. Compared to mineral oils and particularly in the case of screw compressors, the use of polyalphaolefins can reduce the amount of oil mist and oil vapour which collects in oil separators. The amount of oil in the refrigerant vapour can also be reduced to a minimum. As a result of their chemical structure, polyalphaolefins display good viscosity–temperature behaviour (high VI) and thus good cold flowing characteristics. The low pour point and low viscosity of these products guarantee satisfactory oil return even at evaporation temperatures of −50 °C (important especially for plate evaporators).

Polyalphaolefins of the viscosity grade ISO VG 68 are generally used in screw and piston ammonia compressors [10].

12.2.3.6 Fully Synthetic Refrigeration Oils: Polyol Esters (POE)

Polyol ester-based refrigeration oils have been developed for use with HFC and FC refrigerants due to the excellent refrigerant miscibility. The chemical and thermal stability of these lubricants is excellent.

Also for HFO refrigerants, POE oils are suited. Moreover, POE is proved for reliable lubricants for the natural refrigerants CO_2 and hydrocarbons. Products with the suitable viscosities (ISO VG 10-320) are available for industrial and household piston and screw compressors. The manufacturers' viscosity recommendations should be observed.

Like all ester oils, saturated and high-purity polyol ester oils can hydrolyze if they come into contact with water in the compressor (hydrolysis implies splitting of esters with water into partial esters and acidic compounds). It is therefore essential that these products are protected from water, and moisture in general, during storage and use. Ester oils are ultra-dried and are filled into air-tight metal drums with a water content of<30–100 ppm in a nitrogen atmosphere [10,14] (AFEAS (Alternative Fluorcarbons Environmental Acceptability Study), Production, sales of fluorocarbons through 1995, Washington, DC).

The following are the special properties of polyol esters:

- Excellent solubility in FC and HFC refrigerants
- Avoidance of oil build-up in the condenser/evaporator
- Constant thermal conductivity
- High natural viscosity index, good viscosity–temperature behaviour and thus adequate lubrication at high temperatures
- Very good thermal and chemical stability even in the presence of refrigerants
- Excellent flowing properties at low temperatures
- Long oil life
- Compatibility with all commonly used sealing materials such as NBR, HNBR, EPDM and others
- The products are in general ultra-dried to guarantee the chemical stability

Polyol esters are hygroscopic lubricants (i.e. they absorb water) which may hydrolyze over longer periods of time when their water content is >200 ppm. Hydrolysis = splitting the ester into its acidic components [10,14].

Limits for interpretation of used refrigeration oils based on polyolester (POE) RENISO TRITON SE/SEZ.

		Fresh oil	Used oil
Kinematic viscosity @ 40 °C	$mm^2 s^{-1}$	ISO VG +/−10%	ISO VG +/−15%
	$mg\ KOH\ g^{-1}$	<0.05	>0.2 = Increased value! >0.5 = Change oil!

(continued)

(*continued*)

		Fresh oil	Used oil
Neutralization number (with HFC refrigerants, e.g. R134a)			
Neutralization number (with CFC/HCFC refrigerants, e.g. R22)			>0.07 = Increased value! >0.1 = Change oil!
Water content	ppm	<50	>100 = Increased value! > 200 = Change oil!
Wear-elements (e.g. Fe, Al, Cu	ppm	0	>20 = Increased value! >40 = Change oil!

12.2.3.7 Fully Synthetic Refrigeration Oils: Polyalkylene Glycols (PAG) for R134a and HFO-1234 yf

Fully synthetic, polyalkylene glycol-based refrigeration oils for vehicle aircon systems are in worldwide use for vehicle. Although in the past R134a was the standard refrigerant in this application, the situation has changed since 2011. Since then newly homologated passenger cars have to use a refrigerant with GWP ≤150. HFO-1234 yf, a partly fluorinated olefin with GWP = 4 (R134a: GWP = 1300), is one of the alternatives which is already in use. PAG oils for HFO-1234 yf A/C systems have to fulfil high requirements with regard to the thermo-chemical stability with the more reactive refrigerant HFO-1234 yf (compared to R134a). This can be achieved by using defined double-endcapped PAG base fluids with efficient stabilizer additive systems. Polyalkylene glycols are naturally polar and thus miscible with R134a and HFO-1234 yf. These polar characteristics make polyalkylene glycols very hygroscopic and this fact must be taken into consideration when handling these specialized lubricants. When filled, the water content of polyalkylene glycol oils in the system should be <700–1000 ppm (fresh oil according to DIN 51 503-1 – 300 ppm). PAG-based refrigeration oils are ultra-dried before use [10].

12.2.3.8 Fully Synthetic Refrigeration Oils – Polyalkylene glycols for NH_3

Fully synthetic, polyalkylene glycol-based refrigeration oils (ISO VG 68, VG 100) are soluble and partially soluble in ammonia.

By far, most of the present ammonia systems are operated with naphthenic mineral oils, alkylbenzenes or PAO as lubricants. The problems of oil enrichment and oil deposits in such systems are well known. Polyalkylene glycols, on the other hand, display reactively good miscibility in ammonia refrigerants. However, experiences over the last decade showed that it is very challenging to maintain such PAG/NH_3 installations without causing severe problems related to chemical and/or physical reaction in the compressor. It is thus not recommended to use PAG oils in NH_3 systems in general. In any case the compressor manufacturer has to be consulted prior to using PAG oils.

The water content of polyalkylene glycol oils must be kept low (about 300–500 ppm). Mixing or contamination with mineral oil, PAO or alkylbenzenes must be avoided, since PAG is not miscible and compatible with them [15].

12.2.3.9 Other Synthetic Fluids

In the past, polysilicic acid, ester-based synthetic fluids, were used for evaporation temperatures below −120 °C. Products based on low-viscosity silicone oils (polydimethylsiloxane (PDMS)) are also used. Observe manufacturer's recommendations for such applications alternatives can be low-viscosity polyolester oils.

12.2.3.10 Refrigeration Oils for CO_2

The natural refrigerant CO_2 is gaining more and more acceptance among users. In principle, one can say that the miscibility of CO_2 in POE is better than in PAG-based refrigeration oil. Apart from miscibility, thermal stability is also an important factor (outlet temperatures approximately 160–180 °C) [16,17].

The refrigerant CO_2 is increasingly used in supermarket cooling furniture, in deep-freezing systems (low-temperature cascades), for climate control, in heat-pump systems, and in container refrigeration units. These units can run transcritically and/or subcritically.

Special synthetic polyolester oils (POE – ISO VG 55, 85, 170) with selected anti-wear extreme pressure additives (to reduce and prevent wear problems) are used in industrial and commercial refrigeration applications. The solubility of these polyolester oils in CO_2 is controlled (reduction of the viscosity of the POE–CO_2 mixture must be checked) and the miscibility is excellent (especially important for very-low-temperature systems). This excellent miscibility (no miscibility gap down to −40 °C) guarantee oil flowability and oil return from the evaporator to the compressor. Transcritical air-conditioning systems in buses also run with special polyolesters.

Carbon dioxide may also be an option for R134a in car air-conditioning systems in the near future. In these subcritical CO_2 car air-conditioning systems, selected and double-endcapped polyalkylene glycols (PAG) with anti-wear and extreme pressure additives will be used to guarantee the lifetime of the compressor under the severe conditions of the CO_2 transcritical process. These special PAG-based lubricants are also in use in other transcritical processes like heat pump applications.

12.2.3.11 New Refrigeration Oils for HFO Refrigerants

The use of environmentally acceptable refrigerants, so-called Low GWP Refrigerants (GWP = Global Warming Potential), is becoming even more important. In the meantime with the EU Regulation No. 517/2014 the legal framework to reduce the impact of HFC refrigerants to the worldwide greenhouse effect is given. The application of refrigerants with a high GWP value will become more and more difficult. Beside natural refrigerants like carbon dioxide, ammonia and hydrocarbons, the use of partly fluorinated olefins, so-called HFO (Hydrogenated, Fluorinated Olefin) refrigerants will increase.

The refrigerant R1234yf (GWP = 4) is already in use in air conditioning systems of new vehicle types as successor refrigerants for R134a (GWP = 1300). But R1234yf is at least disputed due to its flammability (classification A2L).

R1234ze (GWP = 6) which has the same chemical composition, but a different molecular structure, has also turbo-dynamic properties which enable a use as refrigerant. But the volumetric refrigeration capacity is approximately 25% below the capacity of 1234yf respectively R134a.

Beside these pure substances mixtures of HFO refrigerants can appear. Initial promising experiences with these new HFO refrigerants and HFO/HFC refrigerant mixtures do already exist. For vehicle/car air conditioning systems new developed PAG lubricating oils are recommended.

In stationary applications POE-based refrigeration oils are proven to be reliable lubricants for HFO refrigerants and their mixtures. The HFO refrigerants can have slightly different miscibility with polyolester lubricants compared to R134a and R404A systems. Also the solubility of HFO and/or HFO/HFC refrigerants can be different to polyolesters (POE). This has to be checked and compared [20].

12.2.3.12 Copper Plating

This phenomenon can occur under unsuitable conditions in compressors. Copper from the refrigeration circuit is dissolved in the lubricating oil and transferred around the system where it is primarily deposited on hot metal surfaces. If the mechanical components concerned run at close tolerances, this can lead to failure (bearings, slip rings). Copper plating is not a subject specifically connected with the oil, although certain properties of the oil may assist its occurrence. It is reasonable to assume that the oil resin and sulfur contents of a good refrigeration oil are below the threshold above which these values could begin to encourage copper plating. Factors likely to promote copper plating are insufficient oil–refrigerant stability, moisture in the system (high water content in the oil), impurities of various kinds, oil acidification by the refrigerant and, last but not least, oil ageing due to contact with oxygen [5,6].

12.2.4
Types of Compressor

Compressors, as the core element of refrigerant systems, pump gaseous refrigerant around the circuit and compress the vaporized refrigerant to the liquefaction pressure required to release heat.

Figure 12.6 classifies refrigeration compressors according to their construction. Compressors are divided into two groups: The displacement types which discontinuously compress the refrigerant into increasingly small spaces and the dynamic types which continuously accelerate the refrigerant to increase its pressure [18].

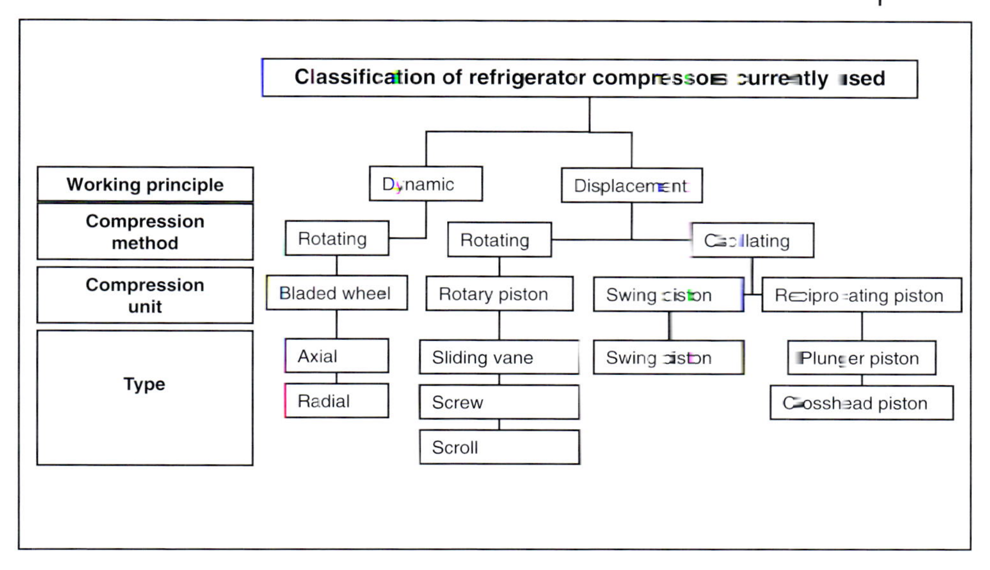

Figure 12.6 Classification of refrigeration compressors according to their construction.

12.2.5
Viscosity Selection

12.2.5.1 General Overview

The guidelines for the selection of refrigeration compressor lubricants are much the same as those applied to engineering in general, that is fast running machines permit the use of lower viscosities than slow running machinery. High bearing loads require higher viscosities than low bearing loads. In addition, refrigerator compressors require much lower viscosities than those calculated from the hydrodynamic lubrication theory. This fact has been proved by years of practical experience and by the elastohydrodynamic theory of lubrication. The influence of refrigerants on the operational viscosity of a refrigeration oil must be considered when a selection is made. In the case of piston compressors, the viscosity of the oil depends on the crankcase pressure and in the case of screw compressors, on the outlet pressure (pressure in the oil separator).

In the past, many refrigerator systems were operated with chlorinated refrigerants in order to provide some reliability reserves. Chlorine compounds are excellent EP additives which protect against wear. This is why previously used refrigeration oils used with CFCs were considered to contain 'anti-wear' agents when the refrigerant was dissolved in the oil. Since the introduction of chlorine-free refrigerants, this function must be performed by the refrigeration oil and/or additives (e.g. isobutane applications).

Some of the important interrelationships to be considered when selecting the correct lubricating oil for a refrigeration compressor are discussed in the following

Reference should also be made to refrigerant oil data sheets which contain important information for both compressor manufacturers and users [19].

The most important parameter for determining the lubricity of oils or oil–refrigerant mixtures is viscosity. The viscosity of oil–refrigerant mixtures should be viewed as the viscosity of a pure oil when bearing load calculations are being made. This applies to the hydrodynamic lubrication of cylindrical plain bearings. In the field of piston and screw lubrication, additional factors are the mixture's boundary friction phenomena. In general, reciprocating piston compressors are lubricated with oils ISO VG 32, VG 46, and VG 68 and screw compressors are lubricated with oils ISO VG 150, VG 170, VG 220, and VG 320 – depending on the refrigerant, temperature, pressure and the solubility of the refrigerant in the oil.

Normally used refrigeration oils are listed in Table 12.8.

Table 12.8 Overview of normally used refrigeration oils.

Compressor type	HFC[a]/FC (e.g. R134a, R404A)	Ammonia[b] (NH_3)	Hydrocarbons[c] (e.g. R290 R600a)	Carbon dioxide (CO_2)	HCFC/CFC (e.g. R12, R22)
Hermetic compressors (e.g. piston compressors	POE ISO VG 10-32	—	MO AB ISO VG 7-32	—	MO AB (MO/AB) ISO VG 15-46
Open-piston compressors	POE ISO VG 32-68	MO AB PAO, PAG ISO VG 32-68	MO AB PAO ISO VG 46-100	POE PAG ISO VG 46-170	MO AB MO/AB ISO VG 32-100
Semi-hermetic compressors	POE ISO VG 32-68	—	MO AB PAO ISO VG 46-100	POE PAG ISO VG 50-150	MO AB MO/AB ISO VG 32-100
Scroll compressors	POE ISO VG 32-68	—	MO AB ISO VG 46-100	—	MO AB ISO VG 32-100
Screw compressors	POE ISO VG 100-320	MO AB PAO, PAG ISO VG 32-68	MO AB PAO, PAG ISO VG 68-220	POE PAG ISO VG 170-220	MO AB ISO VG 68-220
Turbo compressors	POE ISO VG 68-150	—[d]	MO PAO, PAG ISO VG 68-100	—	MO ISO VG 68-100

The viscosity recommendations of the compressor manufacturers must be taken into consideration.
MO: mineral oil; AB: alkylbenzene; MO/AB: mineral oil alkylbenzene mixture; PAO: polyalphaolefin; PAG: polyalkylene glycol; POE: polyol ester.

a) PAG lubricants are used in R134a air-conditioning systems of cars and trucks (PAG 46, PAG 100).
b) MO, AB and PAO are not soluble with ammonia, PAG is (partly) soluble with ammonia, ISO VG 68 is used in piston compressors, up to ISO VG 220 is used in screw compressors.
c) PAG lubricants are partly soluble with hydrocarbons (low-viscosity reduction); MO, AB and PAO are highly soluble with hydrocarbons (high viscosity reduction).
d) Normally oil-free.

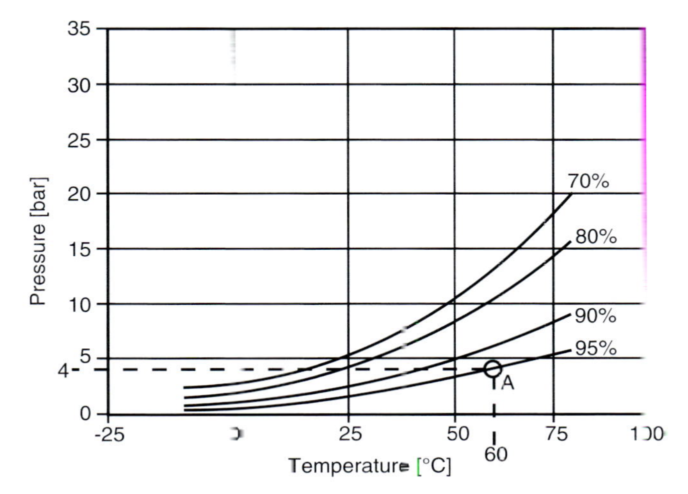

Figure 12.7 Mixture concentration in relationship to temperature and pressure (RENISO Triton SE 55 – R134a), at 4 bar and 60 °C, ~ 5% R134a is dissolved in the oil.

12.2.5.2 Mixture Concentration in Relation to Temperature and Pressure (RENISO Triton SE 55 – R134a)

Figure 12.7 shows how much refrigerant is dissolved in a refrigeration oil when saturated at a defined operating condition (pressure, temperature). As saturation is time related, the refrigerant concentration shown by the diagram is, as a rule, greater than the actual value. It can be viewed as the maximum concentration at any given operating condition. The viscosity which can be taken from the mixture concentration adds a safety margin to any bearing loading calculations.

In this diagram, the refrigerant concentration can be allocated to a point with a defined pressure and a defined temperature [10].

12.2.5.3 Mixture Viscosity in Relation to Temperature, Pressure and Refrigerant Concentration (RENISO Triton SE 55 – R134a)

The precise refrigerant concentration in a system is given at a defined pressure and temperature, as shown in Figure 12.7. Figure 12.8 can be used to read-off the kinematic viscosity of the oil–refrigerant mixture at a defined temperature, defined pressure and defined refrigerant concentration from the left scale (unit of kinematic viscosity is $10^{-6}\,m^2\,s^{-1} = 1\,mm^2\,s^{-1}$). The diagram shows the viscosity in relation to the temperature of the oil–refrigerant mixture at various concentrations.

If the viscosity of the mixture (again only valid when in a state of equilibrium) is required and the refrigerant concentration derived from pressure and temperature is unknown, refer to the diagram. This results from the competing influences of viscosity gains when the temperature of the oil falls and viscosity drops caused by greater quantities of dissolved refrigerant at lower temperatures. This

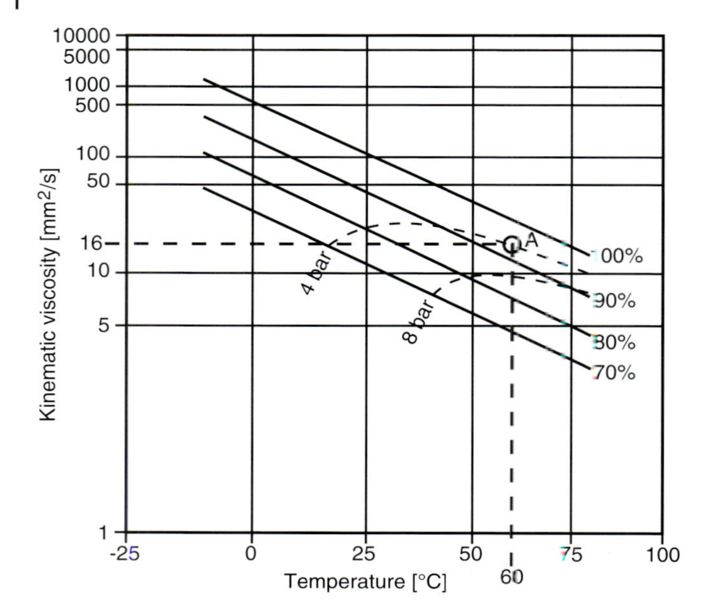

Figure 12.8 Mixture viscosity in relation to temperature, pressure, and refrigerant concentration (RENSIO Triton SE 55 – R134a), at 4 bar and 60 °C, ~ 5% R134a is dissolved, the viscosity of the POE/R 134a mixture is 16 $mm^2 s^{-1}$.

fact is of considerable importance to the design and operation of a refrigerant compressor. Care should therefore be taken that the oil does not reach its maximum viscosity at problematic points in the oil return circuit (e.g. upward flows, evaporators). It is also important that the conditions in the compressor crankcase do not approach those indicated by the falling left-hand part of the viscosity–temperature graph because the smallest temperature fluctuations can have a significant effect on viscosity in such conditions [10].

Mixture Density in Relation to Temperature and Refrigerant Concentration (RENISO Triton SE 55 – R134a) (Figure 12.9)
The density of an oil–refrigerant mixture depends on the density–temperature behaviour of the oil and the refrigerant (Figure 12.9).

Miscibility Gap, Solubility Threshold (RENISO Triton Series with R134a) (Figure 12.10)
Generally, not all refrigerants are miscible in refrigeration oils at all temperatures and in all concentrations. If, for example, one cools a fully dissolved oil–refrigerant mixture, a point will be reached when the fully dissolved mixture separates into two fluid phases. This area of phase separation is called the miscibility gap. The miscibility gap depends on the type of refrigerant and also to a large extent on the type of refrigeration oil. Refrigerant miscibility is defined statically in DIN 51 514. For normal applications, miscibility gap (with alkylbenzenes) is not a

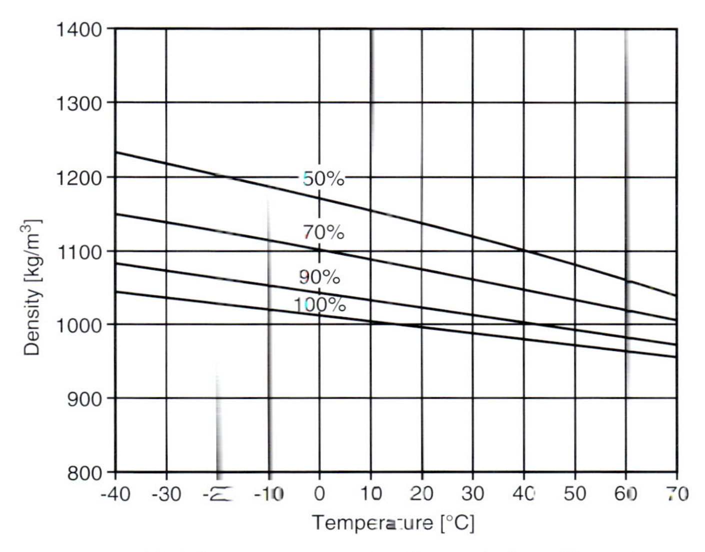

All %-figures represent the oil mass in the refrigerant

Figure 12.9 Mixture density in relation to temperature and refrigerant concentration (RENISO Triton SE 55 – R134a), POE lower density compared with R134a.

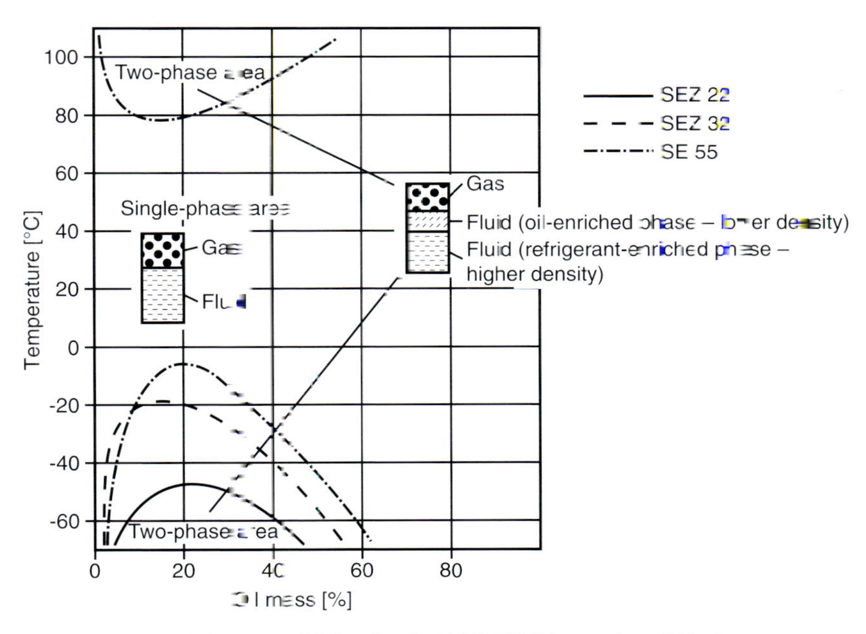

Figure 12.10 Miscibility gap, solubility threshold (RENISO Triton series – R134a).

problem for a series of refrigerants such as R22. Some other refrigerants display pronounced miscibility thresholds. The miscibility gap is of great importance to the refrigerant circuit. If the oil–refrigerant ratio is within the miscibility gap, problems can occur caused by oil-enriched fluid phases depositing in collectors, condensers, evaporators and the crankcase. Flooded evaporators require the largest possible quantity of refrigerant to be dissolved at evaporating temperatures without the phase separation. Figure 12.10 shows various examples of solubility thresholds.

12.2.6 Summary

The selection of an optimum refrigeration oil depends on the specifications of the compressor, the system as a whole and the refrigerant itself. Particularly important factors are the lubricity of the refrigeration oil and any interactions with the refrigerant, evaporation behaviour and solubility and mixture behaviour. Apart from traditional mineral oil-based refrigeration oils, polyol esters (POE) for chlorine-free refrigerants represent the most important group. Polyalkylene glycols (PAG) are used in R134a and HFO-1234 yf vehicle air-conditioning systems. Polyalphaolefins are increasingly being used for ammonia systems. In industrial and commercial CO_2 systems, special synthetic AW/EP polyolester oils (POE) are in use. In car air-conditioning systems, special synthetic polyalkylene glycols will be used.

References

1 Dubbel, H. (1987) Chapter P22, in *Taschenbuch für den Maschinenbau*, 16th edn, Springer, Berlin.
2 Uniti Organization (1986) *Schmierstoff-Tabelle für Verdichter, Kompressoren und Vakuumpumpen*, 3rd edn, Technischer Dienst UNITI, Hamburg, Germany.
3 Deutsches Institut für Normung e.V. (1985) DIN 51 506 – Schmieröle VB und VC ohne Wirkstoffe und Schmierol VDL, Einteilung und Anforderungen, September.
4 International Organization for Standardization (1988) ISO 6743-3A – Compressors (1987), ISO 6743-3B – gas and refrigeration compressors.
5 Planck, R. (1956) *Karlsruhe*, Handbuch der Kältetechnik, Springer, Berlin.
6 Steimle, H. (1950) *Kältemaschinenöle*, Springer, Berlin.
7 Deutsches Institut für Normung e.V. (1998) DIN 51 503-1 Kältemaschinenöle – Mindestanforderungen, August 1997, DIN 51 503-2 Kältemaschinenöle – gebrauchte Kältemaschinenöle, Beuth, Berlin, November.
8 American Society of Heating, Refrigerating and Air-Conditioning Engineers, (1992) ASHRAE Standard ANST/ASHRAE 34–1992, Atlanta.
9 Solkane Taschenbuch, *Kälte- und Klimatechnik*, 1st edn, Solvay Fluor u Derivate GmbH, 1997, Hanover, Germany.
10 Product Management 4 (1999) Fuchs Refrigeration Oils, Fuchs Petrolub AG, Mannheim, Germany.
11 Frigen Fibel (1988), Hoechst Aktiengesellschaft, Verkauf Chemikalien, Frankfurt/Main.

12 Klamann, D. (1982) *Schmierstoffe und verwandte Produkte*, Verlag Chemie, Weinheim.

13 Muri, P. (2005) *Kälte Warme Klima Taschenbuch 2005*, CF Muller Verlag, Heidelberg.

14 Forschungsrat Kältetechnik e.V. (1996) Forschungsvorhaben, Erprobung von Kältemittelgemischen im Verdichterkreislauf, Frankfurt/Main, Germany.

15 Forschungsrat Kältetechnik e.V., (1993) NH_3-losliche Kältemaschinenöle.

16 Reichelt, J. (1996) *Fahrzeugklimatizierung mit naturlichen Kältemitteln*, 1st edn, CF Muller, Heidelberg.

17 Kohlendioxid-Besonderheiten und Einsatzchancen als Kältemittel (1998) Statusbericht des DKV Nr 20, DKV, Stuttgart, Germany, November.

18 von Cube, H.L., Steimle, F. and Lotz, H. (1997) *Lehrbuch der Kältetechnik, Band 1,2, 4., überarbeitete Auflage*, CF Muller, Heidelberg.

19 Mang, T. (1974). Lubricants for refrigerators. *Mineralöltechnik*, (8–9) 1–46.

20 Fuchs Refrigeration Oils (2005) Fuchs Schmierstoffe GmbH, Mannheim Germany.

13 Turbine Oils

Wolfgang Bock

13.1 Introduction

Steam turbines have existed for more than 90 years. Steam turbines are motors with rotating elements which transform the energy of steam into mechanical energy in one or more stages. A steam turbine is normally connected to a driven machine, sometimes through a gearbox [1,2]. Steam temperatures can reach about 560 °C and pressures about 130–240 bar. The greater efficiency made possible by higher steam temperatures and pressures were fundamental factors in the further development of steam turbines. However, higher temperatures and pressures also increased the demands made on the lubricants used. The turbine oils without additives of the early years were unable to cope with these increased demands. Oils with additives have been used in steam turbines for about 50 years now [2,3]. These turbine oils which contain anti-ageing inhibitors and corrosion protection agents offer outstanding reliability in practice if certain basic rules are observed. Steam turbines are used by power stations to drive electrical power generators. In conventional power stations, their output is about 700 MW, while in nuclear plants this figure is about 1300 MW [2,3]. Gas turbines also require turbine oils with improved thermal and oxidative stability. These demands are covered with the development of high-temperature turbine oils [4,5,6,6a].

Turbo compressors and gas compressors are very often lubricated with turbine oils. The oil requirements of these machines are comparable to the requirements of gas and steam turbines. Often oils without EP components are recommended for turbo and gas compressors.

Lubricants and Lubrication, Third Edition. Edited by Theo Mang and Wilfried Dresel.

13.2
Demands on Turbine Oils – Characteristics

The demands on turbine oils are defined by the turbines themselves and their specific operating conditions. The oil in the lubricating and control circuits of steam and gas turbines has to fulfil the following objectives [3,7]:

- Hydrodynamic lubrication of all bearings as well as the lubrication of gearboxes
- Heat dissipation
- Functional fluid for control and safety circuits
- Avoidance of friction and wear on gear tooth flanks in turbine gearboxes and when the turbine is spooled-up.

Apart from these mechanical–dynamic requirements, the following physical–chemical specifications also have to be fulfilled by turbine oils [3,7]:

- Ageing stability for long operating periods
- Hydrolytic stability (especially of the additives used)
- Corrosion protection even if water, steam and/or condensation is present
- Reliable separation of water vapour
- Rapid air release and low foaming
- Good filterability and purity

These stringent demands on steam and gas turbine oils are met with carefully selected base oils and the inclusion of special additives.

13.3
Formulation of Turbine Oils

Turbine oils used today contain special paraffinic base oils with good viscosity–temperature characteristics as well as antioxidants and corrosion inhibitors. If geared turbines require a degree of load-bearing capacity (e.g. FZG failure load stage 9 – DIN 51 345), some mild EP additives are included.

Nowadays, turbine base oils are created exclusively by extraction and hydration. Refining and subsequent selective high-pressure hydrating significantly determine and influence turbine oil characteristics such as oxidation stability, water separation, air release and foaming. This applies in particular to water separation and air release because these features cannot be subsequently improved with additives. Steam turbine oils are generally made of special paraffinic base oil cuts [2,3,7].

Phenolic antioxidants combined with aminic antioxidants (synergies) are added to turbine oils to improve their oxidation stability [3]. To improve corrosion protection, non-emulsifiable corrosion protection agents and non-ferrous metal passivators are used. These are not negatively affected by water or steam contamination during operation and remain suspended. If standard turbine oils are used in geared turbines, mild, long-life and temperature- and oxidation-stable EP additives (organo-phoshpor and/or sulfur compounds) are included.

Furthermore, small quantities of silicone-free anti-foamers and pourpoint depressors are also used in turbine oils [3,7,8].

Great care must be taken to ensure that the defoamer used is absolutely free of silicone. Moreover, these additives must not detrimentally affect the air release (very sensitive) of the oil. All additives should be free of ash (e.g. zinc-free). The purity of a turbine oil according to ISO 4406 should be 15/12 in the tank [4,6,9,10]. No circuits, wires, cables or insulation-containing silicone should come into contact with turbine oils (observe when manufacturing and during application).

13.4 Physical and Chemical Data of Turbine Oils

The following typical data are used to characterize the turbine oils

13.4.1 Colour According to DIN ISO 2049

The colour is product specific and can vary between crystal-clear (colour code 0) and dark brown (colour code 5) (see Figure 13.1).

13.4.2 Density According to DIN 51757

Density refers to the mass of a fluid in relation to its volume. In general, to characterize the turbine oil, the density at 15 °C is reported. The density of turbine

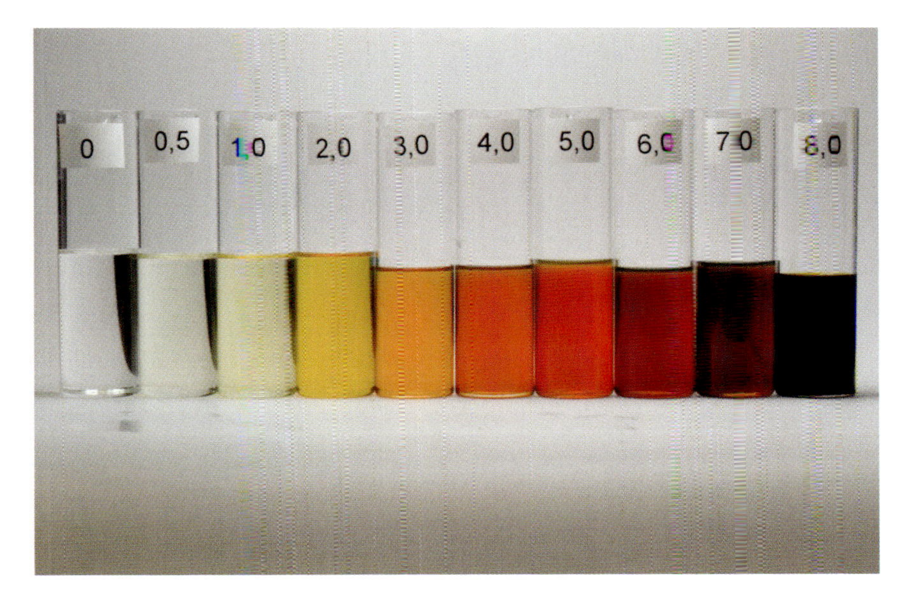

Figure 13.1 Colour according to DIN ISO 2049.

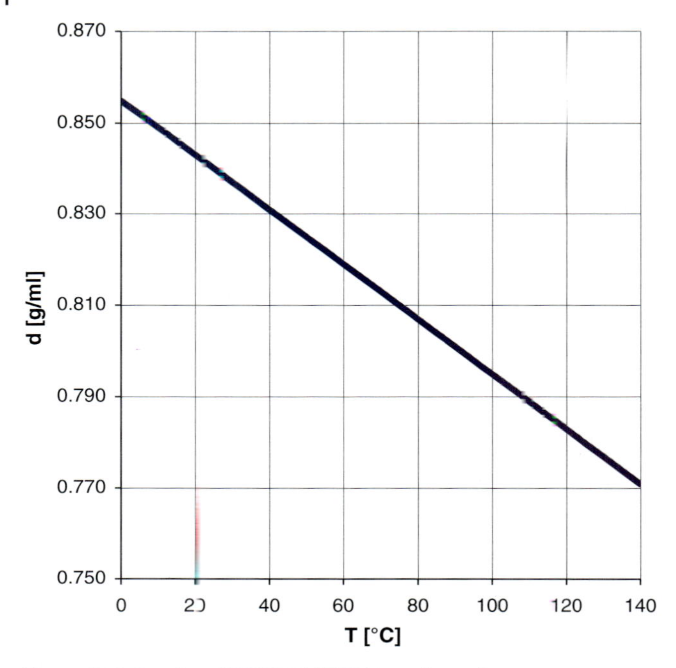

Figure 13.2 Density of RENOLIN ETERNA 46 depends on temperature.

oil is largely dependent on the temperature of the fluid because the volume expands with higher temperature (Figure 13.2).

13.4.3 Kinematic Viscosity According to DIN EN ISO 3104

Viscosity (the thickness of the oil) is the most important characteristic describing the load-carrying capacity of an oil. Turbine oils along with other industrial lubricants are classified according to their kinematic viscosity into ISO Viscosity Grades. The reference temperature is 40 °C and the official unit of kinematic viscosity is $m^2\,s^{-1}$ but in the lubrication sector, the units cSt or $mm^2\,s^{-1}$ are more common. DIN 51519 lays down 18 different viscosity grades from 2 to $1000\,mm^2\,s^{-1}$ at 40 °C for fluid industrial lubricants. Every viscosity grade is described by the mean viscosity at 40 °C and the permissible deviation of ±10% of this value.

The thickness or viscosity of an oil decreases with increasing temperature. The Viscosity Index (VI) describes this temperature dependence and is calculated according to DIN ISO 2909 from the kinematic viscosity at 40–100 °C. A suitably high lubricant viscosity is necessary to form a load-carrying lubricating film in the bearings, cylinders and so on of the turbine (Figure 13.3).

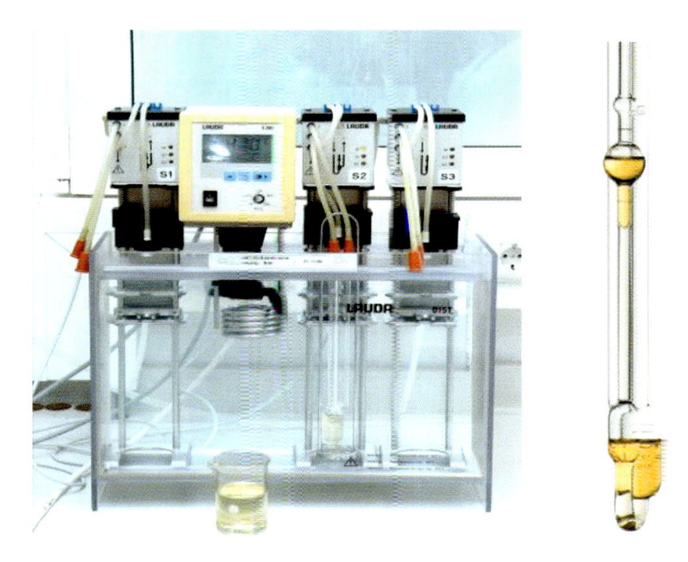

Figure 13.3 Measurement of kinematic viscosity – test equipment.

ISO VG	32	46	68
Viscosity index (VI)	126	132	128
	Kinetic viscosity at		
0°	230.0	445.1	770.1
10°	143.2	221.2	365.2
20°	80.6	121.4	192.2
30°	49.2	72.2	110.3
40°	**32.0**	**46.0**	**68.0**
50°	22.0	30.9	44.5
60°	15.8	21.9	30.6
70°	11.8	16.0	22.0
80°	9.1	12.2	16.4
90°	7.2	9.5	12.5
100°	**5.8**	**7.6**	**9.9**
110°	4.8	6.2	8.0
120°	4.0	5.2	6.5

13.4.4
Flashpoint According to DIN ISO 2592

The flashpoint of a turbine oil provides information about its base oils or mixtures. The flashpoint can be used to determine the vapour pressure of mineral oil-based fluids.

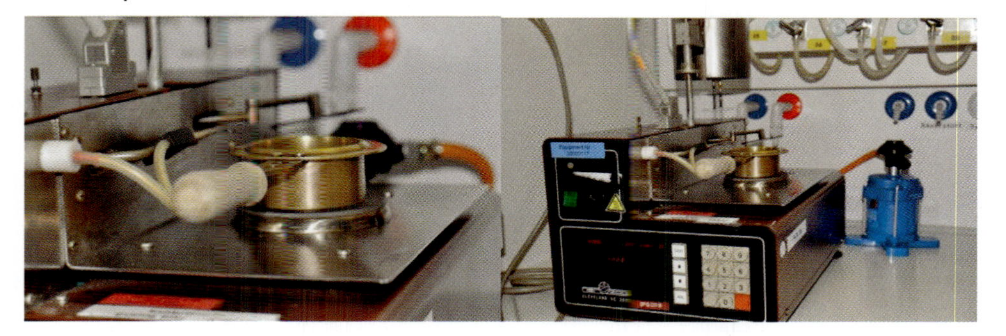

Figure 13.4 Flashpoint measurement – test equipment.

A crucible is filled with the fluid to be tested and its temperature is increased. A small naked flame is passed over the surface of the fluid in defined temperature stages. The lowest temperature at which the flame ignites the vapours above the surface of the fluid is known as the flashpoint. The flame point is the temperature at which the flame ignites the oil, which then burns for at least 5 s.

Flashpoint test methods: ASTM D 92/DIN ISO 2592 (COC) and ASTM D 93 (PM) (Figure 13.4).

13.4.5
Pourpoint According to DIN ISO 3016

The pourpoint shows the lowest temperature at which an oil still flows when it is cooled down under defined conditions (Figure 13.5). According to DIN ISO 3016, the sample is cooled and its flowing behaviour is tested in 3 °C steps. The pourpoint and boundary viscosity define the lowest temperature at which an oil can be used normally.

13.4.6
Foaming According to ASTM D 892

Surface foaming occurs when the air release speed is greater than the speed at which the bubbles at the surface of the fluid collapse, that is more bubbles are created than collapse. In the worst case, this foam can be forced out of tank vents or openings or can be drawn into the pump. Additives, silicone-based or silicone-free defoamers, can accelerate the collapse of bubbles by reducing the surface tension of the foam. However, they negatively influence the air release properties of the fluid, and this can cause compression and cavitation problems. Defoamers are thus used in very small concentrations (about 0.001%). The concentration of a defoamer in a fluid can progressively decline as a result of ageing and settling on metal surfaces and foaming problems often affect old, used fluids. Foaming behaviour is defined by DIN 51 566. This involves defined air being

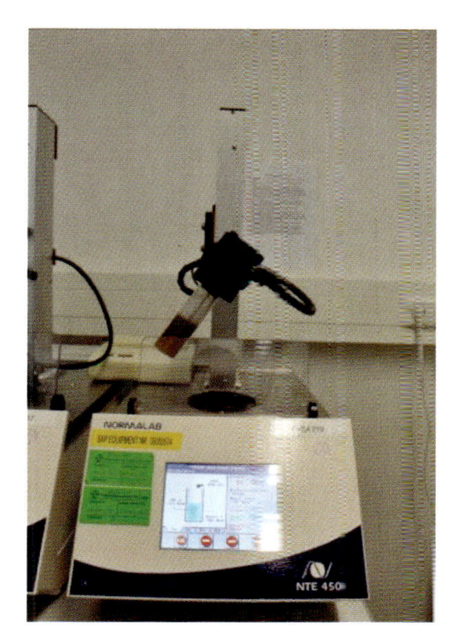

Figure 13.5 Pourpoint measurement – test equipment.

passed through a porous stone which is submerged in the fluid. The volume of foam which gathers on the surface of the fluid is measured against time (at once, after 10 min) and at different temperatures:

Sequence 1: at 24 °C, Sequence 2: at 93.5 °C, Sequence 3: at 24 °C after 93.5 °C.

Surface-active substances, detergent or dispersing additives, contamination in the form of grease, corrosion preventives, cleaners, cutting fluids, ageing by-products and so on can all have a negative effect on foaming behaviour.

Foam behaviour test method: ISO 6247, ASTM D 892 (old DIN 51 566) (Figure 13.6).

13.4.7
Neutralization Number According to DIN 51558

If the neutralization number of a turbine oil increases as a result of ageing, overheating or oxidation, the ageing products formed can attack system components such as pumps and bearings. This, in turn, can cause serious system failures. The neutralization number is therefore an important criterion for evaluation of the condition of a turbine oil. The neutralization number indicates the amount of acidic or alkaline substances in a lubricant. Acids in mineral oils can attack materials. A high acid content which can be the result of oxidation is therefore undesirable (Figure 13.7).

Figure 13.6 Foaming – test equipment.

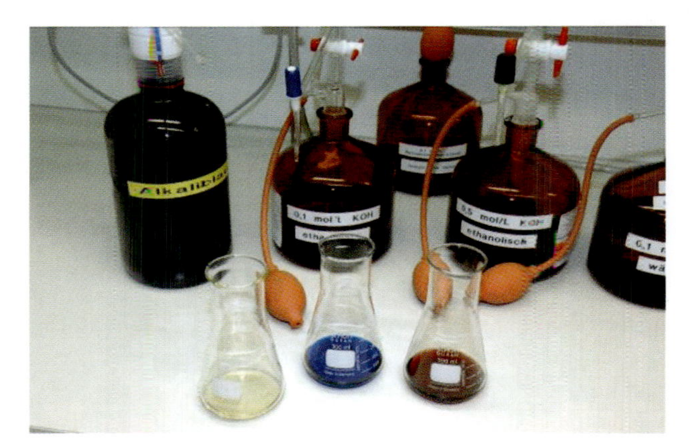

Figure 13.7 Measurement of the neutralization number – test equipment.

13.4.8
FZG Mechanical Gear Test Rig According to DIN ISO 14635-1

Turbine oils, particularly EP (extreme pressure)-containing grades, are used as hydraulic and lubricating oils in combined systems. Dynamic viscosity is the key wear protection factor in hydrodynamic lubrication. At low sliding speeds or high pressures under boundary friction conditions, the wear protection offered by a fluid depends on the additives used (reactive layer formation). These boundary conditions are replicated by the FZG test (Figure 13.8).

Figure 13.8 FZG mechanical gear test rig

The test is primarily used to determine the boundary performance of lubricants. Defined gear wheels turning at a defined speed are either splash- or spray-lubricated with an oil whose initial temperature is recorded. The tooth flank load is increased in stages and the appearance of the tooth flanks is recorded. This is repeated until the final 12th load stage.

- Load stage 07 = Hertzian pressure at the pitch point; 1.080 N mm^{-2}.
- Load stage 08 = Hertzian pressure at the pitch point; 1.232 N mm^{-2}.
- Load stage 09 = Hertzian pressure at the pitch point; 1.386 N mm^{-2}.
- Load stage 10 = Hertzian pressure at the pitch point; 1.538 N mm^{-2}.
- Load stage 11 = Hertzian pressure at the pitch point; 1.691 N mm^{-2}.
- Load stage 12 = Hertzian pressure at the pitch point; 1.841 N mm^{-2}.

- Starting temperature at load stage 4: 90 °C
- Peripheral speed: 8.3 m s^{-1}
- Upper temperature: not defined
- Gear geometry: A

The failure load stage for hydraulic fluids as defined by DIN 51 524-2 is at least 10.

ISO VG 46 turbine oils which do not contain anti-wear EP additives normally achieve load stage 6 (ca. 929 N/mm^2).

EP-containing turbine oils for use in turbine gear sets should reach minimum failure load stage 8 according to DIN 51515, parts 1 and 2.

13.4.9
Air Release at 50 °C According to DIN ISO 9120

The test involves air being blown into the oil. The air release figure is the time the air (down to 0.2 vol%) takes to escape from the oil at 50 °C under given conditions. The proportion of the dispersed air is determined by measuring the density of the oil–air mixture (Figures 13.9 and 13.10).

13.4.10
Water Content According to DIN 51777

Determining the water content according to Karl Fischer, DIN 51777, Part 1 – direct method, Part 2 – indirect method (Figure 13.11).

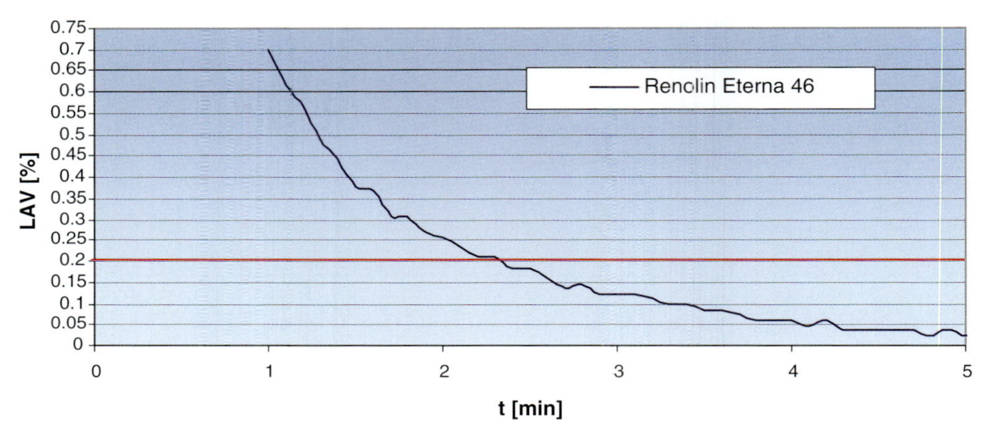

Figure 13.9 Air release properties (LAV) – RENOLIN ETERNA 46.

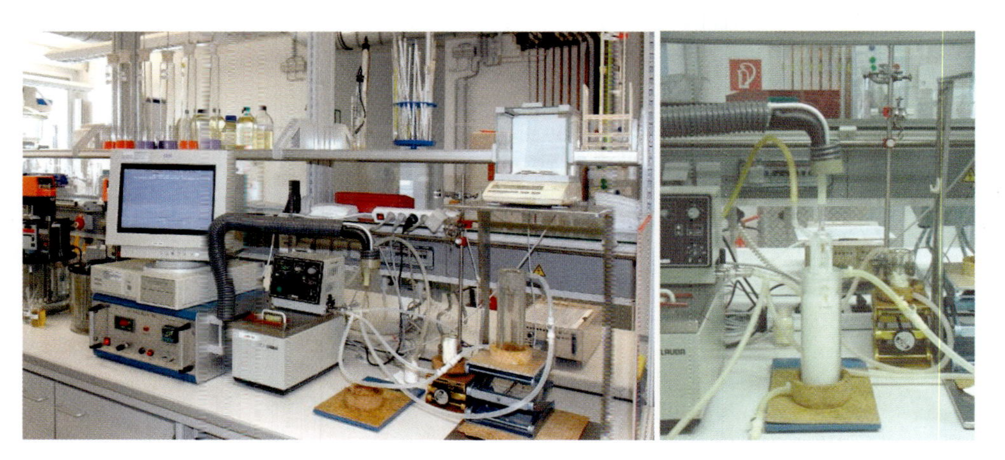

Figure 13.10 Air release – test equipment.

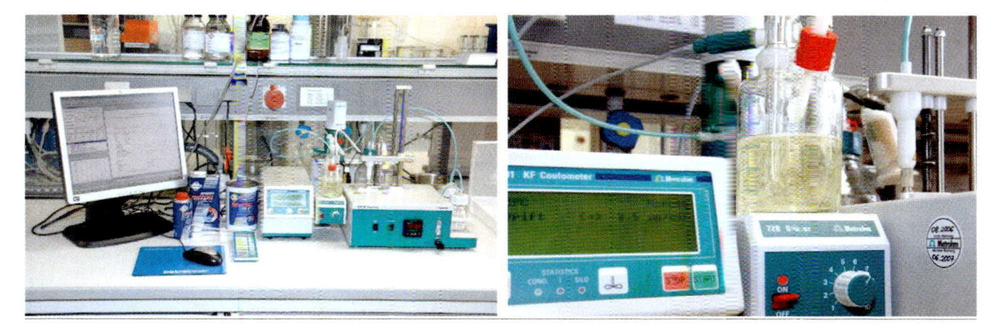

Figure 13.11 Measurement of the water content – test equipment.

The water content according to Karl Fischer, shown as mg kg^{-1} (= ppm: parts per million), is determined by titration. The quantity of dissolved water in refrigeration oils can only be determined with this method. Undissolved water (free water) can also be determined using the Water-Xylol method (ISO 3733/IP 74).

13.4.11
Water Separation According to DIN 51 589

Testing of lubricants and related products – determination of water separability after steam treatment (for lubricating oils and fire-resistant hydraulic fluids) (Figure 13.12).

A constant steam flow is conducted in 100 ml oil at 65 °C over a period of 20 min. The formation of an oil–water mixture (oil and water bubbles) will occur. After the steam treatment, the separation time (the separation velocity) of the oil–water phase is measured. The water separation time (in seconds) of a turbine oil should be less than 150 s, used oil will have higher values.

Maximum viscosity of the tested oil: ISO VG 100.

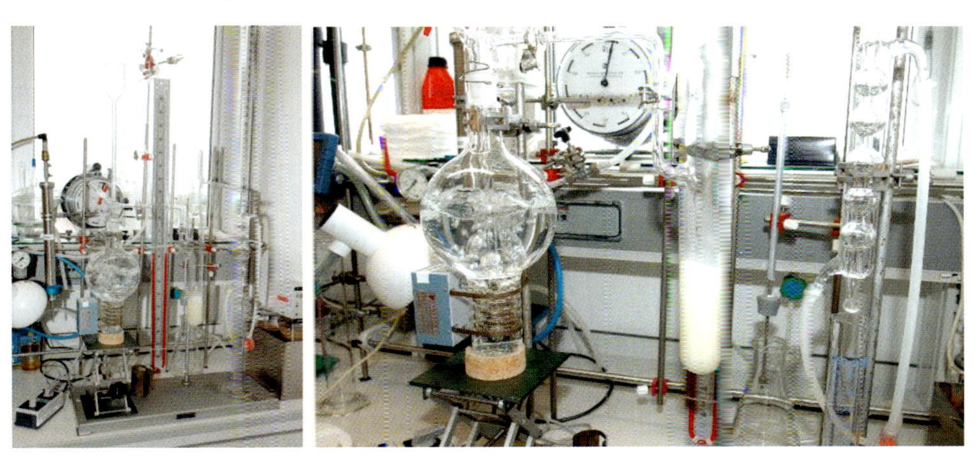

Figure 13.12 Water separation – test equipment.

13.4.12 Demulsifying Power at 54°C According to DIN ISO 6614

Demulsification is the time taken to separate an oil–water mixture (Figure 13.13). The parameters are as follows:

- Viscosity up to 95 $mm^2 s^{-1}$ at 40 °C: test temperature = 54 °C
- Viscosity >95 $mm^2 s^{-1}$: test temperature = 82 °C.

The RENOLIN ETERNA series guarantees quick water separation.

13.4.13 Steel/Ferrous Corrosion Protection Properties According to DIN ISO 7120

Turbine oils often contain dispersed, dissolved or free water, and a turbine oil must provide corrosion protection to all wetted components under all operating conditions, including contamination with water. This test examines the performance of the corrosion protection additives under different operating conditions (Figure 13.14).

Figure 13.13 Demulsification – test equipment.

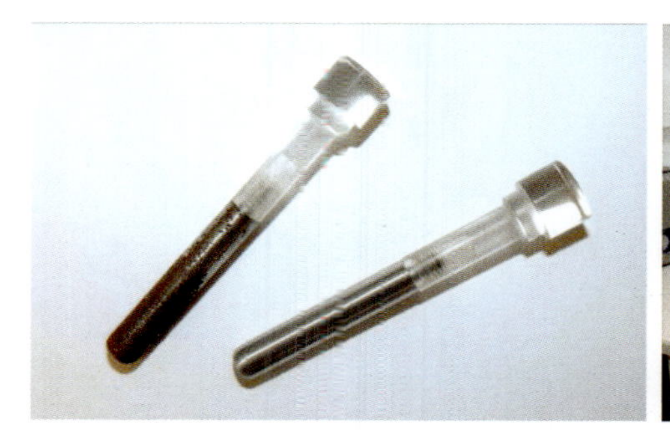

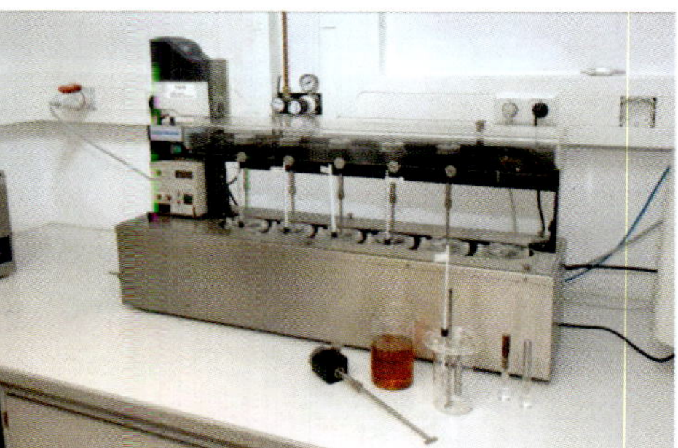

Figure 13.14 Steel/ferrous corrosion protection – test equipment.

The oil to be tested is mixed with distilled water (method A) or artificial sea-water (method B) and stirred continuously for 24 h at 60 °C while a steel rod is submerged in the mixture. After 24 h, the steel rod is examined for corrosion. The results enable assessment of the corrosion protection offered by the oil to steel components which are in contact with water or water vapour.

- Degree of corrosion 0 – no corrosion
- Degree of corrosion 1 – little corrosion
- Degree of corrosion 2 – moderate corrosion
- Degree of corrosion 3 – heavy corrosion

13.4.14
Copper Corrosion Protection Properties According to DIN EN ISO 2160

Copper or copper-containing materials are often used in turbine systems. Materials such as brass, cast bronze or sintered bronze are found in the bearing elements, guides or control units or sliding blocks of hydraulic pumps and motors. Copper pipes are used in cooling systems. Because copper corrosion can cause an entire lubrication and/or hydraulic system to fail, the copper strip test provides information about the corrosive effect of base fluids and additives on copper-containing materials.

According to DIN EN ISO 2160, corrosion to a copper strip can take the form of discolouration or flaking. A ground copper strip is submerged in the fluid to be tested for a given time at a given temperature. Turbine oils and other lubricating oils are normally tested for 3 h at 100 °C. The degrees of corrosion are shown in Figure 13.15.

13.4.15
RPVOT 150 °C According to ASTM D2272

The method according to ASTM D 2272 determines the oxidation stability of steam turbine oils in an oxygen pressure rotating steel chamber/vessel (Figure 13.16).

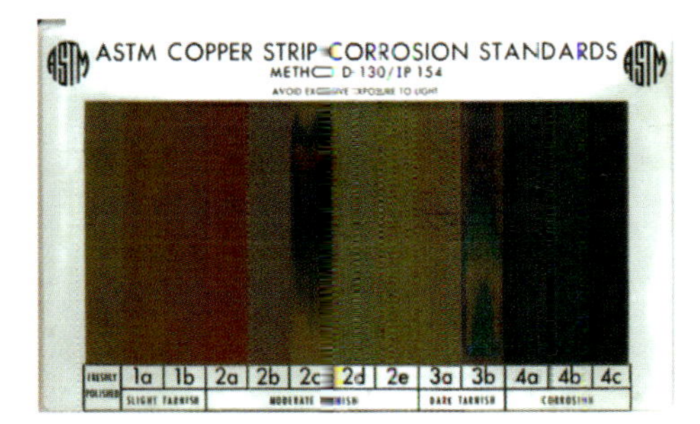

Figure 13.15 Copper corrosion – degrees of corrosion. 1: slightly discoloured, 2: moderately discoloured, 3: heavily discoloured, 4: corrosion (dark discolouration).

Figure 13.16 Rotating pressure vessel oxidation test – test equipment.

A beaker filled with test oil, water and a copper coil (as a catalyst) are placed in the steel chamber (which is filled with a pressure gauge). The vessel is charged to 6 bar oxygen in an oil bath at a temperature of 150 °C. The vessel is rotated at 100 rpm. The time is measured in minutes that is required to reach a specific pressure drop (reaction of O_2 with oil). This is the oxidation stability of the oil. Especially aminic antioxidants will improve the oxidation stability values in this test.

13.4.16
TOST Lifetime According to DIN EN ISO 4263-1

The ageing stability of steam turbine and hydraulic oils containing additives (Figure 13.17).

Figure 13.17 TOST test – test equipment.

The TOST test has been used for many years to test mineral oil-based turbine and hydraulic oils. In a modified form (without water), the dry TOST test is used. The ageing of a lubricant is characterized by an increase in the neutralization number when the oil is exposed to oxygen, water, steel and copper for a maximum of 1000 h at 95 °C (neutralization ageing curve). The maximum permissible increase in the neutralization number is 2 mg KOH after 1000 h.

13.4.17
Thermal Stability

The exposure of lubricating oils to high temperatures over longer periods of time can lead to the formation of decomposition products and these can cause serious problems. Ageing stability is thus an important lubricant selection criterion. Decomposition processes are generally complex chemical reactions which are catalyzed by metals such as copper, iron or aluminium. Also, water in the system can lead to the formation of decomposition products.

Experience shows that an increase in temperature of 10 K doubles the speed of ageing.

Well-known indicators of oil ageing are an increase in neutralization number (acid number) (Figure 13.18).

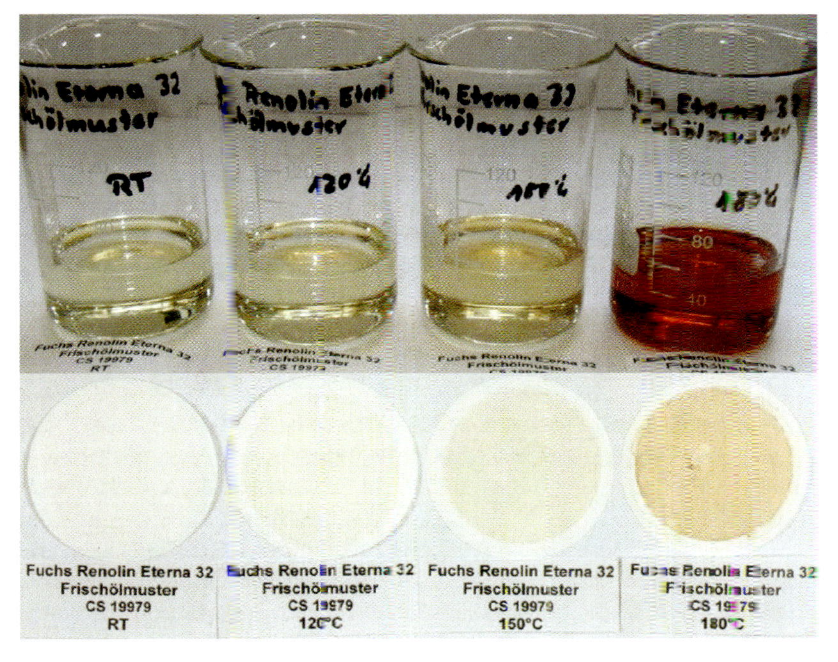

Figure 13.18 MAN high temperature (HT) test.

RENOLIN ETERNA 32 meets and exceeds the requirements of the MAN HT test with excellent results.

RENOLIN ETERNA 32 is characterized by very good wear protection, high oxidation stability and excellent thermal stability.

13.4.18
Thermal Conductivity

Thermal conductivity is the property of a material to conduct heat. The (specific) thermal conductivity of materials is a temperature-dependent absolute term. Its measurement unit is $W\,m^{-1}\,K^{-1}$.

RENOLIN ETERNA 46 – measurement of thermal conductivity:

Oil	RENOLIN ETERNA 46
$\lambda_{25\,°C}$ ($W\,m^{-1}\,K^{-1}$)	0.1500
β ($W\,m^{-1}\,K^{-2}$)	-1.4×10^{-4}
Temperature (C°)	Thermal conductivity ($W\,m^{-1}\,K^{-1}$)
20°	0.150
30°	0.149
40°	0.147
50°	0.146
60°	0.145
70°	0.143
80°	0.142
90°	0.140
100°	0.139
110°	0.137
120°	0.136

$\lambda(\nu) = \lambda_{25\,°C} + \beta \cdot \nu - 25\,°C$
(λ in ($W\,M^{-1}K^{-1}$); ν in °C)

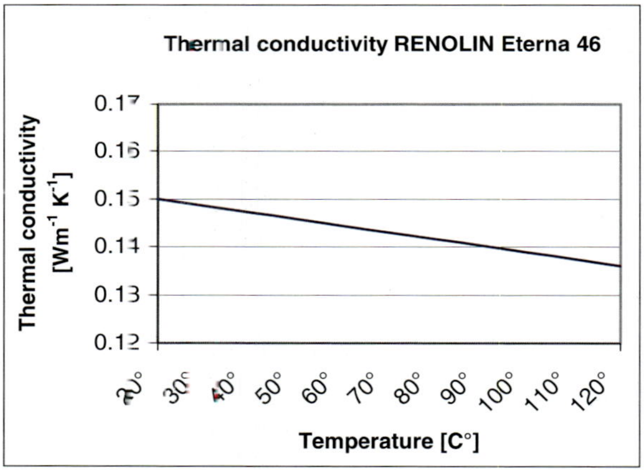

Thermal conductivity RENOLIN Eterna 46

Thermal conductivity [$Wm^{-1}\,K^{-1}$]

Temperature [C°]

13.4.19
Specific Heat

The specific heat is the heat required to increase the temperature of 1 kg of a substance by 1 K. Its measurement unit is $kJ\,kg^{-1}\,K^{-1}$.

RENOLIN ETERNA 46 – measurement of the specific heat:

Oil	RENOLIN ETERNA 46
a ($\mathrm{J\,g^{-1}\,K^{-1}}$)	0.9[illegible]58
b ($\mathrm{J\,g^{-1}\,K^{-1}}$)	0.0038
Temperature (C°)	Specific heat ($\mathrm{J\,g^{-1}\,K^{-1}}$)
20°	2.0[illegible]6
30°	2.053
40°	2.091
50°	2.128
60°	2.166
70°	2.203
80°	2.241
90°	2.278
100°	2.316
110°	2.3[illegible]3
120°	2.3[illegible]1

$C_p = a + b{\cdot}T$
(C_p in $\mathrm{J\,g^{-1}K^{-1}}$); T in K)

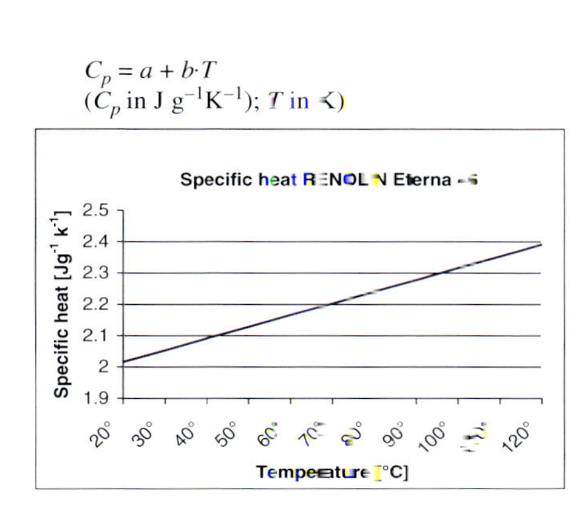

13.4.20
Vapour Pressure

The vapour pressure is a substance- and temperature-dependent gas pressure and denotes the ambient pressure, below which a liquid, at constant temperature, begins to pass into the gaseous state. Its measurement unit is mbar.

RENOLIN ETERNA 46 – measurement of the vapour pressure:

Oil	RENOLIN ETERNA 46
A	4.[illegible]5
B (K)	−[illegible]9678
Temperature (C°)	Vapour pressure (mbar)
20°	6.[illegible]
30°	8.[illegible]
40°	1[illegible]
50°	13[illegible]
60°	17[illegible]
70°	20[illegible]
80°	24[illegible]
90°	29[illegible]
100°	34[illegible]
110°	40[illegible]
120°	47[illegible]

$\mathrm{Lg}\,p = A + B{\cdot}1000/T$
(p in mbar; T in K)

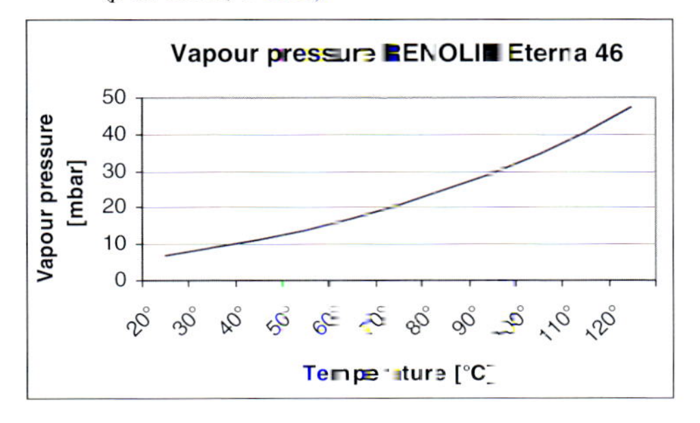

13.4.21
Surface Tension

The surface tension is the interface tension of solids and liquids towards gas, the air. It describes the vertical force action at an imaginary line with 1 m length which is on the surface. Its measurement unit is N m^{-1}.

RENOLIN ETERNA 46 – measurement of the surface tension:

Oil	RENOLIN ETERNA 46
$\sigma_{25\,°C}$ (mNm)	29.7200
α (mN m^{-1} K^{-1})	−0.0762
Temperature (C°)	Surface tension (mN m^{-1})
0°	30.1
30°	29.3
40°	28.6
50°	27.8
60°	27.1
70°	26.3
80°	25.5
90°	24.8
100°	24.0
110°	23.2
120°	22.5

$\sigma(\nu) = \sigma_{25\,°C} + \sigma \cdot (\nu - 25\,°C)$
(σ in mN m^{-1}; σ in °C)

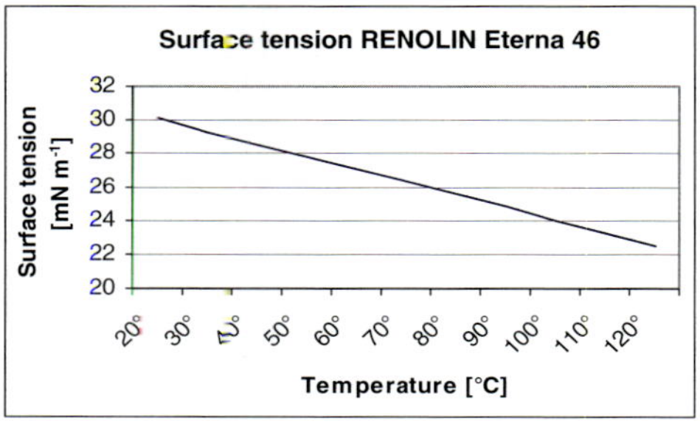

Surface tension RENOLIN Eterna 46

13.4.22
Remaining Useful Life Evaluation Routine (RULER Method)

This method was developed to measure the remaining useful life of, for example, gas turbine lubricants. The RULER measurement gives information about the remaining amount of antioxidants in used oils (monitoring of aminic and phenolic antioxidants, depletion of antioxidants) (Figure 13.19). The method is based on linear sweep voltammetry (electrochemical analysis). The used oil samples are taken in regular intervals and compared with fresh oil values. The oil sample is mixed with a specific RULER solvent and a solid substrate. The solvent separates the antioxidants from the oil. The substrate is used to adhere oil to the substrate surface which settles to the bottom of the vial. Then the RULER electrode inserts into the vial (only the solvent with the antioxidants is analysed).

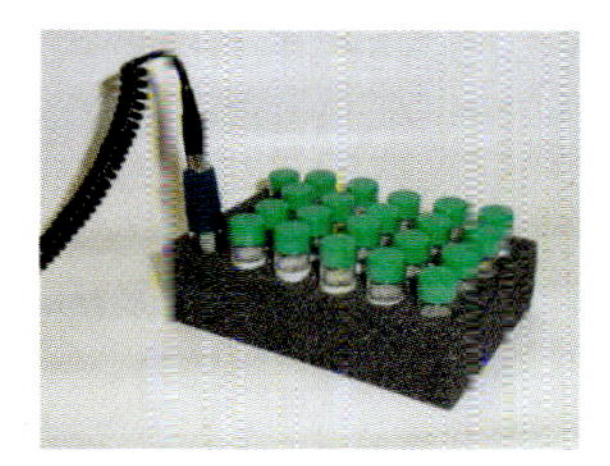

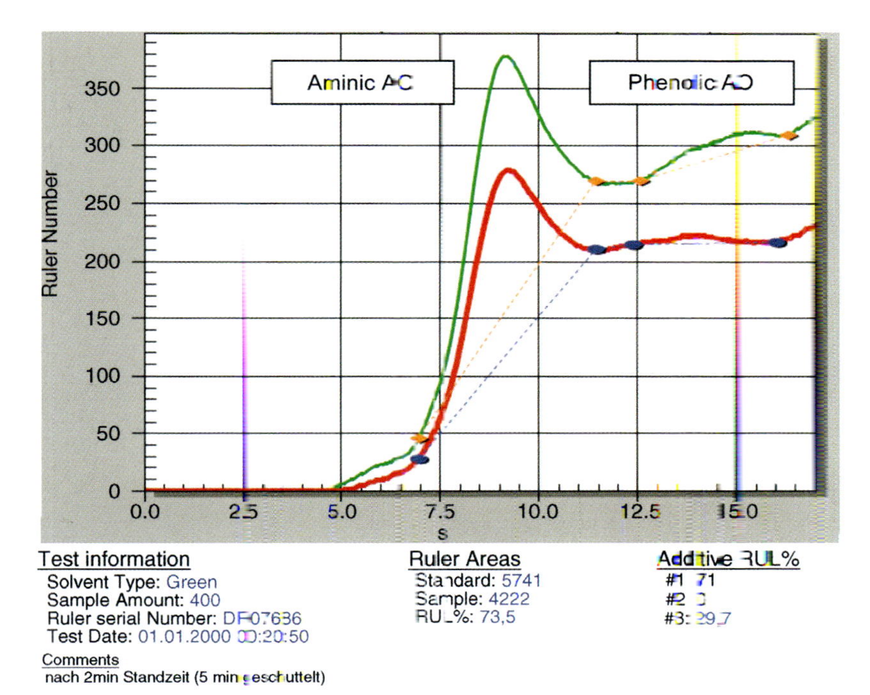

Figure 13.19 RULER test – test equipment and results (RULER together with other used oil values should be discussed and compared).

The specific RULER test solutions:

- green – general application
- yellow – for R&D application
- red – for gas turbine jet oil

The principle: Linear voltage is applied via the electrode (voltage ramp 0–1.7 V); when the oxidation potential of the antioxidant is approached, the antioxidant will oxidize at the carbon electrode; current begins to flow.

For the RULER method, only a small oil sample volume is needed (50–400 µl). It is a very fast method (the measurement itself: less than 1 min), and it will be an ASTM-approved method.

In general, we measure at RENOLIN ETERNA Series the content of aminic antioxidants with the solvent type green, and we compare with fresh oil values. Low phenolic antioxidant contents cannot be detected accurately in the RENOLIN ETERNA Series. The phenolic antioxidant content should be measured by IR-spectrum for comparison.

13.4.23
Membrane Patch Colorimetry (MPC) Test

The MPC test (ASTM draft) is a relatively new method to determine insoluble soft contaminants from a used turbine oil sample (Figure 13.20). These soft contaminants are extracted onto a patch. The colour of the membrane patch is analysed using a spectrophotometer.

Figure 13.20 MPC test – test equipment and results.

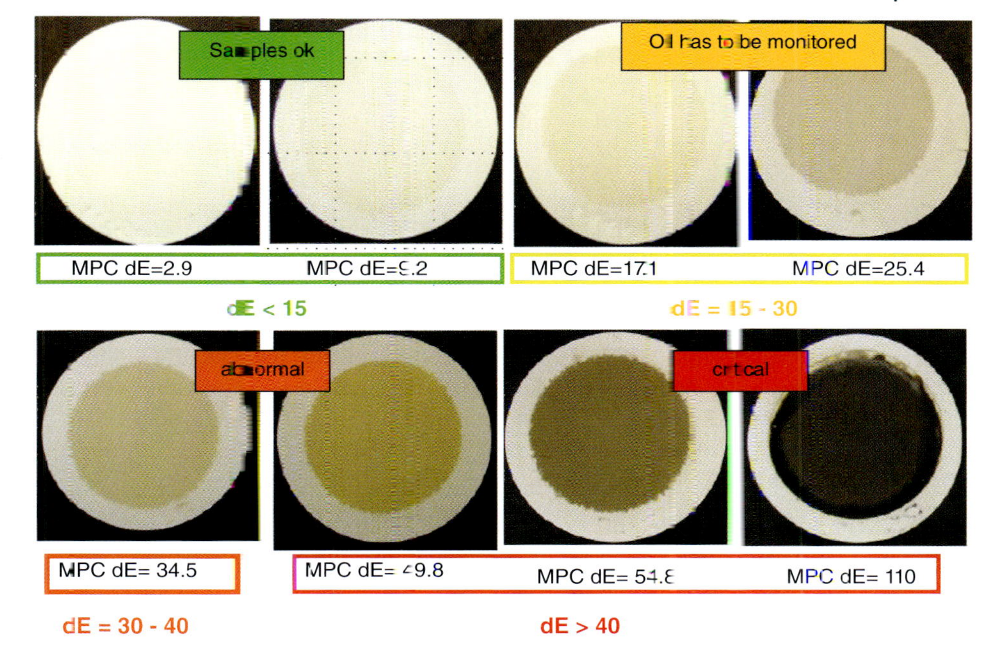

Figure 13.20 (*Continued*).

A 50 ml oil sample is mixed with 50 ml heptane (agglomeration of polar contaminants). The mixture is filtered through a cellulose 0.45 µm filter. After drying the filter, spectral analysis can be performed (unused filter as reference). Calculation of a MPC index is possible via spectral analysis – MPC test equipment. The MPC index can show values between 1 and 100:

- MPC index < 15: sample OK
- MPC index 15–30: sample has to be monitored
- MPC index 30–40: abnormal
- MPC index > 40: critical

MPC values can differentiate and can demonstrate the varnish potential of used turbine oils (which do not show any other critical values) together with other analytical values.

MPC test results together with other used oil values should be discussed and compared.

13.5 Turbine Lubricants: Description According to DIN 51515, Parts 1 and 2 [11]

Description of a turbine oil, viscosity grade ISO VG 46:

Turbine oil DIN 51515-TD 46/DIN 51515-TG 46

If the requirements for higher wear protection are fulfilled, code letter P is added to the abbreviation TD/TG:

Turbine oil: DIN 51515-TDP 46/DIN 51515-TGP 46

13.6
Turbine Lubricants: Specifications

See Tables 13.1 and 13.2.

A turbine oil for increased wear protection must achieve a scuffing load of at least 8 according to DIN ISO 14635-1.

For determining whether a turbine oil meets the requirements of the standard, DIN EN ISO 4259 is applied. This definition applies to all test results that are obtained according to the test standards listed in the second last column.

Special paraffinic mineral oils with additives are normally used for gas and steam turbine lubricants. These serve to protect all shaft bearings in the turbine and the generator as well as the gearbox in corresponding configurations. They can also be used as hydraulic fluids in control and safety systems. Fire-resistant, synthetic type HFD-R fluids [7] are normally used in hydraulic systems which operate at pressures around 40 bar, if there are separate lubricating oil and control oil circuits (twin circuit systems).

Table 13.1 DIN 51 515 – Part 1, lubricants and governor fluids for turbines, minimum requirements – Part 1: turbine oils TD for normal service (February 2010) [11].

Test		Limits				Testing according to[a)]	Comparable ISO[b)] standards
Lubricating oil group		TD 32	TD 46	TD 68	TD 100		
ISO viscosity grade[c)]		ISO VG 32	ISO VG 46	ISO VG 68	ISO VG 100	ISO 3448	ISO 3448
Kinematic viscosity[d)] at 40 °C in $mm^2\ s^{-1}$	min. max.	28.8 35.2	41.4 50.6	61.2 74.8	90.0 110	DIN 51 562-1 or DIN EN ISO 3104	ISO 3104
Viscosity index	min.			90		DIN ISO 2909	ISO 2909
Flash point (COC), °C	min.	185	185	205	215	DIN ISO 2592	ISO 2592
Air release at 50 °C, minutes	max.	5	5	6	no value specified	DIN ISO 9120	ISO 9120
Foam: Seq. I at 24 °C Seq. II at 93 °C Seq. III at 24 °C after 93 °C	 max. ml max. ml max. ml			450/0 50/0 450/0		ISO 6247	ISO 6247
Density at 15 °C, $g\ ml^{-1}$	max.			To be specified by supplier		DIN 51 757 or DIN EN ISO 12185[e)]	ISO 3675 or ISO 12185[e)]

Table 13.1 (*Continued*)

Test		Limits				Testing according to[a]	Comparable ISO[b] standards
Pour point, °C	max.	−6				DIN ISO 3016	ISO 3016
Neutralization number	mg KOH/g	To be specified by supplier				DIN 51 558 Part 1	ISO 6618 or ISO 6619
Ash (oxide ash)	% mass	To be specified by supplier				DIN EN ISO 6245	ISO 6245
Water content,[f] mg kg^{-1}	max.	150				DIN 51 777-1 or DIN EN ISO 12937	ISO 12937
Cleanliness level[g]	min.	20/17/14[h]				DIN ISO 5884 with DIN ISO 4406	ISO 5884 with ISO 4406
Water separation ability (after steam treatment), s.	max.	300				DIN 51 589 Part 1	—
Copper corrosion corrosiveness (3 h/100 °C)	max.	2				DIN EN ISO 2160	ISO 2160
Steel corrosion protection, procedure A[i] corrosiveness	max.	Pass				DIN ISO 7120	ISO 7120
Ageing behaviour[j] (TOST) time in hours to reach delta NZ of 2.0 mg KOH g^{-1}	min.	3000	3000	2500	2000	DIN EN ISO 4263 part 1	ISO 4263-1

a) The oil sample has to be stored without any contact to light before performing the test.
b) International Organization for Standardization.
c) Middle point viscosity at 40 °C in $mm^2\ s^{-1}$.
d) The SI unit of the kinematic viscosity is $m^2\ s^{-1}$.
e) Along with ISO 12185 Technical Corrigendum 1.
f) Turbine oil must be bright and clear.
g) Automatically particle counting method within ISO 11500 is recommended, calibration according to ISO 11171.
h) Requirements to the cleanliness are specific to each installation. The declared values are state of technology. Other values can be agreed between supplier and consumer.
i) At risk of ingress of seawater to the lubrication system, a test according to DIN ISO 7120 with synthetic sea water (procedure B) can be agreed.
j) The ageing behaviour test (TOST-test) has to be done as a type test procedure because of the long testing time.

Additional requirement for EP turbine oils:
If the turbine oil supplies also a turbine gear set, the oil has to reach a failure load stage of min. 8 according to DIN ISO 14635, part 1 (FZG).

Table 13.2 DIN 51 515 – Part 2: lubricants and governor fluids for turbines, minimum requirements – Part 2: turbine oils TG for high-temperature service (February 2010) [11].

Lubricating oil group		Limits TG 32	TG 46	Testing according to[a)]	Comparable ISO[b)] Standards
ISO viscosity grade[c)]		VG 32	VG 46	ISO 3448	ISO 3448
Kinematic viscosity[d)] at 40 °C in $mm^2 s^{-1}$	Min. Max.	28.8 35.2	41.4 50.6	DIN 51 562-1 or DIN EN ISO 3104	ISO 3104
Viscosity index	Min.	90	DIN ISO 2909	ISO 2909	
Flash point (COC), °C	Min.	185	185	DIN EN ISO 2592	ISO 2592
Air release at 50 °C, minutes	Max.	5	5	DIN ISO 9120	ISO 9120
Foam: Seq. I at 24 °C Seq. II at 93 °C Seq. III at 24 °C after 93 °C	Max. ml Max. ml Max. ml	450/0 50/0 450/0		ISO 6247	ISO 6247
Density at 15 °C, $g\ ml^{-1}$	Max.	To be specified by supplier		DIN 51 757 or DIN EN ISO 12185[e)]	ISO 3675 or ISO 12185[e)]
Pour point, °C	Max.	−6		DIN ISO 3016	ISO 3016
Neutralization number	mg KOH g^{-1}	To be specified by supplier		DIN 51 558-1	ISO 6618 or ISO 6619
Ash (oxide ash)	% Mass	To be specified by supplier		DIN EN ISO 6245	ISO 6245
Water content[f)], mg kg^{-1}	Max.	150		DIN 51 777-1 or DIN EN ISO 12937	ISO 12937
Cleanliness level[g)]	Min.	20/17/14[h)]		DIN ISO 5884 with ISO 4406	ISO 5884 with ISO 4406
Water separation ability (after steam treatment), s.	Max.	300		DIN 51 589 Part 1	—
Copper corrosion corrosiveness (3 hrs/125 °C)	Max.	2		DIN EN ISO 2160	ISO 2160
Steel corrosion protection, procedure A[i)] corrosiveness	Max.	Pass		DIN ISO 7120	ISO 7120
Ageing behaviour[j)] Time in hours to reach delta NZ of 2.0 mg KOH g^{-1}	Min.	3.500	3.500	DIN EN ISO 4263-1	ISO 4263-1

Table 13.2 (*Continued*)

Lubricating oil group		Limits TG 32	TG 46	Testing according to[a)]	Comparable ISO[b)] Standards
Rotating pressure vessel oxidation test (RPVOT), min	Min.	750		ASTM D2272	–
RPVOT (modified) – see Annex A	Minutes % of time in unmodified test	85			

a) The oil sample has to be stored without any contact to light before performing the test.
b) International Organization for Standardization.
c) Middle point viscosity at 40 °C in mm²/s.
d) The SI unit of the kinematic viscosity is m²/s.
e) Along with ISO 12185 Technical Corrigendum 1.
f) Turbine oil must be bright and clear.
g) Automatically particle counting method within ISO 11500 is recommended, calibration according to ISO 11171.
h) Requirements to the cleanliness are specific to each installation. The declared values are state of technology. Other values can be agreed between supplier and consumer.
i) At risk of ingress of sea water to the lubrication system, a test according to DIN ISO 7120 with synthetic sea water (procedure B) can be agreed.
j) The ageing behaviour test has to be done as a type test procedure because of the long testing time.

Annex A (normative) – Modified RPVOT
The oil sample for the modified RPVOT (Rotating Pressure Vessel Oxidation Test) is pretreated. For this purpose, the oil sample is washed with nitrogen (48 h, nitrogen flow rate 3 l h⁻¹) in a glass vessel (inner diameter about 38 mm, height about 300 mm) in an oil bath of 121 °C. The oil pretreated this way will be retested according to ASTM D2272.

Additional requirement for EP turbine oils:
If the turbine oil also supplies a turbine gear set, the oil has to reach a failure load stage of min. 8 according to DIN ISO 14635, part 1 (FZG).

13.6.1
Lubricants for Turbines: ISO 8068 Specifications

ISO 6743-5 in conjunction with ISO 8068 (dated 2006) classifies turbine lubricants according to whether they are used in steam or gas turbines and if they contain EP agents [5,12]:

	Normal turbine oils	High-temperature turbine oils
Without EP	ISO-L – TSA (steam) ISO-L – TGA (gas)	ISO-L – TGB (gas) ISO-L – TGSB (steam)
With EP	ISO-L – TSE (steam) ISO-L – TGE (gas)	ISO-L – TGF (gas) ISO-L – TGSE (steam)

Furthermore, synthetic fluids such as PAO and phosphoric acid ester are also listed (see Table 13.3) [5,12].

Table 13.3 ISO 8068 (2006), lubricants, industrial oils and related products (Class L) – classification.

	Part 5 Family T (Turbines): Table – classification of lubricants for turbines			
	General application	**Composition and properties**	**Symbol ISO-L**	**Typical application**
1.	Steam turbines, directly coupled or geared to the load, normal service	Highly refined petroleum oil with rust protection and oxidation stability *(no EP)*	TSA	Power generation and industrial drives and their associated control systems, marine drives, where improved load carrying capacity is not required for the gearing
2.	Gas turbines, directly coupled or geared to the load, normal service		TGA	
3.	Steam turbines, directly coupled or geared to the load, high load carrying capacity	Highly refined petroleum oil with rust protection, oxidation stability and *enhanced load carrying capacity (with EP)* FZG: FLS 8/9/10	TSE	Power generation and industrial drives and marine geared drives and their associated control systems, where the gearing requires improved load carrying capacity.
4.	Gas turbines, directly coupled or geared to the load, high load carrying capacity		TGE	
5.	Gas turbines, directly coupled or geared to the load, higher temperature service	Highly refined petroleum oil with rust protection and *improved oxidation stability (no EP)*	TGB	Power generation and industrial drives and their associated control systems, where high-temperature resistance is required due to hot-spot temperatures
6.	Steam turbines, directly coupled or geared to the load, higher temperature service		TGSB	
7.	Gas turbines, directly coupled or geared to the load, higher temperature service	Highly refined petroleum oil with rust protection, *improved oxidation stability and enhanced load carrying capacity (with EP)* FZG: FLS 8/9/10	TGF	Power generation and industrial drives and marine geared drives and their associated control systems, where the gearing requires improved load carrying capacity.
8.	Steam turbines, directly coupled or geared to the load, higher temperature service		TGSE	

Table 13.3 (Continued)

Part 5 Family T (Turbines): Table – classification of lubricants for turbines			
General application	Composition and properties	Symbol ISO-L	Typical application
9. Other lubricants (a) TGCH – Synthetic steam turbine fluids with no specific fire-resistant properties (e.g. PAO) (b) TSD/TGD – Synthetic steam turbine fluids based on phosphate esters with fire resistant properties (c) THCH – Synthetic fluids, polyalphaolefins and related hydrocarbons (d) THCE – Synthetic fluids, synthetic ester types (according to ISO 15380)			

Different engineering designs make different demands on the lubricant used. The main differences between these demands relate to the ageing and/or oxidation tests and/or FZG performance and other important test requirements.

13.7 Turbine Oil Circuits

The oil circuits play an especially important role in the lubrication of power station turbines. Steam turbines are normally fitted with pressurized oil circuits. Separate lubricating and control circuits as well as combined lubricating and control oil tanks are used.

Under normal operating conditions, the turbine shaft-driven main oil pump draws oil from the tank into the control and bearing supply circuits. Pressure in control circuits is normally between 10 and 40 bar (separate header pump: turbine shaft header pressure about 100–200 bar) [1–3,7]. Oil tank temperatures range from 40 to 60 °C. The velocity of the oil in the feed circuit is about 1.5–4.5 m s^{-1} (about 0.5 m s^{-1} in return circuits) [3]. Cooled and passing through reduction valves, the oil reaches the turbine, generator and possible gearbox bearings at a pressure of about 1–3 bar. The individual oil feeds return to the oil tank at atmospheric pressure. The main turbine and generator shaft bearings are mostly white-metal shell bearings. Axial loads are normally absorbed by trunnion bearings. The lubricating oil circuit of a gas turbine is largely similar to that of a steam turbine. However, gas turbines sometimes use roller bearings as well as plain bearings [1,3,7].

Larger oil circuits are fitted with centrifugal bypass filtering systems. These ensure that the finest of contaminants are removed along with ageing by-products and sludge. Depending on the size of the turbine, the oil in bypass systems with separate pumps should pass through the filters about once every 1–5 h. The oil should be removed from the lowest point in the tank and be filtered directly before the oil is returned to the tank. If the oil is taken from the main oil stream, the flow rate should be reduced to about 2–3% of the main pump's flow capacity. The following equipment is often used: oil centrifuges, fine-paper filters, cellulose cartridge fine-filters and filter units with separators. An additional magnetic filter is also highly recommended. Sometimes bypass and main flow filters are fitted with coolers to reduce the temperature of the filtered oil to about 25 °C. It must also be possible to remove the oil from the tank with a mobile filter unit or centrifuge if water, steam or other contaminants enter the system. For this, the lowest point of the oil tank is normally fitted with a corresponding connector, which can also be used to draw oil samples [3,7].

Oil ageing is also influenced by the frequency with which the oil is pumped through the circuit. If the oil is pumped too fast, excessive amounts of air are either dispersed or dissolved (problem: cavitation in bearings, premature ageing, etc.). Oil tank foaming can also occur but this generally collapses rapidly. Engineering design measures can positively influence air release and tank foaming. These include oil tanks with larger surface areas and larger return circuit pipe cross sections. Simple measures such as returning the oil to the tank through an inverted U-tube can produce astonishing benefits. Fitting baffles to the tank also positively influences air release. These have the effect of prolonging the time in which air, water and solid contaminants can be released from the oil [2,3,7].

13.8
Flushing Turbine Oil Circuits

Before commissioning, all oil circuits should be mechanically cleaned and finally flushed. Every effort should be made to remove all contaminants such as cleaners and corrosion preventives from the system. The oil can now be added for flushing purposes. About 60–70% of the total oil volume is required for flushing [4,7]. The flushing pump should be operated at full power. It is recommended that the bearings are removed and temporarily replaced with blanks (to avoid the penetration of contaminants into the gap between shaft and bearing shells). The oil should be repeatedly heated to a maximum of 70 °C and then cooled to about 30 °C. The expansion and contraction in the pipework and fittings are designed to dislodge dirt in the circuit [7]. The shaft bearing shells should be flushed in sequence to keep the flow rate high. After at least 24 h of flushing, oil filters, oil sieves and bearing oil sieves can be fitted. Mobile filtering units, which may be used, should work with a mesh size of ≤5 µm. All parts of the oil supply chain, including reserve machinery, should be extensively flushed.

Finally, the flushing oil should be drained from the oil tank and coolers. All system components should be thoroughly cleaned externally. The flushing oil may be reused after very fine filtering. However, a careful oil analysis should first be performed and care should be taken that the oil still fulfils DIN 51 515 or the equipment-specific specifications [7]. Flushing should be performed until all solid contaminants are removed from the filter and/or no measurable increase in pressure is recorded in bypass filters after 24 h. It is recommended that a few days of flushing and a subsequent oil analysis follows any system modifications or repair work [7].

13.9 Monitoring and Maintenance of Turbine Oils – General

In normal circumstances, oil monitoring intervals of 1 year are perfectly acceptable [2,4,7]. As a rule, these should be performed in the oil manufacturer's laboratories. In addition, a weekly visual inspection of the oil should be performed to spot contamination and impurities in the oil in good time. Filtering the oil with a centrifuge in a bypass circuit is a reliable method. The contamination of the air surrounding a turbine with gases and other particles should be considered when operating a turbine. The refreshing of additive levels by topping-up lost oil has to be taken into consideration when interpreting the used oil values. Filters, sieves as well as oil temperature and oil level should be checked regularly. In cases of longer shutdowns (longer than two months), the oil should be circulated on a daily basis and the water content should be checked regularly [7].

The control of used fire-resistant fluids, lubricants and turbine oils in turbines can be done in the laboratory of the oil supplier or an independent qualified laboratory. The analysis and the warning values of the different properties and their following-up are described in the VGB-Kraftwerkstechnik Merkblätter, Germany (VGB = Association of German Power Plants) [13].

13.10 Turbine Oils: Evaluation of Used Oil Values – Parameters and Warning Values/Limits According to VGB Recommendation [13]

Parameters	Test method	Warning values/limits
Visual assessment	Visual	Cloudiness, impurities, phase separation, emulsion oil/water
Colour	DIN ISO 2049	≥5 Delta >2/year
Kinematic viscosity at 40 °C	DIN 51562-1	±10% Delta >5%/year

(continued)

(continued)

Parameters	Test method	Warning values/limits
Neutralization number (NZ) acid number	DIN 51558-1/ISO 6618/ISO 6619	≥0.5 mg KOH g^{-1} Delta > (0.1 mg KOH g^{-1})/year
Air release (at 50 °C)	DIN ISO 7120	≥8 min
Water separation (after steam treatment at 6[illegible] °C)	DIN 51589-1	>600 s
Demulsification (at 54 °C)	DIN ISO 6614	40/37/3 after >30 min Oil/water emulsion
Water content according to Karl Fischer (DIN 51777-1 and DIN 51777-2)	DIN 51777-1 DIN EN ISO 12937	>300 mg kg^{-1}; >300 ppm
Degree of cleanliness – Particles/contaminants	DIN ISO 5884 ISO 4406	Lubricating oil circuit: 21/18/15 For additional steering oil circuit: 20/17/14
Decomposition of antioxidants – Phenolic inhibitor[a)] (IR, 3650 cm^{-1}) – RPVOT[a)] – Voltammetry (RULER)[a)] – DSC/PDSC – HPLC	ASTM D 2272 ASTM D 6971	<0.1% absolute (depending on the formulation) <100 min <25% <25% <0.1% absolute
Foaming behaviour	ISO 6247	Seq. I >450 ml per 0 ml
Metals (dissolved in oil)		Zn*, Cu: ≥10 mg kg^{-1} Fe, Ca, Si, Al: ≥20 mg kg^{-1} Sn, Pb, Sb: ≥10 mg kg^{-1} (*only zinc-free formulations)
Solid contaminants	DIN ISO 5884	>300 mg kg^{-1}
Corrosion protection, steel	DIN ISO 7120	≥1 – A

a) The total amount of antioxidant levels is an important criterion of the ageing characteristic. Often the aminic antioxidant content is more important for the lifetime of the fluid.

13.11 Turbine Oils: Evaluation of Used Oil Values – Causes and Measures [13]

Exceeding of the following warning values	Recommended tests	Possible causes and measures
Visual assessment • Cloudiness • Impurity • Phase separation	Water content Solid contaminants Air release	If there is free water (phase separation), remove water at tank sump If there are solid contaminants, clean oil by filtration or separation.

(*continued*)

Exceeding of the following warning values	Recommended tests	Possible causes and measures
• Emulsion (oil/water)	Water separation NZ/acid number	If an emulsion has formed, an oil change should be considered (further test results should be observed)
Colour • 5	Additive content	Thermal stress is too high, eliminate temperature influence/stress
• Delta >2/year	NZ/acid number Air release Water separation	If the NZ and the oxidation inhibitor are in normal range, the oil is ready to operate in a good working condition
Viscosity at 40 °C • Deviation >10% to ISO VG-class • Delta >5%/year		Investigate a possible contamination with lower viscosity or higher viscosity oils. Consultation with the turbine manufacturer if the oil is ready to continue operating
Neutralization number • 0.5 mg KOH g^{-1} • Delta > (0.10 mg KOH g^{-1})/year	Water content Additive content Water content Additive content Metal content	After consultation with the oil supplier, turbine manufacturer or the testing laboratory, possibly accomplish an oil change If the bearing temperature is too high, investigate and eliminate the cause Eliminate leakage in the cooling system ⇒ Water content Investigate if a mixture with other oils has happened If the NZ/acid number is <0.5 mg KOH g^{-1} ⇒ Additive content If the NZ/acid number is <0.5 mg KOH g^{-1} and the additive content is in normal range, the oil is ready to continue operating

13.11.1 Turbine Oils: Used Oil Values

Causes and measures [13].

Exceeding of the following warning values	Recommended tests	Possible causes and measures
Air release (at 50 °C) • >8 min	Visual assessment Foaming properties	Oil contamination, for example silicon oil, detergents, different oils, solid contaminants. Causes for the contamination must be investigated and eliminated

(*continued*)

(*continued*)

Exceeding of the following warning values	Recommended tests	Possible causes and measures
	Water separation	If there are problems with the oil pressure or pump noise due to high circulating rates, an oil change should be accomplished If there are no unusual pump noises and the oil pressures are in a normal range, the oil is ready to continue operating
Water separation/ demulsification: WAV (at 54 °C) • >600 s (DIN 51589-1) WAV • 40/37/3 (>30 min) (ISO 6614)	Visual assessment Water content Air release Metal content	Oil contamination, for example detergents, other oils, solid contaminants. Causes for the contamination must be investigated and eliminated If water and steam contamination are avoided, the oil is ready to continue operating
Water content according to Karl Fischer (DIN 51777-1 and DIN 51777-2) • >300 mg kg^{-1} • >300 ppm	Additive content Water separation (WAV) Metal content	Investigate and eliminate water entry. Drain the water at the bottom of the tank If all the other parameters are in normal range, the oil is still fully operational
Cleanliness – Particles/contaminants Lubrication oil circuit: 21/18/15 For additional Steering oil circuit: 20/17/14	Metal content	Eliminate cause and clean oil, if applicable identification of the solid contaminants If all the other parameters are in normal range, the oil is still fully operational

13.11.2 Turbine Oils: Used Oil Values – Causes and Measures [13]

Exceeding of the following warning values	Recommended tests	Possible causes and measures
Additives (Antioxidants) • <0.1% absolute (IR, HPLC) • <100 min (RPVOT) • <25% (voltammetry, RULER) (DSC/PDSC)	Visual assessment Water NZ/acid number Metal content	If water is present → water If the bearing temperature is too high, investigate and eliminate the cause If all the other parameters are in normal range, the additive content can be raised by partly renewal or top up of additives A top-up of additives should be performed only by professionals (Lube oil supplier). A consultation with the oil manufacturer might be necessary

(continued)

Exceeding of the following warning values	Recommended tests	Possible causes and measures
Foaming behaviour • 450 ml per 0 ml (Seq I (24 °C))	Air release, solid contaminants	If there are solid contaminants perform oil cleaning Aspirate foam and, if applicable add defoamer (Take Care!!!) *Attention, when adding defoamer, especially silicon, the air release increases* A consultation with the oil manufacturer might be necessary If the increased foaming does not lead to operating difficulties, the oil is still ready to operate
Metal content (dissolved in oil) • Zn*, Cu ≥10 mg kg (*only zinc-free formulations) • Fe ≥20 mg kg^{-1} • Ca, Si, Al ≥20 mg kg^{-1} • Sn, Pb, Sb ≥ 10 mg kg^{-1}	Corrosion protection, steel	Corrosion in the oil system, investigate the oil cooler and eliminate the cause. Entry through the ambient air, if possible filter the air The elements Sn/Pb/Sb are an indication of a bearing defect, which can have both corrosive and mechanical reasons. Observe bearing temperature. Check the bearings and eliminate the cause.

13.11.3
Turbine Oils: Used Oil Values – Causes and Measures [13]

Exceeding of the following warning values	Recommended tests	Possible causes and measures
Solid contaminants • <300 mg kg^{-1}	Cleanliness Metal content	Eliminate cause and clean oil, if applicable identification of the solid contaminants If all the other parameters are in normal range, the oil is still fully operational
Corrosion protection steel • 1 – A		If all the other parameters are in normal range, the additive content can be raised by partly renewal or top up of additives A top-up of additives should be performed only by professionals (lubricant supplier). A consultation with the oil manufacturer might be necessary

13.12
Lifetime of (Steam) Turbine Oils

Oil life of 100 000 h is not uncommon in large steam turbines [2,3]. However, the antioxidant level in the oil can fall to about 20–40% of the fresh oil (oxidation, ageing). The life of turbine oil depends heavily on operating conditions such as temperature and pressure, oil circulation speed, filtering and the quality of maintenance and finally, the amount of oil topped-up (this helps maintain adequate additive levels).

The temperature of the oil in a turbine depends on the bearing loading, speed, bearing dimensions and the oil's flow rate. Radiated heat can also be an important parameter. The oil circulation factor, that is the ratio between flow volume/hour and tank volume should be between 8 and 12 h^{-1} [3,7,8]. Such relatively low oil circulation factors ensure that gaseous, fluid and solid impurities can be efficiently separated while air and other gases can be released. Furthermore, low oil circulation factors reduce the thermal loads on oil (with mineral oils, oxidation speed doubles when the temperature increases by 8–10 K). During operation, turbine oils are exposed to considerable oxygen enrichment. Turbine lubricating oils are exposed to air at a number of points around a turbine. The temperatures of bearings can be monitored using thermoelements. High bearing temperatures can be around 100 °C [2,8] and sometimes even more in the lubrication gap. The temperature of bearings can reach up to 200 °C if localized overheating takes place. Such conditions can only be countered by large oil volumes and rapid circulation. The oil draining from plain bearings can be about 70–75 °C [3] and the oil in the tank can be about 60–65 °C. Depending on the oil circulation factor, the oil remains in the tank for between 5 and 8 min [3,7]. During this time, any trapped air can be released, solid contaminants can settle and water can be separated. If the temperature of the oil in the tank is higher, additive components with high vapour pressures can evaporate. This evaporation problem is worsened by the installation of oil vapour extraction units. The maximum temperature of plain bearings is limited by the threshold temperatures of the white metal bearing shells. These are about 120 °C. The development of alternative bearing shell metals, which are less heat sensitive, is currently underway [1,3,7].

13.13
Gas Turbine Oils: Application and Requirements

Gas turbine oils are used in stationary gas turbines. These produce either electricity or heat. The compressor fans generate pressures of up to 30 bar which vent into the combustion chambers where gas is injected [3]. Depending on the type involved, combustion temperatures of up to 1000 °C are reached (generally 800–900 °C) [3,14]. Exhaust gas temperatures can reach about 400–500 °C. Gas turbines with capacities ranging up to about 250 MW are used in urban and suburban steam heating systems, in the paper industry and in the chemical industry. The advantages of gas turbines are compact size, rapid start-ups (<10 min) as well as small oil and water requirements [1,3,7].

Common mineral oil-based steam turbine oils are used for conventional gas turbines. However, it should be remembered that the temperature of some bearings in gas turbines is higher than that in steam turbines so that premature oil ageing can be expected. Moreover, hotspots can occur around some turbine bearings and localized temperatures can reach 200–280 °C [3] whereby the temperature of the oil in the tank remains at about 70–90 °C (hot air and hot gases can accelerate the ageing process). The temperature of the oil reaching a bearing is mostly between 50 and 55 °C and the exit temperature is about 70–75 °C [3]. As the volume of gas turbine oils is generally smaller and they circulate more rapidly, their life is somewhat shorter. The volume of oil for a 40–50 MW generator (GE) is about 6000–7000 l and its life is between 20 000 and 30 000 h (in the case of a 40–60 MW Siemens, 14 000 l and 40 000–80 000 h) [4,6]. Semi-synthetic turbine oils (multiple hydrated base oils) or fully synthetic turbine oils based on synthetic PAOs are recommended for these applications [3,7,10].

In civil and military aviation, gas turbines are used for propulsion. Because of the high temperatures encountered special low-viscosity (ISO VG 10, 22) synthetic oils based on saturated esters (e.g. neopolyesters) are used in these aircraft engines or turbines [14]. These synthetic esters have a high viscosity index, good thermal stability, oxidation resistance as well as excellent low-temperature characteristics. Some of these lubricants can contain additives and some not. The pour-points of these oils are between −50 and <−60 °C. And finally, all relevant civilian and military product specifications must be fulfilled. The lubricants used in aircraft turbines can in some cases also be used in helicopter, ship and stationary industrial turbines. Aviation turbine oils containing special naphthenic base oils (ISO VG 15–32) with good low-temperature characteristics are also used [14].

13.14 Fire-Resistant, Water-Free Fluids for Power Station Applications

For safety reasons, fire-resistant fluids are used in control and governor circuits, which are exposed to ignition and fire hazards. In power stations, this applies in particular to hydraulic systems in high-temperature zones such as near- to super-heated steam pipes. Fire-resistant fluids should not spontaneously ignite when they contact hot surfaces. The fire-resistant hydraulic fluids used in power stations are generally water-free, synthetic fluids based on phosphoric acid esters (type HFD-R according to DIN 51 502 or ISO 6743-0, ISO VG 32–68). These fire-resistant HFD-R fluids based on phosphoric acid esters offer the following features [7]:

- Fire-resistance
- Self-ignition temperature over 500 °C
- Auto-oxidation stable at surface temperatures up to 300 °C
- Good lubricity
- Good protection against corrosion and wear
- Good ageing stability
- Good demulsification
- Low foaming
- Good air release and low vapour pressure

Additives to improve oxidation stability (possibly foam inhibitors) as well as rust and corrosion inhibitors are sometimes used. According to the 7th Luxembourg Report, the maximum permissible temperature of HFD fluids in hydrodynamic systems is 150 °C. Continuous temperatures of 80 °C should not be exceeded in hydraulic systems. These phosphoric acid ester-based synthetic fluids are generally used for control circuits, but in some special cases, also for the lubrication of plain bearings in turbines as well as other hydraulic circuits in steam and gas turbine installations. However, these systems must be designed for these fluids (HFD-compatible elastomers, paints, varnishes and coatings). (E) DIN 51 518 lists the minimum requirements which power station control circuit fluids have to fulfil. Further information can be found in guidelines and specifications relating to fire-resistant fluids, for example in the VDMA Sheet 24 317 and in the CETOP recommendations R 39 H and RP 97 H. Information relating to the change of one fluid to another is contained in VDMA Sheet 24 314 and CETOP RP 86 H [7].

13.15 Lubricants for Water Turbines and Hydroelectric Plants

Hydroelectric power stations have to pay particular attention to the handling of water-polluting substances, that is lubricants. Lubricants with or without additives are used in hydroelectric power stations. The oils are used to lubricate the bearings and gearboxes of principal and ancillary machinery as well as for hydraulic functions in control and governing equipment. The specific operating conditions of the hydroelectric plant need to be considered when selecting lubricants. The lubricants must display good water and air release, low foaming, good corrosion protection, FZG wear protection > 12 in gearboxes, good ageing resistance and compatibility with standard elastomers [7]. As there are no established standards for water turbine oils, the existing product specifications for general turbine oils are adopted as basic requirements. The viscosity of water turbine oils depend on the type and design of the turbine as well as on its operating temperature and can range from 46 to 460 $mm^2 s^{-1}$ at 40 °C. Type TD and LTD lubricating and control oils according to DIN 51 515 are used. In most cases, the same oil can be used for bearings, gearboxes and control equipment. In many cases the viscosity of these turbine and bearing oils is between 68 and 100 $mm^2 s^{-1}$ at 40 °C. When starting up, control and gearbox oil temperatures should not fall below 5 °C and bearing oil temperatures should not fall below 10 °C. In the case of machinery located under cold ambient conditions, the installation of oil heaters is strongly recommended. Water turbine oils are subject to little thermal stress, and as oil tank volumes tend to be high, the life of water turbine oils is very long. In hydroelectric power stations, the oil sampling and analysis intervals can be correspondingly long. Particular care should be taken when sealing the turbine's lubricating oil circuit from possible water ingress. In recent years, rapidly biodegradable water turbine oils based on saturated esters have proven successful in practice. Compared to mineral oils, these products are more rapidly biodegradable and are allocated to a lower water pollution category. In addition, type HLP 46 hydraulic oils type HEES 46 rapidly biodegradable fluids and NLGI grade 2 and 3 greases are used in hydroelectric plants [7].

Often special Zn-free hydraulic oils based on phosphorus sulfur additive technology are used in hydroelectric plants for the lubrication of Kaplan and/or Pelton turbines (mainly ISO VG 68).

References

1 Society of Tribologists and Lubrication Engineers (1983) Turbine Lubrication, 840 Dusse Highway, Parkridge, IL.

2 Bartz, J.W. (1993) Arbeitskreis Schriftentumauswertung Schmierungstechnik, Denkendorf: Dampfturbinenöle – Stand der Technik und Betriebsverhalten, E. Sommer, Deutsche Shell AG, Hamburg.

3 Leugner, L. Tex. (1993) Maintenance, in *The Practical Handbook of Lubrication, Section 17, Turbines and Turbine Oils*, Maintenance Technology International Incorporation, Cochrane, Alberta, Canada.

4 Siemens, A.G. (2010) Sector Energy: Technische Liefervorschrift TLV 901304: Turbinenöle mit normaler thermischer Stabilität (Turbine oils with normal thermal stability), September 2010, Technische Liefervorschrift TLV 901305: Turbinenöle mit erhöhter thermischer Stabilität (Turbine oils with higher thermal stability), May.

5 ISO 6743-5 (1988) (E), Part 5 Family T (Turbines), Classification.

6 General Electric Company (1999) Lubricating Oil Recommendations for Gas Turbines with Bearing Ambients above 500°F (260°C), GE Power Systems, Gas Turbine, GEK 32568 E, revised May 1999 General Electric Company, (2002), Lubricating Oil Recommendations for Gas Turbines with Bearing Ambients above 500°F (260°C), GE Power Systems, Gas Turbine, GEK 32568f, revised February 2002.

7 VDEW Ölbuch (1996) Schmierstoffe und Steuerflüssigkeiten, Band 1 VDEW-Verlag, Frankfurt/Main.

8 Klamann (1982) *Dieter: Schmierstoffe und verwandte Produkte, Kapitel 11.4.1: Dampfturbinenöle* Verlag Chemie, Weinheim.

9 General Electric Company (1999) Lubricating Oil Recommendations with Antiwear Additives for Gas Turbines with Bearing Ambients above 500°F (260°C), GE Power Systems, Gas Turbine: GEK 101941A, November.

10 ABB – Asea Brown Boveri (1994) Schmier- und Steueröle für Turbinen, Spezifikation und Überwachung, HTGD 90117 D, June; ALSTOM (2009) Lubricating and Control Oils for Gas and Steam Turbines, Specification HTGD 90117 V0001 W.

11 DIN 51515 (2010) Part 1 and Part 2 – Lubricants and Governor Fluids for Turbines – Minimum Requirements Part 1: Turbine Oils TD for Normal Service February 2010, Part 2: Turbine Oils TG for High Temperature Service.

12 ISO 8068 (2006) Petroleum Products and Lubricants – Petroleum Lubricating Oils for Turbines (Categories ISO-L TSA and ISO-L TGA) – Specifications, Second edition, 15 September, 2006.

13 VGB PowerTech (2010) Merkblatt VGB-M416, Teil A und Teil B, März 2010, ISBN 978-3-86875-194-9.

14 Zaretsky, E.V. (1997) *Tribology for Aerospace Applications*, STLE Publication SP-37, 1999, Library of Congress Catalog Number 97–061830.